COMPLETE HOME PLUMBING & HEATING HANDBOOK

COMPLETE HOME PLUMBING & HEATING HANDBOOK

JEANNETTE T. ADAMS

ARCO PUBLISHING COMPANY INC.
219 Park Avenue South, New York, N.Y. 10003

Published by Arco Publishing Company, Inc.
219 Park Avenue South, New York, N.Y. 10003

Library of Congress Cataloging in Publication Data
Adams, Jeannette T
Complete home plumbing and heating handbook.

Includes index.
1. Plumbing—Amateurs' manuals. 2. Heating—Amateurs' manuals. I. Title.

TH6124.A3 696'.1 76-28519
ISBN 0-668-03939-6 (Library Edition)

Printed in the United States of America

TABLE OF CONTENTS

Preface

THIS BOOK has been produced to satisfy a demand for a complete book on home plumbing and heating, and covers every phase of this important subject.

It has been prepared to serve as a guide for you to have good and economical plumbing and heating systems in your home.

The subject matter of the *Complete Home Plumbing and Heating Handbook* is divided into the following four sections:

Section 1 describes tools and how they are used, materials, pipe and tubing tools and how to use them, plumbing fixture arrangement, septic tank—soil absorption system, what you should know about plumbing, planning your bathroom, planning your kitchen and workroom, and the water works and hot water heating systems.

Section 2 describes systems, fuels, and controls; plenum heating systems; and peripheral circulation system in old houses.

Section 3 describes chimneys and fireplaces; flashing, gutters and downspouts; protection against fire; and soldering with soft solders.

Section 4 describes plumbing and water system care and repair, heating and ventilating care and repair, and piping installation, repair, and maintenance.

All of the procedures and the projects described in this book are well within the abilities of the average worker. This book was written in the simplest language possible that would keep it within the range of accepted technical phraseology.

Illustrations are copious. To facilitate quick reference several tables have been included throughout the text.

In addition you will find in the Appendices a *Glossary of Terms, Standard Plumbing Symbols, Suggested Specifications for Watertight Concrete*, and *Drainage Fixture Unit Values.*

The *Complete Home Plumbing and Heating Handbook* has

been produced with the conviction that it will be a welcome and valued aid not only to apprentices and beginners, but to experienced plumbers, homeowners, teachers, architects, and others.

J.T.A.

Acknowledgments

THE AUTHOR desires to acknowledge with thanks the assistance of the following firms and national organizations that have cooperated in the production of this book:

American Brass Co. (Anaconda), American Gas Assoc., American Radiator & Standard Sanitary Corp., American Radiator Co., American Society for Testing Materials, American-Standard Plumbing/Heating, American Standards Assoc., American Water Works Assoc., A. O. Smith Corp., Armstrong Manufacturing Co., Better Heating-Cooling Council, Bohn Aluminum & Brass Corp., Borg-Warner Corp.-Plumbing Products Division, Brown & Sharpe Co., Bryant Heater Co., Canada's National Home Builders Assoc., Canadian Hydronics Council, Chicago Pump Co., Cleveland Twist Drill Co., Copper Development Assoc., Crane Co., Dale Valve Co., Delta Faucet Co.-Division of Masco Corp., Duro Co., Electric Energy Assoc., Eljer Fixtures and Fittings—Plumbingware Division Wallace-Murray Corp., E. M. Dart Mfg. Co., Enterprise Brass Works, Emerson Electric Co., Fitzgibbons Boiler Co., Florida Engineering and Industrial Experiment Station, Fulton Sylphon Division—Robertshaw-Fulton Controls Co., General Motors Corp., George W. Ludwig, Globe Brass Mfg. Co., Harvey F. Ludwig, Hays Manufacturing Co., H. E. Robertson, Hoffman Specialty Co., Hot Stream Heater Co., Housing and Home Finance Agency, Humphreys Manufacturing Co., Hydronics Institute, Illinois Engineering Co., Illuminating Engineering Society, Imperial Brass Mfg. Co., Ingersoll-Rand Corp., Institute of Boiler and Radiator Manufacturers, Jenkins Bros., J. H. Williams Co., John E. Kiker, Jr., John Stewart, Johnson Service Co., Jones and Loughlin Steel Corp., Kansas Engineering Experimental Station, Kennedy Valve Mfg. Co., Kenny Mfg. Co., Kewanee Boiler Co., Kohler of Kohler, Kroeschell

Boiler Co., Lead Industries Assoc., Minneapolis-Honeywell Regulation Co., National Bureau of Standards, National Fire Protection Assoc., National Lumber Manufacturers Assoc., National Plumbing and Heating Assoc., New York State Education Department, Owens-Corning Fiberglass Corp., Pennsylvania Electric Switch Co., Pennsylvania State University, Plumbing-Heating-Cooling Information Bureau, Plumbing Contractors Association of the City of New York, Portland Cement Assoc., Robert A. Taft Sanitary Engineering Center, Rockwell Mfg. Corp., Rund Manufacturing Co., Sloan Valve Co., Spear Water and Sewerage Supplies, Stack Heater Co., Steel Boiler Institute, Strauss Electric Appliance Co., Structural Clay Products Institute, University of Wisconsin, U.S. Department of Agriculture—Forest Products Laboratory, U.S. Department of Commerce, U.S. Department of Health, Education, and Welfare, U.S. Radiator Corp., W. A. Case & Son Mfg. Co., Wallace Murray Corp., Washington Metal Products Co., Washington State University, Weil-McLain Co.—Hydronic Division.

Section I

Plumbing in the Home

THE PLUMBING SYSTEM of a house includes the water supply system, the fixtures and fixture traps, the soil, waste and vent pipe, the house drain and sewer and the storm-water-drainage pipe, together with their valves, valve specialties and fittings, all within or adjacent to the home.

Chapter 1

Plumbing Tools and How They Are Used

Tools are designed to make a job easier and enable you to work more efficiently. They are a worker's best friend, regardless of the type of work to be done. This chapter describes the tools needed for plumbing purposes, heating maintenance, installation, and repair work and the correct use and proper care of the more general tools.

Plumbing tools are divided into two classifications: (1) those used for cutting and threading the pipe prior to installation, and (2) those used for actual installation or repair work.

VISES AND CLAMPS

Vises are used for holding work when it is being planed, sawed, drilled, shaped, sharpened, or riveted, or when wood is being glued. *Clamps* are used for holding work which cannot be satisfactorily held in a vise because of its shape and size, or when a vise is not available. Clamps are generally used for light work.

The most common bench vises used are the machinist's bench vise, the bench and pipe vise, the clamp base vise, the blacksmith's vise, and the pipe vise.

A *machinist's bench vise* (Fig. 1) is a large steel vise with rough jaws that prevent the work from slipping. Most of these vises have a swivel base with jaws that can be rotated, while others cannot be rotated. A similar light duty model (Fig. 1) is equipped with a cutoff. These vises are usually bolt-mounted onto a bench.

The *bench and pipe vise* (Fig. 1) has integral pipe jaws for holding pipe from ¾ inch (1.905 cm.) to 3 inches (7.62 cm.)

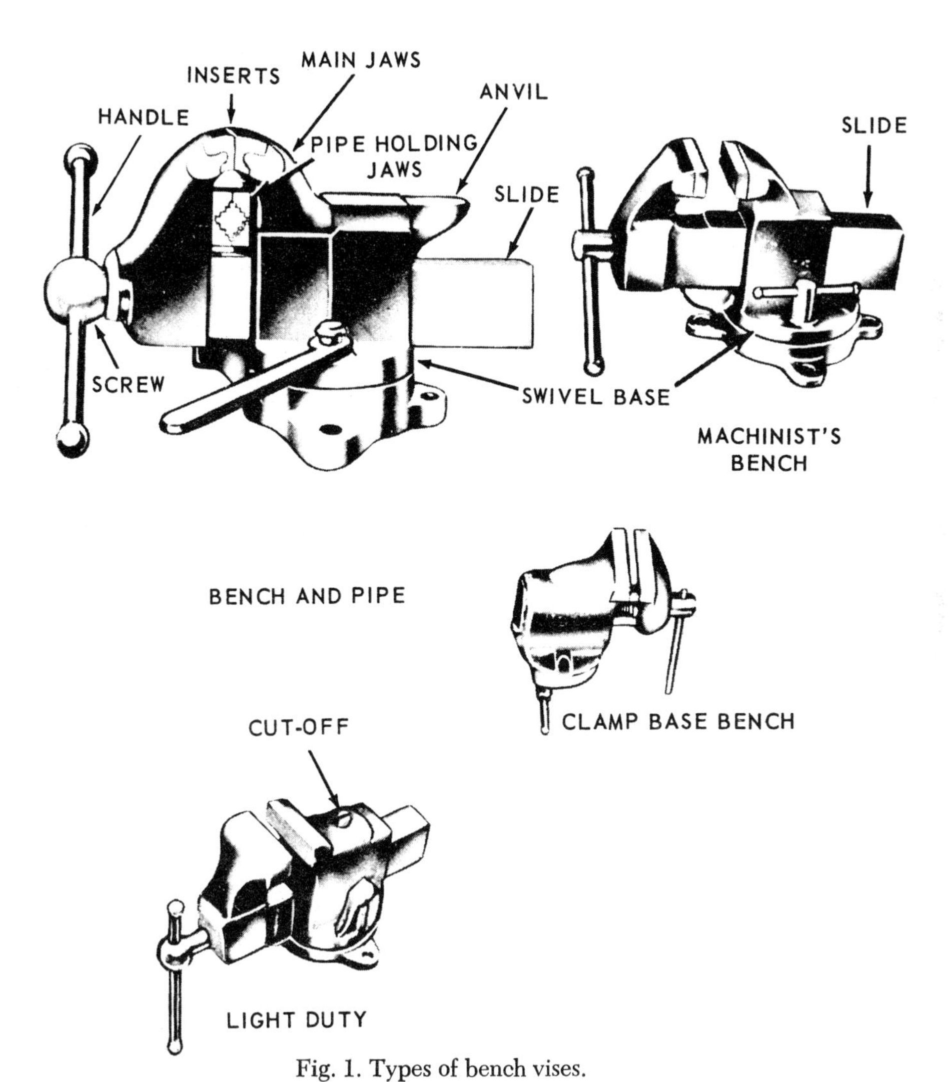

Fig. 1. Types of bench vises.

in diameter. The maximum working main jaw opening is usually 5 inches (12.7 cm.) with a jaw width of 4 to 5 inches (10.16 to 12.7 cm.). The base can be swiveled to any position and locked. These vises are equipped with an anvil and are bolted onto a workbench.

The *clamp base vise* usually has a smaller holding capacity than the machinist's or the bench and pipe vise and is usually clamped to the edge of a bench with a thumbscrew. These types of vises can be obtained with a maximum holding capacity varying between 1½ inches (3.81 cm.) and 3 inches (7.62 cm.). They normally do not have pile holding jaws.

The *blacksmith's vise* (Fig. 2) is used for holding work that must be pounded with a heavy hammer. It is fastened to a sturdy workbench or wall, and the long leg is secured into a solid base on the floor.

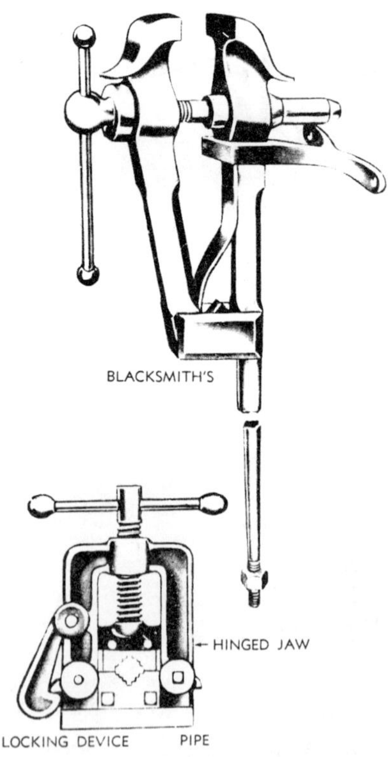

Fig. 2. Blacksmith's and pipe vises.

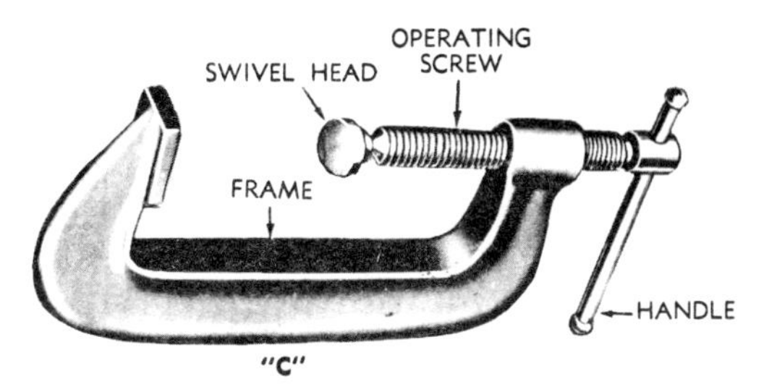

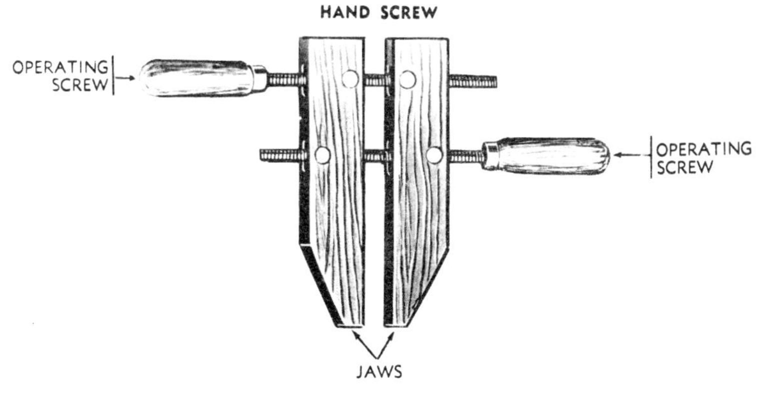

Fig. 3. C clamp and handscrew clamp.

The *pipe vise* (Fig. 2) is specifically designed to hold round stock or pipe. The vise shown in Fig. 2 has a capacity of 1 to 3 inches (2.54 to 7.62 cm.). One jaw is hinged so that the work can be positioned and then the jaw brought down and locked. This vise is also used on a bench. Some pipe vises are designed to use a section of chain to hold down the work. *Chain pipe* vises range in size from 1/8 (.3175 cm.)- to 2½-inch (6.35 cm.) pipe capacity up to ½ (1.27 cm.)- to 8-inch (20.32 cm.) pipe capacity.

A C clamp (Fig. 3) is shaped like the letter *C*. It consists of a steel frame threaded to receive an operating screw with a swivel head. It is made for light, medium, and heavy service in a variety of sizes.

A *hand screw clamp* (Fig. 3) consists of two hard maple jaws connected with two operating screws. Each jaw has two metal inserts into which the screws are threaded. The hand screw clamp is also available in a variety of sizes.

Care of Vises and C Clamps

Keep vises clean at all times. They should be cleaned and wiped with light oil after using. Never strike a vise with a heavy object and never hold large work in a small vise, since these practices will cause the jaws to become sprung or otherwise damage the vise. Keep jaws in good condition and oil the screws and the slide frequently. Never oil the swivel base of a swivel jaw joint as its holding power will be impaired. When the vise is not in use, bring the jaws lightly together or leave a very small gap. (The movable jaw of a tightly closed vise may break due to the expansion of the metal in heat.) Leave the handle in a vertical position.

Threads of C clamps must be clean and free from rust. The swivel head must also be clean, smooth, and grit free. If the swivel head becomes damaged, replace it as follows: pry open the crimped portion of the head and remove the head from the ball end of the screw. Replace with a new head and crimp.

Safety Precautions

When closing the jaw of a vise or clamp, avoid getting any portion of your hands or body between the jaws or between one jaw and the work.

When holding heavy work in a vise, place a block of wood under the work as a prop to prevent it from sliding down and falling on your foot.

Do not open the jaws of a vise beyond their capacity, as the movable jaw will drop off, causing personal injury and possible damage to the jaw.

PIPE AND TUBE CUTTERS AND FLARING TUBES

Pipe cutters (Fig. 4) are used to cut pipe made of steel, brass, copper, wrought iron, and lead. *Tube cutters* (Fig. 5) are used to cut tubing made of iron, steel, brass, copper, and aluminum. The essential difference between pipe and tubing is that tubing has considerably thinner walls. *Flaring tools* (Fig. 6) are used to make single or double flares in the ends of tubing.

Two sizes of hand pipe cutters are generally used. The No.

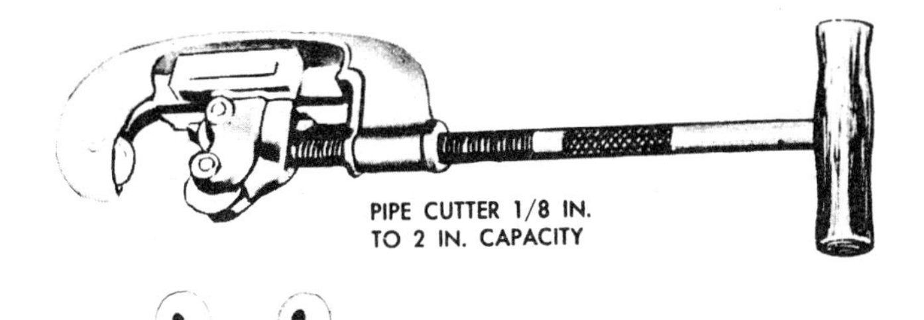

Fig. 4. Pipe cutter.

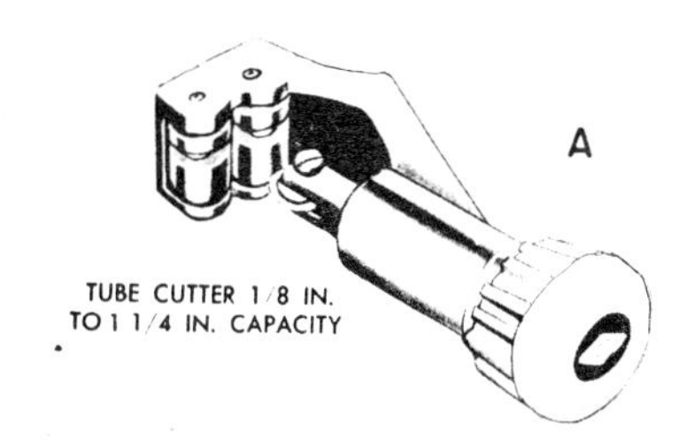

B

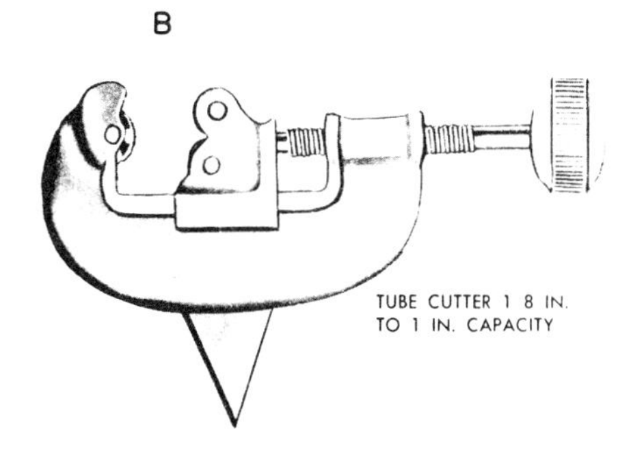

C

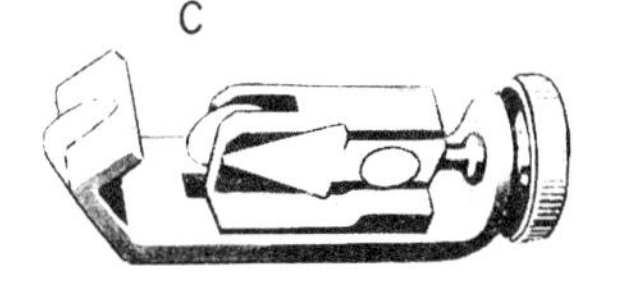

TUBE CUTTER 1/8 IN.
TO 3/4 IN. CAPACITY

Fig. 5. Tubing cutters.

1 pipe cutter has a cutting capacity of $1/8$ to 2 inches (.3175 to 5.08 cm.), and the No. 2 pipe cutter has a cutting capacity of 2 to 4 inches (5.08 to 10.16 cm.). The pipe cutter (Fig. 4) has a special alloy steel cutting wheel and two pressure rollers which are adjusted and tightened by turning the handle.

Most *tube cutters* closely resemble pipe cutters, except that they are of lighter construction. A hand screw feed tubing cutter of $1/8$-inch to $1\frac{1}{4}$-inch (.3175- to 3.175-cm.) capacity (A, Fig. 5) has two rollers with cutouts located off center so that cracked flares may be held in them and cut off without waste of tubing. It also has a retractable cutter blade that is adjusted by turning a knob. The other tube cutter shown in B, Fig. 5 is designed to cut tubing up to and including $3/4$ (1.905 cm.)- and 1-inch (2.54-cm.) outside diameter. The tube cutter shown in C, Fig. 5 cuts tubing from $1/8$-inch (.3175-cm.) to $3/4$-inch (1.905-cm.) in capacity. Rotation of the triangular portion of the tube cutter within the tubing will eliminate any burrs.

Flaring tools (Fig. 6) are used to flare soft copper, brass, or aluminum. The single flaring tool consists of a split die block that has holes for $3/16$ (.47625 cm.)-, $1/4$ (.635 cm.)-, $5/16$ (.79375 cm)-, $3/8$ (.9525 cm)-, $7/16$ (1.11125 cm.)-, and $1/2$-inch (1.27 cm.) outer diameter (o.d.) tubing, a clamp to lock the tube in the die block, and a yoke that slips over the die block and has a compressor screw and a cone that forms a 45-degree flare or a bell shape on the end of the tube. The screw has a T handle. A double flaring tool has the additional feature of adapters that turn in the edge of the tube before a regular 45-degree double flare is made. It consists of a die block with holes for $3/16$ (.47625-cm.)-, $1/4$ (.635-cm.)-, $5/16$ (.79375-cm.)-, $3/8$ (.9525-cm.)-, and $1/2$-inch (1.27-cm.) tubing, a yoke with a screw and a flaring cone, plus five adapters for different size tubing, all carried in a metal case.

REAMERS

Reamers are used to enlarge and true a hole. The reamer consists of three parts—the *body*, the *shank*, and the *blades*. The shank has a square tang to allow the reamer to be held with a wrench for turning. The main purpose of the body is to support the blades.

The blades on a reamer are made of steel and hardened to

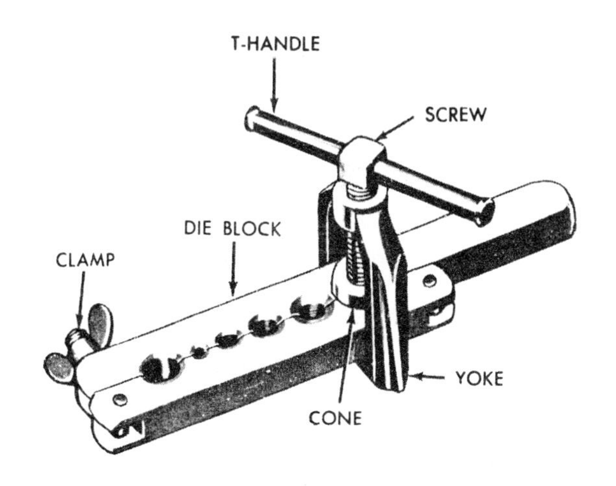

SINGLE FLARING TOOL

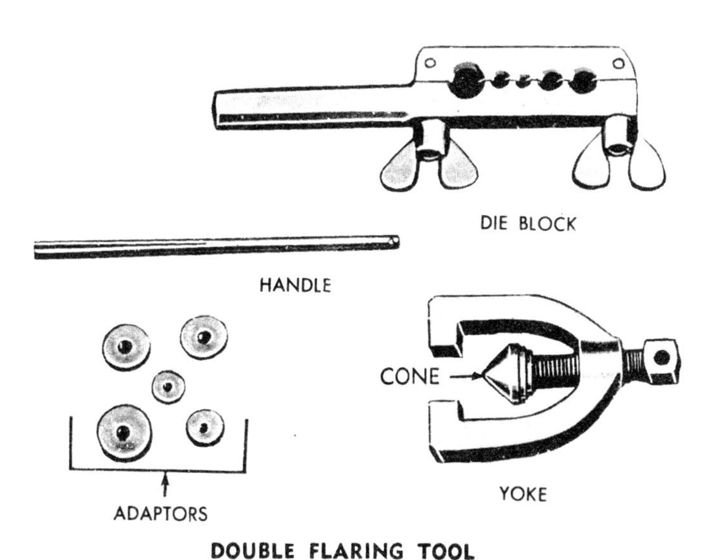

DOUBLE FLARING TOOL

Fig. 6. Flaring tools.

such an extent that they are brittle. For this reason you must be careful when using and storing the reamer to protect the blades from chipping. When you are reaming a hole, turn the reamer in the cutting direction only. This will prevent chipping or dulling of the blades. Great care should be used to assure even, steady turning. Otherwise the reamer will chatter, causing the hole to become marked or scored. To prevent damage to the reamer while not in use, wrap it in an oily cloth and keep it in a box.

Reamers shown in Fig. 7 are available in any standard size. They are also available in size variations of .001″ for special work. A *solid straight flute reamer* (A, Fig. 7) lasts longer and is less expensive than the expansion reamer. However, the *solid spiral flute reamer* (B, Fig. 7) is preferred by workers because it is less likely to chatter.

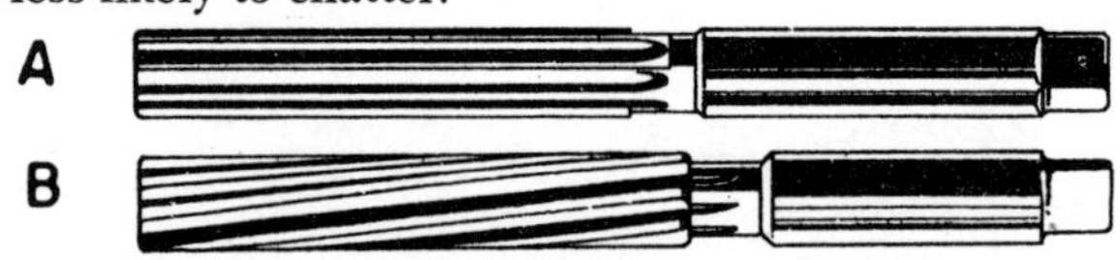

Fig. 7. Reamers. A, Solid straight flute reamer; B, solid spiral flute reamer.

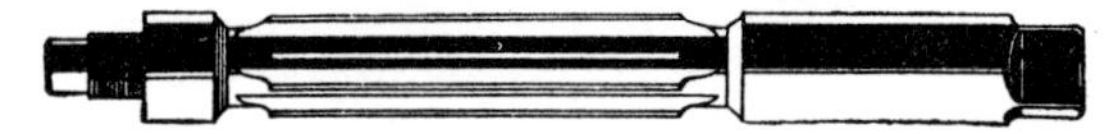

Fig. 8. Expansion reamer.

For general purposes, an *expansion reamer* (Fig. 8) is the most practical. This reamer can usually be obtained in standard sizes from ¼ inch (.635 cm.) to 1 inch (2.54 cm.), by 32nds. (.079375 cm.). It is designed to allow the blades to expand 1/32 inch (.079375 cm.). For example, the ¼-inch (.635-cm.) expansion reamer will ream a ¼-inch (.635-cm.) to a 9/32-inch (.714375-cm.) hole. A 9/32 inch (.714375-cm.) reamer will enlarge the hole from 9/32 inch (.714375-cm.) to 5/16 inch (.793750-cm.). This range of adjustment allows a few reamers to cover sizes up to 1 inch (2.54-cm.).

Reamers are made of carbon steel and high-speed steel. In general, the cutting blades of a high-speed reamer lose their keenness more quickly than a carbon steel reamer. However, after that keenness is gone, it will last longer than the carbon reamer.

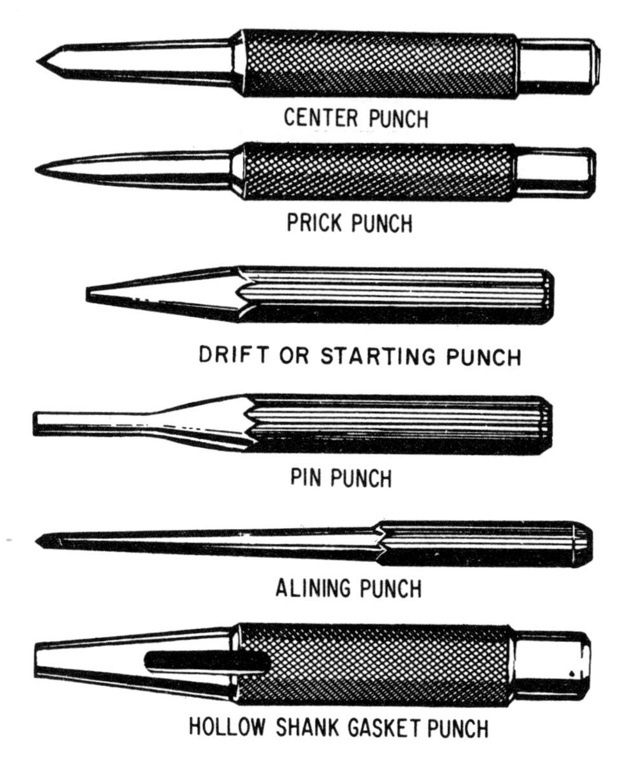

Fig. 9. Punches.

PUNCHES

A *hand punch* is a tool that is held in the hand and struck on one end with a hammer. There are many kinds of punches designed to do a variety of jobs. Figure 9 shows several types of punches. Most punches are made of tool steel. The part held in the hand is usually octagonal, or it may be knurled. This prevents the tool from slipping around in the hand. The other end is shaped to do a particular job.

When you use a punch, there are two things to remember.

1. When you hit the punch you do not want it to slip sideways over your work.

2. You do not want the hammer to slip off the punch and strike your fingers. You can eliminate both of these troubles

by holding the punch at right angles to the work, and striking the punch squarely with your hammer.

The center punch, as the name implies, is used for marking the center of a hole to be drilled. If you try to drill a hole without first punching the center, the drill will "wander" or "walk away" from the desired center.

Another use of the center punch is to make corresponding marks on two pieces of an assembly to permit reassembling in the original positions. Before taking a mechanism apart, make a pair of center punch marks in one or more places to help in reassembly. To do this, select places, staggered as shown in Fig. 10, where matching pieces are joined. First clean the places selected. Then scribe a line across the joint with single and double marks as shown to eliminate possible errors. In reassembly, refer first to the sets of punch marks to determine the approximate position of the parts. Then line up the scribed lines to determine the exact position.

Automatic center punches are useful for layout work. They are operated by pressing down on the shank by hand. An inside spring is compressed and released automatically, striking a blow on the end of the punch. The impression is light, but adequate for marking, and serves to locate the point of a regular punch when a deeper impression is required.

The point of a center punch is accurately ground central with the shank, usually at a 60–90-degree angle, and is difficult to regrind by hand with any sort of accuracy. It is, therefore, advisable to take care of a center punch and not to use it on extremely hard materials. When extreme accuracy is required a prick punch is used. Compare the point angle of the center and prick punches.

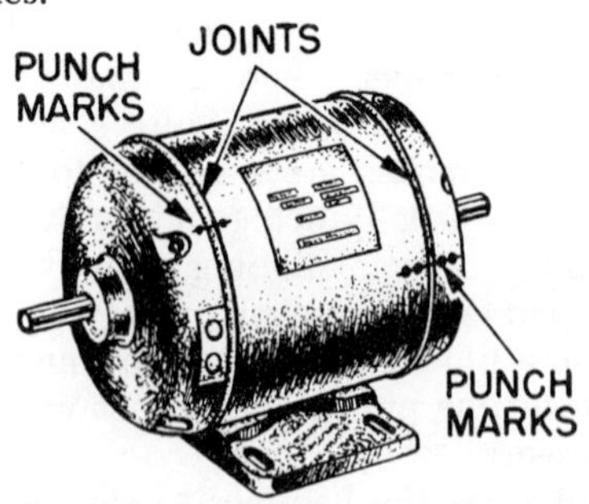

Fig. 10. Punching mating parts of a mechanism.

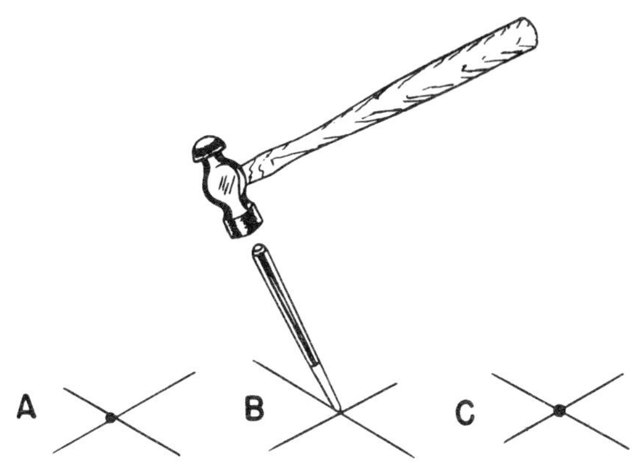

Fig. 11. Marking the intersection of lines with a prick punch.

To make the intersection of two layout lines, bring the point of the prick punch to the exact point of intersection and tap the punch lightly with a hammer. If inspection shows that the exact intersection and the punch mark do not coincide, as shown at A, Fig. 11, slant the punch as shown at B, Fig. 11 and again strike with the hammer, thus enlarging the punch mark and centering it exactly. When the intersection has been correctly punched, finish off with a light blow on the punch held in an upright position. The corrected punch mark is shown at C, Fig. 11.

Drift punches, sometimes called *starting punches*, have a long taper from the tip to the body. They are made that way to withstand the shock of heavy blows. They may be used for knocking out rivets after the heads have been chiseled off, or for freeing pins which are "frozen" in their holes.

After a pin has been loosened or partially driven out, the drift punch may be too large to finish the job. The follow-up tool to use is the *pin punch.* It is designed to follow through the hole without jamming. Always use the largest drift or pin punch that will fit the hole. These punches usually come in sets of three to five assorted sizes. Both of these punches will have flat points, never edged or rounded.

To remove a bolt or pin that is extremely tight, start with a drift punch that has a point diameter that is slightly smaller than the diameter of the object you are removing. As soon as it loosens, finish driving it out with a pin punch. Never use a pin punch for starting a pin because it has a slim shank and a hard blow may cause it to bend or break.

For assembling units of a machine an *alinement* (alining) punch is invaluable. It is usually about 1 foot (30.48 cm.) long and has a long gradual taper. Its purpose is to line up holes in mating parts.

Hollow metal cutting punches are made from hardened tool steel. They are made in various sizes and are used to cut holes in light gage sheet metal.

Other punches have been designed for special uses. One of these is the *soft-faced drift*. It is made of brass or fiber and is used for such jobs as removing shafts, bearings, and wrist pins from engines. It is generally heavy enough to resist damage to itself, but soft enough not to injure the finished surface on the part that is being driven.

For cutting holes in gasket materials a hollow shank *gasket punch* may be used (Fig. 9). Gasket punches come in sets of various sizes to accommodate standard bolts and studs. The cutting end is tapered to a sharp edge to produce a clean uniform hole. To use the gasket punch, place the gasket material to be cut on a piece of hard wood or lead so that the cutting edge of the punch will not be damaged. Then strike the punch with a hammer, driving it through the gasket where holes are required.

HAMMERS, MALLETS, AND SLEDGES

Hammers, mallets, and *sledges* are used to apply a striking force. The tool you select (Fig. 12) will depend upon the intended application.

Hammers

A tool kit would not be complete without at least one hammer. In most cases, two or three are included since they are designated according to weight (without the handle) and style or shape. The shape (Fig. 12) will vary according to the intended work. The carpenter's hammer is designed for one purpose while the machinist's hammer has other primary functions.

Carpenter's Hammer. The primary use of the carpenter's hammer is to drive or draw (pull) nails. Note the names of the various parts of the hammer shown in Fig. 12. The carpenter's

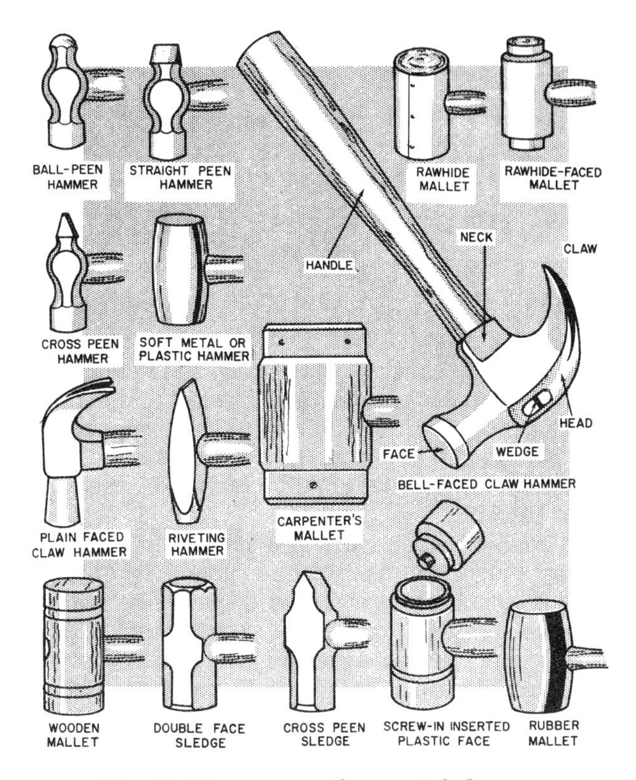

Fig. 12. Hammers, mallets, and sledges.

hammer has either a curved or straight claw. The face may be either bell-faced or plain-faced, and the handle may be made of wood or steel. The carpenter's hammer generally used has a curved claw, bell face, and wooden handle.

Machinist's Hammer. Machinist's hammers are used mostly by people who work with metal or around machinery. These hammers are distinguished from carpenter's hammers by a variable shaped peen, rather than a claw, at the opposite end of the face (Fig. 12).

The *ball-peen hammer*, as its name implies, has a ball which is smaller in diameter than the face. It is therefore useful for striking areas that are too small for the face to enter.

Ball-peen hammers are made in different weights, usually 4, 6, 8, and 12 ounces and 1, 1½, and 2 pounds. For most work a 1½-pound and a 12-ounce hammer should be used. However, a 4- or 6-ounce hammer will often be used for light work such as tapping a punch to cut gaskets out of sheet gasket material.

Machinist's hammers may be further divided into *hard-face* and *soft-face* classifications. The hard-face hammer is made of forged tool steel while the soft-faced hammers have a head made from brass, lead, or a tightly rolled strip of rawhide. Plastic-tipped hammers, or solid plastic with a lead core for added weight, are becoming increasingly popular.

Soft-faced hammers (Fig. 12) should be used when there is danger of damaging the surface of the work, as when pounding on a machined surface. Most soft-faced hammers have heads that can be replaced as the need arises. Lead-faced hammers, for instance, quickly become battered and must be replaced, but have the advantage of striking a solid, heavy nonrebounding blow that is useful for such jobs as driving shafts into or out of tight holes. If a soft-faced hammer is not available, the surface to be hammered may be protected by covering it with a piece of soft brass, copper, or hard wood.

Fig. 13. Correct way to use a ball-peen hammer.

Using Hammers

Simple as the hammer is, there is a right and wrong way of using it. (*See* Fig. 13.) The most common fault is holding the handle too close to the head. This is known as choking the hammer, and reduces the force of the blow. It also makes it harder to hold the head in an upright position. Except for light blows, hold the handle close to the end to increase the lever arm and produce a more effective blow. Hold the handle with the fingers underneath and the thumb alongside or on top of the handle. The thumb should rest on the handle and never overlap the fingers. Try to hit the object with the full force of the hammer. Hold the hammer at such an angle that the face of the hammer and the surface of the object being hit will be parallel. This distributes the force of the blow over the full face and prevents damage to both the surface being struck and the face of the hammer.

Mallets and Sledges

The *mallet* is a short-handled tool used to drive wooden-handled chisels, gouges, wooden pins, or form or shape sheet metal where hard-faced hammers would mar or injure the finished work. Mallet heads are made from a soft material, usually wood, rawhide, or rubber. For example, a rubber-faced mallet is used for knocking out dents in an automobile. It is cylindrically shaped with two flat driving faces that are reinforced with iron bands. (*See* Fig. 12.) Never use a mallet to drive nails, screws, or any object that may cause damage to the face.

The *sledge* is a steel-headed, heavy-duty driving tool that can be used for a number of purposes. Short-handled sledges are used to drive bolts, driftpins, and large nails, and to strike cold chisels and small hand rock drills. Long-handled sledges are used to break rock and concrete, to drive spiles, bolts, or stakes, and to strike rock drills and chisels.

The head of a sledge is generally made of a high carbon steel and may weigh from 6 to 16 pounds. The shape of the head will vary according to the job for which the sledge is designed.

Care of Hammers, Sledges, and Mallets

Hammers, sledges, and mallets should be cleaned and repaired if necessary before they are stored. Before using, ensure that the faces are free from oil or other material that would cause the tool to glance off nails, spikes, or stakes. The heads should be dressed to move any battered edges.

Never leave a wooden or rawhide mallet in the sun, as it will dry out and may cause the head to crack. A light film of oil should be left on the mallet to maintain a little moisture in the head.

The *hammer handle* should always be tight in the head. If it is loose the head may fly off and cause an injury. The eye or hole in the hammer head is made with a slight taper in both directions from the center. After the handle, which is tapered to fit the eye, is inserted in the head, a steel or wooden wedge is driven into the end of the handle that is inserted into the head. This wedge expands the handle and causes it to fill the opposite taper in the eye. Therefore the handle is wedged in both directions as shown in Fig. 14. If the wedge starts to come out, it should be driven in again to tighten the handle. If the wedge comes out, replace it before continuing to use the hammer. If another wedge is not available, you may file one out of a piece of flat steel, or cut one from a portion of the tang of a worn out file. The tang is the end of the file that fits into the handle.

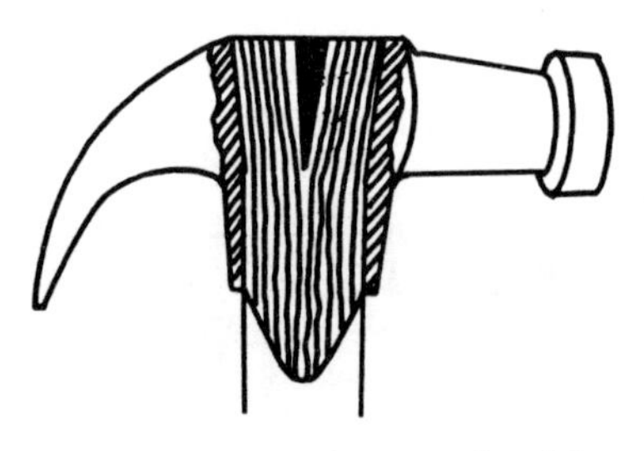

Fig. 14. Handle expanded in hammer head by wedges.

Safety Precautions

Hammers are dangerous tools when used carelessly and without consideration. Some things to remember when using a hammer or mallet are as follows:

1. Do not use a hammer handle for bumping parts in assembly, and never use it as a pry bar. Such abuses will cause the handle to split, and a split handle can produce bad cuts or pinches. When a handle splits or cracks, do not try to repair it by binding with string or wire. Be sure to replace it.
2. Make sure the handle fits tightly on the head.
3. Do not strike a hardened steel surface with a steel hammer. Small pieces of steel may break off and injure someone in the eye or damage the work. However, you can strike a punch or chisel directly with the ball-peen hammer because the steel in the heads of punches and chisels is slightly softer than that of the hammer head.

WRENCHES

A *wrench* is a basic tool that is used to exert a turning or twisting force on bolt heads, nuts, studs, and pipes. The special wrench is designed to do certain jobs are, in most cases, variations of the basic wrenches that will be described in this chapter.

The best wrenches are made of chrome vanadium steel. Wrenches made of this material are light in weight and almost unbreakable. Most common wrenches are made of forged carbon steel or molybdenum steel. These latter materials make good wrenches, but they are generally built a little heavier and bulkier in order to achieve the same degree of strength as chrome vanadium steel.

The *size* of any wrench used on bolt heads or nuts is determined by the size of the opening between the jaws of the wrench. The opening of a wrench is made slightly larger than the bolt head or nut that it is designed to fit. Hex-nuts (six-sided) and other types of nut or bolt heads are measured across opposite flats (Fig. 15). A wrench that is designed to fit a ⅜-inch (.9525-cm.) nut or bolt usually has a clearance of from 5 to 8 thousandths of an inch. This clearance allows the wrench to slide on and off the nut or bolt with a minimum of "play." If the wrench is too large, the points of the nut or bolt head will be rounded and destroyed.

There are many types of wrenches. Each type is designed for a specific use.

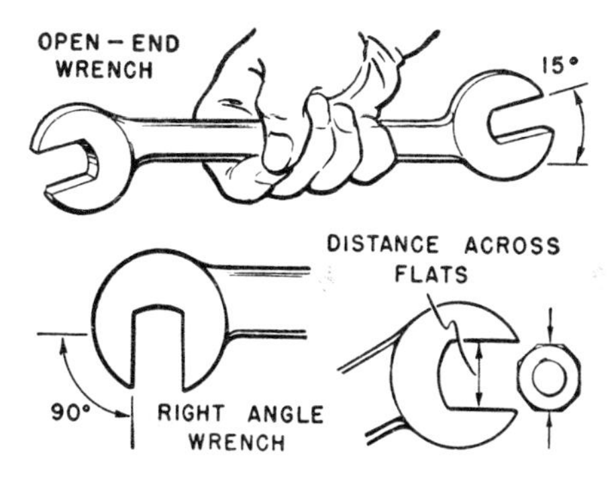

Fig. 15. Open-end wrenches.

Open-end Wrenches

Solid, nonadjustable wrenches with openings in one or both ends are called *open-end wrenches* (Fig. 15). Usually they come in sets of from 6 to 10 wrenches with sizes ranging from $\frac{5}{16}$ to 1 inch (.79375 to 2.54 cm.). (*See* Fig. 15a.) Wrenches with small openings are usually shorter than wrenches with large openings. This proportions the lever advantage of the wrench to the bolt or stud and helps prevent wrench breakage or damage to the bolt or stud. During certain phases of hydraulic maintenance it may be impossible to swing an ordinary wrench due to its length. Ordinary wrenches that are normally available increase in length as their size increases. Therefore when a large size wrench is needed, the length of the wrench sometimes prevents its use, due to the space available to swing the wrench. The Bonney wrench shown in Fig. 16 is an open-end wrench that may be used to great advantage due to its thickness and short length. This wrench is normally procured in the larger sizes, although it is available in a range of sizes to fit all hydraulic fittings.

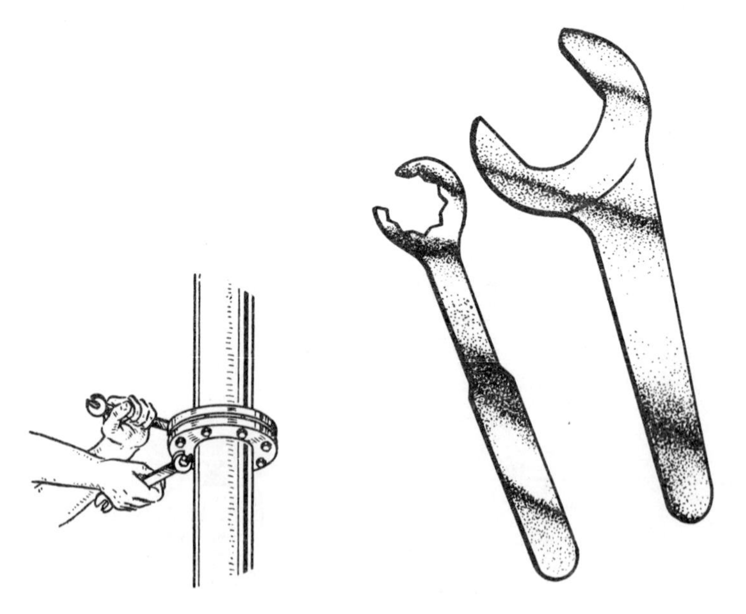

Fig. 15a. Using open-end wrenches. Fig. 16. Hydraulic wrenches.

Open-end wrenches may have their jaws parallel to the handle or at angles anywhere up to 90 degrees. The average angle is 15 degrees (Fig. 15). This angular displacement variation permits selection of a wrench suited for places where there is room to make only a part of a complete turn of a nut or bolt. If the wrench is turned over after the first swing, it will fit on the same flats and turn the nut farther. After two swings on the wrench, the nut is turned far enough so that a new set of flats are in position for the wrench as shown in Fig. 17: (1) Wrench, with opening sloping to the left, about to be placed on nut. (2) Wrench positioned and ready to tighten nut. Note that space for swinging the wrench is limited. (3) Wrench has been moved clockwise to tighten the nut and now strikes the casting which prevents further movement. (4) Wrench is removed from nut and turned counterclockwise to be placed on the next set of flats on nut. But corner of casting prevents wrench from fitting onto the nut. (5) Wrench is being flopped over so that wrench opening will slope to the

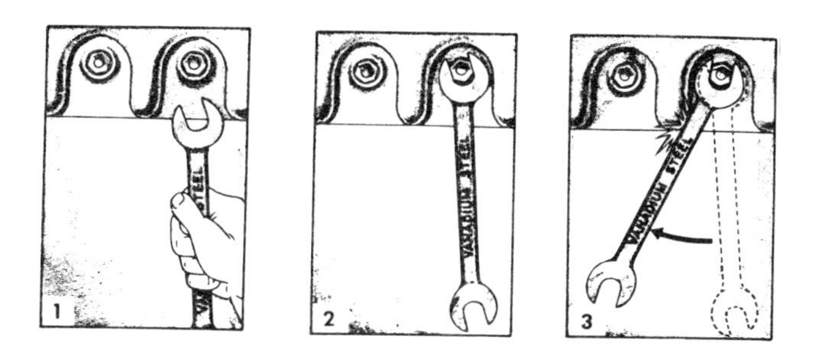

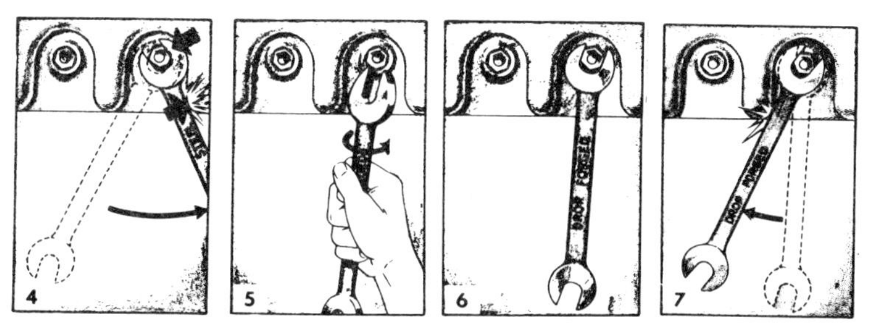

Fig. 17. Use of open-end wrench.

right. (6) In this flopped position, the wrench will fit the next two flats on the nut. (7) Wrench now is pulled clockwise to further tighten nut until wrench again strikes casting. By repeating the flopping procedure, the nut can be turned until it is tight.

Handles of open-end wrenches are usually straight, but may be curved. Those with curved handles are called S *wrenches.* Other open-end wrenches may have offset handles. This allows the head to reach nut or bolt heads that are sunk below the surface.

Fig. 18. Box-end wrench (12-point).

Box Wrenches

Box wrenches (Fig. 18) are safer than open-end wrenches since there is less likelihood they will slip off the work. They completely surround or box a nut or bolt head.

The most frequently used box wrench has 12 points or notches arranged in a circle in the head and can be used with a minimum swing angle of 30 degrees. Six- and eight-point wrenches are used for heavy, 12 for medium, and 16 for light duty only.

One advantage of the 12-point construction is the thin wall. It is more suitable for turning nuts which are hard to get at with an open-end wrench. Another advantage is that the wrench will operate between obstructions where the space for handle swing is limited. A very short swing of the handle will turn the nut far enough to allow the wrench to be lifted and the next set of points fitted to the corners of the nut.

One disadvantage of the box-end wrench is the loss of time which occurs whenever a worker has to lift the wrench off and place it back on the nut in another position in case there is insufficient clearance to spin the wrench in a full circle.

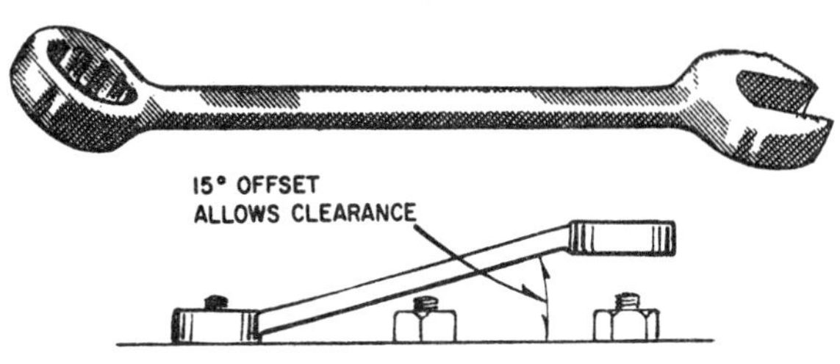

Fig. 19. Combination wrench.

Combination Wrench

After a tight nut is broken loose, it can be unscrewed much more quickly with an open-end wrench than with a box-wrench. This is where a *combination box–open-end wrench* (Fig. 19) comes in handy. You can use the box-end for breaking nuts loose or for snugging them down, and the open-end for faster turning.

The box-end portion of the wrench can be designed with an offset in the handle. Figure 19 shows how the 15-degree offset allows clearance over nearby parts.

The correct use of open-end and box-end wrenches can be summed up in a few simple rules, most important of which is to be sure that the wrench properly fits the nut or bolt head.

When you have to pull hard on the wrench, as in loosening a tight nut, make sure the wrench is seated squarely on the flats of the nut.

Pull on the wrench, do not push. Pushing a wrench is a good way to skin your knuckles if the wrench slips or the nut breaks loose unexpectedly. If it is impossible to pull the wrench, and you must push, do it with the palm of your hand and hold your palm open.

Only actual practice will tell you if you are using the right amount of force on the wrench. The best way to tighten a nut is to turn it until the wrench has a firm and solid "feel." This will turn the nut to proper tightness without stripping the threads or twisting off the bolt.

Socket Wrench

The *socket wrench* is one of the most versatile wrenches in the tool box. Basically, it consists of a handle and a socket type wrench which can be attached to the handle. The "Spin-

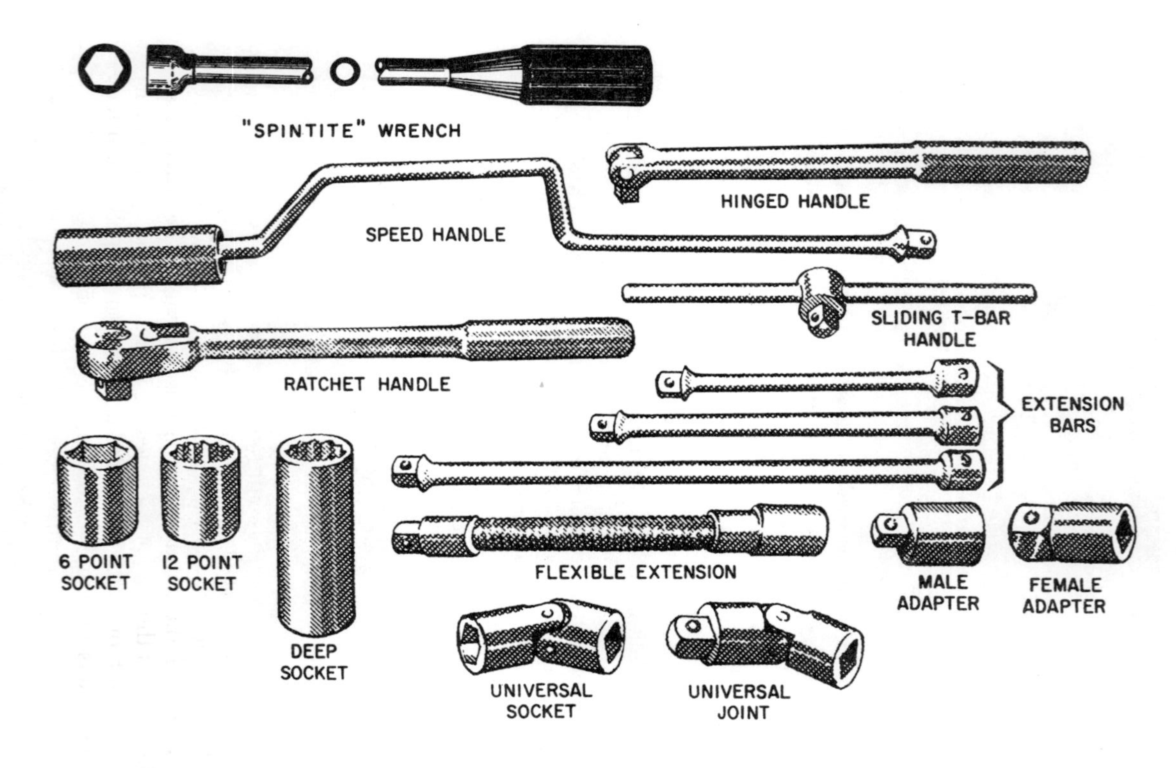

Fig. 20. Socket set components.

tite" wrench shown in Fig. 20 is a special type of socket wrench. It has a hollow shaft to accommodate a bolt protruding through a nut, has a hexagonal head, and is used like a screwdriver. It is supplied in small sizes only and is useful for assembly and electrical work. When used for the latter purpose, it must have an insulated handle.

A complete *socket wrench set* consists of several types of handles along with bar extensions, adapters, and a variety of sockets (Fig. 20).

Sockets. A socket (Fig. 21) has a square opening cut in one end to fit a square drive lug on a detachable handle. In the other end of the socket is a 6-point or 12-point opening like the opening in the box-end wrench. The 12-point socket needs to be swung only half as far as the 6-point socket before it has to be lifted and fitted on the nut for a new grip. It can therefore be used in closer quarters where there is less room to move the handle. A ratchet handle eliminates the necessity of lifting the socket and refitting it on the nut again and again.

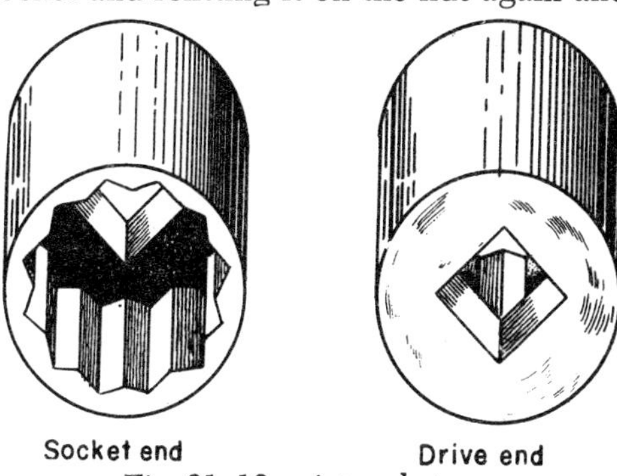

Fig. 21. 12-point sockets.

Sockets are classified for size according to two factors. One is the size of the square opening, which fits on the square drive lug of the handle. This size is known as the drive size. The other is the size of the opening in the opposite end, which fits the nut or bolt. The standard tool box can be outfitted with sockets having ¼ (.635-cm.)-, ⅜ (.9525-cm.)-, and ½-inch (1.27-cm.)-square drive lugs. Larger sets are available. The openings that fit onto the bolt or nut are usually graduated in 1/16-inch (.15895-cm.) sizes. Sockets are also made in deep lengths to fit over sparkplugs and long bolt ends.

Socket Handles. There are four types of handles used with these sockets (Fig. 20). Each type has special advantages, and the worker should choose the one best suited for the job at hand. The square driving lug on the socket wrench handles has a spring-loaded ball that fits into a recess in the socket receptacle. This mated ball-recess feature keeps the socket engaged with the drive lug during normal usage. A slight pull on the socket disassembles the connection.

Ratchet Handles. The ratchet handle has a reversing lever which operates a pawl or dog inside the head of the tool. Pulling the handle in one direction causes the pawl to engage in the ratchet teeth and turn the socket. Moving the handle in the opposite direction causes the pawl to slide over the teeth, permitting the handle to back up without moving the socket. This allows rapid turning of the nut or bolt after each partial turn of the handle. With the reversing lever in one position, the handle can be used for tightening. In the other position, it can be used for loosening.

Hinged Handle. The hinged handle is also very convenient to loosen tight nuts and swing the handle at right angles to the socket. This gives the greatest possible leverage. After loosening the nut to the point where it turns easily, move the handle into the vertical position and then turn the handle with the fingers.

Sliding T-Bar Handle. When using the sliding bar or T handle, the head can be positioned anywhere along the sliding bar. Select the position which is needed for the job at hand.

Speed Handle. The speed handle is worked like the woodworker's brace. After the nuts are first loosened with the sliding bar handle or the ratchet handle, the speed handle can be used to remove the nuts more quickly. In many instances the speed handle is not strong enough to be used for breaking loose or tightening the nut. The speed socket wrench should be used carefully to avoid damaging the nut threads.

Accessories. To complete the socket wrench set, there are several accessory items. Extension bars of different lengths are made to extend the distance from the socket to the handle. A universal joint allows the nut to be turned with the wrench handle at an angle. Universal sockets are also available. The use of universal joints, bar extensions, and universal sockets in combination with appropriate handles makes it possible to form a variety of tools that will reach otherwise inaccessible nuts and bolts.

Another accessory item is an adapter which allows you to use a handle having one size of drive and a socket having a different size drive. For example, a ⅜ (.9525-cm.)- by ¼-inch (.635-cm.) adapter makes it possible to turn all ¼-inch (.635-cm.) square drive sockets with any ⅜-inch (.9525-cm.)-square drive handle.

Torque Wrenches

There are times when, for engineering reasons, a definite force must be applied to a nut or bolt head. In such cases a *torque wrench* must be used. For example, equal force must be applied to all the head bolts of an engine. Otherwise, one bolt may bear the brunt of the force of internal combustion and ultimately cause engine failure.

The three most commonly used torque wrenches are the micrometer setting, dial indicating, and deflecting beam types (Fig. 22). When using the deflecting beam and the dial indicating torque wrenches, the torque is read visually on a dial or scale mounted on the handle of the wrench.

To use the micrometer setting type, unlock the grip and adjust the handle to the desired setting on the micrometer type scale, then relock the grip. Install the required socket or adapter to the square drive of the handle. Place the wrench assembly on the nut or bolt and pull in a clockwise direction with a smooth, steady motion. A fast or jerky motion will result in an improperly torqued unit. When the torque applied reaches the torque value, which is indicated on the handle setting, a signal mechanism will automatically issue an audible click, and the handle will release or "break," and move freely for a short distance. The release and free travel is easily felt, so there is no doubt about when the torquing process is complete.

Manufacturers' instructions generally specify the amount of torque to be applied. To assure getting the correct amount of torque on the fasteners, it is important that the wrench be used properly in accordance with the manufacturers' instructions.

Use that torque wrench which will read about mid-range

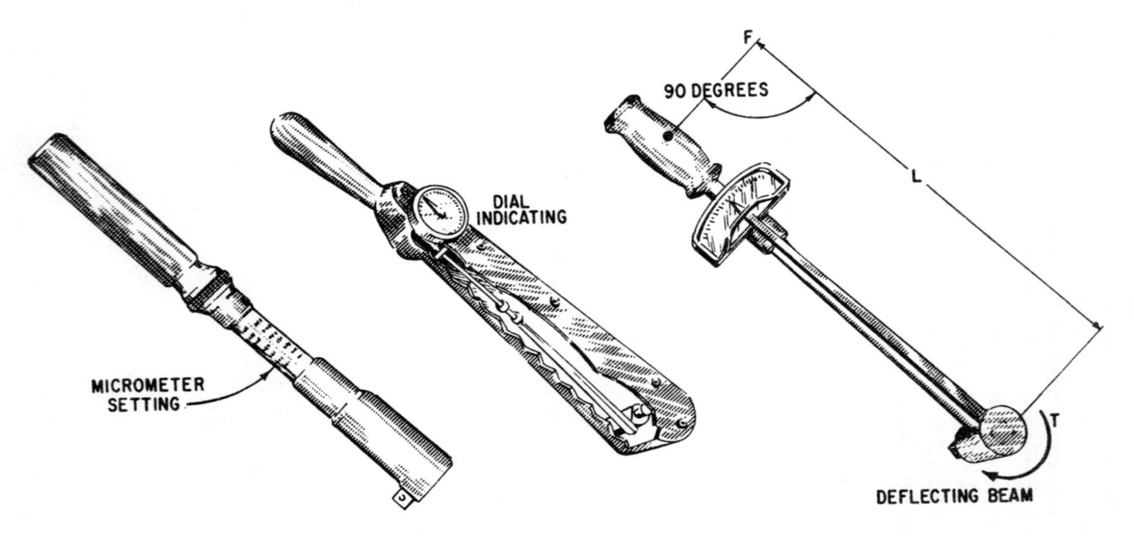

Fig. 22. Torque wrenches.

for the amount of torque to be applied. Be sure that the torque wrench has been calibrated before you use it. Remember that the accuracy of torque measuring depends a lot on how the threads are cut and the cleanliness of the threads. Make sure you inspect and clean the threads. If the manufacturer specifies a thread lubricant, it must be used to obtain the most accurate torque reading. When using the deflecting beam or dial indicating wrenches, hold the torque at the desired value until the reading is steady.

Torque wrenches are delicate and expensive tools. The following precautions should be observed when using them.

1. When using the micrometer setting type, *do not* move the setting handle below the lowest torque setting. However, it should be placed at its lowest setting prior to returning to storage.
2. Do not use the torque wrench to apply greater amounts of torque than its rated capacity.
3. Do not use the torque wrench to break loose bolts which have been previously tightened.
4. Do not drop the wrench. If dropped, the accuracy will be affected.
5. Do not apply a torque wrench to a nut that has been tightened. Back off the nut one turn with a non-torque wrench and retighten to the correct torque with the indicating torque wrench.

Adjustable Wrenches

A handy all-round wrench is the *adjustable open-end* wrench. This wrench is not intended to take the place of the regular solid open-end wrench. Also, it is not built for use on extremely hard-to-turn items. Its usefulness is achieved by being capable of fitting odd-sized nuts. This flexibility is achieved although one jaw of the adjustable open-end wrench is fixed because the other jaw is moved along a slide by a thumbscrew adjustment (Fig. 23). By turning the thumbscrew, the jaw opening may be adjusted to fit various sizes of nuts.

Adjustable wrenches are available in varying sizes ranging from 4 to 24 inches (10.16 to 60.96 cm.) in length. The size of the wrench selected for a particular job is dependent upon the size of nut or bolt head to which the wrench is to be

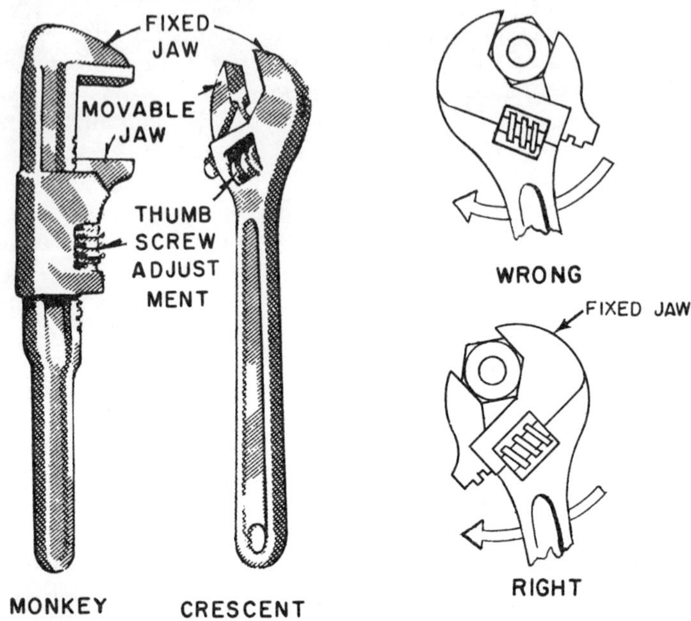

Fig. 23. Adjustable wrenches. Fig. 24. Using a monkey wrench.

applied. As the jaw opening increases, the length of the wrench increases.

Adjustable wrenches are often called "knuckle busters," because mechanics frequently suffer these consequences as a result of improper usage of these tools. To avoid accidents, follow four simple steps. First, choose a wrench of the correct size, that is, do not pick a large 12-inch (30.48-cm.) wrench and adjust the jaw for use on a ⅜-inch (.9525-cm.) nut. This could result in a broken bolt and a bloody hand. Second, be sure the jaws of the correct size wrench are adjusted to fit snugly on the nut. Third, position the wrench around the nut until the nut is all the way into the throat of the jaws. If not used in this manner, the result is apt to be as before. Fourth, pull the handle toward the side having the adjustable jaw (Fig. 24). This will prevent the adjustable jaw from springing open and slipping off the nut. If the location of the work will not allow for all four steps to be followed when using an adjustable wrench, then select another type of wrench for the job. (*See* Fig. 24a.)

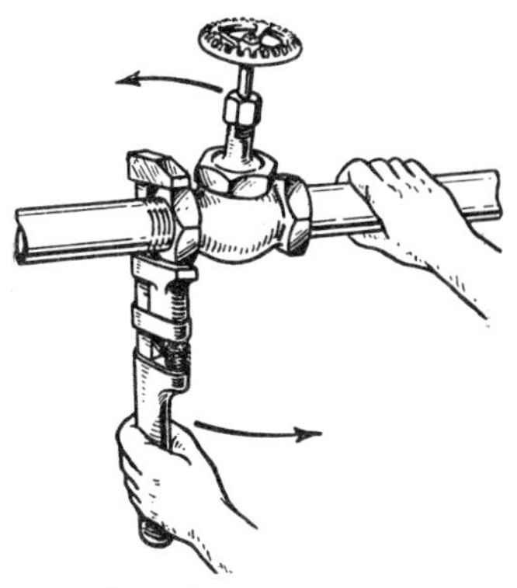

Fig. 24a. Proper procedure for pulling adjustable wrenches.

Pipe Wrench (Stillson). When rotating or holding round work an adjustable pipe wrench (Stillson) may be used (Fig. 25). The movable jaw on a pipe wrench is pivoted to permit a gripping action on the work. This tool must be used with discretion, as the jaws are serrated and always make marks on the work unless adequate precautions are observed. The jaws should be adjusted so the bite on the work will be taken at about the center of the jaws.

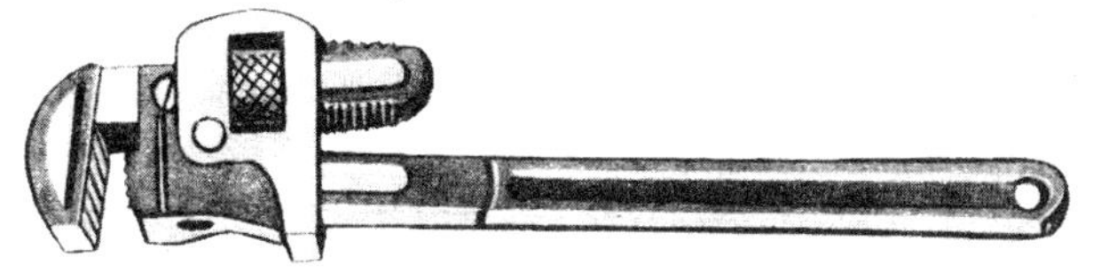

Fig. 25. Adjustable pipe wrench.

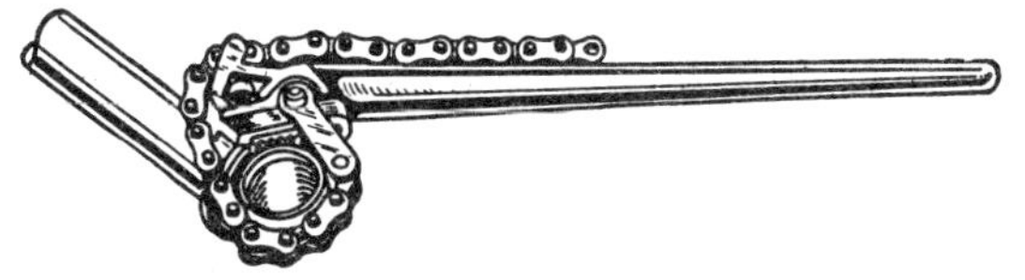

Fig. 26. Chain pipe wrench.

Chain Pipe Wrench. A different type pipe wrench, used mostly on large sizes of pipe, is the chain pipe wrench (Fig. 26). This tool works in one direction only, but can be backed partly around the work and a fresh hold taken without freeing the chain. To reverse the operation the grip is taken on the opposite side of the head. The head is double ended and can

Fig. 27. Using a strap wrench.

be reversed when the teeth on one end are worn out.

Strap Wrench. The strap wrench (Fig. 27) is similar to the chain pipe wrench but uses a heavy web strap in place of the chain. This wrench is used for turning pipe or cylinders where you do not want to mar the surface of the work (Fig. 27). To use this wrench, the webbed strap is placed around the cylinder and passed through the slot in the metal body of the wrench (Fig. 27). The strap is then pulled up tight and as the worker turns the wrench in the desired direction, the webbed strap tightens further around the cylinder. This gripping action causes the cylinder to turn.

Spanner Wrenches

Many special nuts are made with notches cut into their outer edge. For these nuts a *hook spanner* (Fig. 28) is required. This wrench has a curved arm with a lug or hook on the end. This lug fits into one of the notches of the nut and the handle turned to loosen or tighten the nut. This spanner may be made for just one particular size of notched nut, or it may have a hinged arm to adjust it to a range of sizes.

Another type of spanner is the *pin spanner.* Pin spanners have a pin in place of a hook. This pin fits into a hole in the outer part of the nut.

Face pin spanners are designed so that the pins fit into holes in the face of the nut (Fig. 28).

When you use a spanner wrench, you must be sure that the pins, lugs, or hooks make firm contact with the nut while the turning force is transferred from the wrench to the nut. If this is not done, damage will result to either personnel, tools, or equipment.

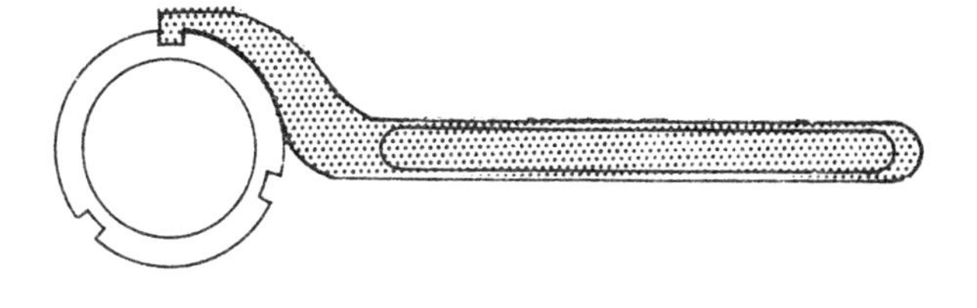

HOOK SPANNER

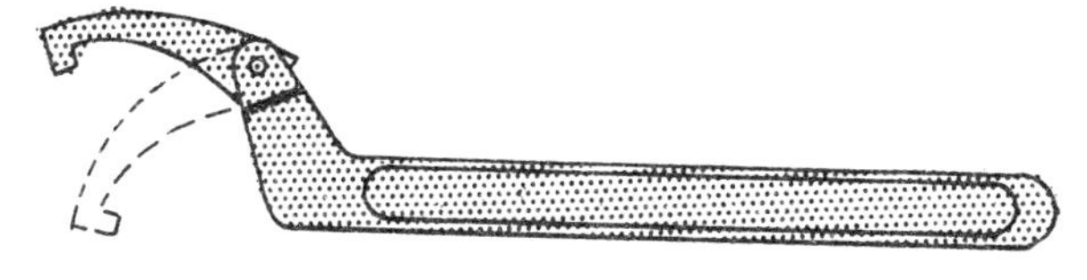

ADJUSTABLE HOOK SPANNER

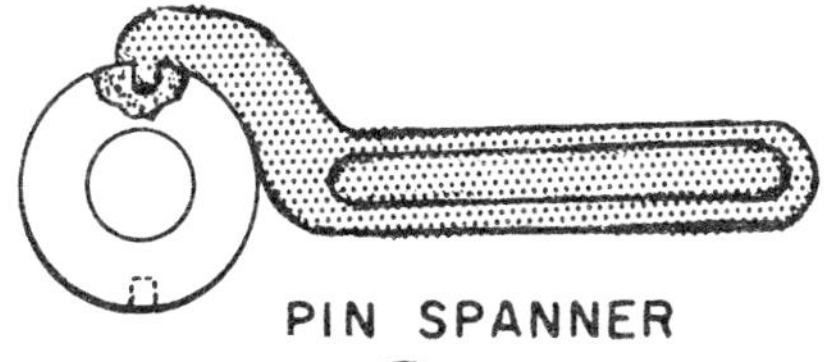

PIN SPANNER

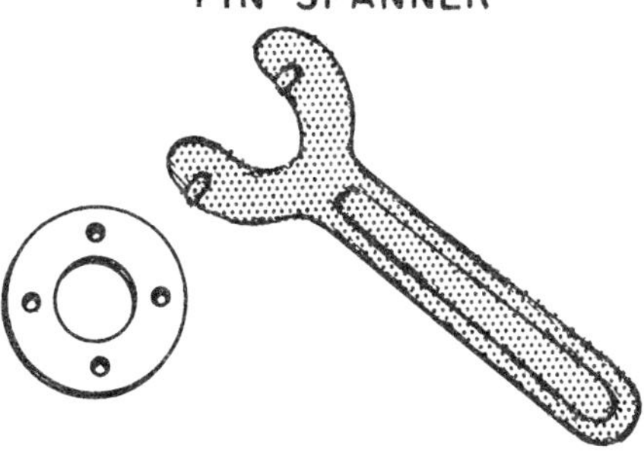

FACE PIN SPANNER

Fig. 28. General-purpose spanner wrenches.

Allen and Bristol Wrenches

In some places it is desirable to use recessed heads on setscrews and capscrews. Recessed head screws usually have a hex-shaped (six-sided) recess. To remove or tighten this type screw requires a special wrench that will fit in the

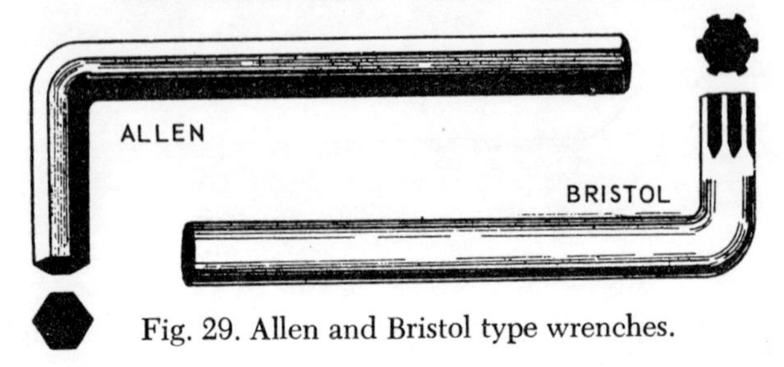

Fig. 29. Allen and Bristol type wrenches.

recess. This wrench is called an *Allen-type wrench.* Allen-type wrenches are made from hexagonal **L**-shaped bars of tool steel (Fig. 29). They range in size up to ¾ inch (1.905 cm.). When using the Allen-type wrench make sure you use the correct size to prevent rounding or spreading the head of the screw. A snug fit within the recessed head of the screw is an indication that you have the correct size.

The *Bristol wrench* is made from round stock. It is also **L** shaped, but one end is fluted to fit the flutes or little splines in the Bristol setscrew (Fig. 29).

Nonsparking Wrenches

Nonsparking wrenches are wrenches that will not cause sparks to be generated when working with steel nuts and bolts. They are generally made from a copper alloy (bronze). However, they may be made from other nonsparking materials.

Nonsparking wrenches must be used in areas where flammable materials are present. These tools are used extensively when working around gasoline carrying vehicles and when working around aircraft or explosives.

Safety Rules for Wrenches

1. Always use a wrench that fits the nut properly.
2. Keep wrenches clean and free from oil. Otherwise they may slip, resulting in possible serious injury to you or damage to the work.
3. Do not increase the leverage of a wrench by placing a pipe over the handle. Increased leverage may damage the wrench or the work.
4. Provide some sort of kit or case for all wrenches. Return them to it at the completion of each job. This saves time and

trouble and facilitates selection of tools for the next job. Most important, it eliminates the possibility of leaving them where they can cause injury or damage to those around or to equipment.

5. Determine which way a nut should be turned before trying to loosen it. Most nuts are turned counterclockwise for removal. This may seem obvious, but even experienced men have been observed straining at the wrench in the tightening direction when they wanted to loosen it.

6. Learn to select your wrenches to fit the type of work you are doing.

SNIPS AND SHEARS

Snips and *shears* are used for cutting sheet metal and steel of various thicknesses and shapes. Normally, the heavier or thicker materials are cut by shears.

One of the handiest tools for cutting light sheet metal (up to 1/16 inch (.15875-cm.) thick) is the *hand snip* (tin snips). The *straight hand snips* shown in Fig. 30 have blades that are

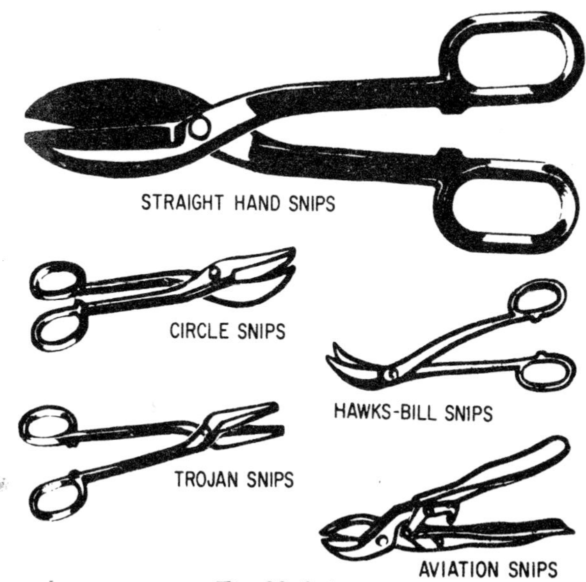

Fig. 30. Snips.

straight and cutting edges that are sharpened to an 85-degree angle. Snips like this can be obtained in different sizes ranging from the small 6-inch (15.24-cm.) to the large 14-inch (35.56-cm.) snip. *Tin snips* will also work on slightly heavier gages of soft metals such as aluminum alloys.

Snips will not remove any metal when a cut is made. There is danger, though, of causing minute metal fractures along the edges of the metal during the shearing process. For this reason, it is better to cut just outside the layout line. This procedure will allow you to dress the cutting edge while keeping the material within required dimensions.

Cutting extremely heavy gage metal always presents the possibility of springing the blades. Once the blades are sprung, hand snips are useless. When cutting heavy material use the rear portion of the blades. This procedure not only avoids the possibility of springing the blades but also gives you greater cutting leverage.

Many snips have small serrations (notches) on the cutting edges of the blades. These serrations tend to prevent the snips from slipping backward when a cut is being made. Although this feature does make actual cutting easier, it mars the edges of the metal slightly. You can remove these small cutting marks if you allow proper clearance for dressing the metal to size. There are many other types of hand snips used for special jobs but the snips discussed here can be used for almost any common type of work.

Cutting Sheet Metal with Snips

It is hard to cut circles or small arcs with straight snips. There are snips especially designed for circular cutting—circle snips, hawks-bill snips, trojan snips, and aviation snips (Fig. 30).

To cut large holes in the lighter gages of sheet metal, start the cut by punching or otherwise making a hole in the center of the area to be cut out. With aviation snips as shown in Fig. 31 or some other narrow-bladed snips, make a spiral cut from the starting hole out toward the scribed circle and continue cutting until the scrap falls away.

To cut a disk in the lighter gages of sheet metal, use combination snips or straight blade snips as shown in Fig. 32. First, cut away any surplus material outside of the scribed

Fig. 31. Cutting an inside hole with snips.

Fig. 32. Cutting a disc out of sheet metal.

circle leaving only a narrow piece to be removed by the final cut. Make the final cut just outside of the layout line. This will permit you to see the scribed line while you are cutting and will cause the scrap to curl up below the blade of the snips where it will be out of the way while the complete cut is being made.

To make straight cuts, place the sheet metal on a bench with the marked guideline over the edge of the bench and hold the sheet down with one hand. With the other hand hold the snips so that the flat sides of the blades are at right angles to the surface of the work. If the blades are not at right angles to the surface of the work, the edges of the cut will be slightly bent and burred. The bench edge will also act as a guide when cutting with the snips. The snips will force the scrap metal down so that it does not interfere with cutting. Any of the hand snips may be used for straight cuts. When notches are too narrow to be cut out with a pair of snips, make the side cuts with the snips and cut the base of the notch with a cold chisel.

Safety and Care of Snips

Learn to use snips properly. They should always be oiled and adjusted to permit ease of cutting and to produce a surface that is free from burrs. If the blades bind, or if they are too far apart, the snips should be adjusted.

Never use snips as screwdrivers, hammers, or pry bars, as they break easily.

Do not attempt to cut heavier materials than the snips are designed for. Never use the snips to cut hardened steel wire or other similar objects. Such use will dent or nick the cutting edges of the blades.

Never toss snips in a toolbox where the cutting edges can come into contact with other tools. This dulls the cutting edges and may even break the blades.

When snips are not in use, hang them on hooks or lay them on an uncrowded shelf or bench.

BOLT CUTTERS

Bolt cutters (Fig. 33) are giant shears with very short blades and long handles. The handles are hinged at one end.

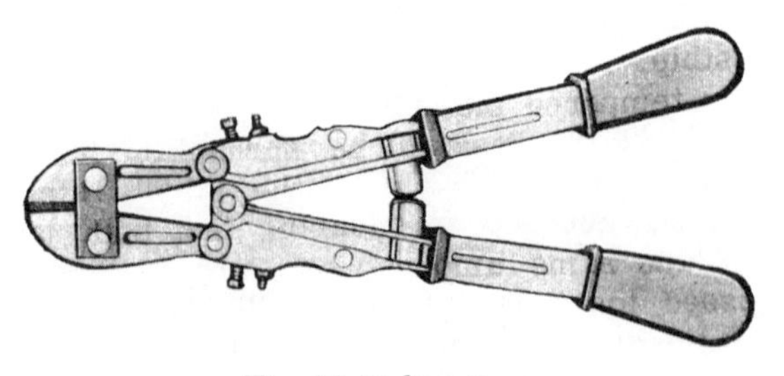

Fig. 33. Bolt cutters.

The cutters are at the ends of extensions which are jointed in such a way that the inside joint is forced outward when the handles are closed, thus forcing the cutting edges together with great force.

Bolt cutters are made in lengths of 18 to 36 inches (45.72 to 91.44 cm.). The larger ones will cut mild steel bolts and rods up to ½ inch (1.27 cm.). The material to be cut should be kept as far back in the jaws as possible. Never attempt to cut spring wire or other tempered metal with bolt cutters. This will cause the jaws to be sprung or nicked.

Adjusting screws near the middle hinges provides a means for ensuring that both jaws move the same amount when the handles are pressed together. Keep the adjusting screws just tight enough to ensure that the cutting edges meet along their entire length when the jaws are closed. The hinges should be kept well oiled at all times.

When using bolt cutters make sure your fingers are clear at the jaws and hinges. Take care that the bolt head or piece of rod cut off does not fly and injure you or someone else. If the cutters are brought together rapidly, sometimes a bolt head or piece of rod being cut off will fly some distance.

HACKSAWS

Hacksaws are used to cut metal that is too heavy for snips or bolt cutters. You can cut metal bar stock with hacksaws.

There are two parts to a hacksaw—the frame and the blade. Common hacksaws have either an adjustable or solid frame (Figs. 34 and 35). Adjustable frames can be made to

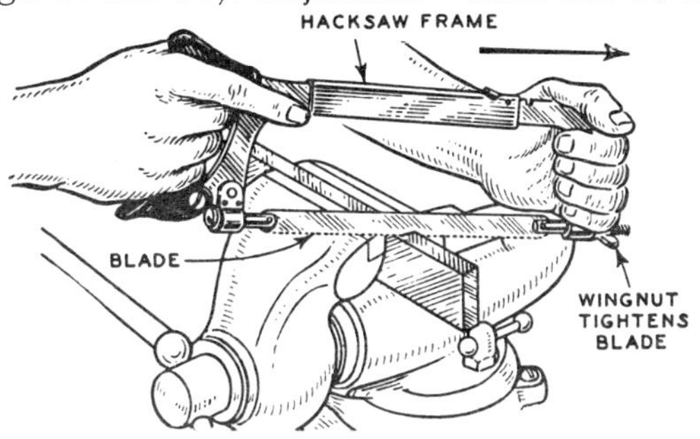

Fig. 34. Proper way to hold an adjustable hacksaw.

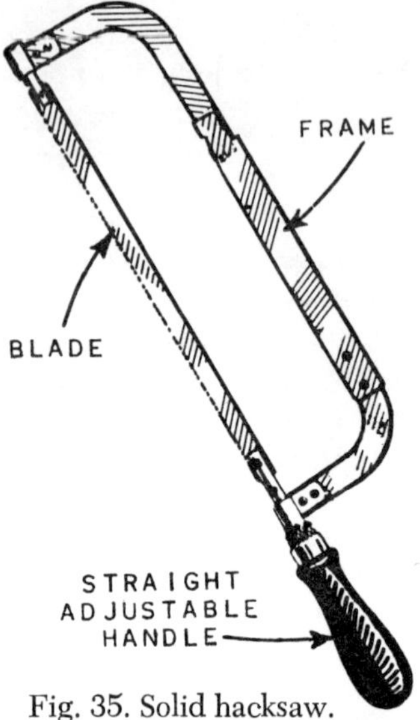

Fig. 35. Solid hacksaw.

hold blades from 8 to 16 inches (20.32 to 40.64-cm.) long, while those with solid frames take only the length blade for which they are made. This length is the distance between the two pins that hold the blade in place.

Hacksaw blades are made of high-grade tool steel, hardened and tempered. There are two types, the all-hard and the flexible. All-hard blades are hardened throughout, whereas only the teeth of the flexible blades are hardened. Hacksaw blades are about ½ inch (1.27 cm.) wide, have from 14 to 32 teeth per inch, and are from 8 to 16 inches (20.32 to 40.64 cm.) long. The blades have a hole at each end which hooks to a pin in the frame. All hacksaw frames which hold the blades either parallel or at right angles to the frame are provided with a wingnut or screw to permit tightening or removing the blade.

The *set* in a saw refers to how much the teeth are pushed out in opposite directions from the sides of the blade. The four different kinds of set are *alternate* set, *double alternate* set, *raker* set, and *wave* set. (*See* Fig. 36.)

The teeth in the alternate set are staggered, one to the left and one to the right throughout the length of the blade. On the double alternate set blade, two adjoining teeth are

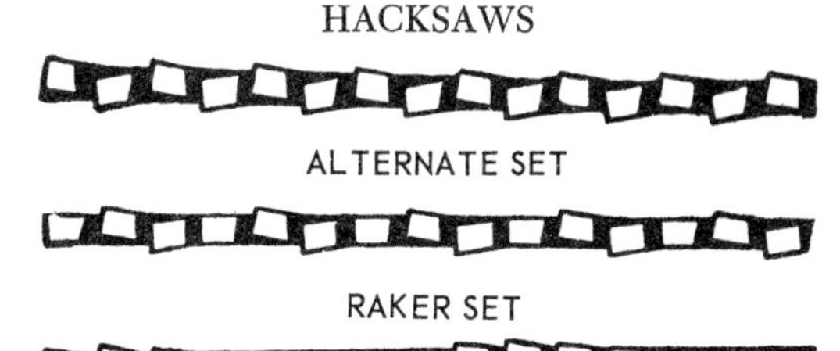

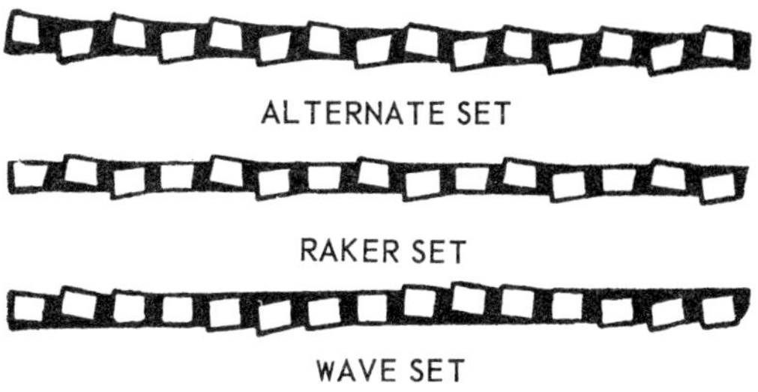

Fig. 36. "Set" of hacksaw blade teeth.

staggered to the right, two to the left, and so on. On the raker set blade, every third tooth remains straight and the other two are set alternately. On the wave (undulated) set blade, short sections of teeth are bent in opposite directions.

Using Hacksaws

Although hacksaws can be used with limited success by an inexperienced person, a little thought and study given to their proper use will result in faster and better work and less dulling and breaking of blades.

Good work with a hacksaw depends not only upon the proper use of the saw, but also upon the proper selection of the blades for the work to be done. Figure 37 will help you select the proper blade to use when sawing metal with a hacksaw. Coarse blades with fewer teeth per inch cut faster and are less liable to choke up with chips. However, finer blades with more teeth per inch are necessary when thin sections are being cut. The selection should be made so that as each tooth starts its cut, the tooth ahead of it will still be cutting.

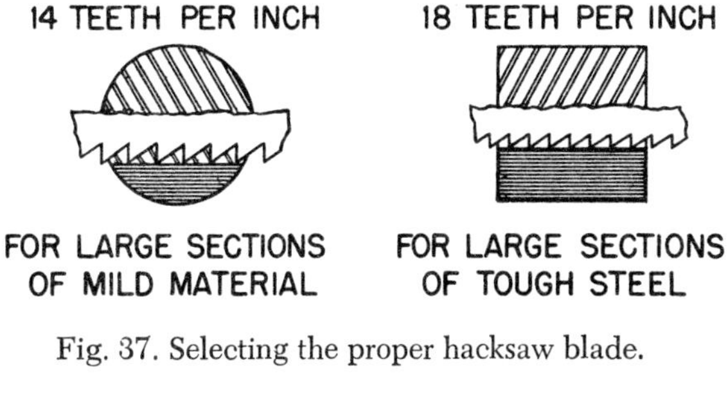

Fig. 37. Selecting the proper hacksaw blade.

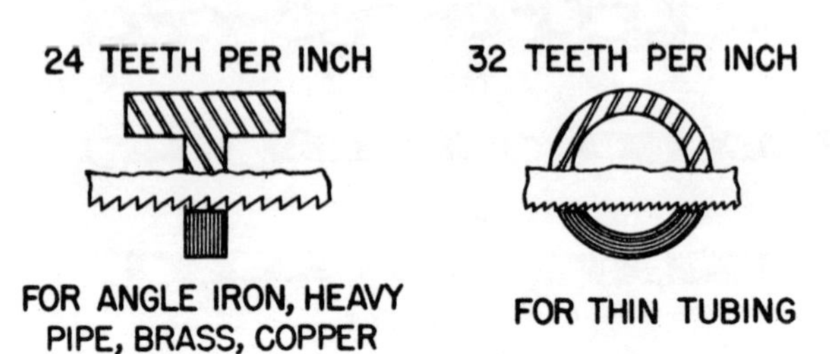

Fig. 37. Selecting the proper hacksaw blade (continued).

To make the cut, first install the blade in the hacksaw frame (Fig. 38) so that the teeth point away from the handle of the hacksaw. (Hand hacksaws cut on the push stroke.) Tighten the wingnut so that the blade is definitely under tension, as this helps make straight cuts.

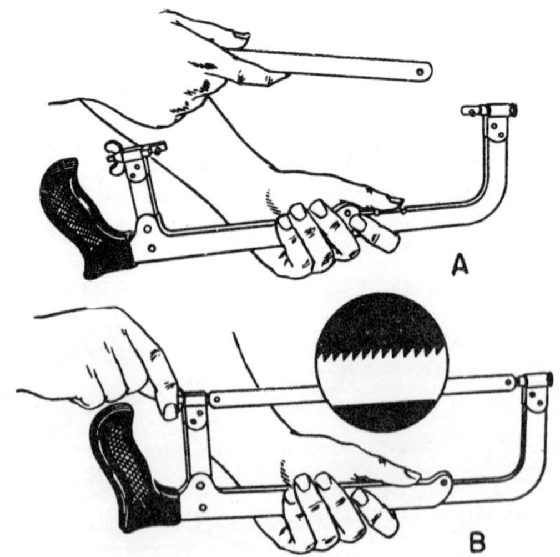

Fig. 38. Installing a hacksaw blade.

Place the material to be cut in a vise. A minimum of overhang will reduce vibration, give a better cut, and lengthen the life of the blade. Have the layout line outside of the vise jaw so that the line is visible while you work.

The proper method of holding the hacksaw is shown in Fig. 34. Note how the index finger of the right hand, pointed forward, aids in guiding the frame.

When cutting, let your body sway ahead and back with each stroke. Apply pressure on the forward stroke, which is the cutting stroke, but not on the return stroke. From 40 to 50 strokes per minute is the usual speed. Long, slow, steady strokes are preferred.

For long cuts (Fig. 39) rotate the blade in the frame so that the length of the cut is not limited by the depth of the frame. Hold the work with the layout line close to the vise jaws, raising the work in the vise as the sawing proceeds.

Saw thin metal as shown in Fig. 40. Note the long angle at which the blade enters the saw groove (kerf). This permits several teeth to be cutting at the same time.

Metal which is too thin to be held, as shown in Fig. 40, can be placed between blocks of wood, as shown in Fig. 41. The

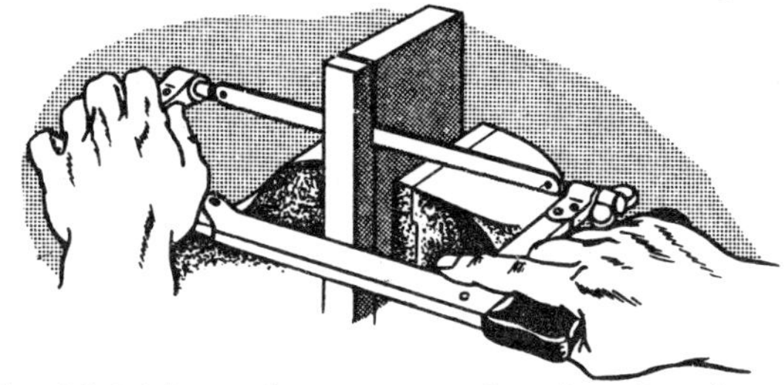

Fig. 39. Making a long cut near the edge of stock.

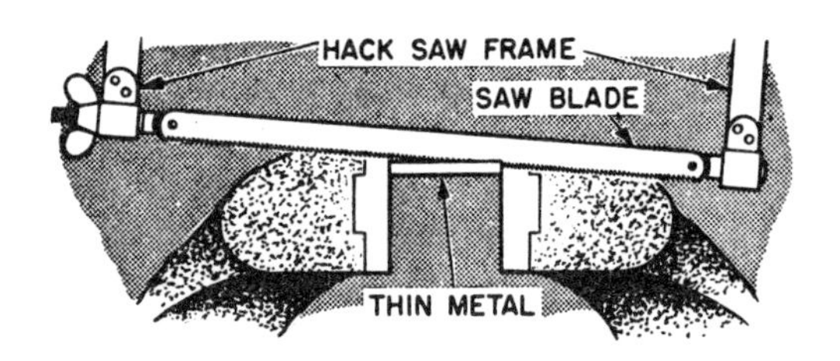

Fig. 40. Cutting thin metal with a hacksaw.

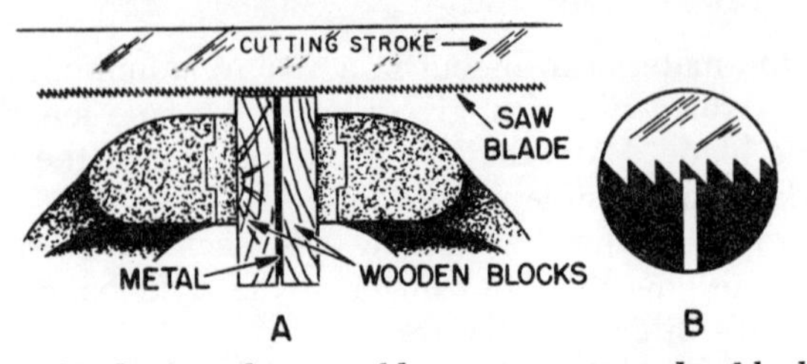

Fig. 41. Cutting thin metal between two wooden blocks.

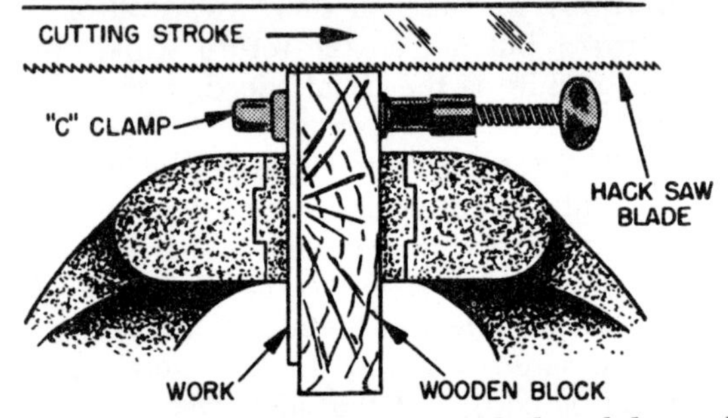

Fig. 42. Cutting thin metal using wood block with layout lines.

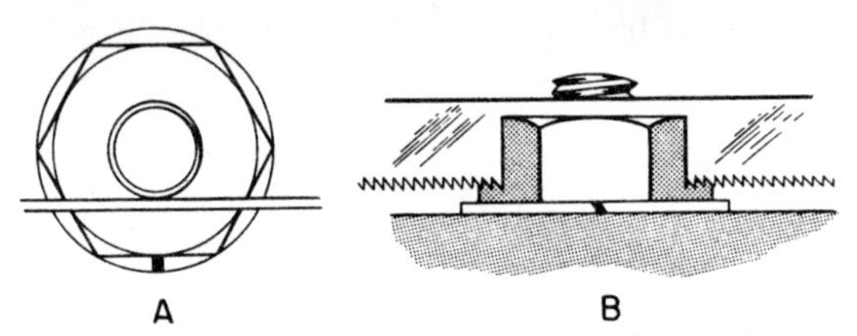

Fig. 43. Removing a frozen nut with a hacksaw.

wood provides support for several teeth as they are cutting. Without the wood, as shown at B, Fig. 41, teeth will be broken due to excessive vibration of the stock and because individual teeth have to absorb the full power of the stroke.

Cut thin metal with layout lines on the face by using a piece of wood behind it (Fig. 42). Hold the wood and the metal in the jaws of the vise, using a C clamp when necessary. The wood block helps support the blade and produces a smoother cut. Using the wood only in back of the metal permits the layout lines to be seen.

To remove a frozen nut with a hacksaw, saw into the nut as shown in Fig. 43, starting the blade close to the threads on the bolt or stud and parallel to one face of the nut as shown at A, Fig. 43. Saw parallel to the bolt until the teeth of the blade almost reach the lockwasher. Lockwashers are hard and will ruin hacksaw blades, so do not try to saw them. B, Figure 43 shows when to stop sawing. Then, with a cold chisel and hammer, remove this one side of the nut completely by opening the saw kerf. Put an adjustable wrench across this new flat and the one opposite, and again try to remove the frozen nut. Since very little original metal remains on this one side of the nut, the nut will either give or break away entirely and permit its removal.

To saw a wide kerf in the head of a capscrew or machine bolt, fit the hand hacksaw frame with two blades side by side, and with the teeth lined up in the same direction. With slow, steady strokes, saw the slot approximately one-third the thickness of the head of the capscrew as shown in Fig. 44. Such a slot will permit subsequent holding or turning with a screwdriver when it is impossible, due to close quarters, to use a wrench.

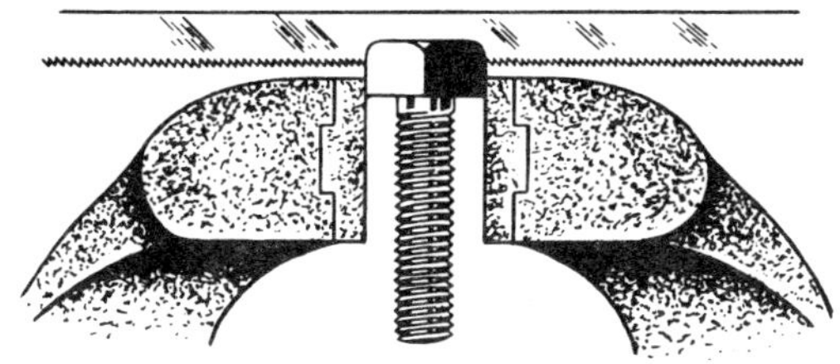

Fig. 44. Cutting a wide kerf in head of a capscrew or bolt.

Hacksaw Safety

The main danger in using hacksaws is injury to your hand if the blade breaks. The blade will break if too much pressure is applied, when the saw is twisted, when the cutting speed is too fast, or when the blade becomes loose in the frame. Additionally, if the work is not tight in the vise, it will sometimes slip, twisting the blade enough to break it.

ROD SAWS

An improvement in industrial technology provides us with a tool that can cut material an ordinary hacksaw cannot even

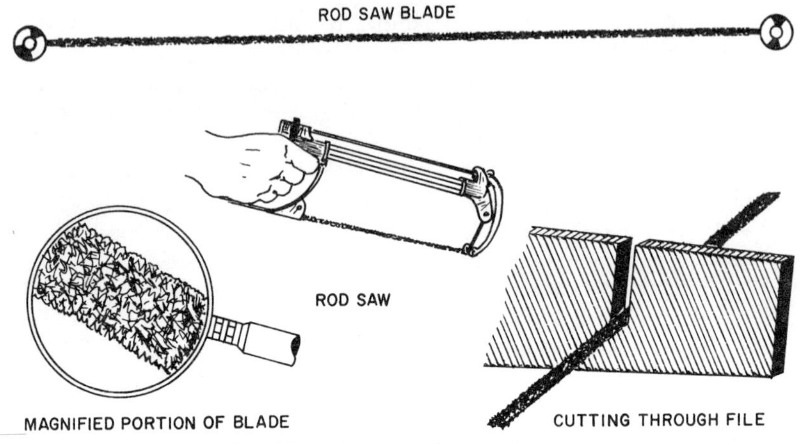

Fig. 45. Rod saw and operations.

scratch. The *rod saw* (Fig. 45) acts like a diamond in its capability of cutting hard metals and materials such as stainless steel, Inconel, titanium, and carbon phenolics.

The rod saw cuts through material by means of hundreds of tungsten-carbide particles permanently bonded to the rod. (*See* magnified portion of blade in Fig. 45.) The rod saw cuts through stainless steel and files with ease. A unique feature of this saw is its capability of cutting on the forward and reverse strokes.

CHISELS

Chisels are tools that can be used for chipping or cutting metal. They will cut any metal that is softer than the materials of which they are made. Chisels are made from a good grade tool steel and have a hardened cutting edge and beveled head. Cold chisels are classified according to the shape of their points, and the width of the cutting edge denotes their size. The most common shapes of cold chisels are flat, cape, round nose, diamond point and side (Fig. 46).

The type chisel most commonly used is the *flat cold chisel*, which serves to cut rivets, split nuts, chip castings, cast iron soil pipe, and cut thin metal sheets. (*See* Figs. 47 and 48.) The *cape chisel* is used for special jobs like cutting keyways,

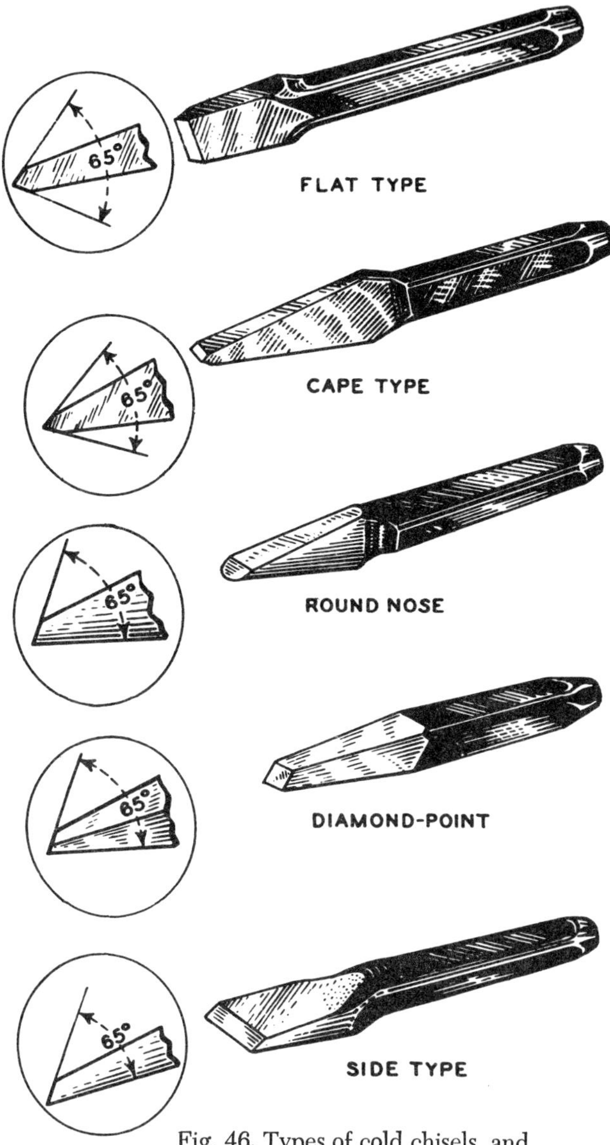

Fig. 46. Types of cold chisels, and average cutting edge angle of each.

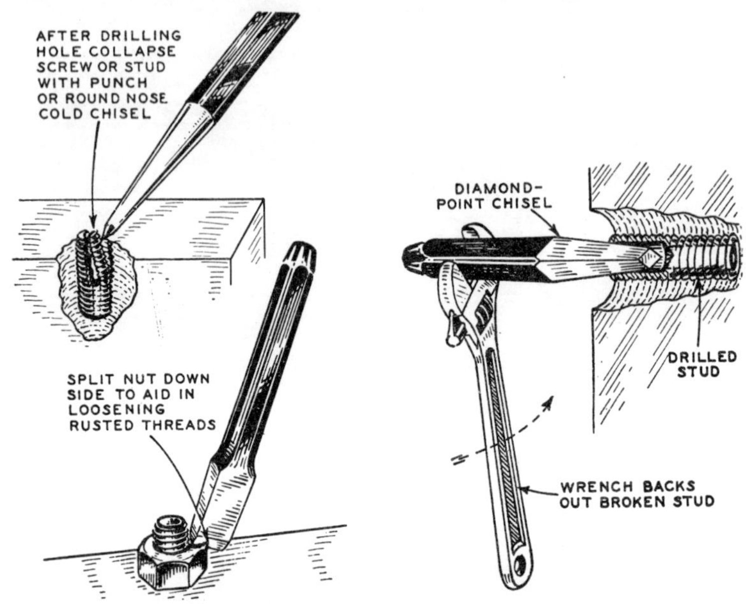

Fig. 47. Always drive cold chisel toward the rear vise jaw.

Fig. 48. Removing broken stud with cold chisel.

narrow grooves and square corners. *Round-nose chisels* make circular grooves and chip inside corners with a fillet. Finally, the diamond-point is used for cutting V-grooves and sharp corners.

As with other tools there is a correct technique for using a chisel. Select a chisel that is large enough for the job. Be sure to use a hammer that matches the chisel—the larger the chisel, the heavier the hammer. A heavy chisel will absorb the blows of a light hammer and will do virtually no cutting.

As a general rule, hold the chisel in the left hand with the thumb and first finger about one inch from the top (Fig. 47). It should be held steadily but not tightly. The finger muscles should be relaxed, so if the hammer strikes the hand it will permit the hand to slide down the tool and lessen the effect of the blow. Keep the eyes on the cutting edge of the chisel, not on the head, and swing the hammer in the same plane as the body of the chisel. If you have a lot of chiseling to do, slide a piece of rubber hose over the chisel. This will lessen the shock to your hand.

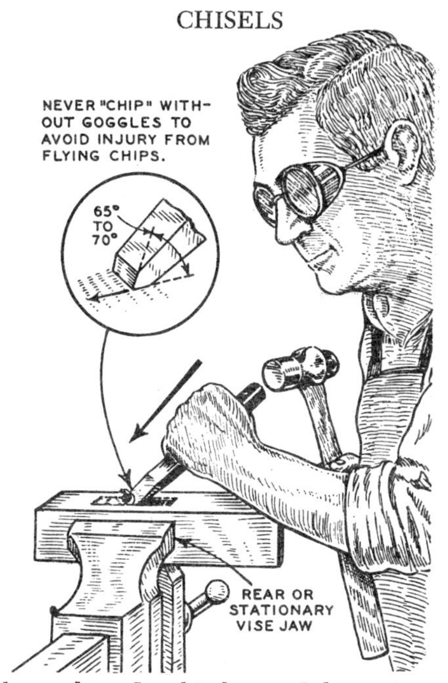

Fig. 49. Always drive flat chisel toward the stationary jaw vise.

When using a chisel for chipping, always wear goggles to protect your eyes (Fig. 49). If other people are close by, see that they are protected from flying chips by erecting a screen or shield to contain the chips. Remember that the time to take these precautions is before you start working on the job.

FILES

A tool kit is not complete unless it contains an assortment of files. There are a number of different types of files in common use, and each type may range in length from 3 to 18 inches (7.62 to 45.72 cm.).

Grades

Files are graded according to the degree of fineness, and according to whether they have single- or double-cut teeth. The difference is apparent when you compare the files in Fig. 50.

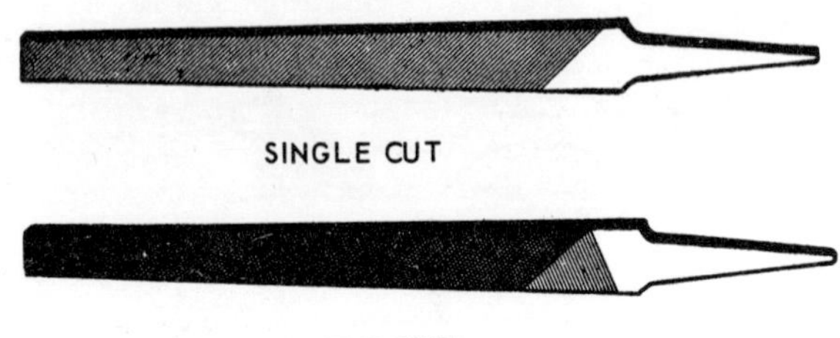

Fig. 50. Single and double-cut files.

Single-cut files have rows of teeth cut parallel to each other. These teeth are set at an angle of about 65 degrees with the centerline. You should use single-cut files for sharpening tools, finish filing, and drawfiling. They are also the best tools for smoothing the edges of sheet metal.

Files with crisscrossed rows of teeth are *double-cut files.* The double cut forms teeth that are diamond-shaped and fast cutting. You should use double-cut files for quick removal of metal and for rough work.

Files are also graded according to the spacing and size of their teeth, or their coarseness and fineness. Some of these grades are shown in Fig. 50. In addition to the three grades shown, you may use some *dead smooth files,* which have very fine teeth, and some *rough* files with very coarse teeth. The fineness or coarseness of file teeth is also influenced by the length of the file. The length of a file is the distance from the tip to the heel, and does not include the tang (Fig. 51). Compare the actual size of the teeth of a 6-inch (15.24-cm.), single-cut smooth file and a 12-inch (30.48-cm.), single-cut smooth file; you will note the 6-inch (15.24-cm.) file has more teeth per inch than the 12-inch (30.48-cm.) file. Design and spacing of file teeth are shown in Fig. 52.

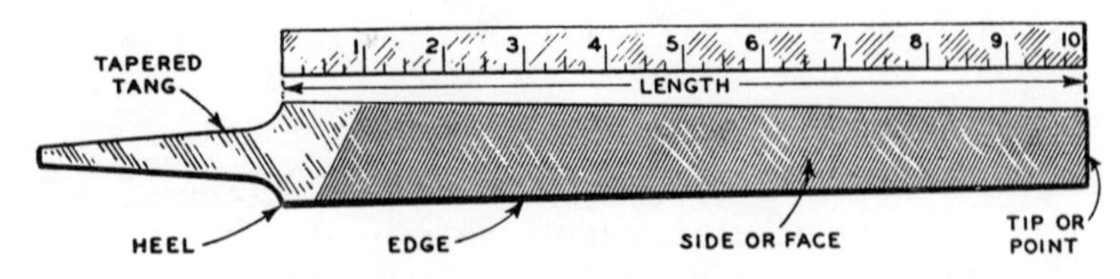

Fig. 51. Parts of a typical file.

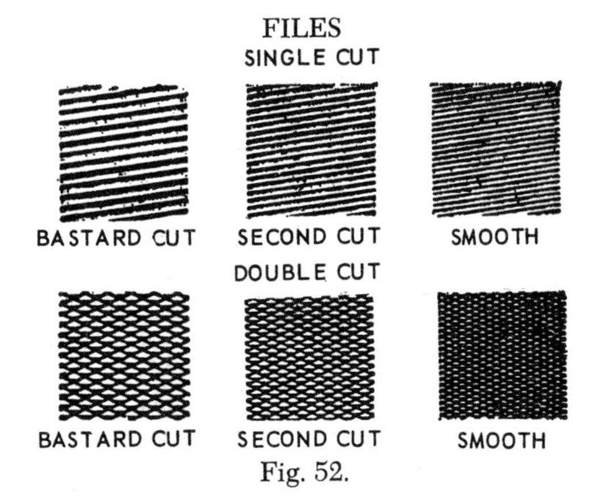

Fig. 52.

Shapes

Files come in different shapes. Therefore, in selecting a file for a job, the shape of the finished work must be considered. Some of the cross-sectional shapes are shown in Fig. 53.

Triangular files are tapered (longitudinally) on all three sides. They are used to file acute internal angles, and to clear out square corners. Special triangular files are used to file saw teeth.

Mill files are tapered in both width and thickness. One edge has no teeth and is known as a *safe edge*. Mill files are used for smoothing lathe work, drawfiling, and other fine, precision work. Mill files are always single-cut.

Flat files are general purpose files and may be either single- or double-cut. They are tapered in width and thickness. *Hard files* are somewhat thicker than flat files. They taper slightly in thickness, but their edges are parallel.

The flat or hard files most often used are the double-cut for rough work and the single-cut, smooth file for finish work.

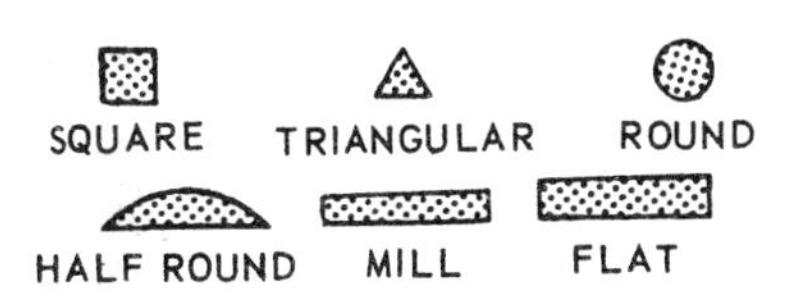

Fig. 53.

Square files are tapered on all four sides and are used to enlarge rectangular shaped holes and slots. *Round files* serve the same purpose for round openings. Small round files are often called *rattail files*.

The *half-round file* is a general purpose tool. The rounded side is used for curved surfaces and the flat face on flat surfaces. When you file an inside curve, use a round or half-round file whose curve most nearly matches the curve of the work.

FILING OPERATIONS

Using a file is an operation that is nearly indispensable when working with metal. You may be crossfiling, drawfiling, using a file card, or even polishing metal. When you have finished using a file it may be necessary to use an abrasive cloth or paper to finish the product. Whether this is necessary depends on how fine a finish you want on the work.

Crossfiling

Figure 54 shows a piece of mild steel being crossfiled. This means that the file is being moved across the surface of the work in approximately a crosswise direction. For best results, keep your feet spread apart to steady yourself as you file with slow, full-length, steady strokes. The file cuts as you push it—ease up on the return stroke to keep from dulling the teeth. Be sure to keep your file clean.

Figure 55 shows the alternate positions of the file when an exceptionally flat surface is required. Using either position first, file across the entire length of the stock. Then, using the

Fig. 54. Crossfiling a piece of mild steel.

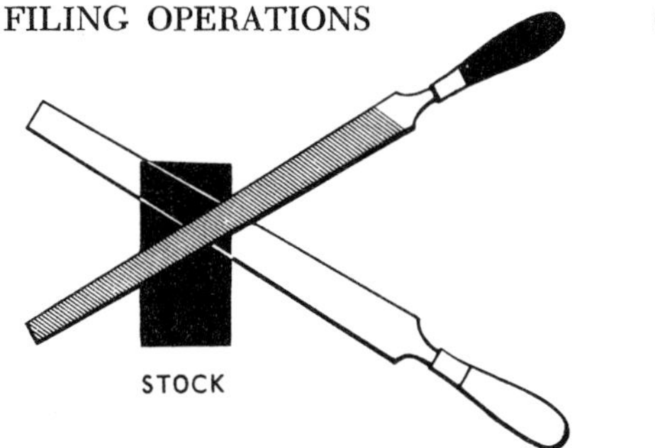

Fig. 55. Alternating positions when filing.

other position, file across the entire length of the stock again. Because the teeth of the file pass over the surface of the stock from two directions, the high spots and low spots will readily be visible after filing in both positions. Continue filing first in one position or direction and then the other until the surface has been filed flat. Test the flatness with a straightedge or with prussian blue and a surface plate.

Drawfiling

Drawfiling produces a finer surface finish and usually a flatter surface than crossfiling. Small parts, as shown in Fig. 56, are best held in a vise. Hold the file as shown in the illustration. Note that the arrow indicates that the cutting stroke is away from you when the handle of the file is held in the right hand. If the handle is held in the left hand, the cutting stroke will be toward you. Lift the file away from the

Fig. 56. Drawfiling a small part.

surface of the work on the return stroke. When drawfiling will no longer improve the surface texture, wrap a piece of abrasive cloth around the file and polish the surface as shown in Fig. 56.

USING FILE CARD

As you file, the teeth of the file may clog up with some of the metal filings and scratch your work. This condition is known as *pinning*. You can prevent pinning by keeping the file teeth clean. Rubbing chalk between the teeth will help prevent pinning, but the best method is to clean the file frequently with a *file card* or *brush* (Fig. 57). A file card has fine wire bristles. Brush with a pulling motion, holding the card parallel to the rows of teeth.

Always keep the file clean, whether you are filing mild steel or other metals. Use chalk liberally when filing nonferrous metals.

FILING ROUND METAL STOCK

Figure 58 shows that as a file is passed over the surface of round work its angle with the work is changed. This results in a rocking motion of the file as it passes over the work. This rocking motion permits all the teeth on the file to make contact and cut as they pass over the work's surface, therefore tending to keep the file much cleaner and thereby doing better work.

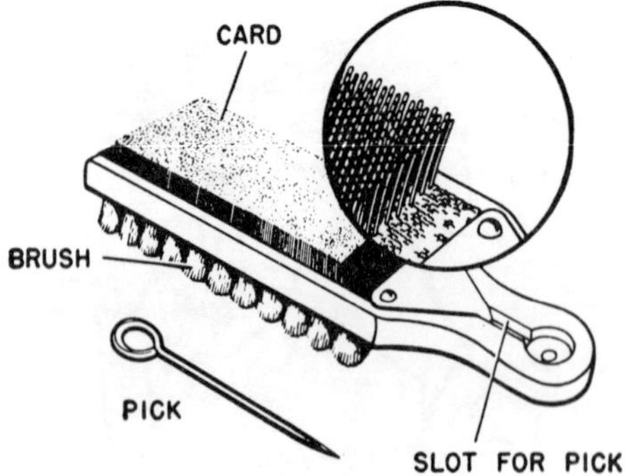

Fig. 57. File cleaner.

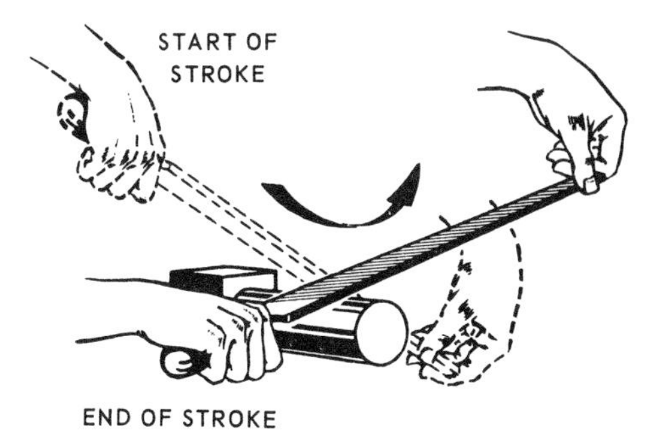

Fig. 58. Filing round metal stock.

Care of Files

A new file should be broken in carefully by using it first on brass, bronze, or smooth cast iron. Just a few of the teeth will cut at first, so use a light pressure to prevent tooth breakage. Do not break in a new file by using it first on a narrow surface.

Protect the file teeth by hanging your files in a rack when they are not in use, or by placing them in drawers with wooden partitions. Your files should not be allowed to rust—keep them away from water and moisture. Also avoid getting the files oily. Oil causes a file to slide across the work and prevents fast, clean cutting. Files that you keep in your toolbox should be wrapped in paper or cloth to protect their teeth and prevent damage to other tools.

Never use a file for prying or pounding. The tang is soft and bends easily. The body is hard and extremely brittle. Even a slight bend or a fall to the floor may cause a file to snap in two. Do not strike a file against the bench or vise to clean it—always use a file card.

Safety Precautions

Never use a file unless it is equipped with a tight-fitting handle. If you use a file without the handle and it bumps something or jams to a sudden stop, the tang may be driven into your hand. To put a handle on a file tang, drill a hole in the handle slightly smaller than a tang. Insert the tang end, and then tap the end of the handle to seat it firmly. Be sure the handle is on straight.

TWIST DRILLS

Making a hole in a piece of metal is generally a simple operation, but in most cases is an important and a precise job. A large number of different tools and machines have been designed so that holes may be made speedily, economically, and accurately in all kinds of material.

In order to be able to use these tools efficiently, it is well to become acquainted with them. The most common tool for making holes in metal is the *twist drill.* It consists of a cylindrical piece of steel with spiral grooves. One end of the cylinder is pointed while the other end is shaped so that it may be attached to a drilling machine. The grooves, usually called *flutes*, may be cut into the steel cylinder, or the flutes may be formed by twisting a flat piece of steel into a cylindrical shape.

The principal parts of a twist drill are the *body*, the *shank*, and the *point* (Fig. 59). The dead center of a drill is the sharp edge at the extreme tip end of the drill. It is formed by the intersection of the cone-shaped surfaces of the point and should always be in the exact center of the axis of the drill. The point of the drill should not be confused with the dead center. The point is the entire cone-shaped surface at the end of the drill.

The lip or cutting edge of a drill is that part of the point that actually cuts away the metal when drilling a hole. It is ordinarily as sharp as the edge of a knife. There is a cutting edge for each flute of the drill.

The lip clearance of a drill is the surface of the point that is ground away or relieved just back of the cutting edge of the drill. The strip along the inner edge of the body is called the margin. It is the greatest diameter of the drill and extends the entire length of the flute. The diameter of the margin at the shank end of the drill is smaller than the diameter at the point. This allows the drill to revolve without binding when drilling deep holes.

The shank is the part of the drill which fits into the socket, spindle, or chuck of the drill press. (*See* Fig. 60.)

A tang is found only on tapered-shank drills. It is designed to fit into a slot in the socket or spindle of a machine. It may bear a portion of the driving torque, but its principal use is to make it easy to remove the drill from the socket of the driving machine.

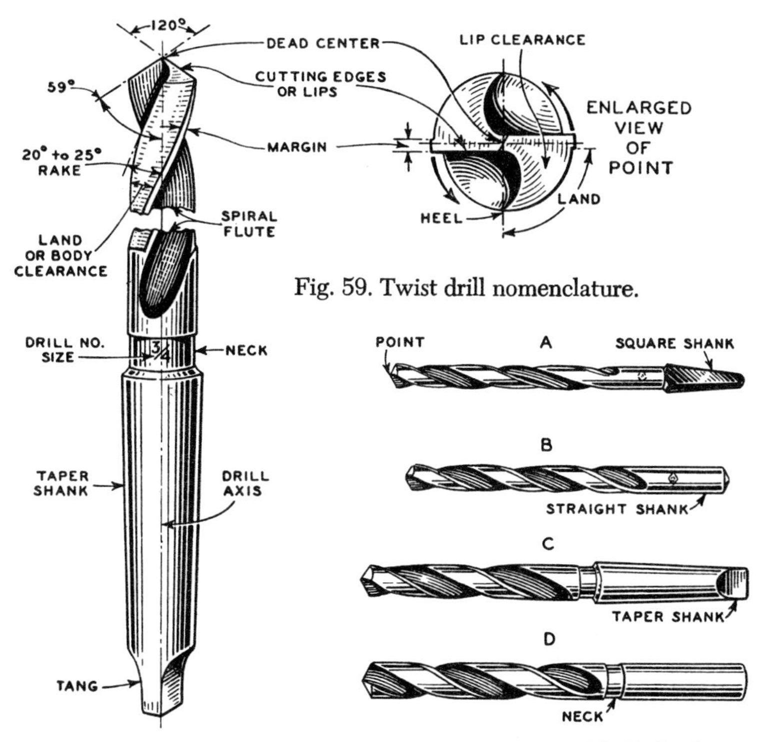

Fig. 59. Twist drill nomenclature.

Fig. 60. Types of drill shanks.

Twist drills are provided in various sizes. They are sized by letters, numerals, and fractions.

Table 1 illustrates the relationship, by decimal equivalents, of all drill sizes (letter, number, and fractional) from number 80 to ½ inch (1.27-cm.). Note how the decimal sizes increase as the number of the drill decreases.

TABLE 1

DECIMAL EQUIVALENTS OF DRILL SIZES.

Drill	Decimal	Drill	Decimal	Drill	Decimal	Drill	Decimal
80	0.0135	49	0.073	20	0.161	H	0.266
79	0.0145	48	0.076	19	0.166	I	0.272
78	0.016	5/64	0.078125	18	0.1695	J	0.277
1/64	0.0156	47	0.0785	11/64	0.171875	K	0.281
77	0.018	46	0.081	17	0.173	9/32	0.28125
76	0.02	45	0.082	16	0.177	L	0.29

TABLE 1 (CONTINUED).

75	0.021	44	0.086	15	0.18	M	0.295
74	0.0225	43	0.089	14	0.182	19/64	0.296875
73	0.024	42	0.0935	13	0.185	N	0.302
72	0.025	3/32	0.09375	3/16	0.1875	5/16	0.3125
71	0.026	41	0.096	12	0.189	O	0.316
70	0.028	40	0.098	11	0.191	P	0.323
69	0.0292	39	0.0995	10	0.1935	21/64	0.328125
68	0.031	38	0.1015	9	0.196	Q	0.332
1/32	0.03125	37	0.104	8	0.199	R	0.339
67	0.032	36	0.1055	7	0.201	11/32	0.34375
66	0.033	7/64	0.109375	13/64	0.203125	S	0.348
65	0.035	35	0.11	6	0.204	T	0.358
64	0.036	34	0.111	5	0.2055	23/64	0.359375
63	0.037	33	0.113	4	0.209	U	0.368
62	0.038	32	0.116	3	0.213	3/8	0.375
61	0.039	31	0.12	7/32	0.21875	V	0.377
60	0.04	1/8	0.125	2	0.221	W	0.386
59	0.041	30	0.1285	1	0.228	25/64	0.390625
58	0.042	29	0.136	A	0.234	X	0.397
57	0.043	28	0.1405	15/64	0.234375	Y	0.404
56	0.0465	9/64	0.140625	B	0.238	13/32	0.40625
3/64	0.046875	27	0.144	C	0.242	Z	0.413
55	0.052	26	0.147	D	0.246	27/64	0.421875
54	0.055	25	0.1495	E	0.25	7/16	0.4375
53	0.0595	24	0.152	1/4	0.25	29/64	0.453125
1/16	0.0625	23	0.154	F	0.257	15/32	0.46875
52	0.0635	5/32	0.15625	G	0.261	31/64	0.484375
51	0.067	22	0.157	17/64	0.265625	1/2	0.5
50	0.07	21	0.159				

Sets of drills are usually made available according to the way the sizes are stated, for example, *sets of letter drills* or *sets of number drills*. However, twist drills of any size (letter, number, or fractional) are available individually if desired.

COUNTERSINKS

Countersinking is the operation of beveling the mouth of a hole with a rotary tool called a *countersink* (Fig. 61). The construction of the countersink is similar to the twist drill. There are four cutting angles, which are taper ground, to the angle marked on the body.

A countersink is used primarily to set the head of a screw or rivet flush with the material in which it is being placed. Countersinks are made in a number of sizes. One size usually takes care of holes of several different sizes. For example, the

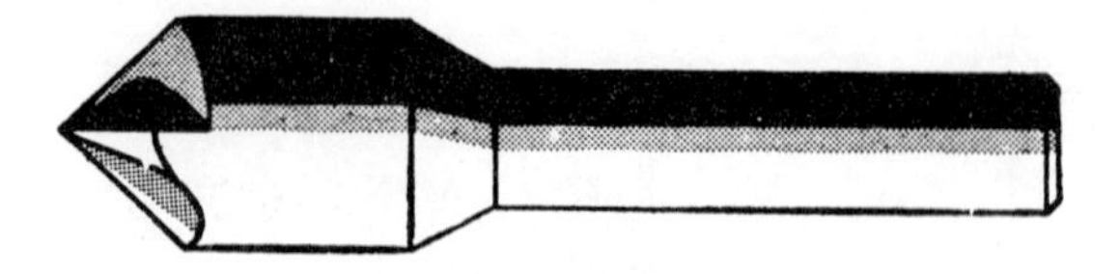

Fig. 61. Countersink.

same countersink can be used for holes from ¼ to ½ inch (.635 to 1.27 cm.) in diameter. Remove only enough metal to set the screw or rivet head flush with the material. If you remove too much material the hole will enlarge and weaken the work.

Select the countersink with the correct lip angle to correspond with the screw or rivet head being used. These countersinks can be turned by any machine that will turn a twist drill.

TAPS AND DIES

Taps and *dies* are used to cut threads in metal, plastics, or hard rubber.

The *taps* are used for cutting internal threads, and the *dies* are used to cut external threads. There are many different types of taps. The most common types are the taper pipe, the taper hand, and the bottoming hand (Fig. 62).

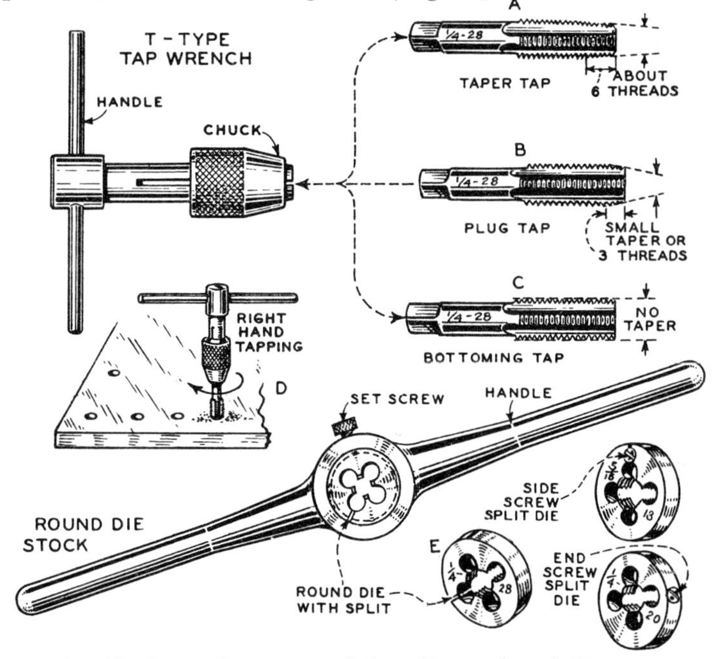

Fig. 62. Types of tapping and threading tools and dies.

The *taper* (starting) *hand tap* has a chamfer length of 8 to 10 threads. These taps are used when starting a tapping operation and when tapping through holes.

Plug hand taps have a chamfer length of 3 to 5 threads and are designed for use after the taper tap.

Bottoming hand taps are used for threading the bottom of a blind hole. They have a very short chamfer length of only 1 to 1½ threads for this purpose. This tap is always used after the plug tap has already been used. Both the taper and the plug taps should precede the use of the bottoming hand tap.

Pipe taps are used for pipe fittings and other places where extremely tight fits are necessary. The tap diameter, from end to end of threaded portion, increases at the rate of ¾ inch (1.905 cm.) per foot (30.48 cm.). All the threads on this tap do the cutting as compared to the straight taps where only the nonchamfered portion does the cutting.

Dies are made in several different shapes and are of the solid or adjustable type. The *square pipe die* (Fig. 63) will cut American Standard Pipe Thread only. It comes in a variety of sizes for cutting threads on pipe with diameters of ⅛ inch to 2 inches (.3175 to 5.08 cm.).

A *rethreading die* (Fig. 63) is used principally for dressing over bruised or rusty threads on screws or bolts. It is available in a variety of sizes for rethreading American Standard Coarse and Fine Threads. These dies are usually hexagonal in shape and can be turned with a socket, box, open-end, or any wrench that will fit. Rethreading dies are available in sets of 6, 10, 14, and 28 assorted sizes in a case.

Round split adjustable dies (Fig. 64) are called *button dies* and can be used in either hand diestocks or machine holders.

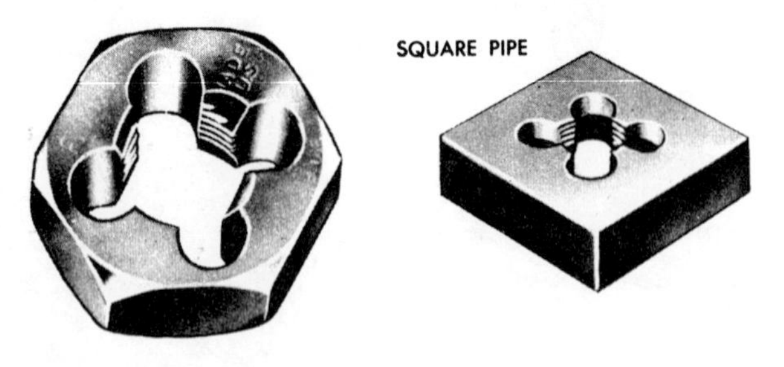

Fig. 63. Types of solid dies.

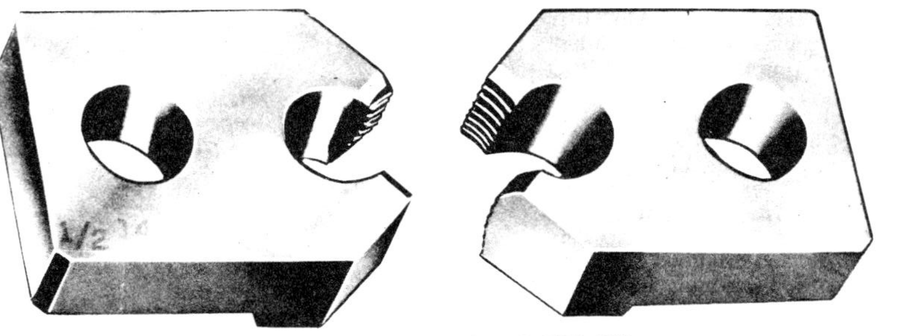

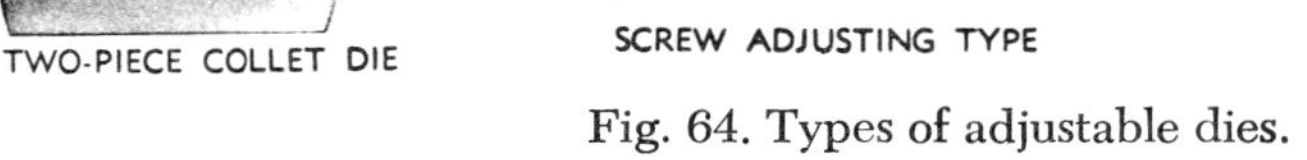

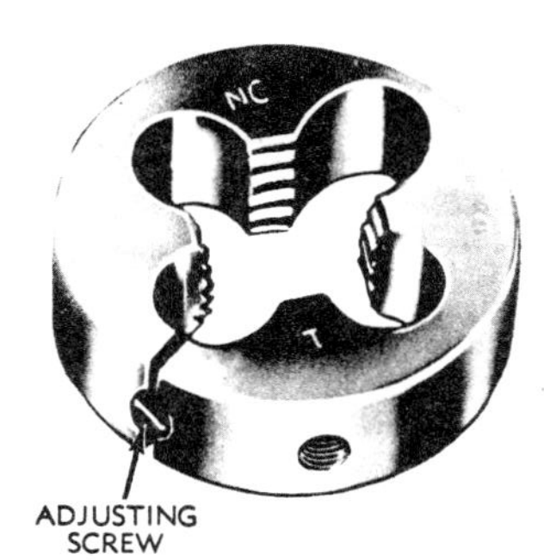

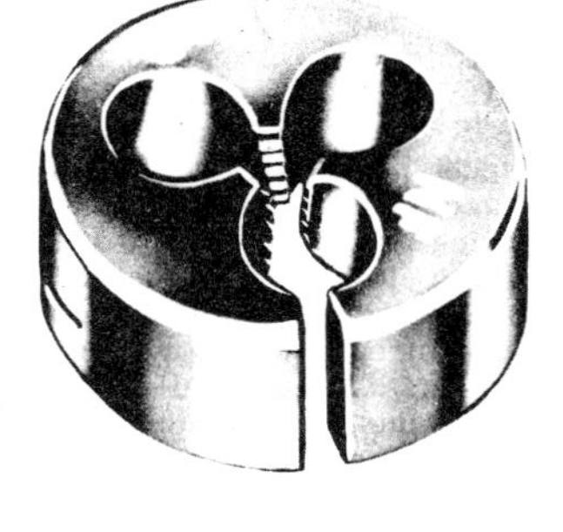

Fig. 64. Types of adjustable dies.

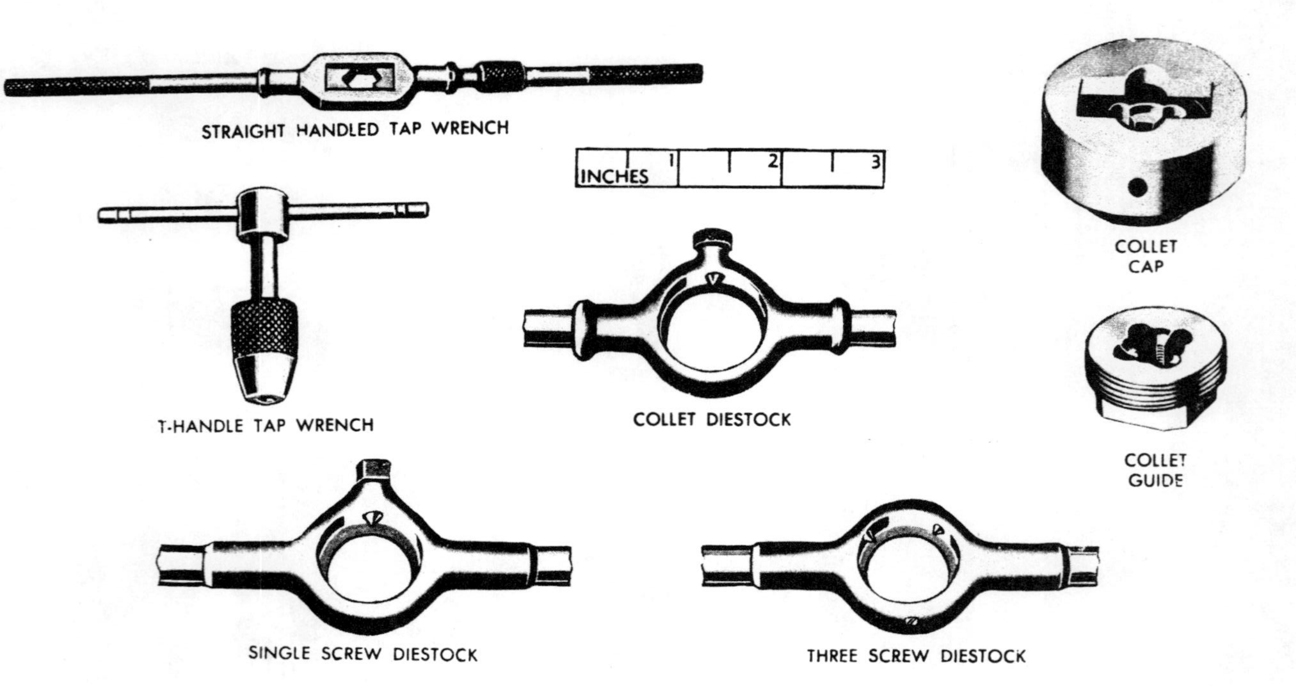

Fig. 65. Diestocks, diecollet, and tap wrenches.

The adjustment in the screw-adjusting type is made by a fine-pitch screw which forces the sides of the die apart or allows them to spring together. The adjustment in the open adjusting types is made by means of three screws in the holder, one for expanding and two for compressing the dies. Round split adjustable dies are available in a variety of sizes to cut American Standard Coarse and Fine Threads, special form threads, and the standard sizes of threads that are used in Britain and other European countries. For hand threading, these dies are held in diestocks (Fig. 65). One type die stock has three pointed screws that will hold round dies of any construction, although it is made specifically for open adjusting type dies.

Two piece collet dies (Fig. 66) are used with a collet cap (Fig. 65) and collet guide. The die halves are placed in the

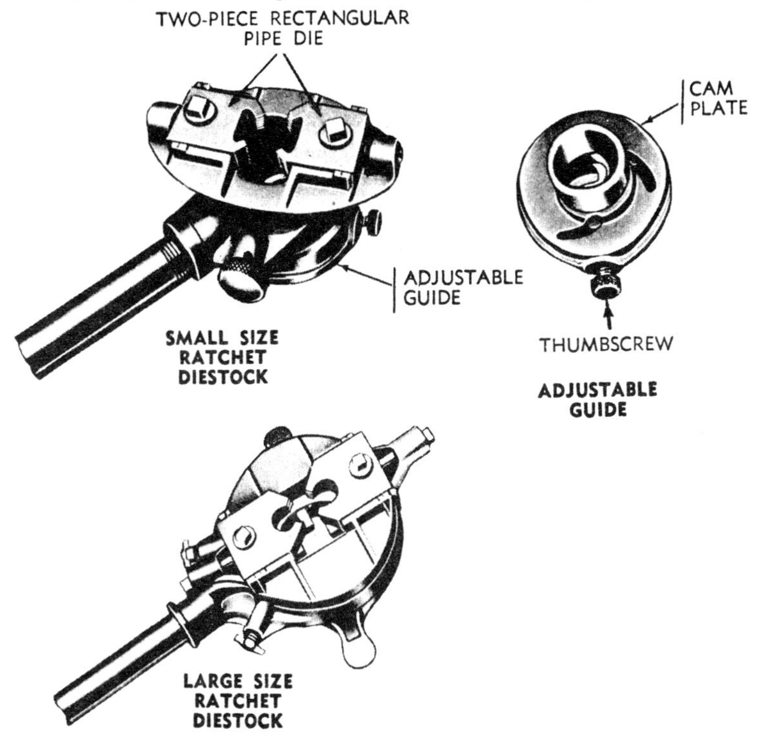

Fig. 66. Adjustable die guide and ratchet diestocks.

cap slot and are held in place by the guide which screws into the underside of the cap. The die is adjusted by means of setscrews at both ends of the internal slot. This type of adjustable die is issued in various sizes to cover the cutting range of American Standard Coarse and Fine and special form threads. Diestocks to hold the dies come in three different sizes.

Two-piece rectangular pipe dies (Fig. 64) are available to cut American Standard Pipe threads. They are held in ordinary or ratchet-type diestocks (Fig. 66). The jaws of the dies are adjusted by means of setscrews. An adjustable guide serves to keep the pipe in alinement with respect to the dies. The smooth jaws of the guide are adjusted by means of a cam plate, and a thumbscrew locks the jaws firmly in the desired position.

Threading sets are available in many different combinations of taps and dies, together with diestocks, tap wrenches, guides and necessary screwdrivers and wrenches to loosen and tighten adjusting screws and bolts. Figure 67 shows a threading set for pipe, bolts, and screws.

Never attempt to sharpen taps or dies. Sharpening of taps and dies involves several highly precise cutting processes which involve the thread characteristics and chamfer. These sharpening procedures must be done by experienced people in order to maintain the accuracy and the cutting effectiveness of taps and dies.

Keep taps and dies clean and well oiled when not in use. Store them so that they do not contact each other or other tools. For long periods of storage, coat taps and dies with a rust preventive compound, place in individual or standard threading set boxes, and store in a dry place.

THREAD CHASERS

Thread chasers are threading tools that have several teeth and are used to rethread (chase) damaged external or internal threads (Fig. 68). These tools are available to chase standard threads. The internal thread chaser has its cutting teeth located on a side face. The external thread chase has its cutting teeth on the end of the shaft. The handle end of the tool shaft tapers to a point.

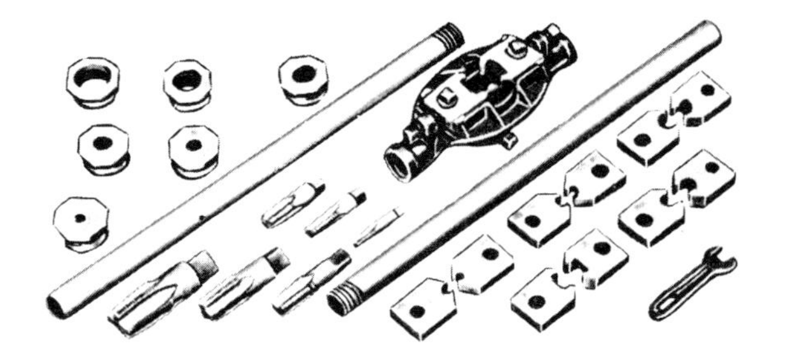

PIPE THREADING SET WITH RECTANGULAR ADJUSTABLE DIES, DIESTOCK, WRENCH, GUIDES AND TAPS

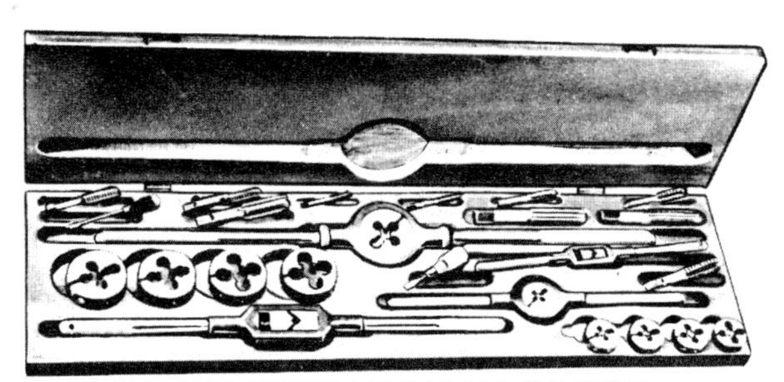

BOLT AND SCREW THREADING SET WITH ROUND ADJUSTABLE SPLIT DIES, DIESTOCKS, TAPS, TAP WRENCHES, AND SCREWDRIVERS

Fig. 67. Threading sets.

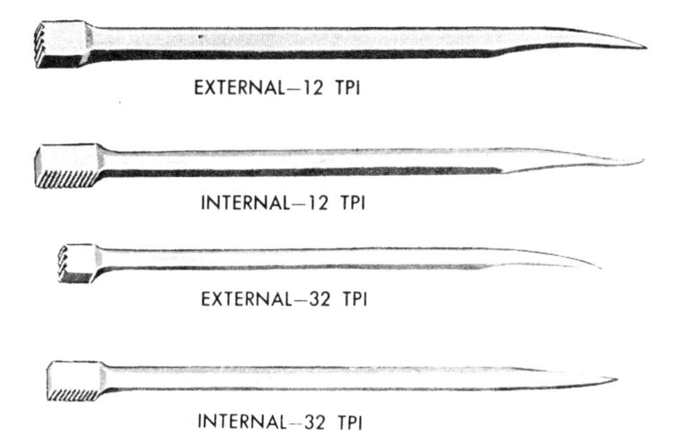

Fig. 68. Thread chasers.

SCREW AND TAP EXTRACTORS

Screw extractors are used to remove broken screws without damaging the surrounding material or the threaded hole. *Tap extractors* are used to remove broken taps.

Some screw extractors (Fig. 69) are straight, having flutes from end to end. These extractors are available in sizes to remove broken screws having ¼- to ½-inch (.635- to 1.27-cm.) outside diameters. Spiral tapered extractors are sized to remove screws and bolts from 3⁄16-inch to 2⅛-inch (.47625- to 5.3975-cm.) outside diameter.

Most sets of extractors include twist drills and a drill guide. Tap extractors are similar to the screw extractors and are sized to remove taps having from 3⁄16- to 2⅛-inches (.47625- to 5.3975-cm.) outside diameter.

To remove a broken screw or tap with a *spiral extractor*, first drill a hole of proper size in the screw or tap. The size hole required for each screw extractor is stamped on it. The extractor is then inserted in the hole, and turned counterclockwise to remove the defective component.

If the tap has broken off at the surface of the work, or slightly below the surface of the work, the *straight tap extractor* shown in Fig. 69 may remove it. Apply a liberal amount of penetrating oil to the broken tap. Place the tap extractor over the broken tap and lower the upper collar to insert the four sliding prongs down into the four flutes of the tap. Then slide the bottom collar down to the surface of the work so that it will hold the prongs tightly against the body of the extractor. Tighten the tap wrench on the square shank of the extractor and carefully work the extractor back and forth to loosen the tap. It may be necessary to remove the extractor and strike a few sharp blows with a small hammer and pin punch to jar the tap loose. Then reinsert the tap remover and carefully try to back the tap out of the hole.

HANDSAWS

The most common *handsaw* consists of a steel blade with a handle at one end. The blade is narrower at the end opposite the handle. This end of the blade is called the *point* or *toe*. The end of the blade nearest the handle is called the *heel*

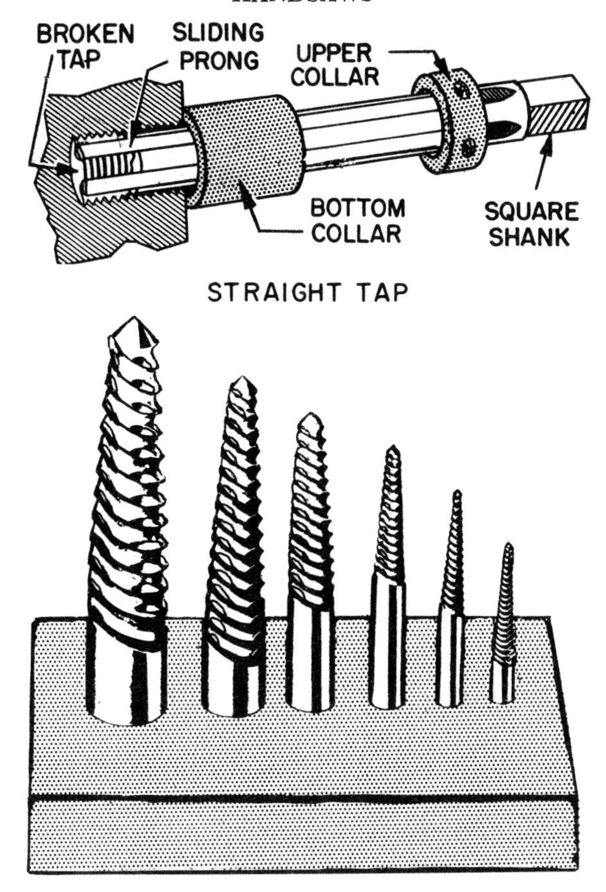

Fig. 69. Screw and tap extractors.

(Fig. 70). One edge of the blade has teeth, which act as two rows of cutters. When the saw is used, these teeth cut two parallel grooves close together. The chips (sawdust) are pushed out from between the grooves (kerf) by the beveled part of the teeth. The teeth are bent alternately to one side or the other, to make the kerf wider than the thickness of the blade. This bending is called the *set* of the teeth (Fig. 71). The number of teeth per inch, the size and shape of the teeth, and the amount of set depend on the use to be made of the

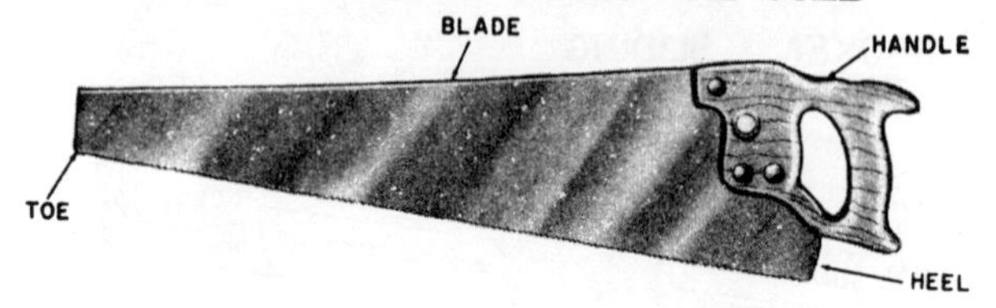

Fig. 70. Nomenclature of a handsaw.

Fig. 71. "Set" of handsaw teeth.

saw and the material to be cut. These handsaws are described by the number of points per inch. There is always one more point than there are teeth per inch. A number stamped near the handle gives the number of points of the saw.

Ripsaws and Crosscut Saws

Woodworking handsaws designed for general cutting consist of *ripsaws* and *crosscut* saws. Ripsaws are used for cutting with the grain and crosscut saws are used for cutting across the grain.

The major difference between a ripsaw and a crosscut saw is the shape of the teeth. A tooth with a square-faced chisel-type cutting edge, like the ripsaw tooth shown in Fig. 72, does a good job of cutting with the grain (called ripping), but a poor job of cutting across the grain (called crosscutting). A tooth with a beveled, knife-type cutting edge, like the crosscut saw tooth shown in Fig. 72, does a good job of cutting across the grain, but a poor job of cutting with the grain.

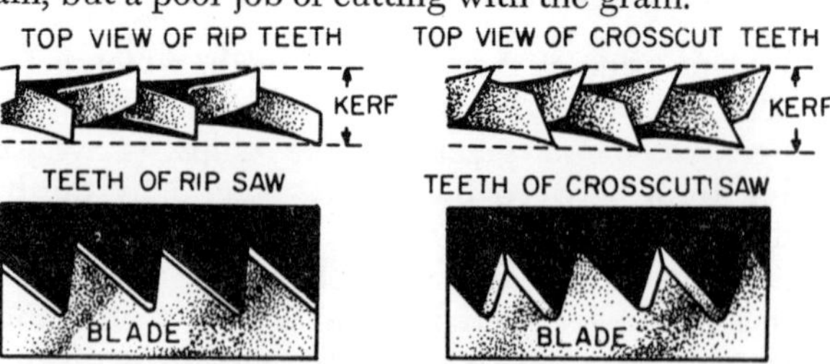

Fig. 72. Comparing rip and crosscut saw teeth.

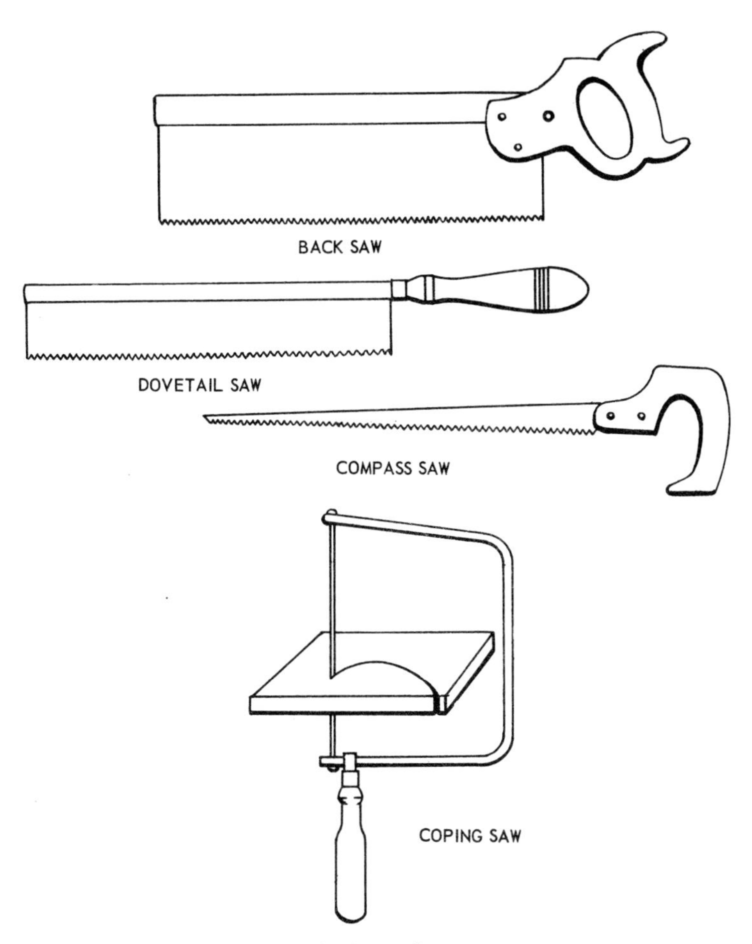

Fig. 73. Special saws.

Special Purpose Saws

The more common types of saws used for special purposes are shown in Fig. 73. The *backsaw* is a crosscut saw designed for sawing a perfectly straight line across the face of a piece of stock. A heavy steel backing along the top of the blade keeps the blade perfectly straight.

The *dovetail saw* is a special type of backsaw with a thin, narrow blade and a chisel-type handle (Fig. 73).

The *compass saw* (Fig. 73) is a long, narrow, tapering ripsaw designed for cutting out circular or other nonrectangular sections from within the margins of a board or panel. A hole is bored near the cutting line to start the saw. A *keyhole saw* is simply a finer, narrower compass saw.

The *coping saw* is used to cut along curved lines as shown in Fig. 73.

Saw Precautions

A saw that is not being used should be hung up or stowed in a toolbox. A toolbox designed for holding saws has notches that hold them on edge, teeth up. Stowing saws loose in a toolbox may allow the saw teeth to become dulled or bent by contacting other tools.

Before using a saw, be sure there are no nails or other edge destroying objects in the line of the cut. When sawing out a strip of waste, do not break out the strip by twisting the saw blade. This dulls the saw and may spring or break the blade.

Be sure that the saw will go through the full stroke without striking the floor or some other object. If the work cannot be raised high enough to obtain full clearance for the saw, you must carefully limit the length of each stroke.

Using a Handsaw

To saw across the grain of the stock, use the crosscut saw, and to saw with the grain, use a ripsaw. Study the teeth in both types of saws so you can readily identify the saw that you need.

Place the board on a sawhorse, or some other suitable object. Hold the saw in the right hand and extend the first finger along the handle as shown in Fig. 74. Grasp the board as shown and take a position so that an imaginary line passing lengthwise of the right forearm will be at an angle of approximately 45 degrees with the face of the board. Be sure the side of the saw is plumb or at right angles with the face of the board. Place the heel of the saw on the mark. Keep the saw in line with the forearm and pull it toward you to start the cut.

To begin with, take short, light strokes, gradually increasing the strokes to the full length of the saw. Do not force or jerk the saw. Such a procedure will only make sawing more difficult. The arm that does the sawing should swing clear of your

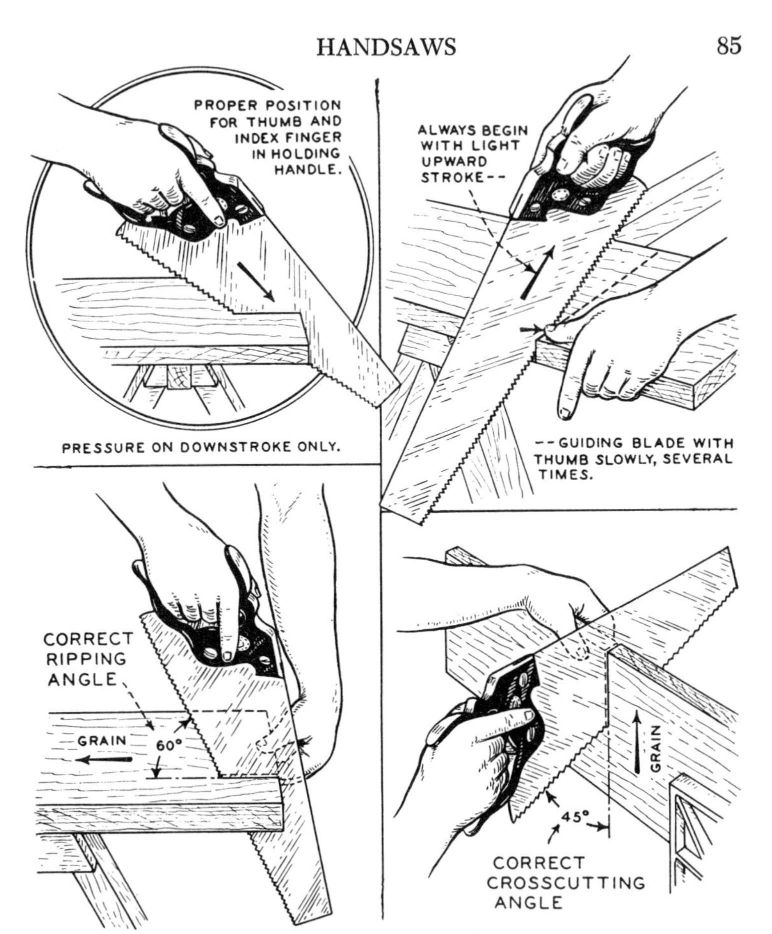

Fig. 74. Proper way to hold a saw.

body so that the handle of the saw operates at your side rather than in front of you.

Use one hand to operate the saw. You may be tempted to use both hands at times, but if your saw is sharp, one hand will serve you better. The weight of the saw is sufficient to make it cut. Should the saw stick or bind, it may be because the saw is dull and is poorly set. The wood may have too much moisture in it, or you may have forced the saw and therefore have caused it to leave the straight line.

Keep your eye on the line rather than on the saw while sawing. Watching the line enables you to see instantly any tendency to leave the line. A slight twist of the handle and taking short strokes while sawing will bring the saw back. Blow away the sawdust frequently so you can see the layout line.

Final strokes of the cut should be taken slowly. Hold the waste piece in your other hand so the stock will not split when taking the last stroke.

Short boards may be placed on one sawhorse when sawing. Place long boards on two sawhorses, but do not saw so your weight falls between them or your saw will bind. Place long boards so that your weight is directly on one end of the board over one sawhorse while the other end of the board rests on the other sawhorse.

Short pieces of stock are more easily cut when they are held in a vise. When ripping short stock it is important that you keep the saw from sticking, so it may be necessary to take a squatting position. The saw can then take upward direction and therefore work easily. When ripping long boards it will probably be necessary to use a wedge in the saw kerf to prevent binding (Fig. 75).

PLANES

The *plane* is the most extensively used of the hand shaving tools. Most of the lumber handled by anyone working with wood is dressed on all four sides, but when performing jobs such as fitting doors and sash, and interior trim work, planes must be used.

Bench and *block planes* are designed for general surface smoothing and squaring. Other planes are designed for special types of surface work.

Fig. 75.

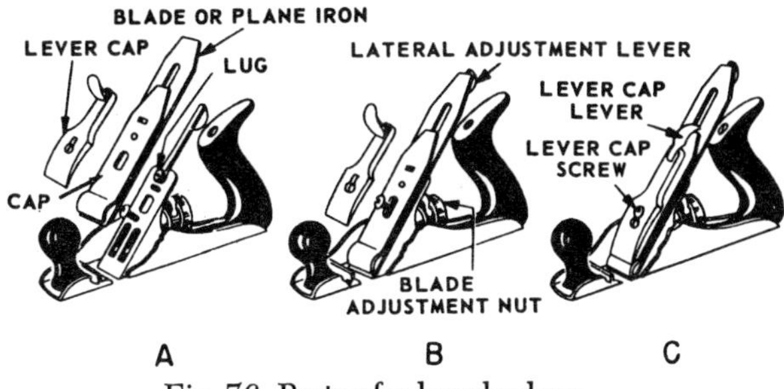

Fig. 76. Parts of a bench plane.

The principal parts of a bench plane and the manner in which they are assembled are shown in Fig. 76. The part at the rear that you grasp to push the plane ahead is called the handle; the part at the front that you grasp to guide the plane along its course is called the knob. The main body of the plane, consisting of the bottom, the sides, and the sloping part which carries the plane iron is called the mouth. The front end of the sole is called the toe; and the rear end is called the heel.

A plane iron cap, which is screwed to the upper face of the plane iron, deflects the shaving upward through the mouth as shown at C, Fig. 77, and therefore prevents the mouth from becoming choked with jammed shavings. The edge of the cap should fit the back of the iron as shown at A, Fig. 77, not as shown at B, Fig. 77. The lower end of the plane iron cap should be set back $\frac{1}{32}$ inch (.079375 cm.) from the edge of the plane top as shown at A, Fig. 77. The iron in a bench plane goes in bevel down.

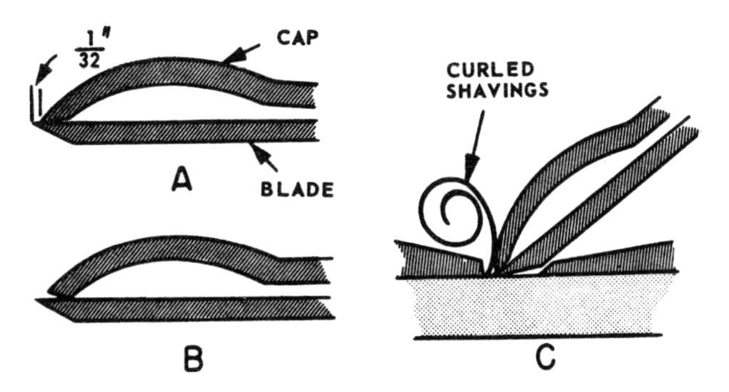

Fig. 77. Plane iron and plane iron cap.

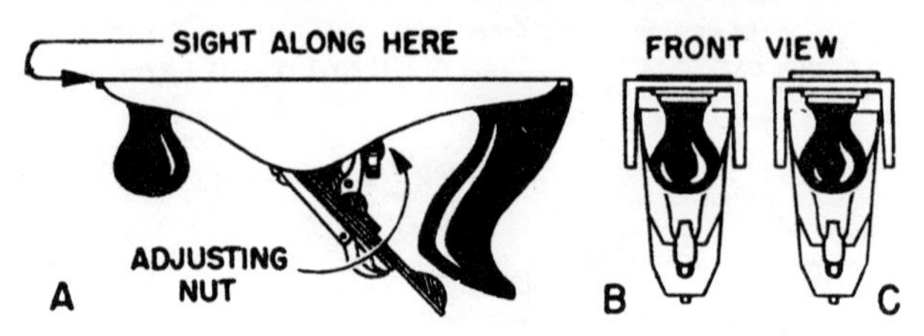

Fig. 78. Manipulation of the adjusting nut moves the plane iron up or down

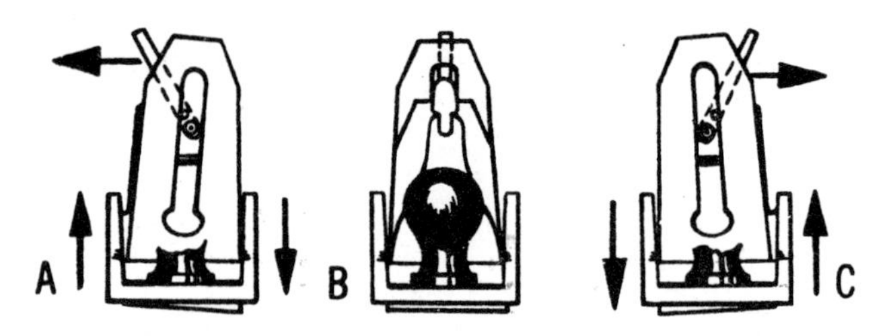

Fig. 79. Effect of manipulation of the lateral adjustment lever.

The edge of the plane iron is brought into correct cutting position by the manipulation of first the *adjusting nut* and next the *lateral adjustment lever* as shown in Figs. 78 and 79. The adjusting nut moves the edge of the iron up or down; the lateral adjustment lever cants it to the right or left. To adjust the plane you hold it upside-down, sight along the sole from the toe, and work the adjusting nut until the edge of the blade appears. Then work the lateral adjustment lever until the edge of the blade is in perfect alinement with the sole as shown in B, Fig. 78 and in B, Fig. 79. Then use the adjusting nut to give the blade the amount of protrusion you want. This amount will depend upon the depth of the cut you intend to make.

There are three types of bench planes (Fig. 80)—the *smooth plane*, the *jack plane*, and the *jointer plane* (sometimes called the *fore plane* or the *gage plane*). All are used primarily for shaving and smoothing with the grain; the chief difference is the length of the sole. The sole of the smooth plane is about 9 inches (22.86 cm.) long, the sole of the jack plane is about 14 inches (35.56 cm.) long, and the sole of the jointer plane is from 20 to 24 inches (55.88 to 60.96 cm.) long.

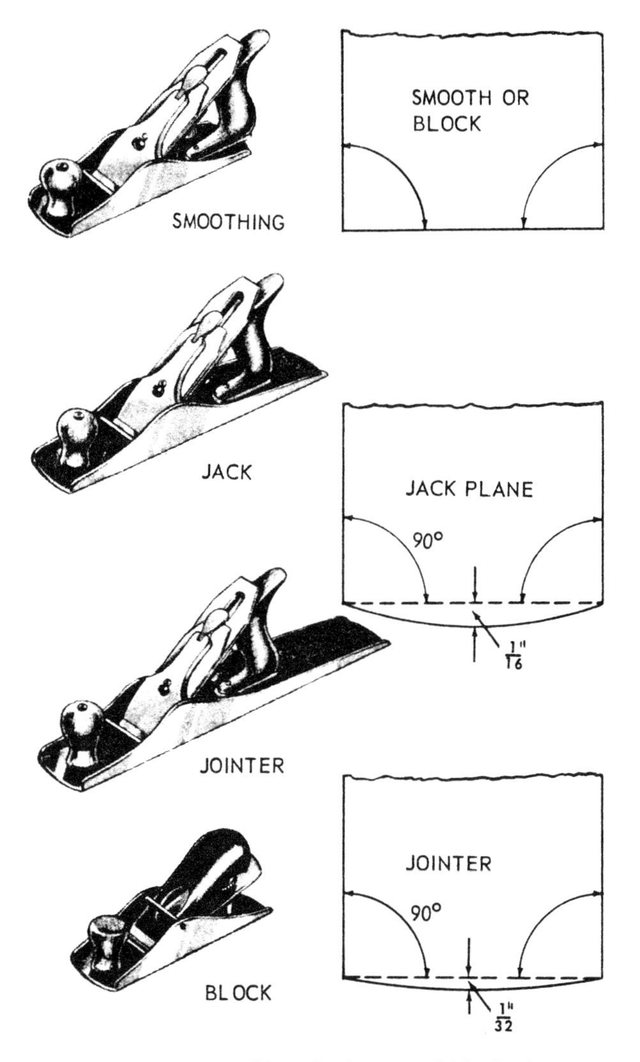

Fig. 80. Types of bench planes and block plane.

The longer the sole of the plane is, the more uniformly flat and true the planed surface will be. Consequently, which bench plane you should use depends upon the requirements with regard to surface trueness. The *smooth plane* is, in general, smoother only. It will plane a smooth, but not an especially true surface in a short time. It is also used for cross grain smoothing and squaring of end stock.

The *jack plane* is the general jack-of-all-work of the bench plane group. It can take a deeper cut and plane a truer surface than the smooth plane. The *jointer plane* is used when the planed surface must meet the highest requirements with regard to trueness.

A *block plane* and the names of its parts are shown in Fig. 81. Note that the plane iron in a block plane does not have a plane iron cap, and also that, unlike the iron in a bench plane, the iron in a block plane goes in bevel up.

The block plane, which is usually held at an angle to the work, is used chiefly for cross grain squaring of end stock. It is also useful for smoothing all plane surfaces on very small work.

BORING TOOLS

When working with wood, you will frequently be required to bore holes. It is important, therefore, that you know the proper procedures and tools used for this job. Auger bits and a variety of braces and drills are used extensively for boring purposes.

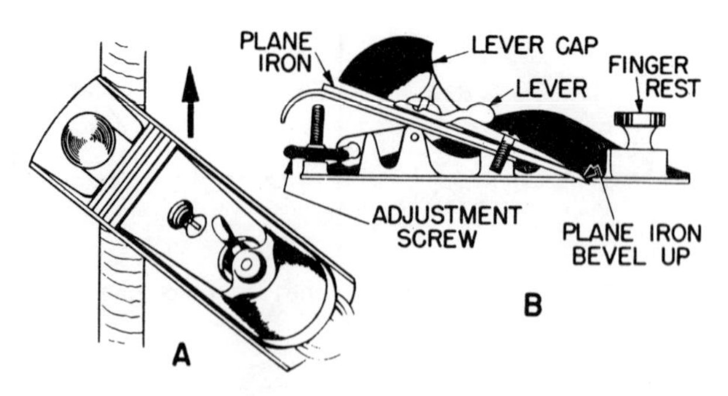

Fig. 81. Block plane nomenclature.

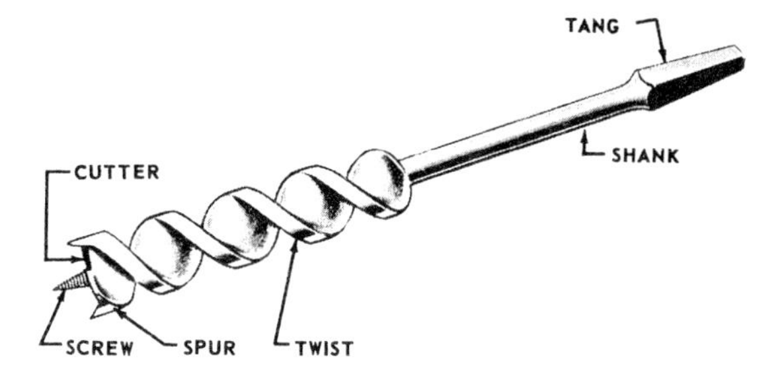

Fig. 82. Nomenclature of an auger bit.

AUGER BITS

Bits are used for boring holes for screws, dowels, and hardware, as an aid in mortising (cutting a cavity in wood for joining members) and in shaping curves and for many other purposes. Like saws and planes, bits vary in shape and structure with the type of job to be done. Some of the most common bits are described in this section.

Auger bits are screw-shaped tools consisting of six parts—the cutter, the screw, the spur, the twist, the shank, and the tang (Fig. 82). The twist ends with two sharp points called the spurs, which score the circle, and two cutting edges which cut shavings within the scored circle. The screw centers the bit and draws it into the wood. The threads of the screw are made in three different pitches—steep, medium, and fine. The steep pitch makes for quick boring and thick chips, and the fine or slight pitch makes for slow boring and fine chips. For end wood boring, a steep or medium pitch screw bit should be used because end wood is likely to be forced in between the fine screw threads, and that will prevent the screw from taking hold. The twist carries the cuttings away from the cutters and deposits them in a mound around the hole.

The sizes of auger bits are indicated in sixteenths of an inch and are stamped on the tang (Fig. 83). A number 10 stamped on the tang means $^{10}/_{16}$ or $^{5}/_{8}$ inch (1.5875-cm.); number 5, $^{5}/_{16}$ inch (.79375 cm.); and so on. The most common woodworker's auger bit set ranges in size from $^{1}/_{4}$ to 1 inch (.635 to 2.54 cm.).

Ordinary auger bits up to 1 inch (2.54 cm.) in diameter are from 7 to 9 inches (17.78 to 22.86 cm.) long. Short auger bits

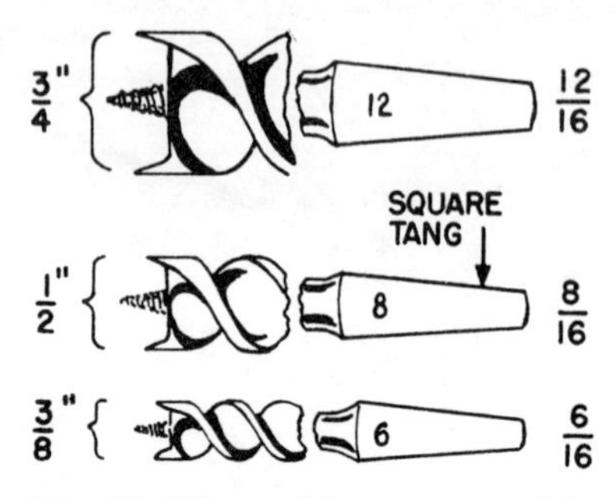

Fig. 83. Size markings on auger bits.

that are about 3½ inches (8.89 cm.) long are called *dowel bits*.

Expansive auger bits have adjustable cutters for boring holes of different diameters (Fig. 84). Expansive bits are generally made in two different sizes. The largest size has three cutters and bores holes up to 4 inches (10.16-cm.) in diameter. A scale on the cutter blade indicates the diameter of the hole to be bored.

BRACES AND DRILLS

The auger bit is the tool that actually does the cutting in the wood. However, it is necessary that another tool be used to hold the auger bit and give you enough leverage to turn the bit. The tools most often used for holding the bit are the *brace,* the *breast drill*, and the *push drill* (Fig. 85).

BORING THROUGH HOLES IN WOOD

To bore a hole in wood with an auger bit, first select the proper fit indicated on or near the square tang. Then insert the auger bit into the chuck (Fig. 86).

To chuck the bit, hold the shell of the chuck (A, Fig. 86) as you turn the handle to open the jaws. When the jaws are apart far enough to take the square tang of the bit, insert it until the end seats in the square driving socket at the bottom of the chuck (B, Fig. 86). Then tighten the chuck by turning the handle to close the jaws and hold the bit in place.

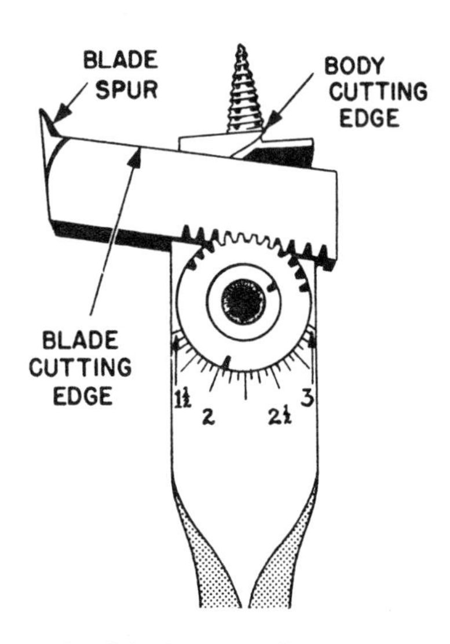

Fig. 84. Expansive bit.

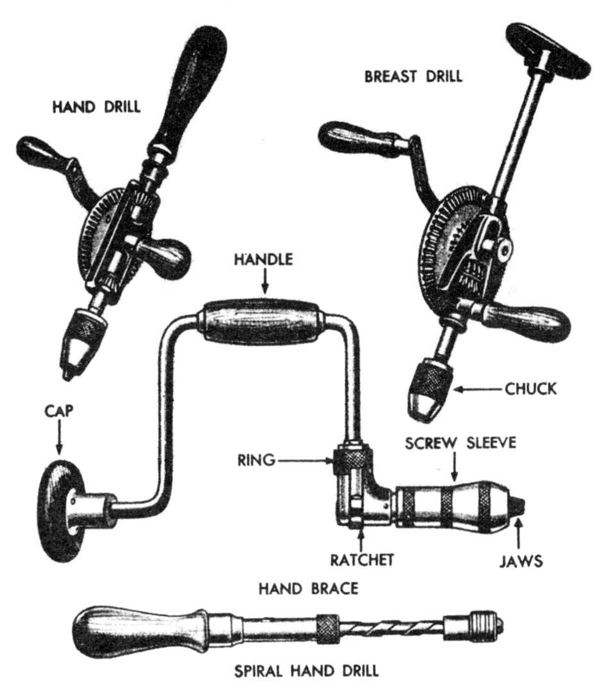

Fig. 85. Drills and hand brace.

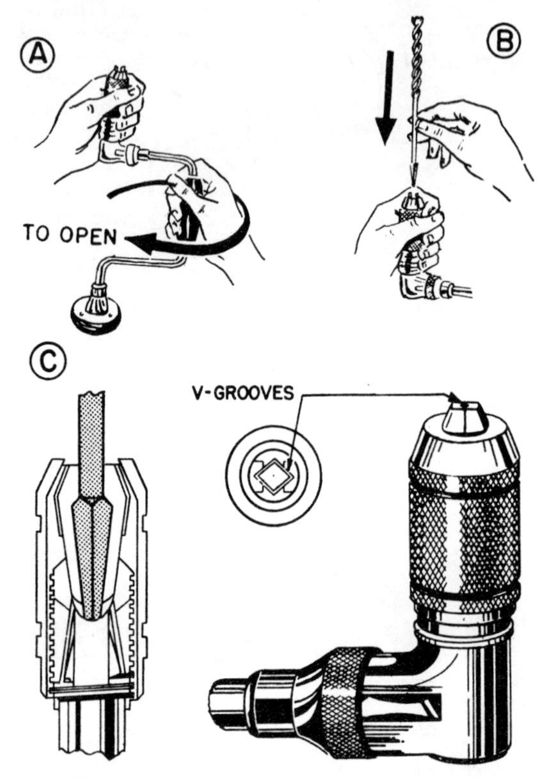

Fig. 86. Placing an auger bit in a chuck.

With a chuck having no driving socket (a square hole which is visible if you look directly into the chuck), additional care must be taken to seat and center the corners of the tapered shank in the grooves of the chuck jaws. (*See* C, Fig. 86.) In this type of chuck the jaws serve to hold the bit in the center and to prevent it from coming out of the chuck.

After placing the point of the feed screw at the location of the center of the hole you will bore, steady the brace against your body, if possible, with the auger bit square with the surface of the work.

To bore a horizontal hole in the stock held in the bench vise, hold the head of the brace with one hand, steadying it against your body, while turning the handle with the other hand. Scrap stock behind the job will prevent splintering (Fig. 88).

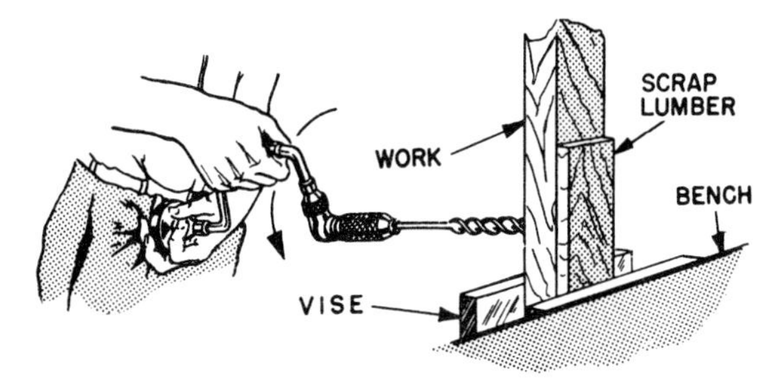

Fig. 87. Using scrap lumber to prevent splintering when boring.

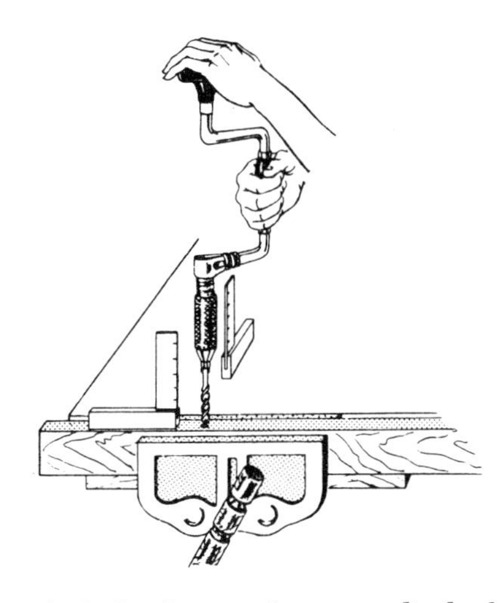

Fig. 88. Method of sighting in for perpendicular hole.

When it is not possible to make a full turn with the handle of the bit brace, turn the cam ring shown in Fig. 85 clockwise until it stops. This will raise one of the two ratchet pawls affording clockwise ratchet action for rotating the bit. For counterclockwise ratchet action, turn the cam ring counterclockwise as far as it will go.

To bore a vertical hole in stock held in a bench vise, hold the brace and the bit perpendicular to the surface of the work. Placing a try square near the bit, alternately in the two positions shown in Fig. 88 will help you sight it in.

Another way to bore a through hole without splitting out on the opposite face is to reverse the bit one or two turns when

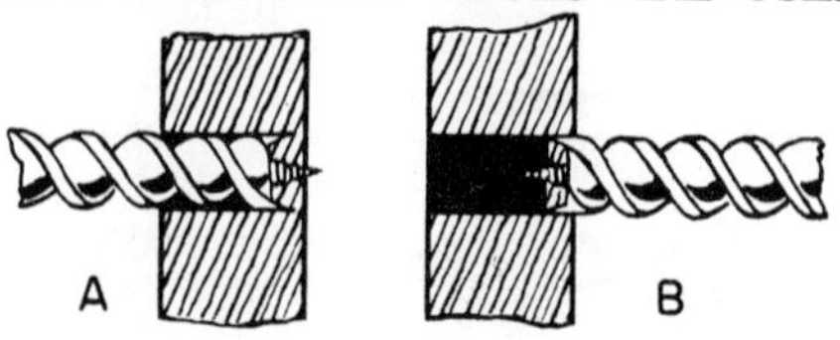

Fig. 89. Boring a through hole by reversing direction.

the feed screw just becomes visible through this opposite face (A, Fig. 89). This will release the bit. Remove the bit while pulling it up and turning it clockwise. This will remove the loose chips from the hole. Finish the hole by boring from the opposite face. This will remove the remaining material which is usually in the form of a wooden disk held fast to the feed screw (B, Fig. 89).

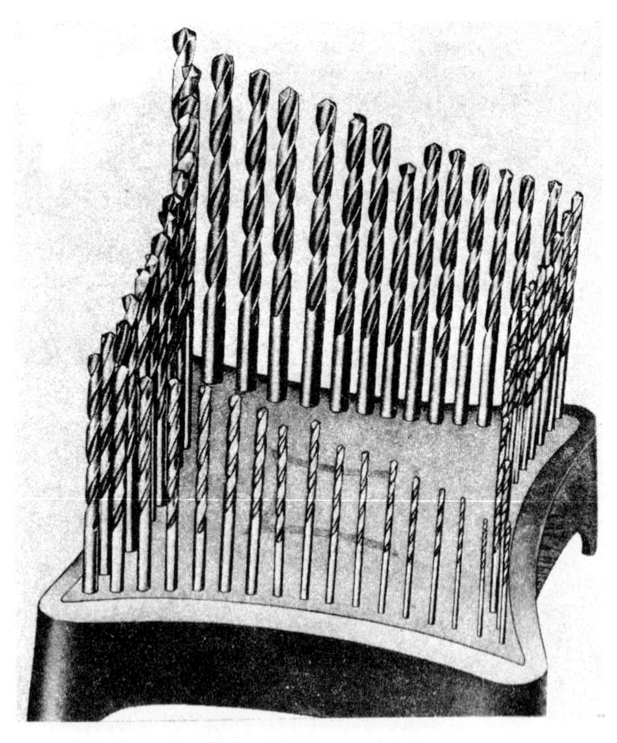

Fig. 90. Twist drills of various sizes.

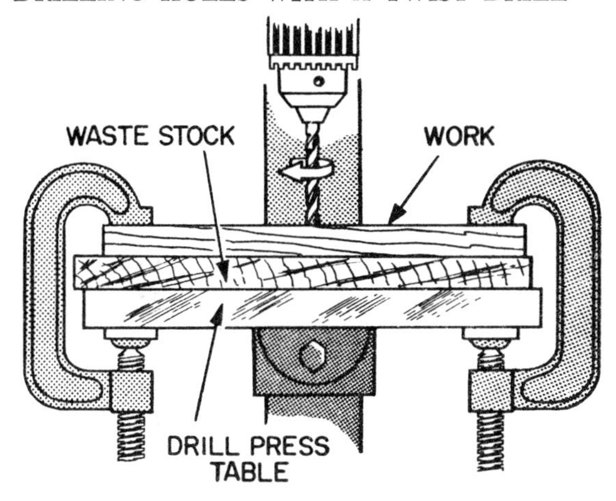

Fig. 91. Drilling a hole in wood with a twist drill.

DRILLING HOLES WITH A TWIST DRILL

An ordinary *twist drill* may be used to drill holes in wood. Select a twist drill of the size required and secure it in the chuck of a drill. Figure 90 shows sizes of twist drills from No. 1 to No. 60.

In Fig. 91 the twist drill has been chucked. Note that the job is secured to the table with a pair of C clamps. Beneath the job is a block of wood. In drilling through wood, a backup block is used to ensure a clean hole at the bottom of the job.

Figure 92 shows a hole being drilled with a breast drill. Turn the crank handle with one hand as you hold the side handle with the other hand. This will steady the breast drill while feed pressure is applied by resting your chest on the breast plate shown in Fig. 92. Note that the breast drill has a high or a low speed available, according to the setting of the speed selector nut. When drilling a horizontal hole, apply feed pressure by resting your body against the breast plate.

In drilling a horizontal hole with the hand drill shown in Fig. 93, operate the crank with the right hand and with the left hand guide the drill by holding the handle which is opposite the chuck end of the drill.

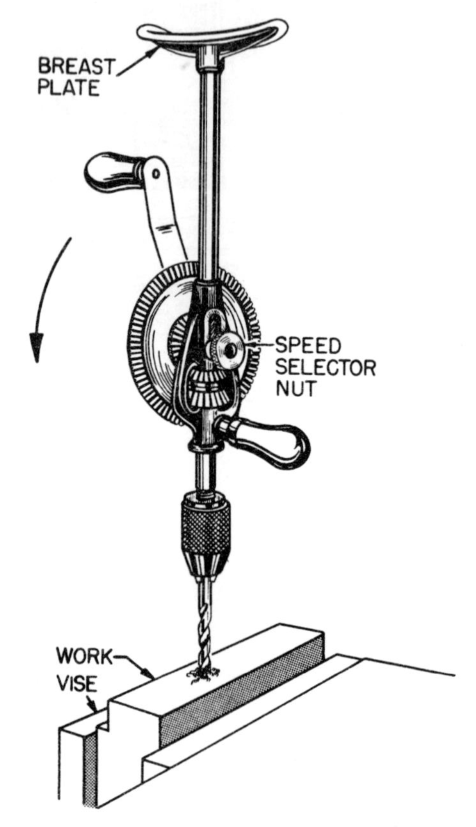

Fig. 92. Drilling a hole with a breast drill.

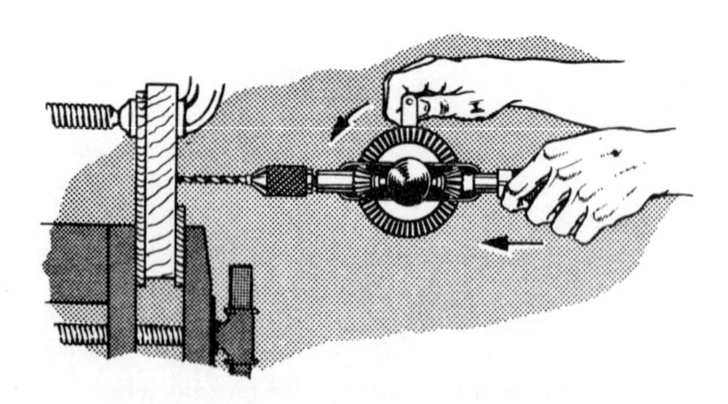

Fig. 93. Drilling a hole with a hand drill.

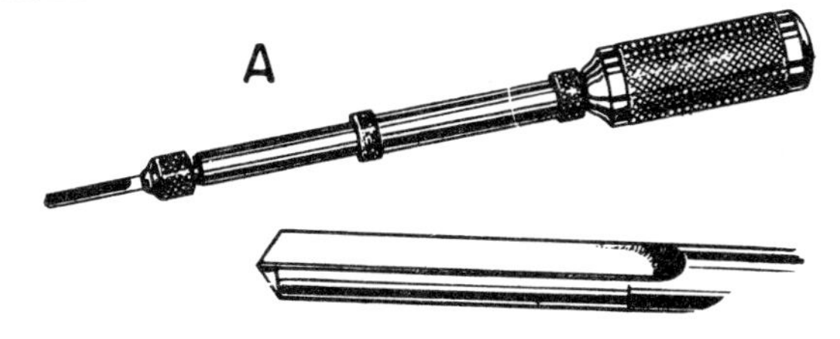

Fig. 94. A, Push drill; B, drill point.

DRILLING HOLES WITH A PUSH DRILL

A, Figure 94 shows the Stanley "Yankee" automatic drill which is often called a *push drill.* This drill can be used to drill either horizontal or vertical holes when the accuracy of the right angle with the work is not critical.

The drill point used in push drills (B, Fig. 94) is a straight flute drill. Sharpen its point on the grinder and provide only slight clearance behind the cutting edge. It will drill holes in wood and other soft materials.

To select a drill for use in a push drill, hold the handle of the drill in one hand and release the magazine by turning the knurled screw as shown in A, Fig. 95. This will permit you to drop the magazine. B, Figure 95 shows the drill magazine lowered to expose the drills from which the proper size can be selected.

To chuck the drill, loosen the chuck several turns and insert the drill as far as it will go. Turn the drill until it seats in the driving socket in the bottom of the chuck. Then tighten the chuck to hold the drill in place (C, Fig. 95).

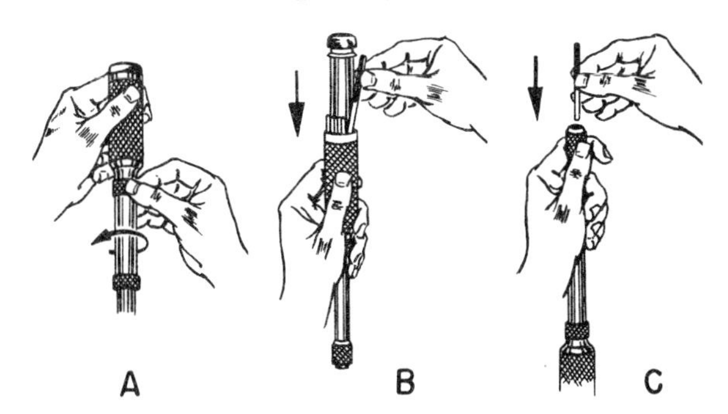

Fig. 95. Selecting a drill for use in a push drill.

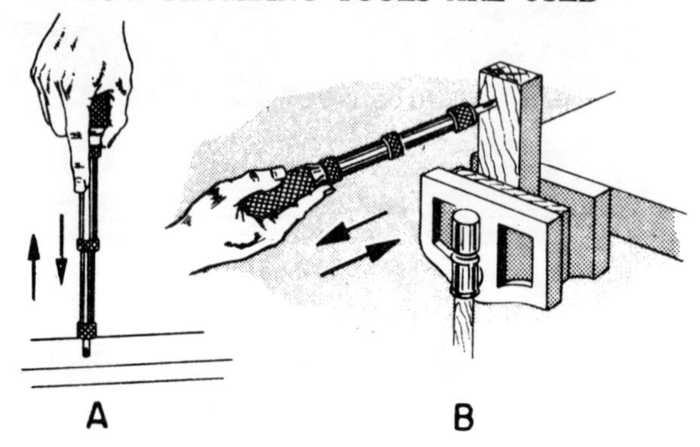

Fig. 96. Drilling horizontal and vertical holes with a push drill.

To drill a vertical hole with this drill (A, Fig. 96), place the job on a flat surface and operate the push drill with alternate strokes up and down. If it is necessary to hold the work in place while it is being drilled, use some mechanical means if you can. If you must hold the job with your hand, grasp the material as far as possible from where the drill is drilling.

In drilling horizontal holes with the push drill, as in B, Fig. 96, secure the job in a vise. The back-and forth strokes rotate the drill, advancing it into the work on the forward stroke as the drilling proceeds. The index finger, extended along the body of the tool, will help to guide the drilling at right angles to the work.

WOOD CHISELS

A *wood chisel* is a steel tool fitted with a wooden or plastic handle. It has a single beveled cutting edge on the end of the steel part, or blade. According to their construction, chisels may be divided into two general classes—*tang chisels,* in which part of the chisel enters the handle, and *socket chisels,* in which the handle enters into a part of the chisel (Fig. 97).

A *socket chisel* is designed for striking with a wooden mallet (never use a steel hammer); while a *tang chisel* is designed for hand manipulation only.

Wood chisels are also divided into types, depending upon their weights and thicknesses, the shape or the design of the blade, and the work they are intended to do.

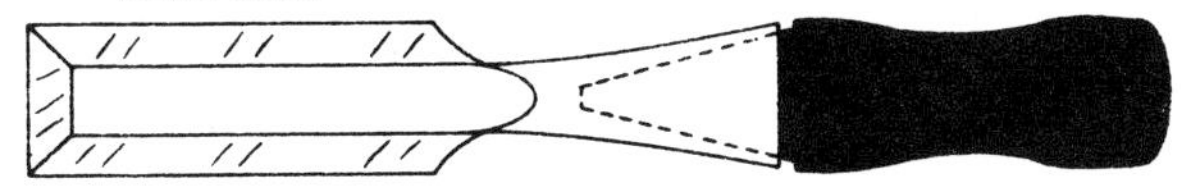

Fig. 97. Tang and socket wood chisels.

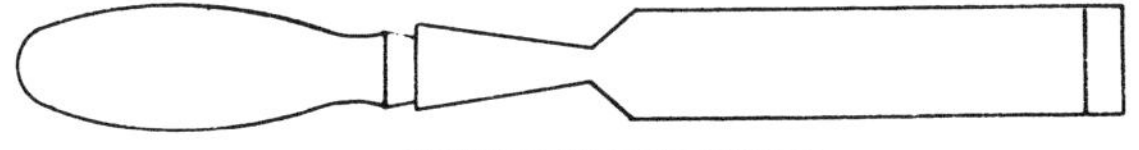

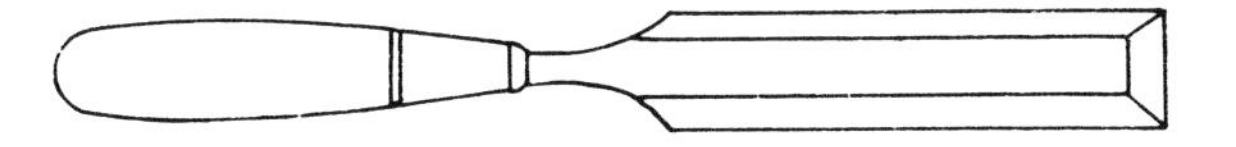

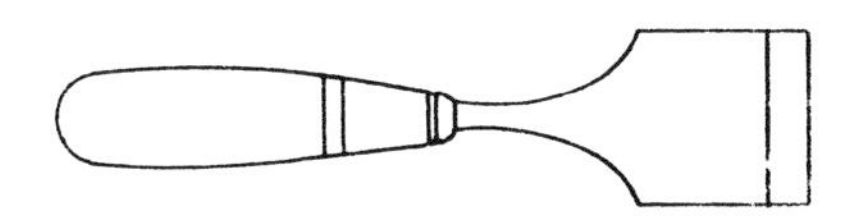

Fig. 98. Common type of wood chisels.

The shapes of the more common types of wood chisels are shown in Fig. 98. The *firmer chisel* has a strong, rectangular cross-section blade, designed for both heavy and light work. The blade of the *paring chisel* is relatively thin, and is beveled along the sides for the fine paring work. The *butt chisel* has a short blade, designed for work in hard to get at places.

The butt chisel is commonly used for chiseling the *gains* (rectangular depressions) for the *butt* hinges on doors (hence the name). The *mortising chisel* is similar to a socket firmer but has a narrow blade designed for chiseling out the deep, narrow *mortises* for mortise-and-tenon joints. This work requires a good deal of levering out of chips; therefore, the mortising chisel is made extra thick in the shaft to prevent breaking.

A *framing chisel* is shaped like a firmer chisel, but has a very heavy, strong blade designed for work in rough carpentry.

A wood chisel should always be held with the flat side or back of the chisel against the work for smoothing and finishing cuts. Whenever possible, it should not be pushed straight through an opening, but should be moved laterally at the same time that it is pushed forward. This method ensures a shearing cut, and with care will produce a smooth and even surface even when the work is cross-grained. On rough work, use a hammer or mallet to drive the socket-type chisel.

On fine work, use your hand as the driving power on tang chisels. For rough cuts, the bevel edge of the chisel is held against the work. Whenever possible, other tools such as saws and planes should be used to remove as much of the waste as possible, and the chisel used for finishing purposes only.

Safety Precautions

1. Secure work so that it cannot move.
2. Keep both hands back of the cutting edge at all times.
3. Do not start a cut on a guideline. Start slightly away from it, so that there is a small amount of material to be removed by the finishing cuts.
4. When starting a cut, always chisel away from the guideline toward the waste wood, so that no splitting will occur at the edge.
5. Never cut toward yourself with a chisel.
6. Make the shavings thin, especially when finishing.

7. Examine the grain of the wood to see which way it runs. Cut with the grain. This severs the fibers and leaves the wood smooth. Cutting against the grain splits the wood and leaves it rough. This type of cut cannot be controlled.

SCREWDRIVERS

Screwdrivers have one main purpose—to loosen or tighten screws. However, they have been used as a substitute for everything from an ice pick to a bottle opener.

Standard Screwdrivers

There are three main parts to a standard screwdriver. The portion you grip is called the handle, the steel portion extending from the handle is the shank, and the end which fits into the screw is called the blade (Fig. 99).

The steel shank is designed to withstand considerable twisting force in proportion to its size, and the tip of the blade is hardened to keep it from wearing.

Standard screwdrivers are classified by size, according to the combined length of the shank and blade. The most common sizes range in length from 2½ inches (6.35 cm.) to 12 inches (30.48 cm.). There are many screwdrivers smaller and some larger for special purposes. The diameter of the shank and the width and thickness of the blade are generally proportionate to the length, but again there are special screwdrivers with long thin shanks, short thick shanks, and extra wide or extra narrow blades.

Screwdriver handles may be wood, plastic, or metal. When metal handles are used, there is usually a wooden hand grip placed on each side of the handle. In some types of wood or plastic handled screwdrivers the shank extends through the handle, while in others the shank enters the handle only a short way and is pinned to the handle. For heavy work, special types of screwdrivers are made with a square shank. They are designed this way so that they may be gripped with a wrench, but this is the only kind on which a wrench should be used.

When using a screwdriver it is important to select the proper size so that the blade fits the screw slot properly. This prevents burring the slot and reduces the force required to

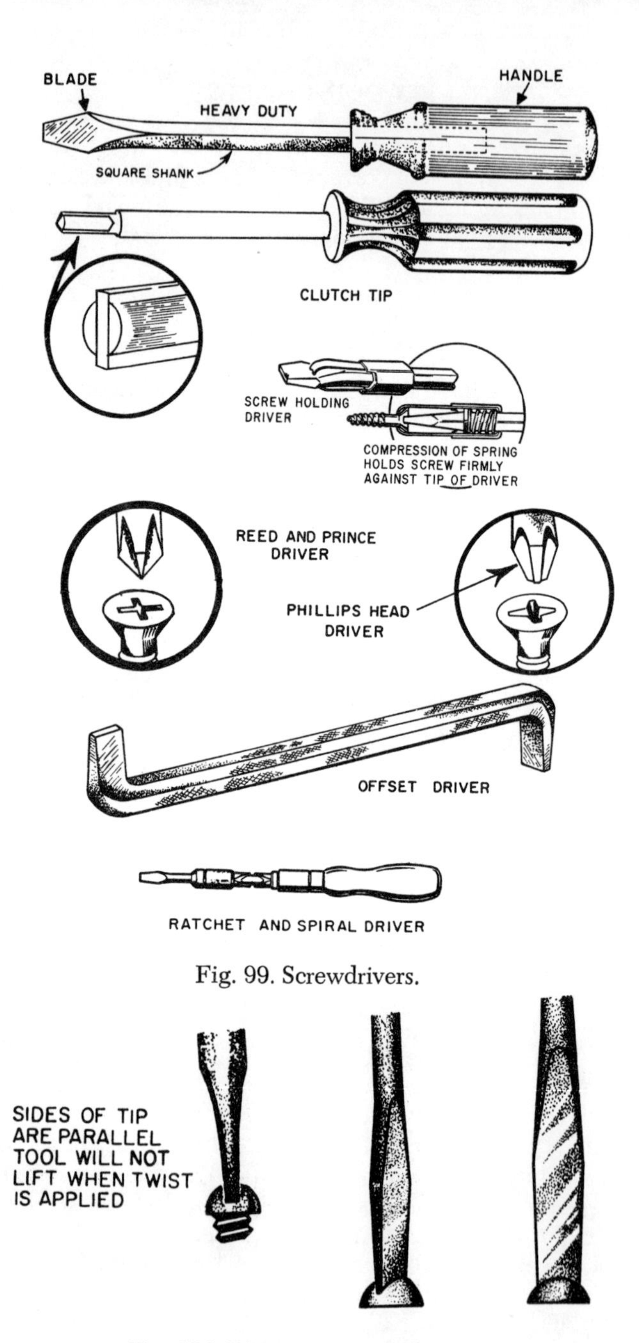

Fig. 99. Screwdrivers.

Fig. 100. Positioning screwdrivers.

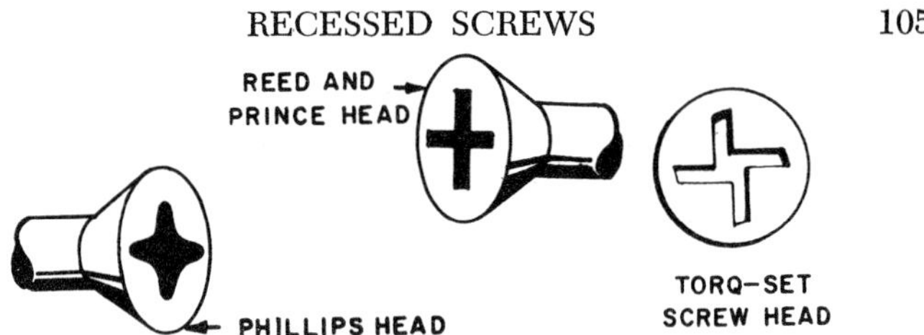

Fig. 101. Phillips, Reed and Prince, and Torq-set screw heads.

hold the driver in the slot. Keep the shank perpendicular to the screw head (Fig. 100).

RECESSED SCREWS

Recessed screws are available in various shapes. They have a cavity formed in the head and require a specially shaped screwdriver. The clutch tip (Fig. 99) is one shape, but the more common include the Phillips, Reed and Prince, and newer Torq-Set types (Fig. 101). The most common type found is the Phillips head screw. This requires a Phillips screwdriver (Fig. 99).

Phillips Screwdriver

The head of a Phillips screw has a four-way slot into which the screwdriver fits. This prevents the screwdriver from slipping. Three standard sized *Phillips screwdrivers* handle a wide range of screw sizes. Their ability to hold helps to prevent damaging the slots or the work surrounding the screw. It is a poor practice to try to use a standard screwdriver on a Phillips screw because both thc tool and screw slot will be damaged.

Reed and Prince Screwdriver

Reed and Prince screwdrivers are not interchangeable with Phillips screwdrivers. Therefore, always use a Reed and Prince screwdriver with Reed and Prince screws and a Phillips screwdriver with Phillips screws, or a ruined tool or ruined screwhead will result. (*See* Fig. 102.)

The Phillips screwdriver has about 30-degree flukes and a blunt end, while the Reed and Prince has 45-degree flukes

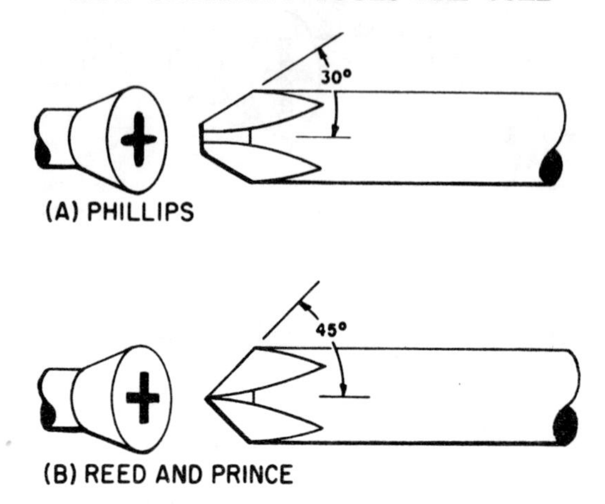

Fig. 102. Matching cross-slot screws and drivers.

and a sharper, pointed end. The Phillips screw has beveled walls between the slots; the Reed and Prince has straight, pointed walls. In addition the Phillips screw slot is not as deep as the Reed and Prince slot.

Additional ways to identify the right screwdriver are as follows.

1. If it tends to stand up unassisted when the point is put in the head of a vertical screw, it is probably the proper one.
2. The outline of the end of a Reed and Prince screwdriver is approximately a right angle, as shown in Fig. 102.
3. In general, Reed and Prince screws are used for airframe structural applications, while Phillips screws are found most often in component assemblies.

"Torq-Set" Screws

"Torq-Set" machine screws (offset cross-slot drive) have recently begun to appear in new equipment. The main advantage of the newer type is that more torque can be applied to its head while tightening or loosening than any other screw of comparable size and material without damaging the head of the screw.

Torq-Set machine screws are similar in appearance to the more familiar Phillips machine screws.

Since a Phillips driver could easily damage a Torq-Set

screwhead, making it difficult if not impossible to remove the screw even if the proper tool is later used, the worker should be alert to the differences (Fig. 101) and be sure to use the proper tool.

OFFSET SCREWDRIVERS

An *offset screwdriver* (Fig. 99) may be used where there is not sufficient vertical space for a standard or recessed screwdriver. Offset screwdrivers are constructed with one blade forged in line and another blade forged at right angles to the shank handle. By alternating ends, most screws can be seated or loosened even when the swinging space is very restricted. Offset screwdrivers are made for both standard and recessed head screws.

RATCHET SCREWDRIVERS

For fast easy work the *ratchet screwdriver* (Fig. 99) is extremely convenient, as it can be used with one hand and does not require the bit to be lifted out of the slot after each turn. It may be fitted with either a standard type bit or a special bit for recessed heads. The ratchet screwdriver is most commonly used by the worker for driving screws in soft wood.

Safety Precautions

1. Never use a screwdriver to check an electrical circuit.
2. Never try to turn a screwdriver with a pair of pliers.
3. Do not hold work in your hand while using a screwdriver, because if the point slips it can cause a bad cut. Hold the work in a vise, with a clamp or on a solid surface. If that is impossible, you will always be safe if you follow this rule—Never get any part of your body in front of the screwdriver blade tip. That is a good safety rule for any sharp or pointed tool.

PLIERS

Pliers are made in many styles and sizes and are used to perform many different operations. Pliers are used for cutting

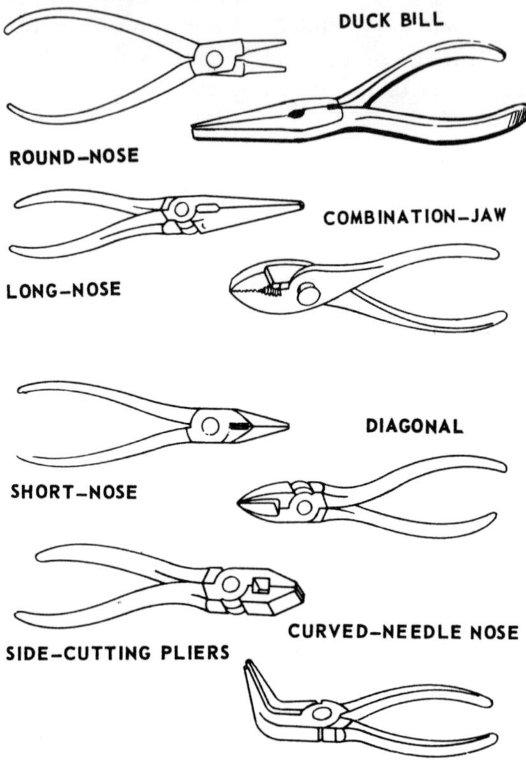

Fig. 103. Pliers.

purposes as well as holding and gripping small articles in situations where it may be inconvenient or impossible to use your hands. Figure 103 shows various types of pliers.

The *combination pliers* are handy for holding or bending flat or round stock. The *long-nosed pliers* are less rugged, and break easily if you use them on heavy jobs. Long-nosed pliers commonly called needle-nose pliers are especially useful for holding small objects in tight places and for making delicate adjustments. The *round-nosed pliers* are handy when you need to crimp sheet metal or form a loop in a wire. The *diagonal cutting pliers*, commonly called *diagonals* or *dikes* are designed for cutting wire and cotter pins close to a flat surface and are especially useful in the electronic and electrical fields. The duckbill pliers are used extensively in aviation areas.

Two important rules when using pliers are as follows.

1. Do not make pliers work beyond their capacity. The long-nosed pliers are especially delicate. It is easy to spring or break them, or nick their edges. If that happens, they are practically useless.

2. Do not use pliers to turn nuts. In just a few seconds a pair of pliers can damage a nut. Pliers must not be substituted for wrenches.

Slip-joint Pliers

Slip-joint pliers (Fig. 104) are pliers with straight, serrated (grooved) jaws, and the screw or pivot with which the jaws are fastened together may be moved to either of two positions, in order to grasp small or large sized objects better.

To spread the jaws of slip-joint pliers, first spread the ends of the handles apart as far as possible. The slip-joint or pivot will now move to the open position. To close the jaws again spread the handles as far as possible, then push the joint back into the closed position.

Slip-joint combination pliers (Fig. 105) are pliers similar to the slip-joint pliers, but with the additional feature of a side

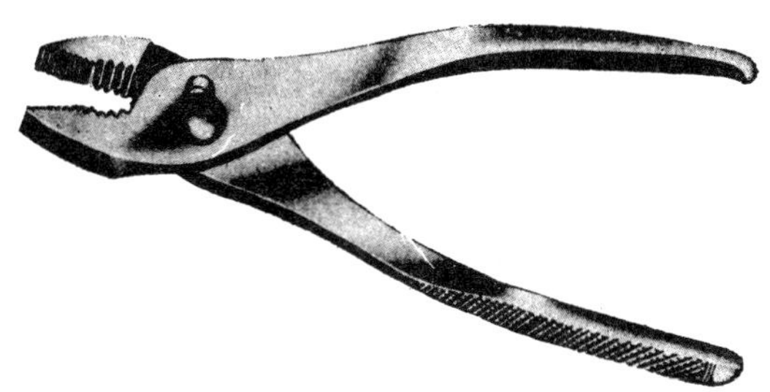

Fig. 104. Slipjoint pliers.

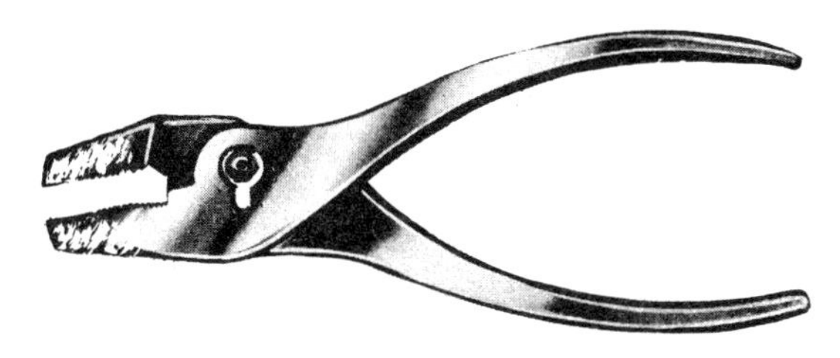

Fig. 105. Slipjoint combination pliers.

cutter at the junction of the jaws. This cutter consists of a pair of square cut notches, one on each jaw, which act like a pair of shears when an object is placed between them and the jaws are closed.

The cutter is designed to cut material such as soft wire and nails. To use the cutter, open the jaws until the cutter on either jaw lines up with the other. Place the material to be cut as far back as possible into the opening formed by the cutter, and squeeze the handles of the pliers together. Do not attempt to cut hard material such as spring wire or hard rivets with the combination pliers. To do so will spring the jaws of the pliers, and if the jaws are sprung it will be difficult thereafter to cut small wire with the cutters.

Vise-grip Pliers

Vise-grip pliers (Fig. 106) can be used for holding objects regardless of their shape. A screw adjustment in one of the handles makes them suitable for several different sizes. The jaws of vise-grips may have standard serrations such as the slip-joint combination pliers or may have a clamp jaw. The clamp jaws are generally wide and smooth and are used primarily when working with sheet metal.

Vise-grip pliers have an advantage over other types of pliers in that you can clamp them on an object and they will stay. This will leave your hands free for other work.

This tool may be used in a number of ways. It may be used as a clamp, speed wrench, portable vise, and for many other uses where a locking plier jaw may be employed. These pliers may be adjusted to various jaw openings by turning the knurled adjusting screw at the end of the handle (Fig. 106). Vise-grips can be clamped and locked in position by pulling the lever toward the handle.

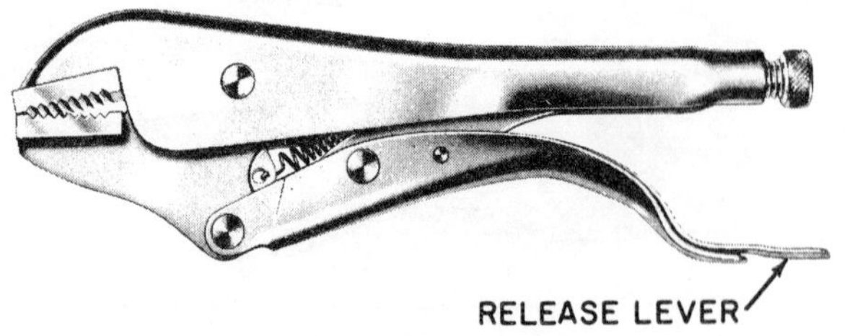

Fig. 106. Vise grip pliers.

Caution: Vise-grip pliers should be used with care since the teeth in the jaws tend to damage the object on which they are clamped. They should not be used on nuts, bolts, tube fittings, or other objects which must be reused.

Water-pump Pliers

Water-pump pliers were originally designed for tightening or removing water pump packing nuts. They are excellent for this job because they have a jaw adjustable to seven different positions. Water-pump pliers (Fig. 107) are easily identified by their size, jaw teeth, and adjustable slip joint. The inner surface of the jaws consists of a series of coarse teeth formed by deep grooves, a surface adapted to grasping cylindrical objects.

Channel-lock Pliers

Channel-lock pliers (Fig. 108) are another version of water-pump pliers easily identified by the extra long handles, which make them a very powerful gripping tool. They are shaped the same as the water-pump pliers, but the jaw opening adjustment is effected differently. Channel-lock pliers have grooves on one jaw and lands on the other. The adjustment is

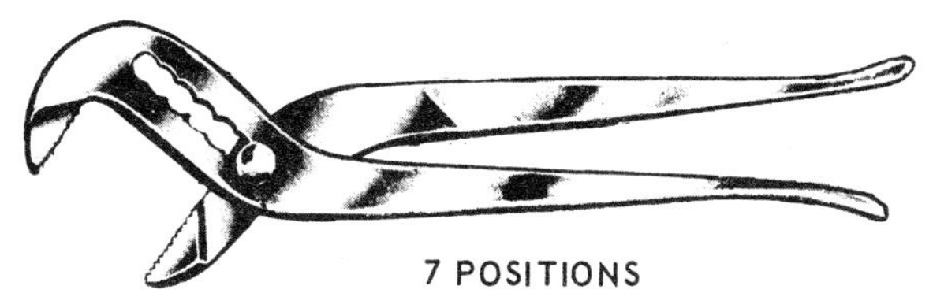

Fig. 107. Water pump pliers.

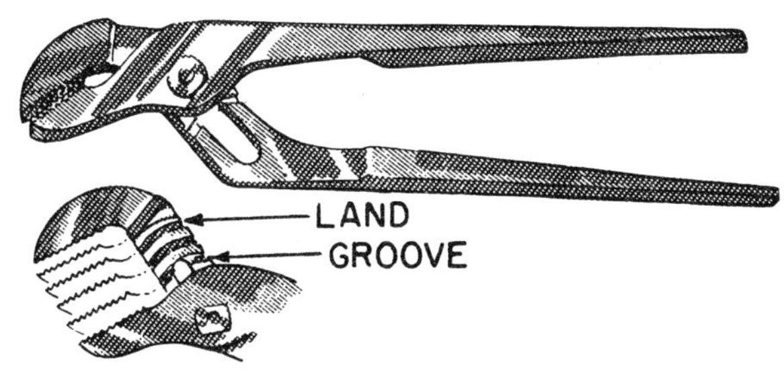

Fig. 108. Channel-lock pliers.

affected by changing the position of the grooves and lands. The channel-lock pliers will only be used where it is impossible to use a more adapted wrench or holding device. Many nuts and bolts and surrounding parts have been damaged by improper use of channel-lock pliers.

Diagonal Pliers

Diagonal cutting pliers (Fig. 103) are used for cutting small, light material, such as wire and cotter pins in areas which are inaccessible to the larger cutting tools. Also, since they are designed for cutting only, larger objects can be cut than with the slip-joint pliers.

As the cutting edges are diagonally offset approximately 15 degrees, diagonal pliers are adapted to cutting small objects flush with a surface. The inner jaw surface is a diagonal straight cutting edge. Diagonal pliers should never be used to hold objects, because they exert a greater shearing force than other types of pliers of a similar size. The sizes of the diagonal cutting pliers are designated by the overall length of the pliers.

Side-cutting Pliers

Side-cutting pliers (sidecutters) are principally used for holding, bending, and cutting thin materials or small gage wire. Sidecutters vary in size and are designated by their overall length. The jaws are hollowed out on one side just forward of the pivot point of the pliers. Opposite the hollowed out portion of the jaws are the cutting edges (Fig. 103).

When holding or bending light metal surfaces, the jaw tips are used to grasp the object. When holding wire grasp it as near one end as possible because the jaws will mar the wire. To cut small diameter wire the side cutting edge of the jaws near the pivot is used. Never use sidecutters to grasp large objects, tighten nuts, or bend heavy gage metal, since such operations will spring the jaws.

Sidecutters are often called electrician or lineman pliers. They are used extensively for stripping insulation from wire and for twisting wire when making a splice.

Duckbill Pliers

Duckbill pliers (A, Fig. 109) are used in the same manner

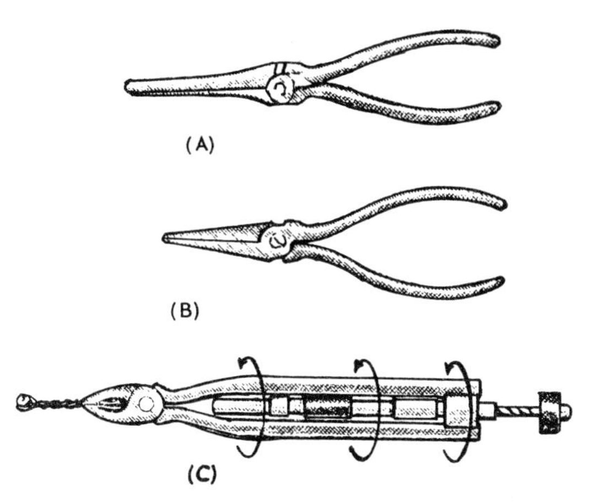

Fig. 109. Pliers. A, Duckbill; B, needle nose; C, wire twister.

as needle-nose pliers. However, there is a difference in the design of the jaws. Needle-nose jaws are tapered to a point which makes them adapted to installing and removing small cotter pins. They have serrations at the nose end and a side cutter near the throat. Needle-nose pliers may be used to hold small items steady, to cut and bend safety wire, or to do numerous other jobs which are too intricate or too difficult to be done by hand alone.

Note: Duckbill and needle-nose pliers (B, Fig. 109) are especially delicate. Care should be exercised when using these pliers to prevent springing, breaking, or chipping the jaws. Once these pliers are damaged, they are practically useless.

Wire-twister Pliers

Wire-twister pliers (C, Fig. 109) are three-way pliers that hold, twist, and cut. They are designed to reduce the time used in twisting safety wire on nuts and bolts. To operate this wire twister grasp the wire between the two diagonal jaws, and the thumb will bring the locking sleeve into place. A pull on the knob twirls the twister, making uniform twists in the wire. The spiral rod may be pushed back into the twister without unlocking it, and another pull on the knob will give a tighter twist to the wire. A squeeze on the handle unlocks the twister, and the wire can be cut to the desired length with the side cutter. The spiral of the twister should be lubricated occasionally.

Maintenance of Pliers

Nearly all side-cutting pliers and diagonals are designed so that the cutting edges can be reground. Some older models of pliers will not close if material is ground from the cutting edges. When grinding the cutting edges never take any more material from the jaws than is necessary to remove the nicks. Grind the same amount of stock from both jaws.

Note: When jaws on pliers do not open enough to permit grinding, remove the pin that attaches the two halves of the pliers so that the jaws can be separated.

PLUNGERS

A *plunger* or "plumber's friend" (Fig. 110) is used to clear minor drain-pipe or trap obstructions. When using the plunger, fill the trap or the drain with water, insert the plunger over the drain and pump until the force of the water clears the obstruction.

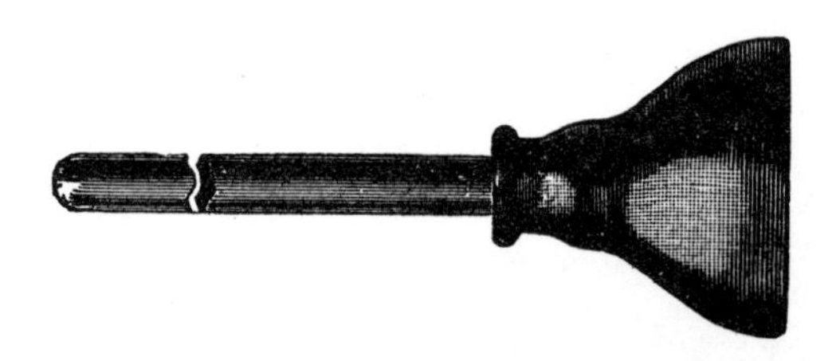

Fig. 110. Plunger or "plumber's friend."

CALKING IRONS

Calking irons are used for packing oakum or special calking compounds and lead in the joints of cast-iron pipes. They are available in various shapes and sizes. The calking iron is held at an angle and the compound forced into the joint by striking the iron with a heavy machinist's hammer.

SHARPENING STONES

Sharpening stones are divided into two groups, natural and artificial. Some of the natural stones are oil treated during and after the manufacturing processes. The stones that are oil treated are sometimes called *oilstones*. Artificial stones are normally made of silicone carbide or aluminum oxide. *Natural stones* have very fine grains and are excellent for putting razor-like edges on fine cutting tools. Most sharpening stones have one coarse and one fine face. Some of these stones are mounted, and the working face of some of the sharpening stones is a combination of coarse and fine grains. Stones are available in a variety of shapes as shown in Fig. 111.

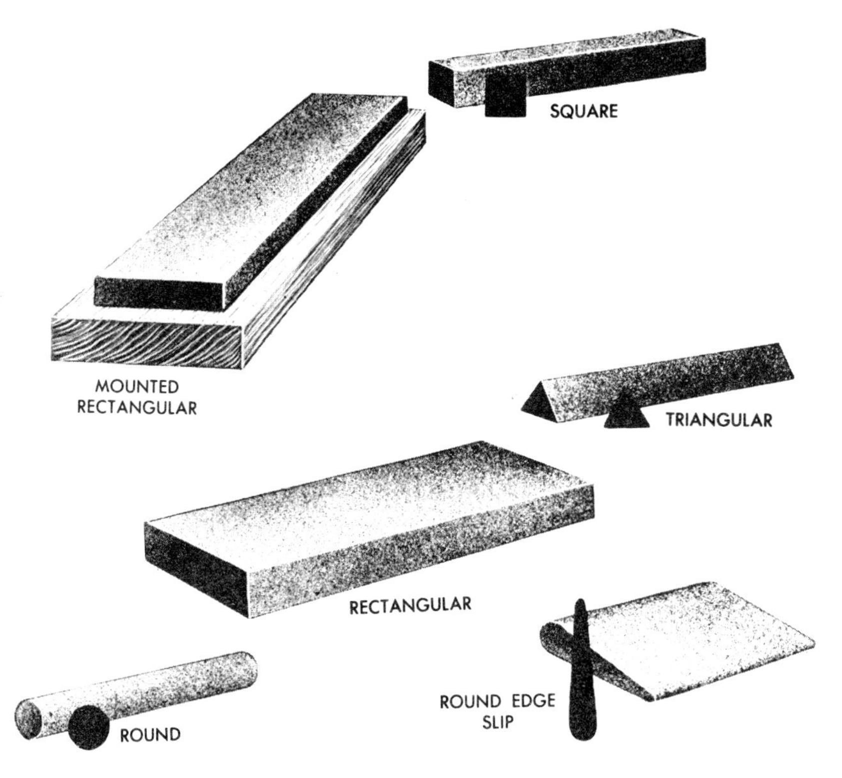

Fig. 111. Shapes of sharpening stones and oilstones.

A fine cutting oil is generally used with most artificial sharpening stones. However, other lubricants such as kerosene may be used. When a tool has been sharpened on a grinder or grindstone, there is usually a wire edge or a feather edge left by the coarse wheel. The sharpening stones are used to hone this wire or feather edge off the cutting edge of the tool. Do not attempt to do a honing job with the wrong stone. Use a coarse stone to sharpen large and very dull or nicked tools. Use a medium grain stone to sharpen tools not requiring a finished edge, such as tools for working soft wood, cloth, leather, and rubber. Use a fine stone and an oilstone to sharpen and hone tools requiring a razorlike edge.

Prevent glazing of sharpening stones by applying a light oil during the use of the stone. Wipe the stone clean with a wiping cloth or cotton waste after each use. If the stone becomes glazed or gummed up, clean with aqueous ammonia or dry cleaning solvent. If necessary, scour with aluminum oxide abrasive cloth or flint paper attached to a flat block.

At times, stones will become uneven from improper use. True the uneven surfaces on an old grinding wheel or on a grindstone. Another method of truing the surface is to lap it with a block of cast iron or other hard material covered with a waterproof abrasive paper, dipping the stone in water at regular intervals and continuing the lapping until the stone is true.

Stones must be carefully stored in boxes or on special racks when not in use. Never lay them down on uneven surfaces or place them where they may be knocked off a table or bench, or where heavy objects can fall on them. Do not store in a hot place.

Sharpening a Pocket Knife

Pocket knives may be sharpened on a medium or fine grade sharpening stone with a few drops of oil spread on the surface. Hold the handle of the knife in one hand and place the blade across the stone. Press down with the fingers of the other hand and stroke the blade following a circular motion as shown in Fig. 112. After several strokes, reverse the blade and stroke the opposite side, following the same type of motion. Use a light even pressure. A thin blade overheats quickly and can lose its temper. The wire edge or burr that may be left on a

Fig. 112. Sharpening a pocket knife.

knife blade after whetting may be removed by stropping both sides on a soft wood block, canvas or leather.

KNIVES

Most knives are used to cut, pare, and trim wood, leather, rubber and other similar materials. The types most frequently used are the *shop knife*, *pocket knife*, and the *putty knife* (Fig. 113).

The *shop knife* can be used to cut cardboard, linoleum, and paper. It has an aluminum handle and is furnished with interchangeable blades stored in the 5-inch (12.7-cm.) handle.

Pocket knives are used for light cutting. They are unsuited for heavy work. *Multi-purpose knives* have an assortment of

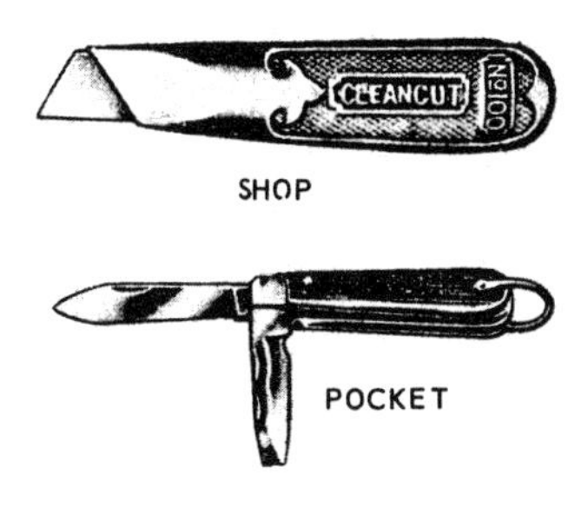

Fig. 113. Knives.

blades designed for forcing holes, driving screws and opening cans, as well as cutting. The blades are hinged and should be contained within the case when not in use. They are spring loaded to keep them firmly in place when open or closed.

Safety with knives is essential. Do not use knives larger than can be safely handled. Use knives only for the purpose for which they were designed. Always cut away from your body. Do not carry open knives in your pocket or leave them where they may come into contact with or cause injury to others. Put knives away carefully after use to protect sharp cutting edges from contacting other hard objects.

FLASHLIGHT

Each toolbox should have a standard vaporproof two-cell *flashlight.* Installed in both ends of this flashlight are rubber seals which keep out all vapors. The flashlight should be inspected periodically for the installation of these seals, the spare bulb, and colored filters which are contained in the end cap. *Note:* Do not throw away the filters as they can be used during night operations.

INSPECTION MIRROR

There are several types of *inspection mirrors* available in a variety of sizes and may be round or rectangular. The mirror is connected to the end of a rod and may be fixed or adjustable (Fig. 114).

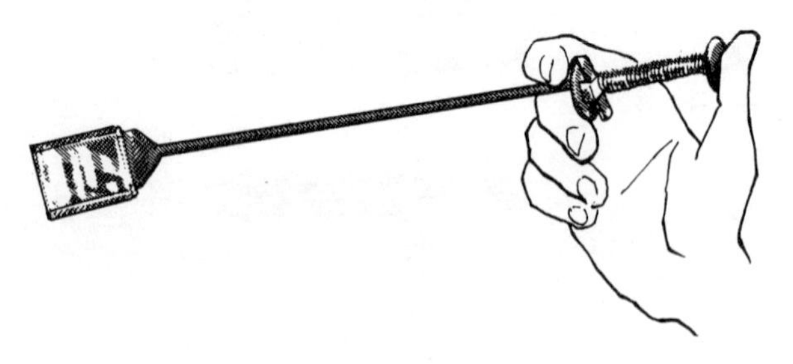

Fig. 114. Inspection mirror.

The inspection mirror aids in making detailed inspection where the human eye cannot directly see the inspection area. By angling the mirror, and with the aid of a flashlight, it is possible to inspect most areas. A late model inspection mirror features a built-in light to aid in viewing dark places where use of a flashlight is not convenient.

FIRE POTS AND BLOWTORCHES

Fire pots and blowtorches are available in a variety of sizes and types, for use with either gasoline or kerosene (Fig. 115). The *fire pot* is used to melt lead or solder when making calked joints, cable and wire splices, and other jobs. The *blowtorch* is used for sweating joints on brass and copper pipe and tubing and for thawing out frozen pipes.

In both the fire pot and blowtorch, air pressure is used to force the gasoline into the generating coil or chamber. The chamber is kept just hot enough to vaporize the gasoline. The gas is then forced through a small opening and draws in the required amount of air to produce a combustible mixture.

Fig. 115. Fire pot and blowtorch.

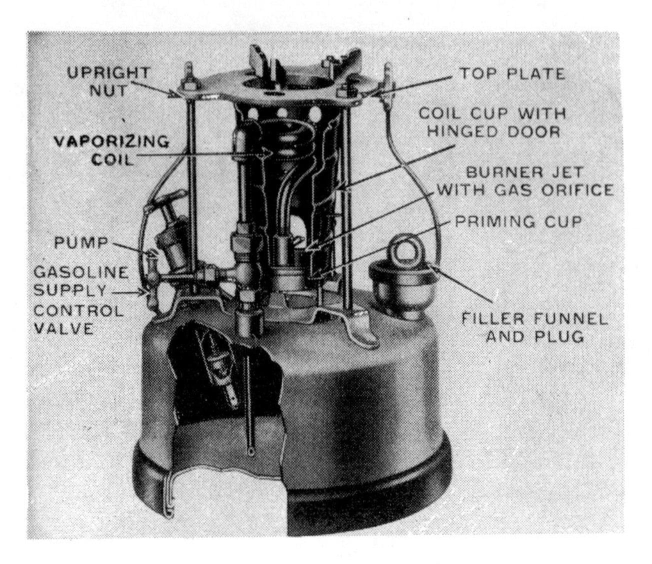

Fig. 116. Cutaway view of a fire pot.

Specific directions for filling and using fire pots and blowtorches are furnished by the manufacturer.

Be sure not to fill either of these tools in a closed room. To fill either the fire pot or the blowtorch, remove the filler plug (Figs. 116 and 117). After filling the tank, close the valve, unscrew pump plunger, and use the pump to build up the required air pressure. Hold the hand over the coil or combustion chamber and turn on the gasoline supply control valve to permit the gasoline to drip into the priming cup. Light the gasoline that has dripped into the priming cup, as this supplies the heat required to preheat the gasoline in the combustion chamber. Preheating the gasoline in the combustion chamber vaporizes it. Before all of the gasoline in the priming cup is consumed, turn on the control valve slowly and carefully to permit the flame from the priming cup to ignite the vapor produced in the combustion chamber.

Be sure to follow directions furnished by the manufacturer for the operation, care, and maintenance of the fire pot and blowtorch.

MECHANICAL FINGERS

Small articles which have fallen into places where they

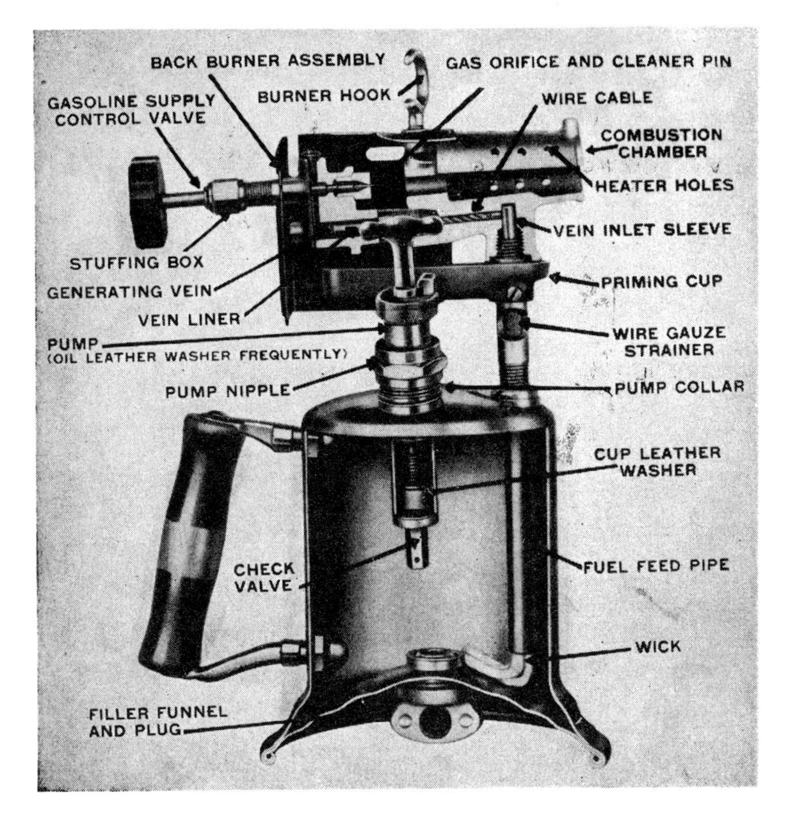

Fig. 117. Cutaway view of a blowtorch.

cannot be reached by hand may be retrieved with the *mechanical fingers* (Fig. 118). This tool is also used when starting nuts or bolts in difficult areas. The mechanical fingers shown in Fig. 118 have a tube containing flat springs which extend from the end of the tube to form clawlike fingers, much like the screw holder. The springs are attached to a rod that extends from the outer end of the tube. A plate is attached to the end of the tube, and a similar plate to be pressed by the thumb is attached to the end of the rod. A coil spring placed around the rod between the two plates holds them apart and retracts the fingers into the tube. With the bottom plate grasped between the fingers and enough thumb pressure applied to the top plate to compress the spring, the tool fingers extend from the tube in a grasping position. When the

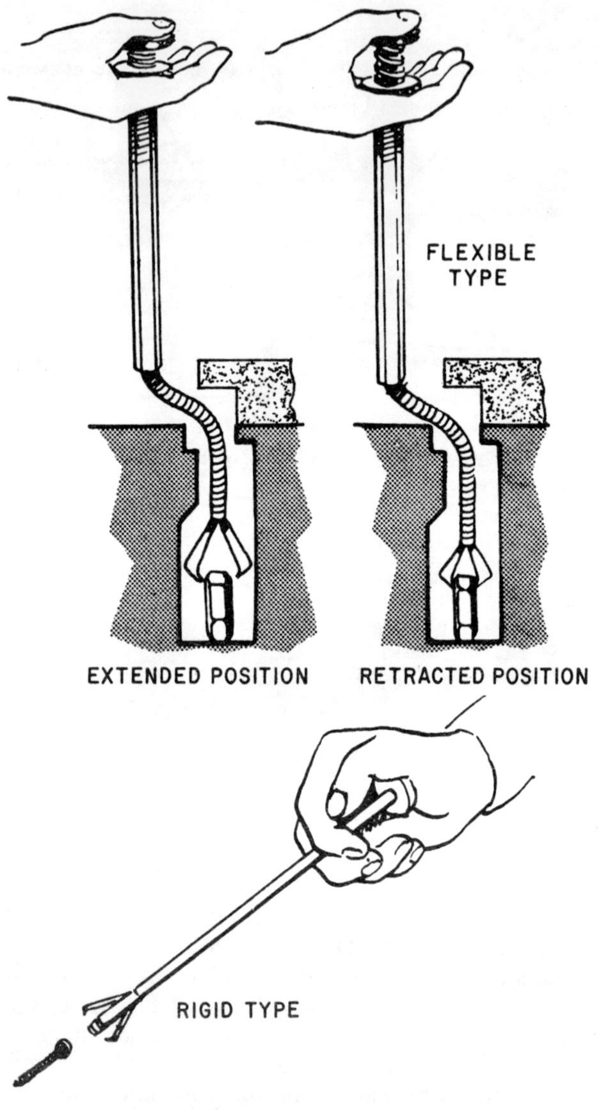

Fig. 118. Mechanical fingers.

thumb pressure is released, the tool fingers retract into the tube as far as the object they hold will allow. Thus, enough pressure is applied on the object to hold it securely. Some mechanical fingers have a flexible end on the tube to permit their use in close quarters or around obstructions (Fig. 118).

Chapter 2

Plumbing Material

The correlation between plumbing and pipe, and valves and fittings is quite natural. Too often we think of plumbing for a residence as consisting chiefly of plumbing fixtures and trim. We must not forget that in one item, specifically pipe, there is more than 300 feet (9144 cm.) of it, which lie within the walls and under the floors of the typical two-story house. Of the ten percent of the new residential building dollar that is used for plumbing (exclusive of heating) about one-half is spent for the "in the wall" arteries. In short, the pipe, the valves and fittings of the plumbing system.

The tremendous progress in the sanitary standards of our people has changed the complexity and scope of plumbing from merely connecting a single pipe faucet and basin in a kitchen to the complete installation in the modern home of multi-baths, complete kitchens, powder rooms, laundries, automatic heat, and air conditioning accessories.

Today, and for the future, the plumbing industry is assuming a vital role in guarding the health of the nation, and a constant challenge is presented to the manufacturers of plumbing material to satisfy the demands of builders for new homes, and for replacement of outmoded residential installations. To this must be added the ever-necessary renovation and improvement, and emergency repairs of old homes.

PLUMBING CLASSIFICATIONS

The *plumbing system* of a building includes the water supply system, the fixtures and fixture traps, the soil, waste and vent pipe, the house drain and sewer and the stormwater-drainage pipe, together with their valves, valve specialties and fittings, all within or adjacent to the building.

Plumbing material can be divided into three main general groupings:

1. Plumbing fixtures.
2. Fixture fittings, sometimes classified in two subgroups (a) trimmings, (b) fittings.
3. Roughing-in.

Plumbing fixtures may be defined as those fixtures which serve as receptacles which receive water, liquid, or water-borne waste and discharge them into a drainage system with which the fixtures are directly or indirectly connected. Closet bowls, bathtubs, shower compartments, lavatories, sinks, and laundry tubs (without fittings or trim) can be classified as plumbing fixtures. They are usually made of vitreous china, enameled iron, Duraclay or enameled pressed steel.

Plumbing trim are those metal parts of a fixture, subject to selection or option, such as faucets, traps, and supply pipe, and those parts regularly supplied with the fixtures, such as wall hangers, closet spuds, lavatory leg rods, bolts for pedestals and flush tank trim.

Roughing-in is the term applied to all the valves, fittings, piping, and appurtenances installed in a building required to carry and control the water to and the discharge from the plumbing fixtures and extending to the wall or floor line of such fixtures. This would include valves and fittings, steel pipe, stacks, vents, drains, soil pipe, and so on.

FIXTURES

Most *fixtures* are made of vitreous china, porcelain enameled cast iron, Duraclay or porcelain enameled sheet metal.

Porcelain Enameled Cast Iron

Enamelware is a product manufactured by actually fusing an opaque glass to a metal surface. In a plumbing fixture, it is the application or fusing of this opaque glass, termed porcelain or vitreous enamel, to a specially prepared cast iron body or shape to cover all visible or water contacting surfaces.

The date of origin of porcelain enamel is unknown, but we do know that hundreds of years ago the art of porcelain making was well established in oriental countries. Porcelain enameled objects in museums, art schools, and in private collections though made hundreds of years ago are still in a state of perfect preservation. Because of the structural hardness of this material it has withstood the test of time.

Component chemicals used in the manufacture of enamelware are divided into three classes—refractories, fluxes and opacifiers. There are many different ingredients that come under each of the classifications, and all of them are not necessarily used together in any one mixture of enamel.

Included in the *refractories* used as ingredients in the manufacture of our dry process enamelware are feldspar, china clay and silica (quartz). Some of the *fluxes* used are fluorspar, calcium carbonate, barium carbonate and borax, and in the *opacifier* group are found tin oxide, zirconium compounds, bone ash and antimony oxides.

Cast Iron Base

Enamel is a glass in every sense of the word, because it is brittle and non-resilient. It is essentially important that the base to which it is applied be as rigid and as inflexible as the law of expansion and contraction will allow.

Cast iron is considered one of the best base materials because of its rigidity and also because of the texture of the iron. The rough and porous surface of cast iron provides natural grip or adherence points to which the enamel can cling.

Regular and Acid-resisting Enamel

Two kinds of enamel are being produced. These are known as *regular* and *acid-resisting. Regular enamel* does not have acid-resisting qualities. *Acid-resisting enamel* fixtures are more costly to manufacture and therefore sell for more. The ingredients used are more expensive, and because these ingredients are more carefully applied and must be fired longer, the manufacturing costs are increased.

All colored ware is made of acid-resisting enamel. Acid-resisting enamel is recommended to resist all acids with the exception of hydrofluoric (the glass solvent). Though it is exceptionally hard, nevertheless it will scratch, and scratches show up prominently (by comparison) just as scratches on a highly polished mahogany table top show up more conspicuously than on a table top of satin finish. Because of its close texture, acid-resisting enamel soils rather than stains, and if soiled it responds to the action of cleaning solutions without losing its high gloss. Acid-resisting porcelain enamel resists attack by acids or alkalis such as in fruit juices, foods, or drugs commonly found in kitchens or bathrooms. It is recommended for kitchen sinks and other fixtures subject to such acids that might attack regular enamel.

Most kitchen sinks are made only of acid-resisting enamel. Sinks are generally subjected to more severe use than other plumbing fixtures in the home. Constant contact with the acids contained in fruits, vegetables and other edible foods as well as those found in cleansing preparations, and deposits from mineral bearing waters will permanently stain regular enamelware. *Note:* The same stains can be scoured from acid-resisting enamel.

Designs

In late years, manufacturers of plumbing ware have developed many new fixtures for home use and for other applications. More emphasis has been directed to styling, and today fixtures are available that are designed in complete accord with the latest styles of equipment being produced in other industries. Many groups of fixtures have been developed for home use, consisting of kitchen sink, bathtub, lavatory and closet in each of which the same style characteristics predominate to make complete ensembles of related designs.

Colored Enamelware

In the manufacture of enamel, *color* is obtained by the addition of color oxides. Colored fixtures cost only slightly more than white, and they represent a fine way to add decorative excitement to your kitchen or bathroom.

Vitreous China

Vitreous china sanitary ware is a glazed fine grain earthenware body, thoroughly vitrified at intense heat. The body is made of selected clays, flints and feldspars and the glaze, a form of glass, is fused into the ware at high temperatures and becomes an integral part of it. The appearance is much the same as the finest china dishes. Vitreous china makes attractive and durable sanitary ware and is especially noted for its brilliant, hard, easily cleaned, impervious to abrasion and acid-resisting qualities.

Vitreous china is more inclusively defined as "a homogenous mixture of ceramic materials bonded by fusion of these materials at high temperatures. Composition and heat treatment is such as to result in a body so impervious to moisture penetration that it will not absorb more than one-half of one percent of its weight of water when subjected to an absorption test as specified therein."

Materials must comply with the requirements. Clay must be selected that is at once plastic and workable and which will vitrify at a reasonable temperature. An impervious bonding agent that will fuse at the desired temperature is also necessary. Clays known as *ball clay* have the desired plasticity and vitrifying range and are the *backbone* of vitreous china. Ball clay is naturally dark in color when fired, so china clay or kaolin is also used to make the body whiter. Feldspar melts at the desired temperatures and forms a glassy bond of all of the particles in the mixture. Flint is incorporated to reduce shrinkage and impart strength.

Clays, Feldspars, and Fints

China as now made, calls for a number of clays, feldspars and flints in its structure. Its principal general ingredients and their purposes are as follows:

Ingredient	*Purpose*
English ball clay	Plasticity
English china clay	Color
Florida clay	Close texture
Georgia clay	Close texture
New Hampshire feldspar	Binding flux
Connecticut feldspar	Binding flux
Domestic flint	To strengthen and reduce shrinkage

Colored Ware

Colored ware is made by introducing the color into the glaze composition. The color pigment is generally the oxide of a metal which on being raised to a high temperature in the presence of the basic glaze materials will produce the desired color.

Duraclay

Several years of research, experimental production, and field tests have resulted in a type of sanitary ware which has all of the desirable properties. It has been introduced to the trade under the name "Duraclay."

Duraclay consists of a blend of carefully selected domestic clays, a previously unused mineral, and a certain percentage of pre-fired clay. This grayish buff body is covered first with a double application of a white vitreous coating, which has a composition similar to vitreous china ware, and then with a brilliant, durable, high fired glaze. The incorporation of this previously unused mineral is responsible for the remarkable properties of Duraclay which is produced by methods otherwise similar to those used in production of vitreous china ware.

During the intense heat treatment of Duraclay ware, there is developed in the body an interlocking crystalline structure which prevents moisture expansion.

The excellent resistance to thermal shock is due to the absence of the glassy bond and to the interlocking crystalline formation developed in this body, which acts as a bond and increases the mechanical strength of the body.

The remarkable and unexcelled resistance of the glazed surface to chemical action and abrasion is due to the chemical composition of the glaze itself. The production of this type of glaze is possible only at the high temperatures to which Duraclay is fired. This glaze is not only impervious to fruit, vegetable and other common acids and alkalis, but is resistant to the wear and tear caused by abrasive scouring materials. Marks from aluminum and other metallic utensils can be made on the glaze surface but they are easily removed with soap and water. The evaporation of rusty or dirty water will leave a stain on the surface which is easily removed with soap and water.

Vitreous Enameled Steel Ware

The basic material used in fabricating vitreous enameled steel bathtubs is the especially prepared and processed sheets of steel, developed exclusively for enameling. The perfection of the method of processing this basic material has contributed much to the success of enameled steel ware since types that were used earlier did not perform satisfactorily. Cold rolled steel was found to have too smooth a surface for proper adhesion of the enamel, but the surface of this enameling steel is of such texture as to insure a firm and permanent grip of the undercoat enamel.

FIXTURE TRIM

Fixture trim is produced in metal. Brass is the material most used for plumbing trim. A complete line of iron trim is available. There are many types of ware in fixtures and trim available.

Brass Trim

Most brass castings of trim are made of a brass composition that has proved satisfactory for the purpose. Many trim parts are forged or pressed, which provide an excellent surface for polishing, buffing, and plating.

Plating

In plating, a high degree of skill and control is essential to a good coat. Various baths or dips are required, such as the cleaning bath, the running bath, the copper bath, and the like. The baths are automatic in operation and thermostatically controlled so that the articles remain in each bath just the required length of time and at the proper temperature.

Design

Almost all plumbing trim may be generally divided into two main groups (supply and waste) each of which is identified with the type fixtures with which it is to be used. This leads to the familiar descriptions, like sink double supply fitting with swinging spout, lavatory pop-up waste fitting, and overrim tub supply. (*See* Figs. 1, 2, and 3.)

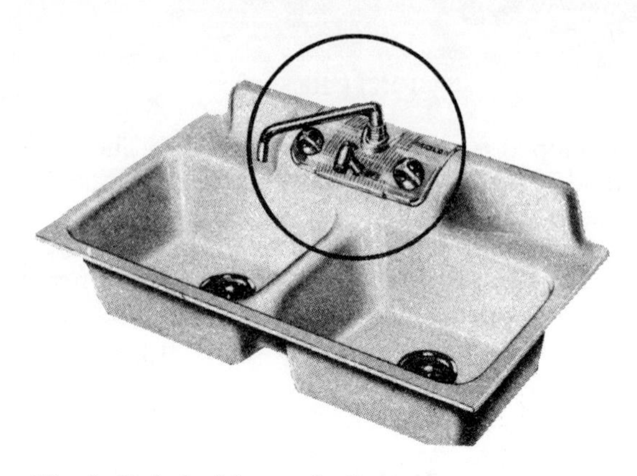

Fig. 1. Sink double supply fitting with swinging spout.

Fig. 2. Lavatory pop-up waste fitting.

Fig. 3. Overrim tub supply.

Fixture Supply Controls

Fixture supply controls are those which control and regulate the water at the fixture. These controls may be classified as concealed (where valve is located behind the wall or beneath the surface of fixture) or as exposed. (*See* Fig. 4.) The concealed controls are especially desirable in new work. In plumbing terminology, the terms *exposed* or *concealed* are used descriptively only on wall-installed items.

The type of supply control at the fixture may embody separate valves for hot and cold water lines which means that water must be led to the fixture from each supply to get the desired temperature for use. A more practical and the more popular type of supply control is the inter-connected double valve discharge to fixture through common spout or the combination supply fixture in which the valves are encased in a housing assembly. (*See* Fig. 4.)

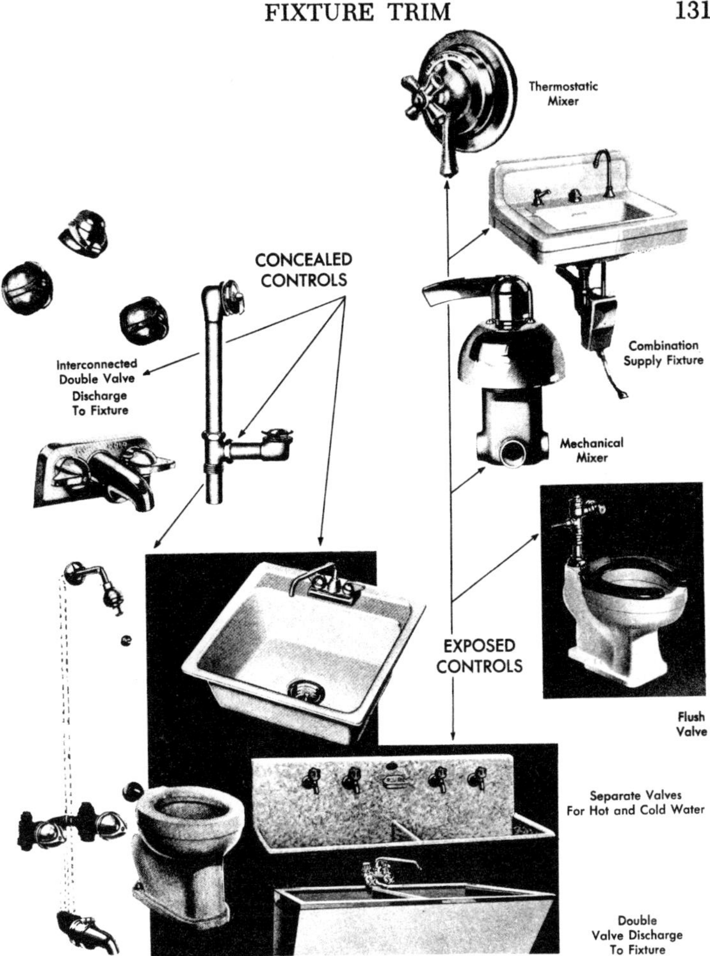

Fig. 4. Fixture supply controls.

Another type of supply is the thermostatic or mechanical type mixing valve, which by means of a single handle control delivers tempered water. (*See* Fig. 4.)

Outlet Discharges

The type of outlet discharge selected depends on a variety of requirements including the personal tastes of the home owner and specific conditions to be overcome. Some of the items which fall in this category are shower heads, spouts for delivery of water, self-closing faucets, and sprays.

The *shower head* selection is essentially a matter of personal choice. There are a variety of types for domestic uses and special types which are designed for a specific purpose.

The *self-closing faucet* is used in public washrooms or in any installation where water conservation is of prime importance.

Special type spouts have various advantages. The gooseneck is excellent for pantry sinks, service sinks, and in various other applications. This spout allows the user to put tall objects underneath its orifice.

Waste Controls

There are several types of waste outlets available for various fixtures.

Bath Wastes. For the bathtub there is available a concealed bath waste and overflow of the mechanical pop-up type and the connected waste and overflow type. The latter type is the simplest waste fixture and consists of a tube with overflow strainer which is attached to a waste spud with rubber stopper and chain. This is connected to the waste pipe. (*See* Figs. 5 and 6.)

The waste unit is made up of three units:

1. Concealed piping unit with spud which holds the waste to the bath.
2. The waste plug unit which can be removed from and replaced in the waste outlet opening.
3. Plug and stopper.

The lever action lavatory waste is noted for its quick draining ability and its quick, easy operating mechanism. The waste plug is directly controlled by lever and the lift rod, and the unusually high lift and the large slots in the side of the plug assure quick draining of the basin. The plug may be removed from the basin by a quarter turn and pull, permitting easy cleaning of the plug and ready accessibility to the outlet for cleaning.

The conventional plug and stopper is also available for lavatory wastes.

There is a fourth type of lavatory waste available. It is an indirect type of lavatory pop-up waste.

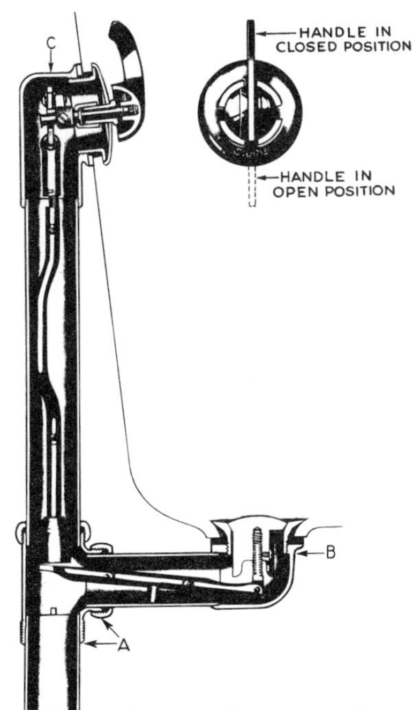

Fig. 5. Cutaway bath waste showing working mechanism.

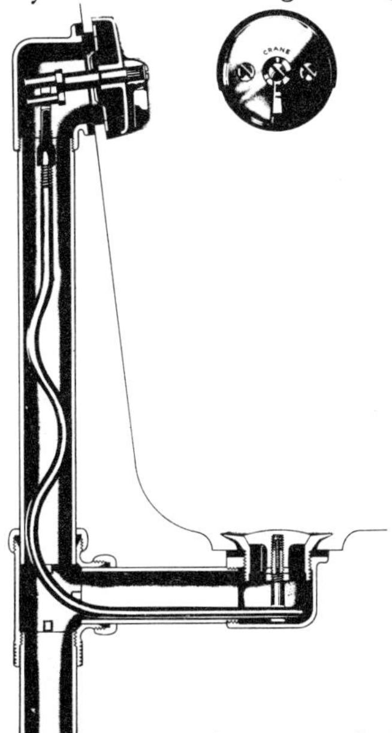
Fig. 6. Cutaway view of automatic bath waste.

Sink Wastes. Great strides have been made in sink waste matter or water disposal since the days of the flat sink strainer. Sinks are now furnished with the waste which is a practically sized, hand operated, removable sink strainer. By a simple turn of the handle the outlet can be stopped and the sink compartment is converted into a dishwashing basin or quickly drained. The large cup strainer may be lifted out and emptied without touching accumulated waste matter.

FITTINGS

Fittings used in making pipe joints (provide outlets, and so on, except where valves are used) comprise a considerable part of any piping system, and in addition constitute a good portion of the aggregate overall cost. Further, and probably most important, is the fact that pipe itself is usually trouble-free, if proper materials for the service have been selected, while joints are likely to be responsible for any unlooked-for maintenance costs which may occur. The working pressure and the nature of the service for which fittings are to be used have an important bearing on their design, and the wide variations in service conditions have necessitated the development of many different types and patterns as well as the use of different metals and alloys.

In order to really know fittings it is best to first familiarize yourself with the various types and kinds of joints which are employed in joining fittings with other elements of piping.

The seven groups into which fittings (joints) generally fall (together with some of their variations) are as follows.

1. *Screwed ends* (Fig. 7) are used more often due to the fact that they make a substantially permanent connection—a very satisfactory metal to metal joint—are adequate for reasonable pressures and temperatures where piping is used and, last but not least, are economical to use. It is important to note that they are most popular in the smaller sizes because the larger the pipe the more difficult it is to make up a screwed joint.

2. *Flanged ends* (Fig. 8) are used where conditions make screwed ends impractical, and bolting up the joint is advantageous. Although some flanged ends are made in sizes as small as ½ inch (1.27 cm.), they are most generally used for lines

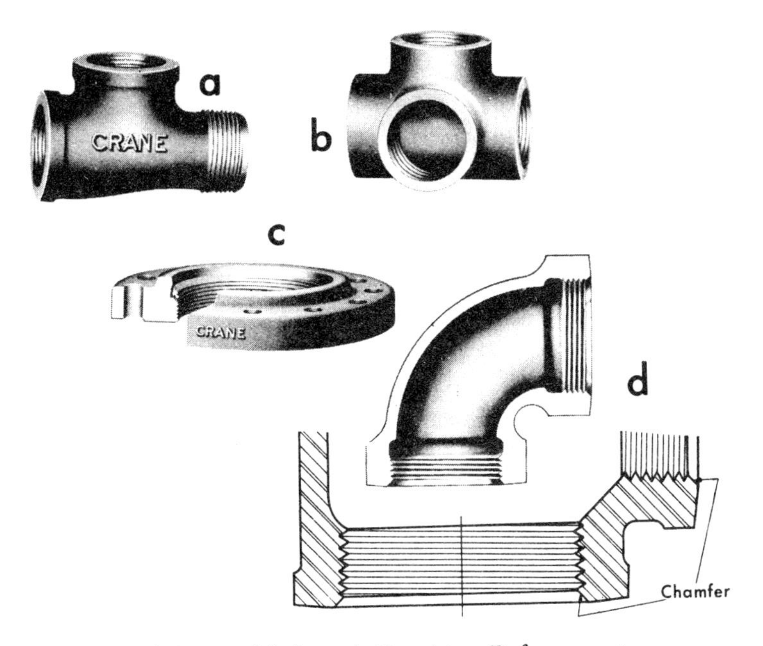

Fig. 7. Screwed fittings. A, Street tee; B, four-way tee; C, flat-band; D, plain or gas-pattern.

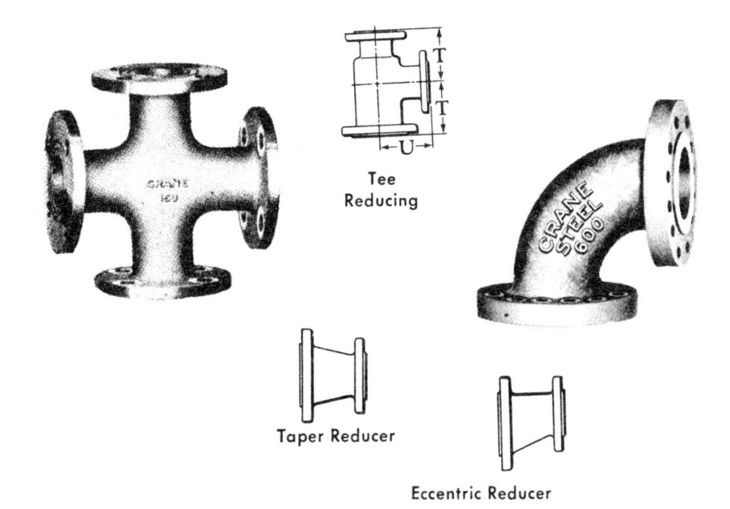

Fig. 8. Flanged fittings.

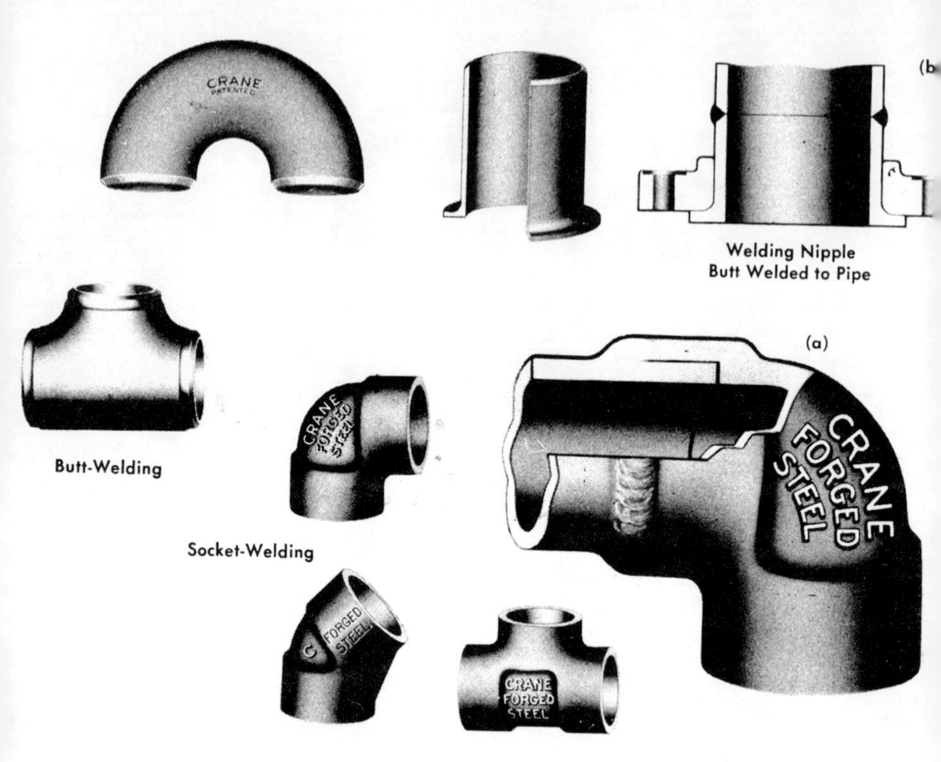

Fig. 9. Welding ends.

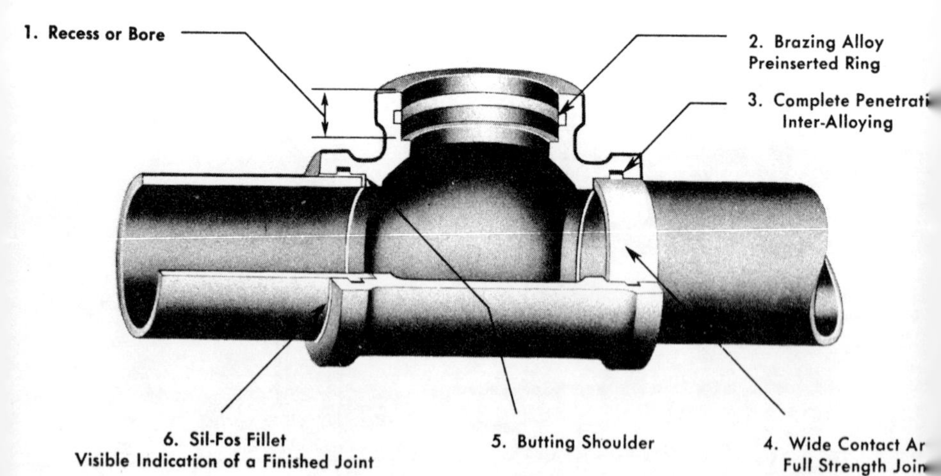

Fig. 10. Brazing ends.

4 inches (10.16 cm.) and larger because of their easier assembly and take-down. Further, flanged joints are made up with much smaller tools than are required for screwed connections of comparable size.

3. *Welding ends* (Fig. 9) are generally grouped under two headings: *Socket welding* primarily used in connection with small diameter piping; and *butt welding* for general all-around use, and for any pressure or temperature to which the piping may be subjected. Also, welding ends are generally used on lines not requiring frequent dismantling.

4. *Brazing ends* (Fig. 10) are made with a brazing ring insert, having end joints similar to a socket weld or solder fitting. These ends are satisfactory for many installations such as steam and water lines, refrigeration, and the like. Like the solder end, they are often preferred as a means of obtaining a more permanently tight joint than is possible with a screwed end.

5. *Solder-joint ends* (Fig. 11) have been perfected for the purpose of forming a smooth, durable and pressure-tight union between copper tubing and the fitting. Fittings of this type are very often used in plumbing and heating lines, lubricating oil lines around machines, and services such as water, air, vacuum, and other non-corrosive and non-hazardous systems.

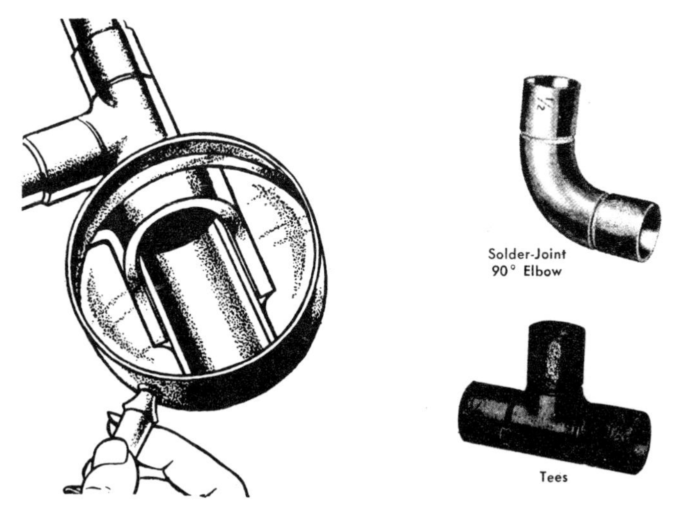

Fig. 11. Solder-joint ends.

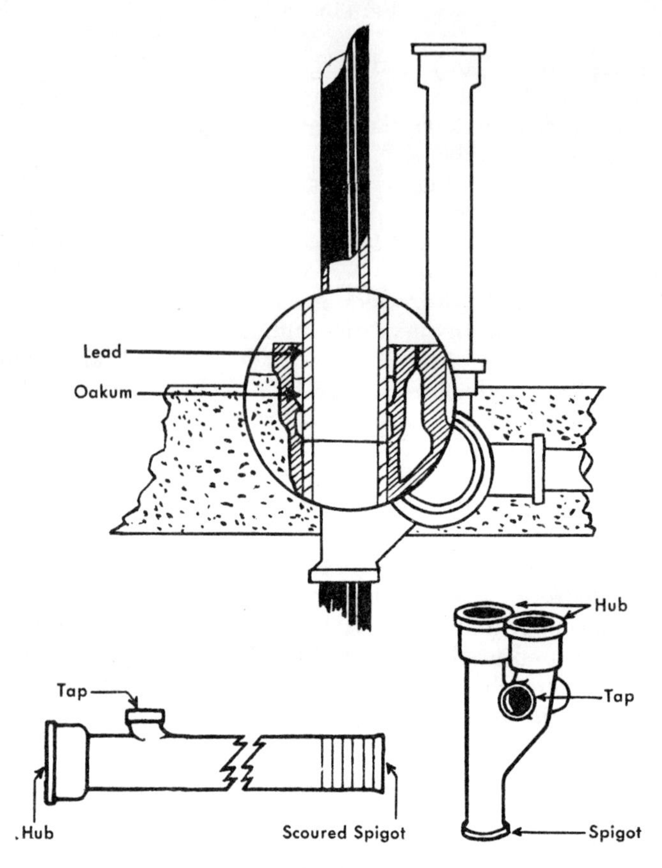

Fig. 12. Hub and spigot ends.

6. *Hub and spigot ends* (Fig. 12) are generally limited to valves and fittings specified for use with gas and water mains, sewage piping or soil pipe. The joint is assembled on the socket principle, with pipe inserted into hub end to valve or fitting, then calked with oakum and sealed with molten lead. Hub end fittings are usually made of cast iron and used with cast iron pipe having diameters which differ from steel pipe. Variation of this joint for use with wrought iron or steel pipe are known as the "Matheson," "Converse," and by other names. Figure 12 shows the hub and spigot end, the most commonly used method of employing calking and packing (lead and oakum) in making up a joint. There are other methods of

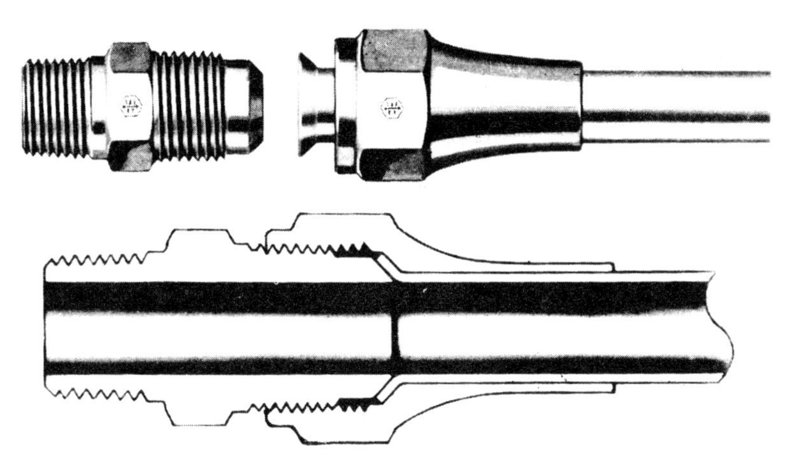

Fig. 13. Flared ends.

calking and by the same token there are various ways of applying bolted, mechanical, threaded, ring, gasket, and welding principles. You will find most of these variations illustrated in this section.

7. *Flared end* (Fig. 13) fittings are commonly used for copper and plastic tubing up to 2 inches (5.08 cm.) in diameter. The end of the tubing is flared or skirted and a ring nut is used to make a union-type joint.

Screwed Fittings

Screwed fittings (Fig. 7) are used generally on small size lines, and are available in a variety of designs and materials, each being applicable to the type of pressure and service for which it is intended, such as steam, water, air, gas, and oil lines; drainage and vacuum systems; railing work; and the like.

All straight and reducing patterns are necessarily well proportioned, of good weight, and entirely suitable for efficient service in connection with the working pressures for which they are recommended. They are easily made up (assembled) in the smaller size lines by average steam fitters.

A, Figure 7 shows the flat band screwed fitting in which the ends are reinforced by the bands as a means of preventing cracking or stretching of the fitting when screwing in a pipe or nipple. The plain or gas pattern (B, Fig. 7) is commonly made in malleable iron and is used in residences and general building work for low pressure gas and plumbing work.

Generally, the inside diameter of a cast screwed fitting is as large as the outside diameter of the pipe. The exception is the drainage fitting which has an inside diameter equal to the inside diameter of the pipe and with the thread chamber designed to eliminate pockets, and create clear flow. (*See* C, Fig. 7.)

Threads in screwed fittings are chamfered for easy entrance and starting of the pipe and as a measure of protection to the threads (D, Fig. 7).

In selecting materials of which screwed fittings are made, importance is given to the personal opinion of the user and to the matter of common practice. Serviceability carries greatest weight in the case of fittings for acid or alkali service, and the like. You will notice that when a pipe line is being dismantled very often cast iron screwed fittings are removed by breaking them with a hammer whereas malleable iron or brass screwed fittings, being more elastic, do not break so easily and therefore must be cut out or unscrewed.

Flanged Fittings

Flanged fittings (Fig. 8) in the larger sizes have a distinct advantage over screwed fittings in that they are comparatively easy to install or dismantle in a pipe line. In addition, their use makes it possible to carry on the practice of fabricating piping by laying out lengths of pipe with flanges attached and the fittings marked at the factory so that field erection becomes simple and fast with all parts fitting as originally laid out. The application of dimensional standards makes it possible to interchange one fitting for another more readily.

Because of their design, flanged fittings are made in various materials and in any size desired, whereas screwed fittings are very seldom made in the larger sizes. In the event that the desired size in a reducing fitting is not obtainable out of stock, it is generally possible (when using flanged fittings) to make up a reducing fitting by simply bolting on a reducing companion flange. In Fig. 8 the straight flanged fitting is shown in comparison to the flanged reducing fitting.

Substantial progress has been made in size standardization of flanged fittings and it is possible to lay out the plan for a new installation without any thought as to which dealer would be most apt to stock the required sizes.

Welding Fittings

Welding fittings (Fig. 9) fall into two general classifications known as socket-welding and butt-welding. Both types have been made possible by the perfection of shop and field welding techniques, and both with their patterns and variations are especially adapted for use where high pressures and high temperatures must be reckoned with. Welding fittings are as strong or stronger than the grade of seamless steel pipe of the thickness for which they are designed, and by the same token can be used with the same degree of safety for the same pressure or temperature for which the pipe is used.

A, Figure 9 illustrates and explains the *socket-welding fitting* which is recommended for use on small size pipe lines. As shown, the fitting (either forged steel or machined bar stock) is designed to slip over the pipe. Since the welding metal is deposited on the outside of the pipe, there are no so-called icicles left on the inner surface to clog the lines or restrict the flow.

There are several advantages which result from the careful designing of the socket-welding fitting and have helped to popularize this type of fitting. They are as follows.

1. The deep socket provides liberal come and go. The welder need not be absolutely accurate in his cutting or measuring procedure. The deep socket speeds up installing, and reduces costs.
2. The thickness of the socket wall is approximately 1¼ times the thickness of pipe with which it is supposed to be used. This conforms to the American Standard on socket-welding fittings and provides suitable space for a weld of ample strength.
3. The inside diameter of all socket-welding fittings is the same as the inside diameter of the pipe with which they are used. Even though the pipe is butted back against the shoulder inside of the fitting there is no ridge or irregularity of surface.

The welding principle is applied to fittings and flanges as

well as steel gate, globe, angle, and check valves in the full range of sizes.

Butt-welding fittings (B, Fig. 9) are designed with a uniform circular section of full diameter and identical center to end dimensions. Ends are accurately beveled and are circular to properly match the end of the pipe to which it is to be welded. Use of butt-welding fittings not only reduces the weight of lines, but minimizes the likelihood of leaky joints.

Brazing

The name "Crane-Seal" on a brass valve or fitting indicates that the item is especially prepared for *oxyacetylene brazing*, and that each product so identified is equipped with Sil-Fox brazing alloy ring inserts. This alloy forms an ideal bond between pipe and fitting that is suitable for reasonably high pressures and temperatures.

Figure 10 shows the design principles and materials which go to make up a *Silbraz joint*. Because of accurate machining, there is a minimum of clearance between pipe and fitting surfaces. Also, the Sil-Fox brazing ring insert (factory installed) assures the user that the correct type and quantity of brazing metal will be used, and that the Sil-Fox actually penetrates into and inter-alloys with the surfaces of the pipe and fittings. The bond becomes stronger than the parts being joined. A square shoulder at the bottom of the bore limits the distance that the pipe can be inserted and, further, provides a dam for the brazing alloy. In the final stages of flow, the molten Sil-Fox forms a fillet at the end of the fitting adjoining the pipe. This continuous fillet appearing around the pipe is visible proof that the joint is completed.

Solder-joint Fittings

Figure 11 illustrates a fairly complete story of the design principles surrounding *solder-joint fittings*.

Brass or wrought copper solder-joint fittings for use with copper tubing (O.D. is ⅛ inch (.3175 cm.) larger than nominal size) have an unusually large range of applications covering practically all types of home, industrial, institutional, food processing and chemical uses. The paramount reason for this wide acceptance is the relative ease of erecting and handling of copper tube lines and fittings. Tests in both the laboratory and under everyday working conditions have proved that they have unusual strength for the services intended and that if failure should occur it is generally in the tubing and not in the fitting or joint. (*See* Figs. 12 and 13.)

TYPES OF FITTINGS

There are many patterns of fittings, each of which is designed as it is for a particular and specific purpose, and the complete assortment is sufficient to care for any known joining problem. Pipe fitters can make up any simple or complex job with a minimum number of fittings and in strict accordance with specifications and codes.

Fittings are made in the same variety of metals as valves (brass, iron, steel, various alloys) and in corresponding pressure classes, and end connections, and in all standard pipe sizes. Note, however, that all types of fittings are not made in all sizes and materials.

Even the regular range of fittings includes a great multiplicity of items and to show them all here would be practically impossible. An acquaintanceship with the most common types as illustrated, and a general knowledge of the typical variations and their uses, will be of the most immediate help to you.

These patterns of fittings are as follows.

Flanges
Elbows
Tees
Crosses
Reducers
Bushings
Couplings
Nipples
Unions
Return Bends
"Y" Bends or Laterals
Manifolds or Headers
Caps and Plugs

When thinking in terms of the multiplicity of items remember that each of the foregoing patterns has its own variations which are designed for specific services such as the circulating boiler elbow, drop tee, circulating boiler coupling, various types of drainage fittings, and the like.

Flanges

The term *companion* is accepted as applying to all types

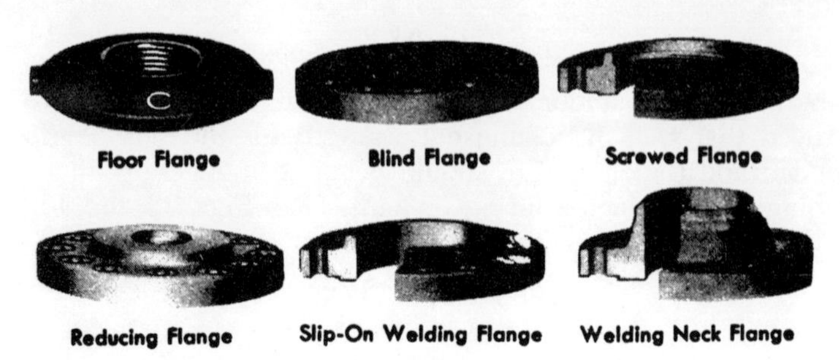

Fig. 14. Flanges.

of loose flanges, just as flanges on the end of a valve or fitting are called *end flanges*. (*See* Fig. 14.)

The more popular and most widely used *companion flanges* usually fall into three groups.

1. Flanges internally threaded to be screwed in place on the end of a threaded pipe.
2. Flanges which are designed for butt-welding to pipe ends, and/or slip-on flanges welded front and back, and refaced.
3. Blind flanges.

Cast iron and brass flanges for the first group are used extensively for relatively low pressure steam and water service. Steel flanges are specified for the higher pressures and temperatures. Screwed flanges are popular in field erecting work, as the operation is greatly simplified by cutting and threading pipe to the required length and screwing on the flange.

The second class companion flange (welding) is employed in welded joints. There are two popular types of welding flanges; the welding neck flange and the slip-on welding. In both types the flange is welded directly to the pipe. Where temperatures and pressures are high, the use of welding for attaching the flange to the pipe (in favor of other types and methods) eliminates the difficulty ordinarily experienced in assuring tightness between the pipe and flange under severe operating conditions. Further, the use of this type of flange has been accelerated by greatly improved welding technique.

The third group of flanges (blind) are solid except for bolt holes and are used as closures.

Elbows

Elbows (commonly called "ells") are usually designed to make a 90-degree or right angle turn in a pipe line, but where

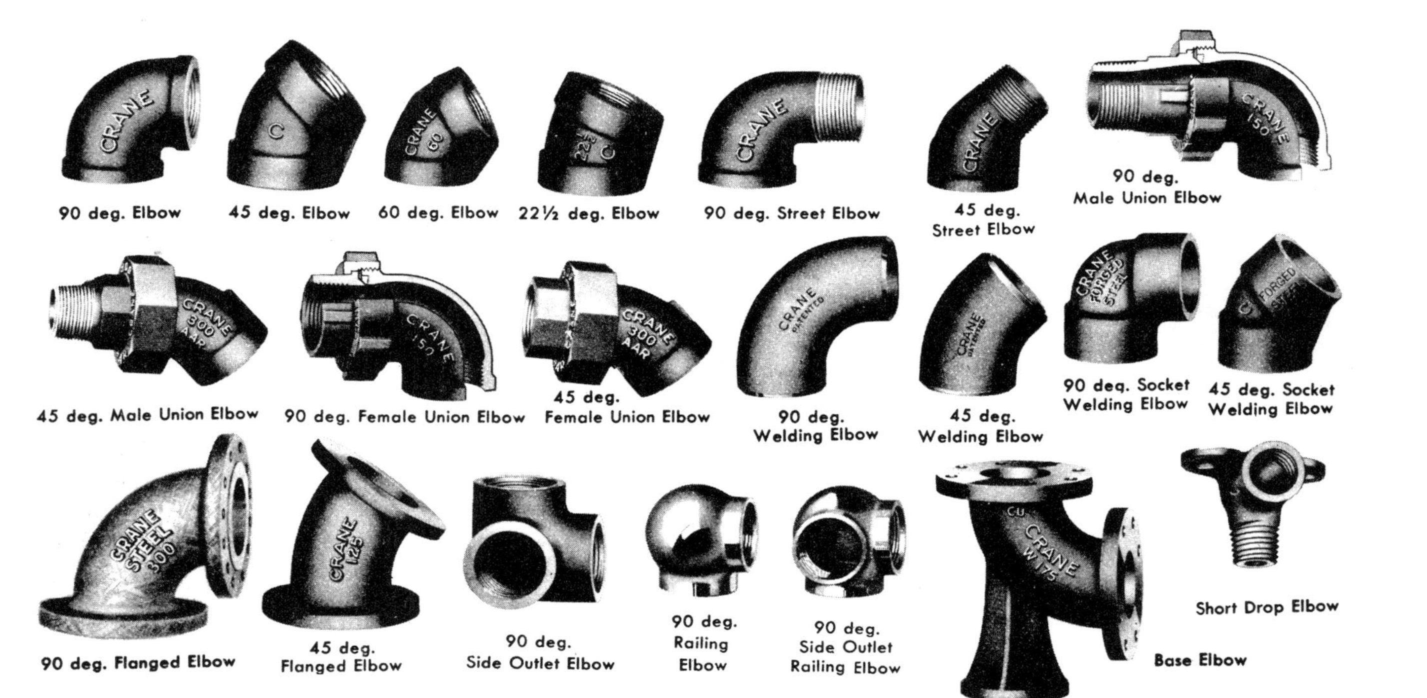

Fig. 15. Elbows.

different angle turns are desired there are 60, 45, 22½, and other degree elbows to be had (Fig. 15).

Another type of elbow is the "street ell" (Fig. 15). In shape it is similar to a regular ell but has one female end and one male end.

Tees

Where it is necessary to take a right angle line from a horizontal or vertical run of pipe, a *tee fitting* shaped like the letter "T" is used. Here again the "street" design may be had and is known as a "street tee," formerly called "service tee." The street tee generally has a male opening at one end and a female opening on the other end of the run; the outlet being female, although in some instances both ends of the run are female and the outlet male.

Crosses

Because they are used to connect intersecting pipe lines, *crosses* have four openings in the same plane at 90-degree angles to each other (Fig. 16). They look like a plus sign.

Reducers

A *reducer* is used for the purpose of permitting the joining of a given size line to one of smaller size. It has a smaller opening at one end than the other, the center lines of which usually coincide (Fig. 17).

There are times (primarily in heating systems) when undesirable water pockets might form, in which case the eccentric reducer should be used in order that the inside bottoms of both openings will be on the same plane or level.

Elbows, tees, and crosses are also made in reducing patterns.

Fig. 16. Crosses.

Fig. 16. Crosses (continued).

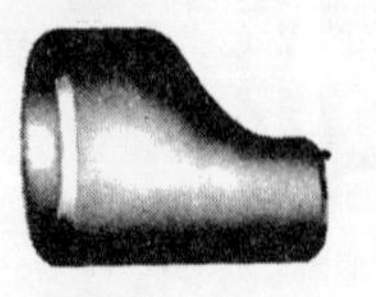

Standard Reducer | Eccentric Reducer | Butt-Welding Reducer | Eccentric Butt-Welding Reducer | Flanged Taper Reduc

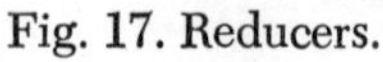

Fig. 17. Reducers.

Outside Hexagon Bushing | Eccentric Bushing | Face Bushing

Fig. 18. Bushings.

Bushings

Bushings (Fig. 18) are made use of when it is impossible to secure a reducing fitting of the proper size. They are inserted into the end of another fitting or valve to reduce the size of the openings.

Caps and Plugs

A *cap* is used to close off the male end of a line. *Plugs* serve the same purpose on female ends. (*See* Fig. 19.)

Couplings

The *coupling* (Fig. 20) connects two lengths of pipe. All couplings (cast and wrought) are made with female ends.

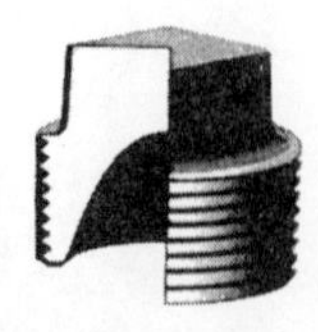

Standard Cap | Countersunk Plug | Square Head Plug | Bar Plug

Fig. 19. Cap and plugs.

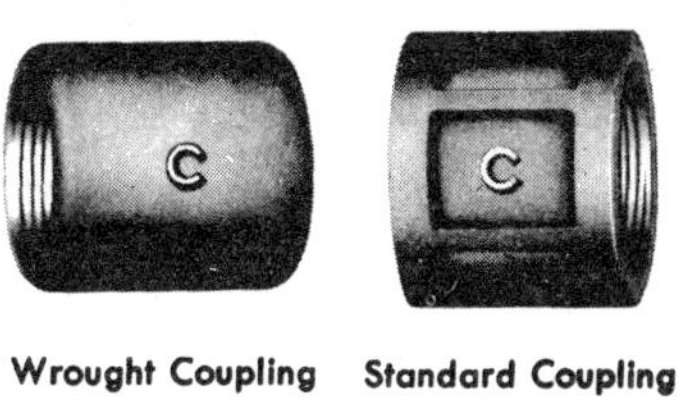

Fig. 20. Couplings.

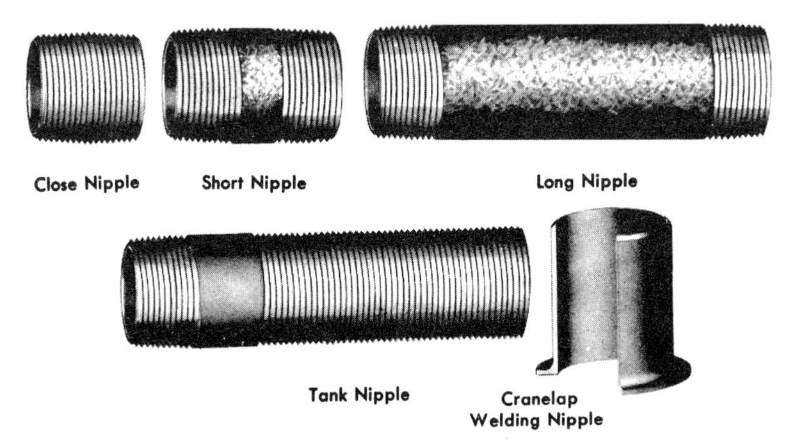

Fig. 21. Nipples.

Nipples

Nipples (Fig. 21) might well be termed "short lengths of thread pipe." They are used to connect two other parts, and come in three types.

1. The *close nipple* usually has its entire surface covered with tapered threads cut from each end to the middle. It is used to make a close connection between two parts. The length of these nipples varies with pipe size, it being determined by the length of thread necessary to make a satisfactory joint.

2. The *short nipple* is longer than the close type and often has a small unthreaded surface between the threaded ends.

3. A *long nipple* is any pipe up to 12 inches (30.48 cm.) long, threaded on both ends. All longer pipe is called "cut pipe."

Also, there are other applications where nipples even shorter (called "butt" nipples) than the "close" type are necessary but must be specially made.

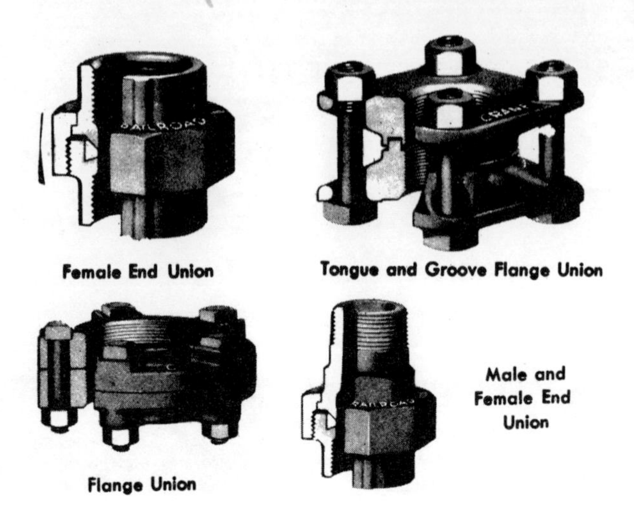

Fig. 22. Screwed and flanged unions.

Screwed and Flange Unions

Unions are designed for use in pipe lines to permit easy assembling, and easy opening or dismantling of the lines. Obviously, their primary purpose is to connect pipe.

Generally the small size unions are designed with a screwed union ring, whereas the larger sizes are of the flange or bolted type (Fig. 22).

A union fitting is a combination of a fitting and a ring type of union which serves the purpose of both. It provides the additional advantage of reducing the number of joints necessary in a line.

Return Bends

A *return bend* (Fig. 23) is a fitting which is shaped like the letter *U* and has the inlet and outlet in the same plane. It returns the flow of fluid to the direction from which it came. It is made to various centers and radii. In installations, return bends are economical because they reduce the number of joints and decrease the makeup time ordinarily consumed in assembling extensive piping systems.

"Y" Bends or Laterals

"Y" bends or laterals (Fig. 24) are rather uncommon because the peculiar shape requires special design to obtain

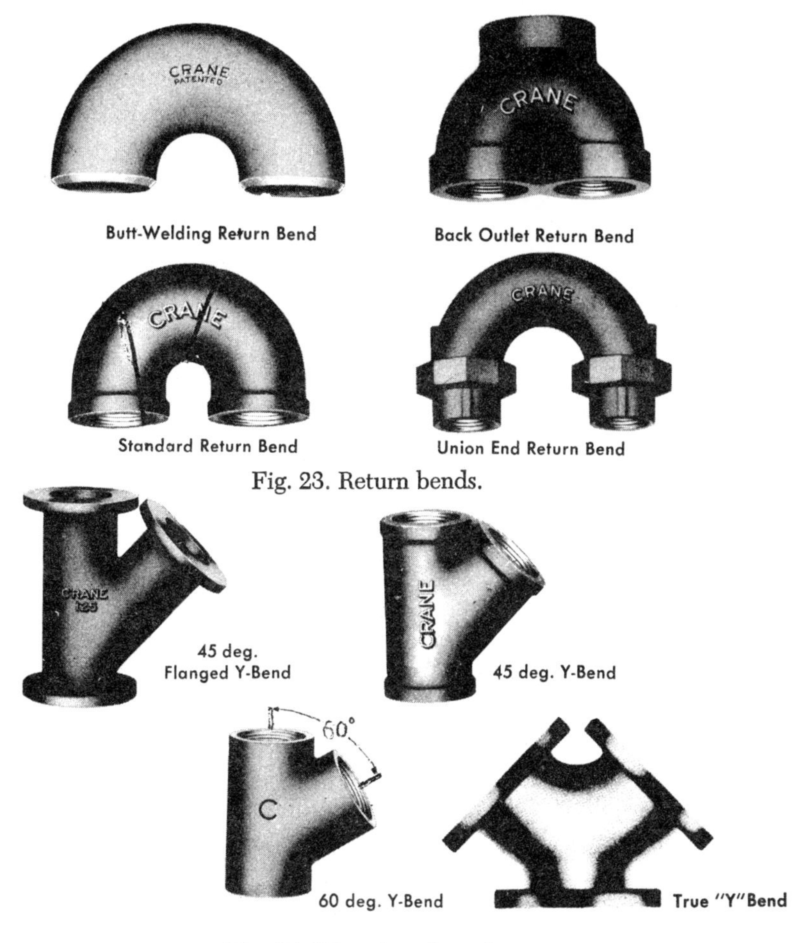

Fig. 23. Return bends.

Fig. 24. Y bends or laterals.

proper strength. The design does reduce pressure drop and there are some installations where this feature or space limitations may demand its being used.

Manifolds or Headers

Manifolds or headers (Fig. 25), which are usually considered a specialty, serve the purpose of diverting flow from a large pipe into one or more smaller lines. They are shop fabricated or cast to specifications furnished by the purchaser or his design engineer.

Fig. 25. Manifolds or headers.

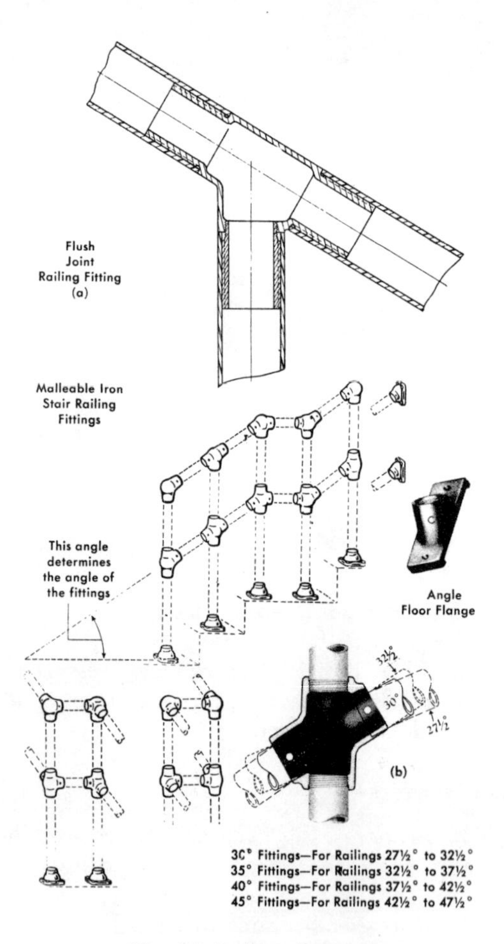

Fig. 26. Railing fittings.

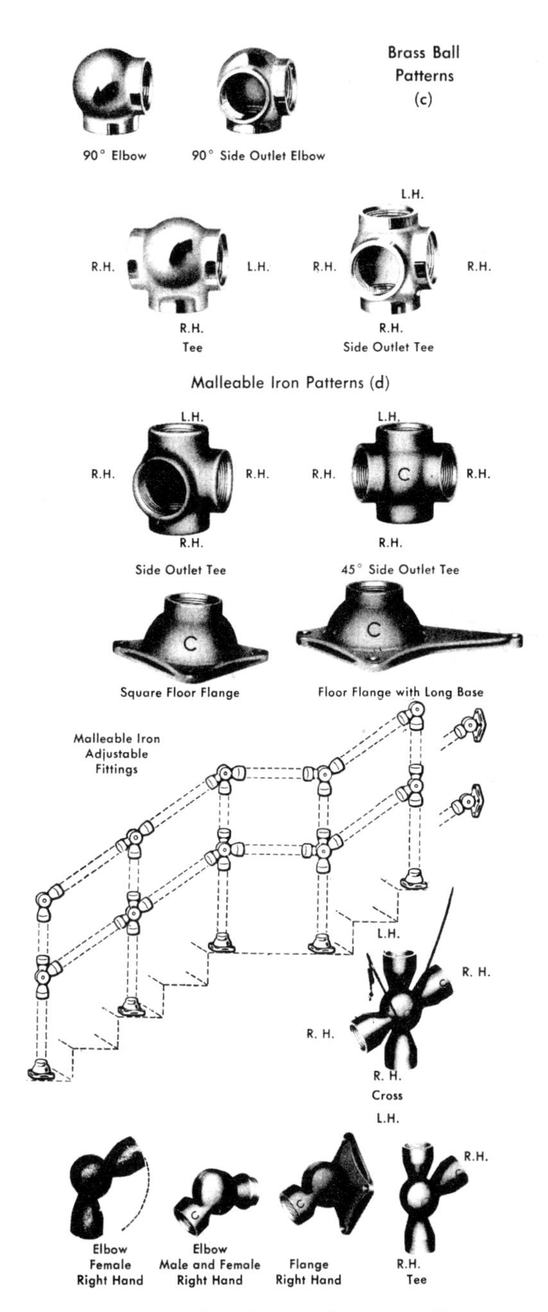

Fig. 26. Railing fittings (continued).

FITTINGS FOR SPECIFIC PURPOSES

Railing Fittings

Railing fittings are made so that pipe can be screwed in, slipped in, or attached in some other manner demanded by the class of work desired. *Screwed railing fittings* are provided with a recess as shown in C, Fig. 26, to protect and conceal the threads on the pipe (D, Fig. 26). The following are the various styles which are most widely used.

The *adjustable fittings* are furnished with screwed ends only and are adaptable to stairs having an angle of 45 degrees or less from the floor line.

Slip and *screw joint variable angle fittings* are made so that all posts or uprights must necessarily have screwed connections for the purpose of securing a solid and substantial job. Note that the run of these fittings is designed to allow an angular variation of the pipe of 2½ degrees on each side of the centerline. After the angle pipe has been inserted into the fitting and is in the correct position, the pipe is drilled, pins are inserted, riveted over, and the ends of the pins filed smooth. A correctly finished railing will make it appear that the fittings had been specially made for the absolute angle required (B, Fig. 26).

Flush joint railing fittings are made with an extension over which the pipe is driven up to the face of the fitting, forming a smooth joint and a continuous rail having the same diameter at all points (A, Fig. 26).

Ammonia Fittings

Ammonia fittings are made in both screwed and flanged patterns. The general shape of the flange is somewhat different from the American Standard for pressure class flanges. The facing is usually tongue and groove and the chamfer in screwed and ammonia fittings is larger than on regular screwed fittings so that the joint can be soldered or sweated if desired.

Drainage Fittings

Drainage fittings are designed so that the inside diameter of the fitting is the same as the inside diameter of the pipe. In short, in drainage service there must not be any obstruction against which solid matter can lodge. Fittings with right angle openings have the threads tapped-pitched so that the horizontal line will pitch ¼ inch per foot (.635 cm. per 30.48 cm.) to assure proper drainage.

SIZES AND ANGLES OF FITTINGS

Fittings are said to be "straight" when all of the openings are of the same size, and "reducing" when their openings are not of the same size. The size of an opening in any fitting is governed by the size of the pipe to which it will be connected with the exception of fittings such as solder-joint and tube fittings used with tubing. A 3-inch (7.62-cm.) nominal size screwed fitting is not 3 inches (7.62 cm.) in diameter. The term merely means that the female openings are threaded to connect to the male threads on a 3-inch (7.62-cm.) pipe which is actually 3½ inches (8.89 cm.) in outside diameter.

In the lower pressure classes, flanged fitting openings are of nominal diameter. For instance, a 3-inch (7.62-cm.) flanged fitting has approximately a 3-inch (7.62-cm.) opening. The higher pressure class steel fittings have openings smaller than nominal because of the heavier metal thicknesses. This has been established by American Standard. This fact should be carefully noted and extreme care exercised when reading the size of a flanged fitting from a casting.

Solder-joint fittings have openings ⅛-inch (.3175-cm.) larger than the nominal size, except those for refrigeration, which have openings the same as nominal size. Flared and compression-tube fittings are used with tubing of the same outside diameter (O.D.) as nominal size. Figure 27 illustrates how to read the size of reducing fittings. While it is necessary to mention the size of only one opening on straight fittings, the size of all the openings on reducing fittings must be given

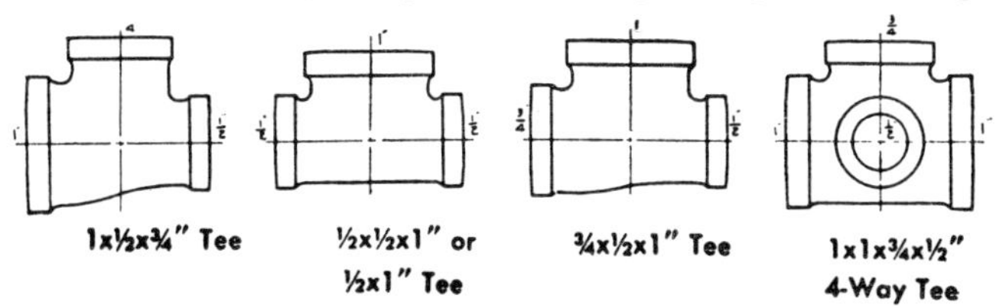

Fig. 27. Reading the size of reducing fittings.

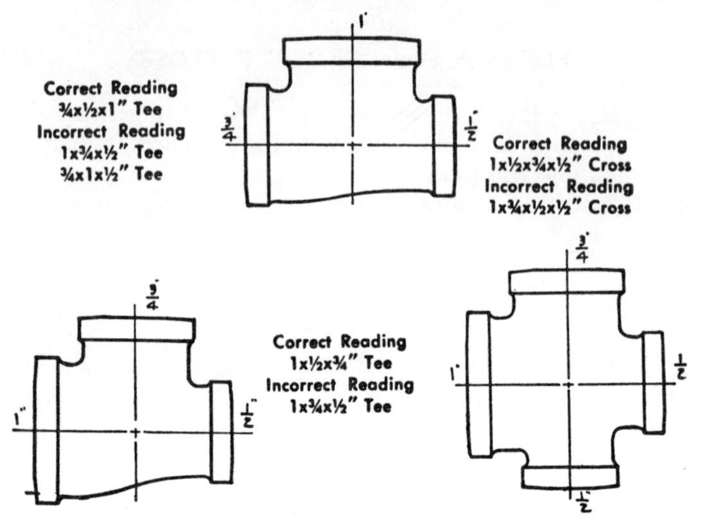

Fig. 28. Correct and incorrect reading of fittings.

when ordering. The exception to this rule is that where the two ends of the run are the same size, this size need be given only once. It is easy to see the confusion that would be caused by incorrect reading of sizes on reducing fittings. Note the comparison of correct and incorrect methods in Fig. 28.

In the case of an eccentric fitting (Fig. 29), the position in which it is to be used in the installation determines the relative position of the openings. *Note:* Position should always be indicated on a sketch accompanying the order.

When figuring or reading the angle of fittings, bear in mind that there are two methods, one of which applies to all fittings (with the exception of railing fittings) and the other to railing fittings only.

Fig. 29. Eccentric fittings.

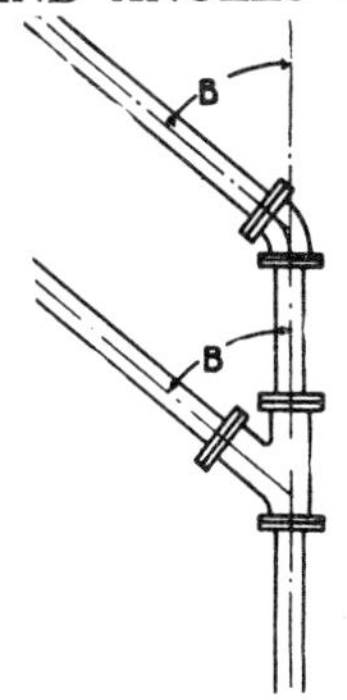

Fig. 30. Railing fitting, B denotes the angle of the fitting.

As shown in Fig. 30, the angle of a fitting (eels, tees, Y branches, and so on) is determined by measuring the angle of the directional change in flow from a straight line.

In the case of railing fittings (Fig. 31), you simply determine the angle of the stairway. This reading becomes the angle of the railing fitting.

Figure 26 illustrates a few special degree railing fittings wherein the angle of the stairway describes the angle of the fitting. Further, it shows other fittings whose angle is equal to that of the railing fittings, but it should be noted that the method of determining the degree of angle is different.

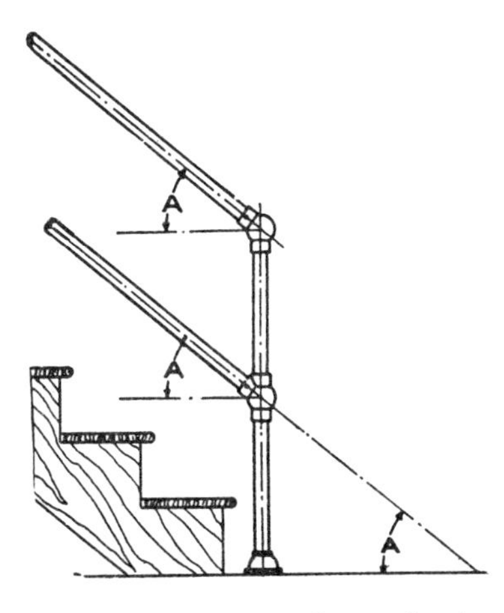

Fig. 31. Railing fitting, A denotes the angle of railing fitting.

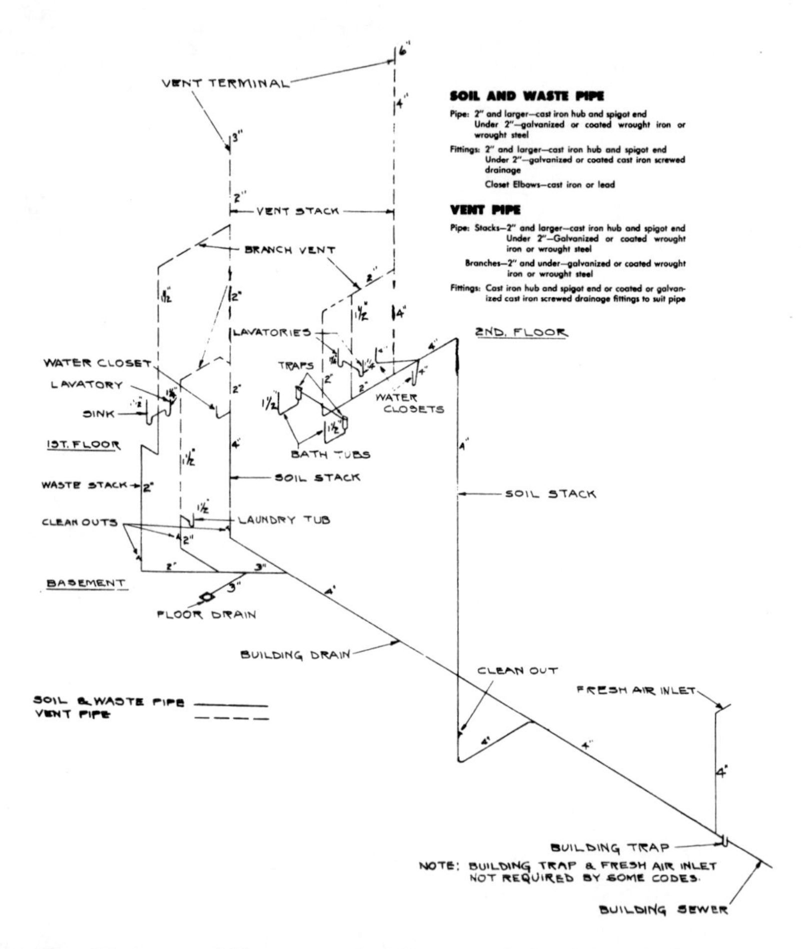

Fig. 32. Pipe and fittings required for typical plumbing installation. (Drainage system.)

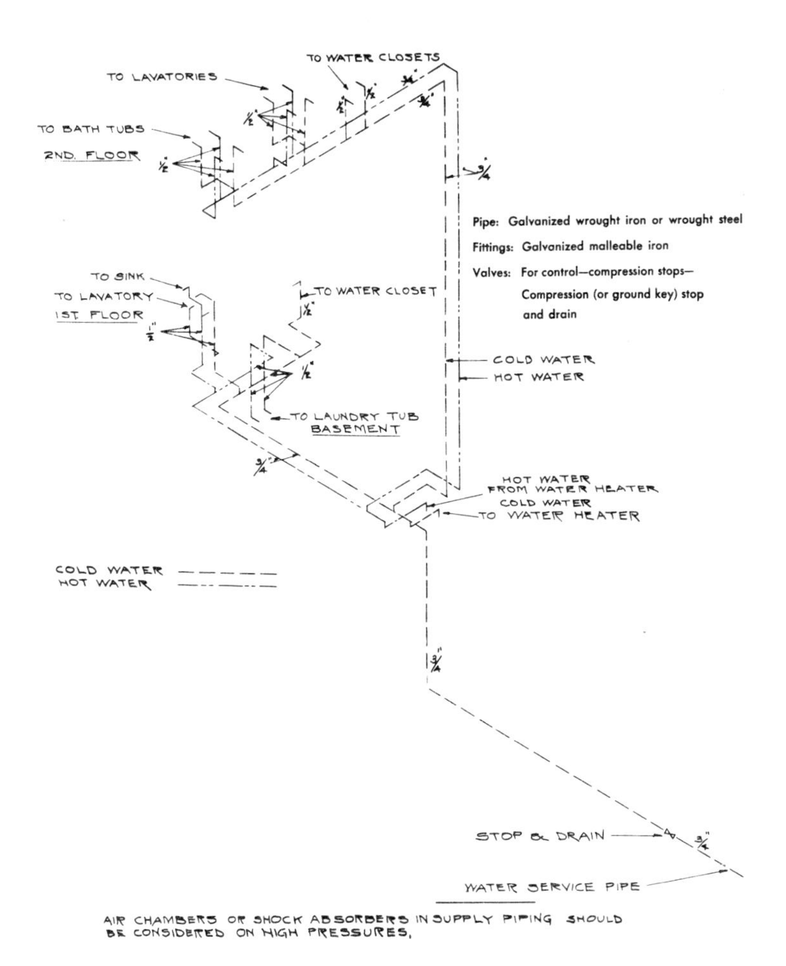

Fig. 33. Pipe and fitting required for typical plumbing installation. (Water supply system.)

ROUGHING-IN

"Roughing-in" is the term usually applied to the many valves, fittings, and lengths of pipe that wind their way from the local water supply and local waste connections through floors, walls, and ceilings on their mission of supplying fresh water to, and removing water-borne waste from all plumbing fixtures throughout the entire building. This also includes all water piping and drainage and vent piping. Roughing-in, which is completed before the installation of the fixtures, includes all piping up to the wall or floor line of the fixtures. (*See* Figs. 32, 33, and 34.)

The plumbing fixtures may be entirely or jointly selected by the homeowner, but the roughing-in materials are most generally left to the discretion of the architect and/or plumbing contractor. Selection of materials and their installation is usually guided by local codes and practices, water conditions, and economy.

The "behind the wall" system consists of the following: cast iron soil pipe, soil fittings, screwed drainage fittings, special fittings, cast iron and malleable iron fittings, stops, valves, and most all kinds of pipe.

PIPE

Pipe is an important item in roughing-in. Domestic piping varies with the locality and structure and includes steel, wrought iron, copper, brass, cast iron, lead and in various degrees substances like vitrified clay, concrete and bituminized fiber. Figures 32, 33, and 34 indicate the type and kind of pipe which would ordinarily be specified for the various domestic services.

Wrought Steel Pipe

Steel is a material consisting basically of iron plus a small amount of carbon and is characterized by its great ability to resist deformation without fracture and be formed by either hot working or cold working processes. From the standpoint of application, *steel pipe* is by far the more generally used.

The outstanding advantage of steel pipe is its great tensile

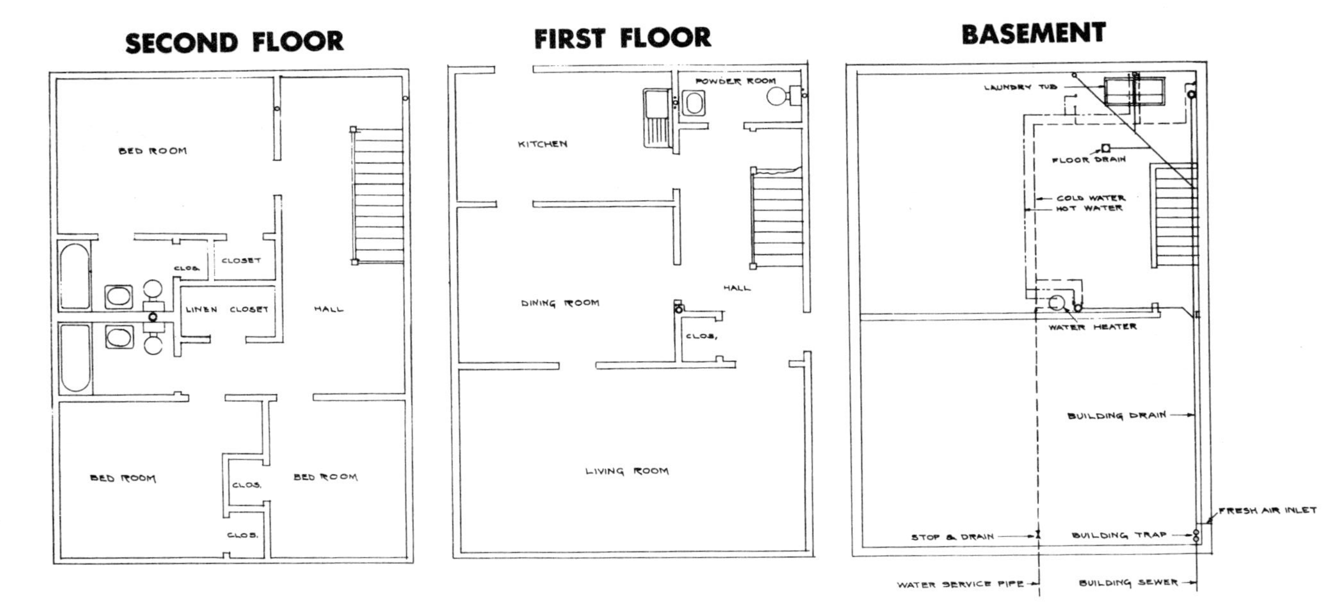

Fig. 34. Typical plumbing installation of residence.

strength combined with ductility and it is especially adaptable to those uses where sudden pressures, vibration, shocks, and water hammer are encountered. It is a material capable of withstanding external forces as well as internal pressures. Steel pipe is used practically everywhere pipe is used, for pressure or gravity flow of steam, air, water, oil, or gas.

There are many grades of steel ranging from mild or low carbon steel to the high alloys. *Mild steel* is the grade most commonly used in steel pipe.

Ordinary steel pipe is produced by welding or the seamless process. Both seamless and welded pipe are available in standard, extra strong and double extra strong weights.

Steel pipe can be joined by welding as well as by screwed joints and fittings. However, welding of pipe used for drainage should be avoided and welding of galvanized pipe is not recommended.

Steel pipe is obtainable in standard sizes and weights and available in both black and galvanized. Black pipe is not normally used on water lines either above or below ground.

The term *wrought pipe* is often used to refer to standard steel pipe and should not be confused with wrought iron pipe. It is advisable to use the term *wrought steel* to avoid any confusion or misunderstanding.

Wrought Iron Pipe

Although *wrought iron* contains a quantity of finely divided slag, it has many of the attributes of steel. The presence of this slag gives *wrought iron* pipe a resistance to corrosion under some conditions.

All wrought iron pipe is produced by welding because its properties do not lend themselves to the seamless process as in the case of steel. In the range of nominal sizes, ⅛ inch to 12 inches (.3175 to 30.48 cm.) inclusive, it is produced in three commercial weights, standard, extra strong and double extra strong. The O.D. sizes, 14 inches (35.56 cm.) and larger, are produced in various wall thicknesses up to 1 inch (2.54 cm.) according to required specification. Black or galvanized wrought iron pipe is available and can be furnished plain end, threaded, and coupled, or threaded only. For wrought iron pipe, as well as for steel pipe, cast iron or malleable iron fittings are usually used.

Wrought iron pipe can be used on water service lines from the main to the building, for raw and hot water lines, drainage lines, heating system return lines, rain leaders, vent stacks, and water or gas lines.

Cast Iron Soil Pipe

Cast iron is a material consisting basically of iron plus carbon and silica in amounts which render it strong but nondeformable. Cast iron does not lend itself to forming and is thereby restricted to forms resulting from molding and machine operations. Its properties are such as to make it ideally suitable for drainage pipe.

Cast iron soil pipe and *fittings*, because of ease of jointing and installation, and the comparative flexibility of joints, are more adaptable to drainage than most pipe and fittings made of other materials. They withstand the wear and tear of service over a long period of time.

Cast iron soil pipe and fittings are made in standard and extra heavy weight, although codes in general require the extra heavy weight. Pipe conforming with the ASA standard must withstand a hydrostatic pressure test of 50 psi. There is no specification or test requirement covering standard weight pipe. Most codes require cast iron soil pipe for underground drain lines and it is also the most commonly used pipe for drainage lines above the ground.

Cast iron soil pipe is made with hub and spigot ends instead of standard threaded connection. Double hub pipe is made but it is used only where cutting is necessary to save scrap. Codes do not allow double hubs in drainage lines.

Where *hub* and *spigot fittings* are used, the joints are made watertight by inserting the spigot end of one pipe into the hub of the second, packing oakum in the joint, pouring hot lead into the hub, and calking. Lead wool can be used instead of the poured lead under certain conditions. The resulting joint is airtight and watertight.

Cast iron soil pipe comes in 5 foot (152.40 cm.) lengths and is available *tarred* or *untarred*.

Brass or Copper Pipe

Brass or *copper pipe* is strong and durable, resists corrosion, and is used extensively in some areas because of its ability to withstand the action of the local corrosive waters. Its qualities particularly meet conditions in certain parts of the country, especially along the various seaboards.

Brass and copper pipe give satisfactory long time service on waters of low permanent hardness, a fair degree of temporary hardness, or low carbonic acid gas content and of relatively high alkalinity.

Commercial lengths of standard pipe regularly available are 12 feet. Intermediate or shorter lengths are cut to order only. Sizes are available from ½ inch to 12 inches (1.27 to 30.48 cm.) for regular and ⅛ inch to 10 inches (.3175 to 25.4 cm.) for extra strong.

Copper Water Tube

Copper water tube is durable, resistant to corrosion, and flexible. It is recommended for many services such as steam, water, air, gas or similar fluids. It is used for the water supply in buildings and in a number of cities for the service pipe from the water main to the house or curb. Copper tubing for general plumbing and heating purposes is available in three weights known as types K, L, and M. All weights are furnished in hard temper and in addition types K and L are furnished annealed.

The three types of copper water tube used in plumbing, depending on the principal uses, are as follows.

Type K. For underground services and general plumbing purposes.

Type L. For general plumbing purposes.

Type M. For use with soldered fittings *only.* (Furnished in 2½ inch to 12 inches (6.35 to 30.48 cm.).)

Hard copper tubing is intended primarily for use in straight lengths. It is not recommended for field bending except with proper bending equipment. *Soft copper tubing* can be bent without special equipment. Twenty-foot lengths for both hard and soft copper tubing and 60-foot coiled lengths for the soft tubing are considered standard. Flared and solder fittings are available and because of the adaptability of copper to jointing by soldering and the permanence of the soldered joint, solder fittings are the ones most commonly used.

Lead Pipe

The use of *lead* as piping material can be traced back to antiquity. It is ductile, durable, flexible and has a low melting point. It is very often used as the service pipe from the main to the house or curb and is required by many water departments for this purpose. It can successfully handle some highly corrosive chemicals, such as concentrated sulphuric acid in certain temperature ranges, but precautions should be taken against the possibility of alkaline corrosion. *Lead pipe* should not be used to conduct soft water or water of swampy or peaty origin, as such waters tend to take the metal into solution.

Due to its lack of elasticity, lead is not too well adapted to mechanical joints, and this plus its low melting point makes the wiped joint the recommended type in joining lead pipe.

Lead is also used for soil and waste pipe, traps, and bends. According to Commercial Standard (set up by the Lead Industries), lead pipe is made in sizes up to 6 inches (15.24 cm.); sizes up to 2 inches (5.08 cm.) are made to carry cold water pressures up to 100 psi.

Cement Lined Pipe

Steel pipe with a *cement lining* has been developed to improve the serviceability of pipe under certain corrosive conditions. The cement lining acts as a chemical resistant covering protecting the pipe. The pipe has good mechanical strength. Fittings of the same type are available.

This type of pipe is intended primarily for carrying waters which rust, corrode, or otherwise attack unprotected metal pipe. Underground supply lines constitute one of its most important uses. When used underground, the exterior surface should be properly protected against corrosion by a suitable coating or wrapping. Present day applications include municipal and suburban water supply, and private water supply lines for industrial establishments, railways, and similar purposes.

Vitrified Clay Sewer Pipe

Vitrified clay sewer pipe is a very popular sewer pipe made of native clay and requires no other material for its production. The glass-like surface of vitrified clay sewer pipe is impervious to the action of acids, alkalis, sewerage gases, industrial wastes, and the abrasive action of silt and gravel.

Sewer pipe comes in laying lengths of 2, 2½, and 3 feet (60.96, 76.20, and 91.44 cm.), with internal diameters of 4 inches to 36 inches (10.16 to 91.44 cm.). Vitrified clay sewer pipe fittings are available, and jointing is accomplished by cement or hot poured bituminous jointing material.

Clay sewer pipe is the most commonly and universally used material for building sewers.

Orangeburg Pipe

Orangeburg is the trade name for a non-metallic pipe of special bituminized fiber composition made in long lightweight lengths. It is a pipe made so as to produce a uniform structure of fibers thoroughly impregnated with coal tar pitch. It is advertised as resistant to corrosion or disintegration by chemical action of sewage and drain cleaners, and easy to connect, install and cut. Installation is aided by taperweld joints made root-proof and permanently tight without cement or jointing compound. Adapters are available for jointing to other kinds of pipe.

It is sometimes used for house-to-sewer (or septic tank) connections where permitted, as conductor pipe for irrigation, for outside drainage lines, outside downspouts or storm drains, industrial waste disposal and for underground protective covering for other pipe. It is also made in the perforated type for use in septic tank leaching beds, foundation footing drains and subsoil drainage.

Orangeburg pipe should not be put in locations where it is exposed to mechanical damage or where it is not well supported as the result of unfirm bedding or soft spots. Stones or other hard objects should not be piled upon it. Orangeburg pipe should not be used where it is exposed to continuous and sustained high temperature in industrial applications and it should never be used with pressure.

Orangeburg pipe should not be used for piping inside of a building or for potable water supply.

Supply Lines or Water Distributing Pipe

The *supply piping* consists of that piping from the curb to the house and throughout the house to the fixtures or any other outlet. The *water pipe* from the water main up to the building (called service pipe) can be furnished of different materials such as copper, lead, galvanized wrought iron, and galvanized steel. The material from the main to the curb is usually determined by the local water department. It is generally the plumbing contractor who is responsible for the piping from the curb into and through the residence. The selection of the water pipe should be determined by the corrosive action of the water and soil when the pipe is underground and a type pipe should be selected to best withstand these conditions.

Inside the building, the supply piping most often used is galvanized wrought steel, although copper, brass, galvanized wrought iron, and lead are also used in certain areas and installations.

Immediately adjacent to the inside wall of the basement at the beginning of the interior supply piping, there is generally installed (as usually required by codes) a stop and drain so the water supply system can be cut off and drained when necessary. In residential work, a compression or ground key stop and drain is used most often, but in larger installations where pipe sizes of 1 inch (2.54 cm.) and up are required, a valve (usually a gate valve) and separate drain are used. Following this the water meter is installed. The water is then piped through various branches to the hot water heater and fixtures throughout the house.

Water pipe should be so graded that the entire system or parts of the system can be properly drained. Adequate size air chambers are recommended for all fixtures.

Stops in the fixture supply pipe are recommended so that any faucet or fixture can be cut off when repairs are necessary without affecting any other fixture. As a minimum, the supply branches to each bathroom or toilet room should be provided with stops so they can be cut off without shutting off the entire house or building. The supply branch to each lawn faucet should include a stop and drain so that the piping will drain properly when the supply is shut off for the winter.

The hot water tank should be provided with approved pressure relief and temperature relief valves as a protection against explosion, particularly in installations which are metered or have a check valve in the cold water supply line to the domestic water heating system. In the case of light copper range boilers, a vacuum valve will afford protection against collapse in the event of a vacuum. A hot water circulating system is recommended for apartment buildings and all large installations. Hot water pipe should be insulated.

Water service pipe should not be laid in the same trench as the building sewer, but should be laid in a separate trench in order to prevent possible contamination of the water supply in case of leakage.

In buildings using *non-potable water*, the *piping* for such water should be kept entirely independent from the *potable water supply piping* and no cross connection with the potable water supply system should be made. The non-potable water piping should be adequately identified by colored paint so that it can readily be distinguished from the potable water piping.

Pipe for Drainage Systems

The *soil* and *waste piping* of a building (known as the drainage system or drainage piping) includes all the piping (within the building) which conveys sewage, rain water, or other liquid waste to the point of disposal. This does not include the main sewer or sewage treatment disposal system.

The piping which conveys sewage is separate from the piping which conveys storm, surface, or ground water. It connects to the sanitary sewer. In some municipalities a combined sewer is used which carries both sewage and storm water. In such cases, the building drain and building storm drain may connect to a combined building sewer, which connects to the main combined sewer.

The differentiation in terminology between soil and waste piping is based on the type of waste or sewage carried in the line. The term *soil pipe* is applied to any pipe which conveys the discharge of water closets or fixtures having similar functions, with or without the discharge from other fixtures. Waste pipe is defined as any pipe which receives the discharge of any fixture, except water closets or similar fixtures, and conveys the same to the building drain, soil or waste stack. When such pipe does not connect direct with the drainage system, it is termed an *indirect waste pipe.*

Cast iron, galvanized or coated wrought iron, or wrought steel screwed pipe are the most commonly used pipe for drainage systems in buildings.

In *residential work*, hub and spigot end cast iron pipe is used most generally, and in *tall buildings*, the drainage piping most often installed is coated iron or steel screwed pipe.

Clay sewer pipe is commonly used for the building sewer and drainage piping outside of buildings.

Drainage piping must be sized, and run in conformity with the plumbing code. *Cleanouts* are recommended at all changes in direction and at sufficiently close intervals in straight runs of pipe to permit easy cleaning in case of stoppage.

Soil or *waste pipe* should not be run over potable water supply tanks or over places where food or drinks are processed, served, or stored.

Area drains, subsoil drains, and *leaders* (downspouts) should connect to the storm sewer system and not to the sanitary sewer system.

Steam exhausts, blowoff or *hot water* higher in temperature than 140° F. should not discharge directly into the building drainage system but should first be properly cooled.

Wastes from ice boxes, refrigerators, food or drink processing equipment, not water connected, should not discharge directly into any building drain.

DRAINAGE FITTINGS

Cast Iron Screwed Drainage Fittings

Cast iron screwed drainage fittings have approximately the same inside diameter as wrought pipe. When a joint is made, the end of the pipe practically meets the shoulder as the back of the thread chamber, forming an unobstructed passageway. Each opening is chamfered to permit easy entrance of the pipe.

Hulbert Drainage Fittings

In addition to the common screwed drainage fittings, a line of fittings is available in both the screwed, and hub and spigot end types for wall hung closets; either single or battery installation. These are called "Hulbert" cast iron fittings. In these fittings the inlet openings are located at progressively different distances above the center of the lateral drain line, thus enabling the drain line to be properly pitched, in keeping with good plumbing practices. All of the closets in a battery can be installed at a uniform height from the floor, the latest drain line can be installed above the floor and the number of fittings and joints are materially reduced.

These fittings are used in conjunction with *supporting flanges* or *supporting chairs*. For iron or brass pipe connection between the closet and the fitting, either the supporting flange or chair may be used. With lead pipe connection, however, the supporting chair should always be used. The chair is also recommended for use with hub and spigot end fittings for battery installations.

The line includes black or galvanized screwed or hub and spigot fittings in a liberal assortment of types and sizes. In addition to the long-turn design shown in Figs. 35 and 36, a variety of short-turn single five-by-four-inch drainage fittings are also available. These are provided for exceptionally compact installations where space restrictions do not permit the use of the long-turn fittings. In addition to these, various types of closet elbows are made with integral flange, and some for separate flanges as well as other fittings to meet structural requirements.

Some special lines of fittings have been specifically designed in conjunction with the various types of soil, waste, and vent stacks, together with horizontal branches and vents, to eliminate as many joints as possible and to facilitate installations.

Cast Iron Soil Drainage Fittings

Cast iron soil fittings are hub and spigot end, and are used with all cast iron soil pipe drainage installations.

The hub and spigot end fittings as used with cast iron soil pipe for drainage installations should not be confused with hub and spigot fittings used with cast iron pipe for gas and water installations since the latter are much heavier in order

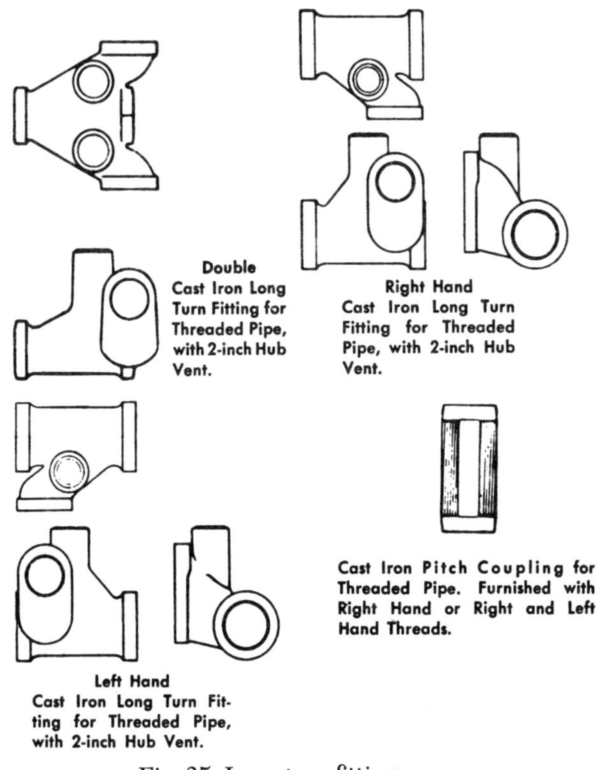

Fig. 35. Long turn fittings.

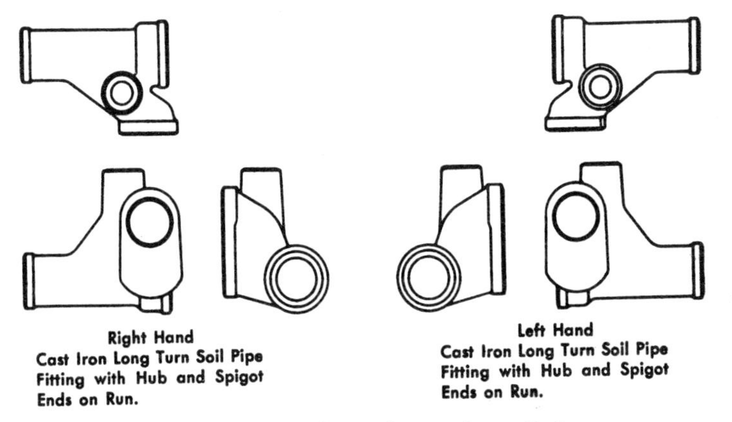

Fig. 36. Long turn soil pipe fittings for wall closets.

to withstand the service to which they are subjected.

The more common standard types of soil pipe fittings are shown in Fig. 37. (*See* Appendix I, Glossary of Terms.)

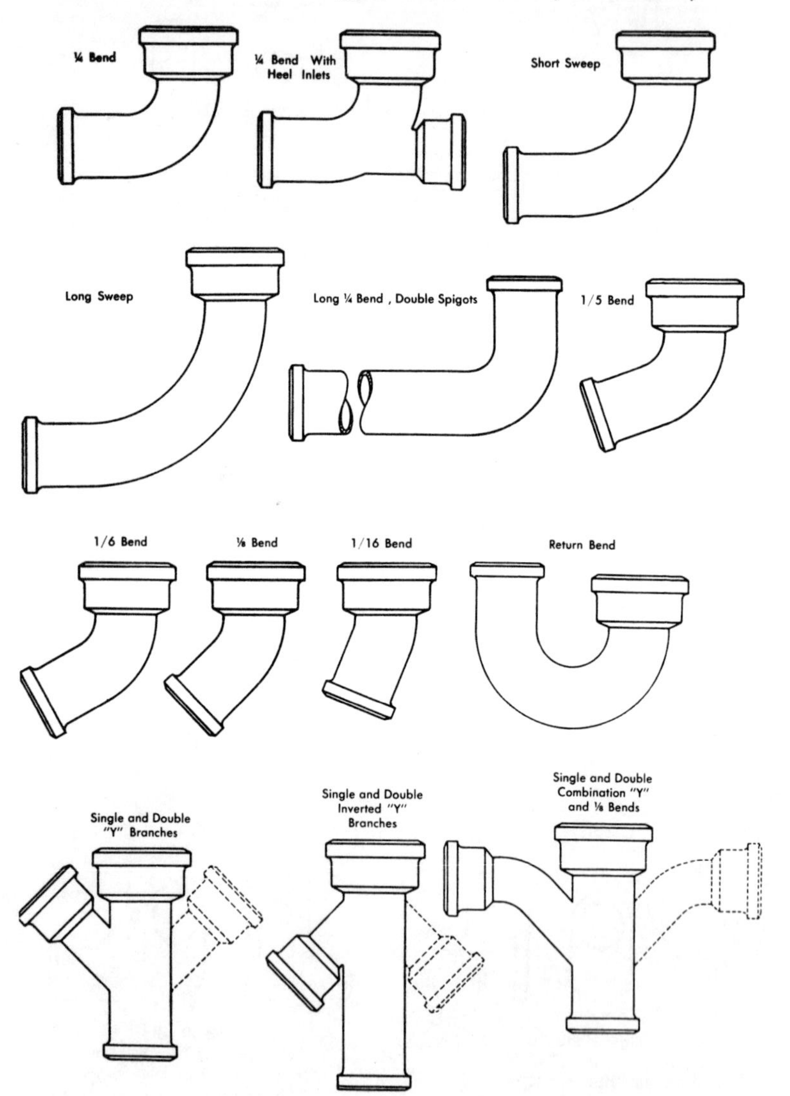

Fig. 37. Cast iron soil pipe fittings.

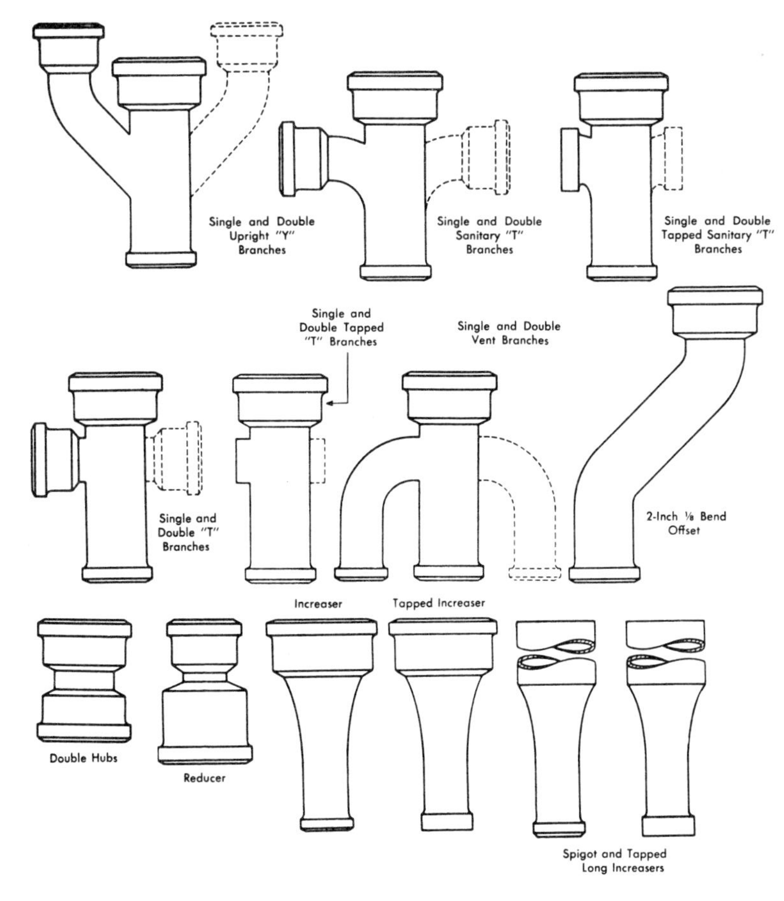

Fig. 37. Cast iron soil pipe fittings (continued).

TRAPS

The purpose or function of a *trap* is to prevent the entrance of sewer gas into buildings. The average code requires a fixture trap with a water seal of two inches to four inches, and every fixture must be trapped.

P traps are almost universally used and required by most codes. They must be sized as specified by the local code. Crown venting of traps is generally prohibited.

S traps are generally prohibited by codes because of their inability to be properly vented thereby permitting possible siphonage. They are used primarily in communities having no codes, even though their use is not recommended.

Drum Traps

Drum traps are commonly used for bathtubs and showers because they have an accessible cleanout. They are required by some codes on shower and bathtub installations.

House or Building Traps

The *house trap*, the function of which is to prevent sewer gas from coming into the drainage system in the house, is required by many eastern cities. It is a running trap with double cleanout, usually installed near the foundation wall in the basement or outside the building. If placed outside of the building it is usually made accessible by being installed in a manhole. This type must be provided with a fresh air inlet which permits the circulation of air through the system. The fresh air inlet pipe connects to the building drain just inside of the building trap and extends upward and through the building wall to the outer air. The open end is protected by a metal perforated grating or vent cap. (*See* Figs. 38, 39, 40, 41, 42 and 43.)

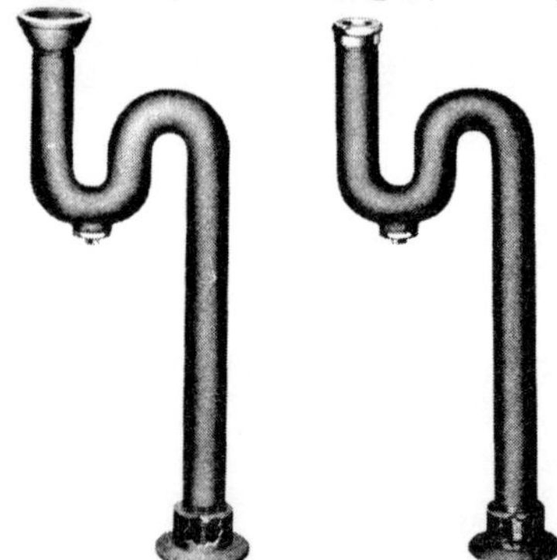

Fig. 38. Cast iron S traps with brass cleanout plug and female outlet.

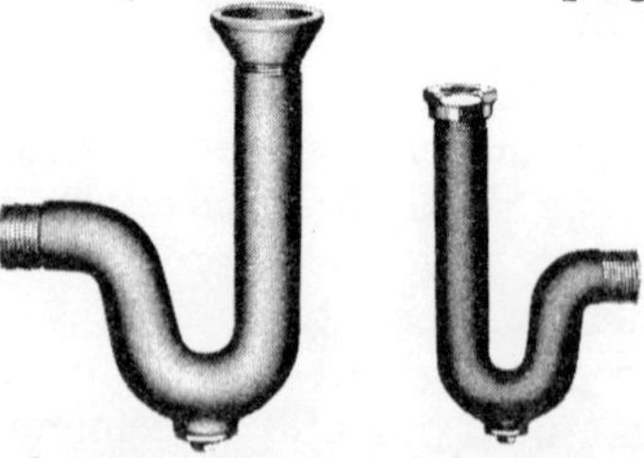

Fig. 39. Cast iron P traps with brass cleanout plug.

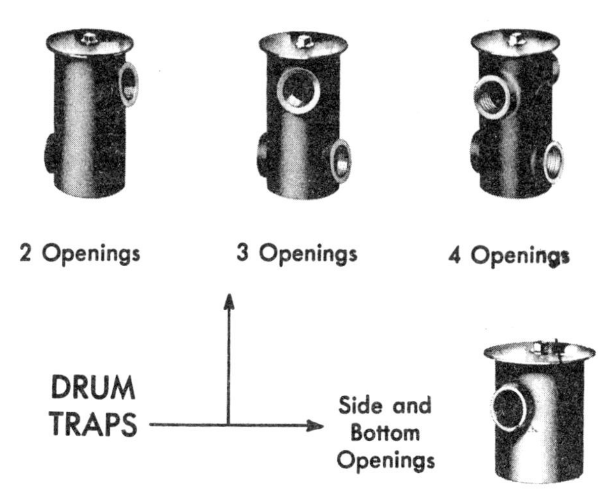

Fig. 40. Drum traps.

Fig. 41. Bath trap ell.

Fig. 42. Rubbing trap with double hub vent.

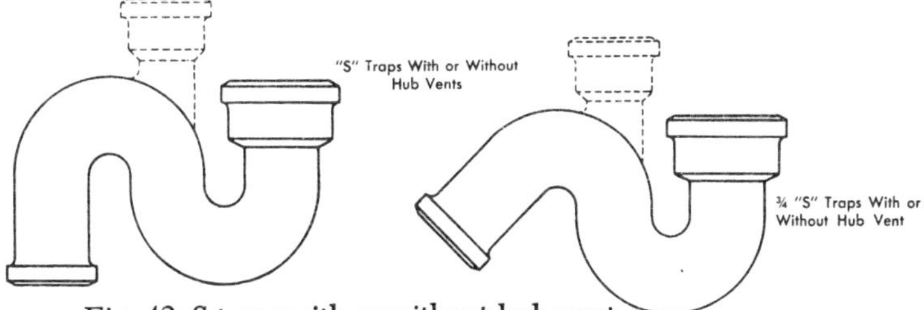

Fig. 43. S traps with or without hub vent.

VENTING

An important part of the "roughing-in" piping and an excellent safeguard to health is the *vent system*. This is composed of one or more pipe lines installed to provide a flow of air to or from the drainage system and to provide air circulation within the system in order to protect trap seals from siphonage and back pressure.

Venting serves as a protection against the breaking of the water seal in each of the fixture traps. Without the benefit of the air furnished through the vent pipe, the slug of water from a closet being flushed or a fixture being drained could create a plus or minus pressure which may break the trap seal or siphon the trap.

Some form of venting is required by all codes and while some require individual venting and others permit loop or circuit venting, all installations should be in conformity with the particular code applying to the locality of the building.

The type of vent used depends on the kind of installation, the governing code, and the preference of the architect, builder, engineer, or owner. In each situation, when a choice is permitted under the code, the type of venting selected should be the best possible installation that can be afforded.

In small residences, the problem of venting is comparatively simple. In this type of installation, wherever fixture traps are located within the minimum distance from the stack and under conditions as specified in the code, no individual venting is usually required. (*See* Fig. 44 showing various kinds of vents. Full lines show soil and waste piping. Dotted lines show vent piping.) (*See* Appendix I, Glossary of Terms.)

Pipe and Fittings for Venting

Although vent stacks in smaller buildings or residences are generally hub and spigot end cast iron pipe, the *vent stack* may be either cast iron, wrought iron, or galvanized wrought steel, depending on the material used for the soil stack. The *vent piping* between the vent stack and the fixtures is usually wrought iron or galvanized wrought steel, the two types of pipe which are most commonly used in venting.

Drainage fittings are the preferred type of fitting for venting because they provide a smooth waterway and prevent moisture

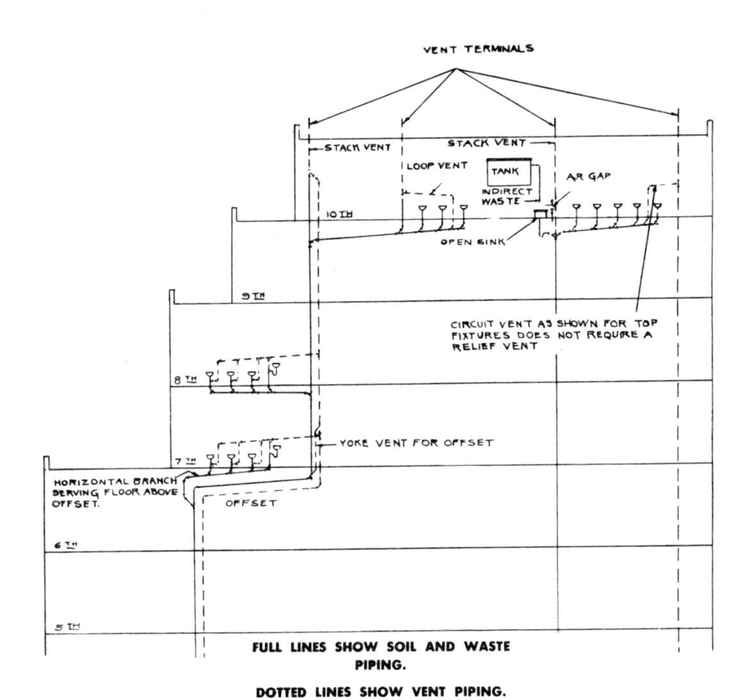

Fig. 44. Various kinds of vents.

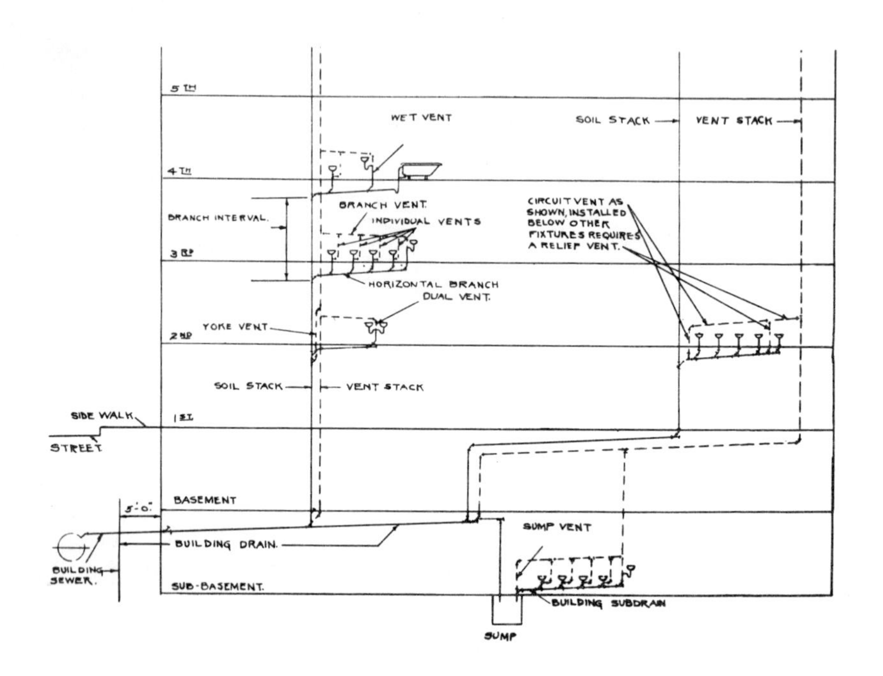

Fig. 44 Various kinds of vents (continued).

condensate from collecting in the fittings. However, ordinary malleable or cast iron screwed fittings are allowed by some codes.

Selecting Correct Pipe Size

In planning a new home, one of the most important items that contribute to the successful operation of the water system is adequate sized piping for maximum requirements. If in doubt, it is much safer to lean toward the larger size since, in using pipe one size larger, successful operation is not impaired and provision is made for reduced flow which occurs in time due to the corrosion or scale forming in pipe or for additional fixtures which might be added at a later date. If too small, however, annoyance at fixtures not producing the desired flow and failure of the system itself may result. The difference in cost between piping that is adequate in size and piping that is too small is estimated to be very little of the total cost of the plumbing. However, should it be necessary to break up the walls and floors to replace inadequate piping, the expense would be substantial.

The soil, waste, and vent pipe together with the traps should be of a size to conform with the code requirements.

Water supply piping should be of a size large enough to supply all fixtures which may be expected to operate simultaneously. If desired, the water supply piping including the water service pipe from the main to the house should have sufficient reserve capacity to permit additional fixtures. When flush valves or other devices using high rates of flow are installed, the size of the water piping should be increased accordingly.

Inadequate size of water piping may sometimes be the cause of a person being scalded when using a shower due to a cold water faucet being opened with a consequent reduction of flow of the cold water to the shower.

Inadequate size water supply piping also presents a back-flow hazard because the pressure drop, due to the use of fixtures on the lower floors, may be sufficient in some instances to produce a vacuum in the piping on the upper floors.

PIPE SUPPORTS

See Fig. 45 for various supports.

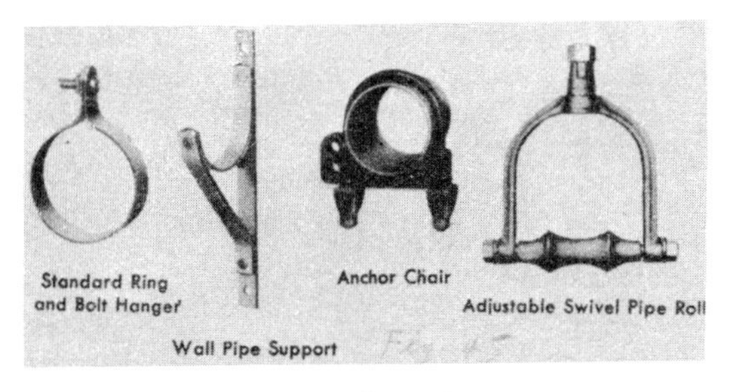

Fig. 45.

INSPECTION AND TESTS

As a safety and uniformity feature and to insure compliance with local codes, all new completed plumbing work is inspected by competent authorities in areas where regulations are in force.

Drainage and vent systems are tested in their entirety or in sections. When tested in sections, no part of the system is subjected to less than 10 feet head except the uppermost 10 feet. All openings in the piping are tightly closed, except the highest opening and the system is filled with water to the point of overflow. The water is kept in the system for at least 15 minutes before the inspector starts to check tightness at all points in the system. All piping remains uncovered until it has passed the required tests.

Air may be used for testing by forcing it into the system at any suitable opening after all inlets and outlets to the system are closed. When a uniform pressure of five pounds per square inch is held for at least 15 minutes without introduction of additional air, the test is considered as being satisfactory.

After the drainage and vent systems are completed and fixtures installed, a final test with smoke, peppermint or air is made. The traps are filled with water and a thick, pungent smoke is introduced into the entire system by means of smoke machines. When smoke appears at stack openings on the roof, they are closed and a pressure equivalent to a column of water 1 inch (2.54 cm.) high (0.036 psi) is built up and maintained for 15 minutes before inspection starts. Where the peppermint test is used, the system is closed as in the smoke test and two ounces of peppermint is introduced for each line or stack. The air test, where used, conforms with the smoke test except that air instead of smoke is introduced into the system.

The water supply is usually tested under a water pressure not less than the pressure on which it is to be used.

VALVES

Two types of valves are most commonly used: (1) *gate valves*, which get their name from their gate-like disc that moves across the path of flow, (2) *globe valves*, which are so named for the globular shape of their body.

Flow through a *gate valve* travels in a straight line. When the valve is wide open, there is very little resistance to flow since seat openings are approximately the same size as the inside diameter of pipe. In a *globe valve*, the direction of flow passing through the body is changed, and this results in greater resistance to the movement of fluids. Each of these designs has certain advantages in specific applications as will be seen later.

Both gate and globe valves are made in varying patterns. While the variants adhere to their respective design principle, their difference is in construction details mainly in the disc, stem, and bonnet.

Angle valves in basic design are the same as globe valves, except that their pipe openings are at right angles to each other. As their name implies, they are used at 90-degree turns in piping where they save a nipple and an elbow. By reducing the number of joints, they cut installation time as well as lessen restriction to flow.

Check valves perform the single function of checking or preventing reversal of flow in pipe lines. They are made in a wide variety of patterns, but in operating principle they conform to two basic designs, (1) swing and (2) lift, which are explained in this section.

There are many other types of valves, including pop safety valves, relief valves, pressure regulators, stop-check valves, quick-opening valves, candle-point valves, cocks, and so on. Each type meets definite service needs.

HEALTH PROTECTING CODES

Today we know that our present plumbing system is the greatest health protection we have. Pure water is delivered where needed and human waste is carried away (by water) and disposed of in a sanitary manner. The contact and contamination of the pure water by disease potent waste matter is the source of all trouble and must be avoided at all costs. This is accomplished by correct plumbing installations. To insure the necessary safeguards for the health of the community, Health Departments and Plumbing Inspection Departments are working very closely together to issue standards and codes to protect public health in the primary functions of removing waste matter, protecting against sewer gases and keeping fresh and pure water from being contaminated.

In almost every community having a water supply and sewage system, the installation of plumbing is subject to municipal or state regulations. Municipal and state plumbing codes, generally speaking, prescribe the minimum requirements for the installation of plumbing, with respect to the material used, the method of their connection or assembly and the design of the system itself. In addition, municipal and state plumbing codes ordinarily include the procedure for examining and licensing plumbers or master plumbers, or both.

Codes also result in a more efficient functioning of the plumbing system and, by prescribing the material that may be used, insure the use of proper materials and proper size drainage pipe.

The applicable code must be complied with since various codes may prescribe different methods of venting, sizing of drains and of piping installations, as well as various material requirements. The principle of sanitary engineering should be applied as far as possible to obtain the most sanitary installations rather than installations which just meet the minimum code requirements. Some of the requirements of codes are governed by local conditions, for example, some codes require a building trap in the building drain near the outside wall, whereas in other localities none is required or used. Some codes require a grease trap or catch basin for all kitchen sinks, while others require them only for sinks in restaurants, hotels and similar places where meals are served.

Some communities in addition to plumbing codes, have health regulations or building codes governing sanitation which must be complied with. Plumbing may come under the Health Department in some localities and in others under the Building Department or Board of Public Works.

State codes cover minimum plumbing regulations effective throughout the state. Municipal codes must be equal to or may be more stringent than the state code. There is no national code for plumbing.

Selecting the correct and most appropriate material and the design and installation of the supply and waste system of the building is of the greatest importance, and correctly done is positive insurance against health hazards.

Chapter 3

Pipe and Tubing Tools and How to Use Them

In commercial usage there is no clear distinction between *pipe* and *tubing*, since the correct designation for each tubular product is established by the manufacturer. If the manufacturer calls a product pipe, it is pipe; if the manufacturer calls it tubing, it is tubing.

Pipes are used to move liquids and gases such as water, steam, fuel, and compressed air. *Tubes* are used in much the same way as pipes, but have thinner walls and are usually of small diameter.

The materials used to make pipe and tubing are copper, brass, steel, cast iron, Monel, stainless steel, wrought iron, and lead. The type of fluid carried by a pipe is labeled on the pipe with a stencil.

PIPING SYSTEMS

The *main piping systems* for water, steam, drainage, and the like, are made up of large pipes, which are joined either by *bolted flanges* or by *welding*.

Pipes up to 2 inches (5.08 cm.) in diameter are usually joined with pipe fittings, such as unions, nipples, couplings, elbows, and others. These fittings are tapped and threaded with standard pipe threads, which taper ¾ inch per foot (1.905 per 30.48 cm.) of thread. The fittings are threaded during manufacture.

PIPE CUTTING

You can cut pipe with an ordinary *hand hacksaw*, a *power*

Fig. 1. Pipe cutter.

hacksaw, or a *pipe cutter*. The pipe cutter is usually used, but the power hacksaw is faster if you have a large number of pieces to cut or if the pipe has a thick wall. The pipe cutter has a special alloy steel cutting wheel and two pressure rollers. These are adjusted and tightened by turning the handle. The whole tool is revolved around the pipe as shown in Fig. 1.

The operation of the pipe cutter leaves a shoulder on the outside of the pipe and a burr on the inside. Always remove the inside burr or the ragged edges will catch dirt and other solid matter, and will block the flow. The *burring reamer* (Fig. 2) is the tool you use to remove the burr.

THREADING

Adjustable and solid hand dies are used to cut standard pipe threads up to 2 inches (5.08 cm.) in diameter. Some installations require *left hand threads*, therefore use *left hand dies* to cut them.

The threading procedure is about the same as that for the N.C. or N.F. threads. Machine oil or lard oil is used as a lubricant and coolant on iron and steel, but copper and brass are threaded dry.

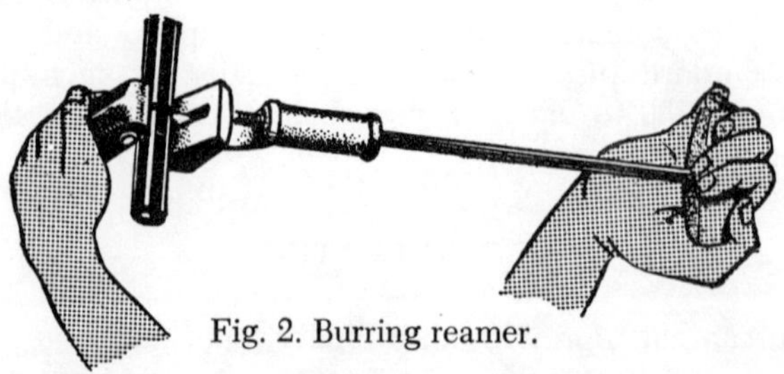

Fig. 2. Burring reamer.

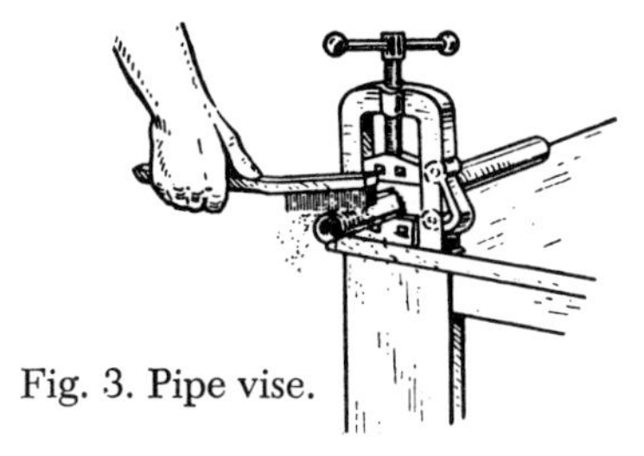

Fig. 3. Pipe vise.

A *pipe vise* (Fig. 3) is used to hold the pipe during the threading operation. The die is adjustable and has a *guide clamp*. This clamp fits over the pipe and is tightened with a thumbscrew. It draws the die on the pipe as the die stock is revolved. The clamp also helps you to get the threads straight.

When you use an adjustable die, cut only one half the depth of the thread at first. Then readjust the die to finish cutting the threads to the full depth.

The number of threads cut should not be greater than the number of threads of the die. The cut is complete when the end of the pipe is flush with the back surface of the die. Do not forget to back up the die frequently to clear the chips.

When it is necessary to cut internal threads, use a *pipe tap*. It cuts standard pipe threads with ¾-inch (1.905-cm.) taper per foot (30.48 cm.). Pipe taps are fluted and are like common screw taps, except for the taper. They are used with tap wrenches.

PIPE BENDING

Pipe bends, when they can be used, have advantages over fittings in that they offer no restriction to the flow and are economical. The smaller sizes of iron pipe may be bent cold, if the radius of the bend is not less than ten times the diameter of the pipe. The seam of the pipe should be on the inside of the bend.

The bend of a pipe should be laid out on paper, on a layout table, or on the floor of the shop. If it is laid out on a metal surface, it is marked with soapstone. A wire template is made and used to check pipe bends.

Brass and copper pipe should be *annealed* before it is formed or bent. Annealing softens the metal and increases its *ductility* (ability to stretch, compress, and bend). You can

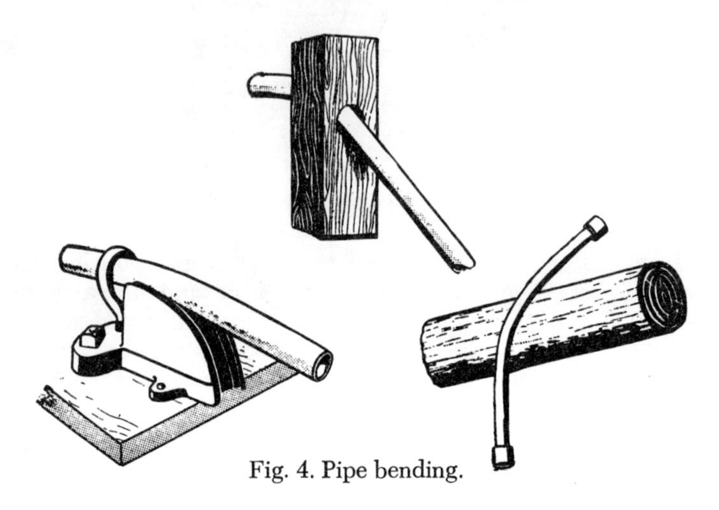

Fig. 4. Pipe bending.

anneal copper tubing by heating it to a cherry-red color and quenching it in cold water. Do not quench steel, aluminum, or brass tubing. Just let it cool in the open air.

The *bending jig* shown in Fig. 4 is a good tool to use in forming a pipe bend, or you can use one of the other methods illustrated. Avoid decreasing the inside diameter of the pipe by using as large a radius as possible for the bend.

Test all pipes that have been bent. There is some danger of leakage, due to seam splitting or through cracks caused by stretching.

If the pipe is large, and a short radius bend is necessary, the pipe should be filled with dry sand, packed tightly, and plugged at both ends. Drill a small hole in one plug to allow the expanded air to escape. The pipe is heated to a bright red and bent to the shape desired. The tightly packed sand prevents the pipe from collapsing (caving in). If you have to do much bending of heavy pipe you will need special bending equipment, such as tables, bending pins, clamps, winches and portable gas heating units.

PIPE ASSEMBLY

Threaded *water pipe joints* are usually made up with *red lead* as a seal. *Steam pipe threads* are sealed with *graphite paint.* Put the sealing compound on the pipe threads only, so it will not get inside the pipe and form a dangerous obstruction. Make sure the threads are clean before you apply the sealing compound.

Threaded joints should be screwed together by hand and tightened with a *pipe wrench* (commonly called a "Stillson"). The pipe should be held in a pipe vise during assembly, but if it is impossible to use a vise the pipe may be held with another pipe wrench.

How tight should you tighten a joint? Experience is the best teacher. Usually you will have two or three unused threads on a properly cut pipe thread. If all the threads are used, the wedging action of the tapered thread may cause the fitting to split.

Pipe wrenches are made in a number of sizes (lengths). Use the following table as a guide for selecting the best size to use.

Wrench Size	for	Pipe Size
6 inch (15.24 cm.)		¼ inch (.635 cm.)
10 inch (25.4 cm.)		⅜ and ½ inch (.9525 and 1.27 cm.)
14 inch (35.56 cm.)		¾ inch (1.905 cm.)
18 inch (45.72 cm.)		1 and 1¼ inches (2.54 and 3.175 cm.)
24 inch (60.96 cm.)		1½ and 2 inches (3.81 and 5.08 cm.)

Chain pipe tongs are often used for turning and holding pipes of all sizes. One type of pipe vise utilizes the chain clamping principle.

TUBING

Tubing serves in many places to convey water, fuel, lubricating oil, hydraulic fluid, and so on. Copper tubing is used extensively, but some tubes are made of brass, stainless steel, Monel, or aluminum alloys.

Because tubing is usually thin walled, it is seldom threaded. Special threaded *fittings* and *couplings* are generally used with tubing. These are either soldered to the tubing or held by a flared end. Pieces of tubing may be soldered together without the use of a fitting. Hard solders (usually silver solders) are best. Soft solders are not strong enough.

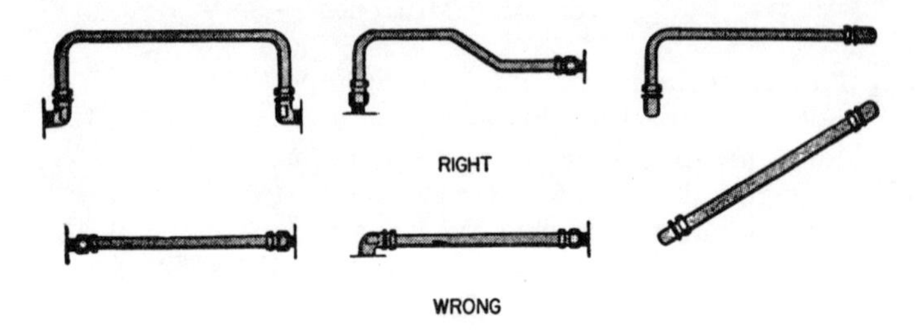

Fig. 5. Use the right installation method.

Tubing is usually supplied in coils in the soft (annealed) condition.

A straight line may be the shortest distance between two points but it is not the best distance for tubing lines. It is almost impossible to cut and flare tubing so that it will be exactly the required length. And a straight tube would be easily damaged or pulled loose by an accidental blow, or by expansion or contraction resulting from temperature changes. Figure 5 illustrates the correct method of tubing installation.

CUTTING TUBING

You can do a good job of cutting tubing with a hand hacksaw if you use the setup shown in Fig. 6. Always use a blade with 32 teeth per inch (2.54 cm.) for sawing thin

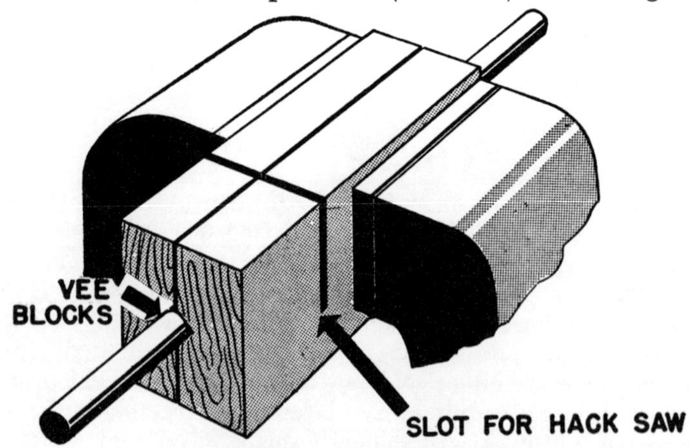

Fig. 6. Set-up for sawing tubing.

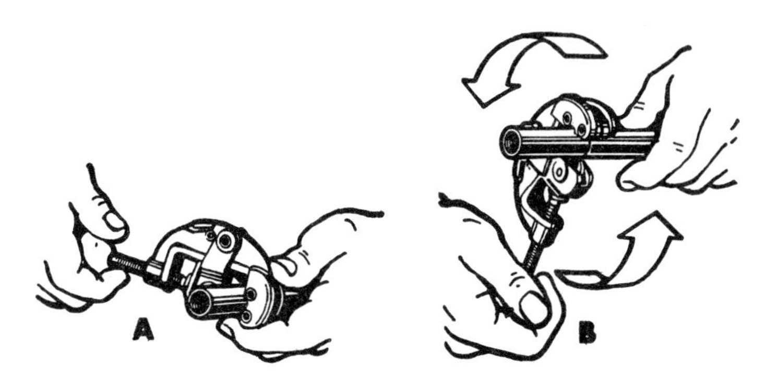

Fig. 7. Tube cutter.

tubing. But use a *tube cutter* if one is available. The tube cutter is operated as illustrated in A and B, Fig. 7. Be sure you operate the cutter in the direction indicated by the arrows in B, Fig. 7. To avoid crushing the tubing, turn the handle only a short distance after each revolution.

After the tubing is cut to length, square up the ends and remove the burrs. The outside burrs may be removed with a file. The inside burrs may be removed with a bearing scraper, a knife, or a small file. Avoid nicks and scratches—they are the beginnings of potential cracks. Screw the cutting wheel as shown at A, Fig. 7; rotate the cutter, keeping a slight pressure against the cutting wheel with the screw adjustment as shown at B, Fig. 7. Be sure to get out all the filings, chips, dirt, and the like. Stubborn dirt or scale can be removed by running a piece of "fuzzy" wire cable through the tubing.

BENDING TUBING

After the tubing has been cut, squared, burred, and cleaned, it is ready to be formed (bent) to the desired shape. You can bend tubing by hand, but you will do a better job if you use some kind of bending device.

Small sizes of tubing, like those used for engine fuel and oil lines, may be formed with the *spring coil tool* (Fig. 8). A coil is selected that just fits over the tubing. The tool helps to prevent the collapse of the tube and produces a smooth curve.

Medium size tubes (¼ to ⅝ inches, .635 to 1.5875 cm.) may be formed with the *hand tube bender*. The use of the hand tube bender is illustrated in Fig. 9.

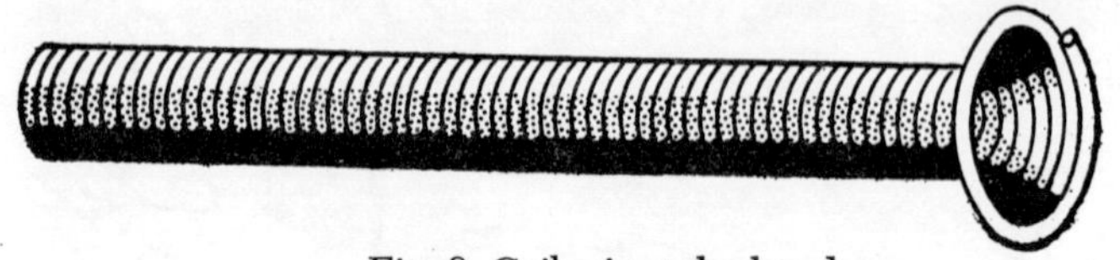

Fig. 8. Coil wire tube bender.

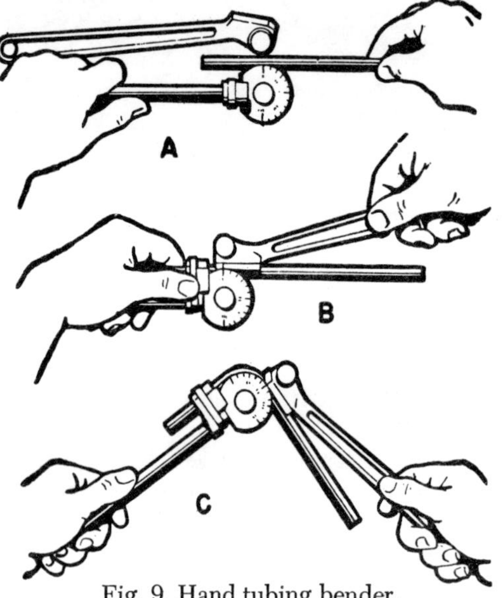

Fig. 9. Hand tubing bender.

Tubing over ½ inch (1.27 cm.) in diameter will partially collapse during the bending process if it is not filled with some kind of easily removable material. Filler materials include sand, resin, and special bending alloys which have very low melting points.

Sand is a commonly used filler. It must be fine and dry, and be packed tightly in the tube. It is not necessary to heat the tubing, but it should be in the annealed condition.

Commercial *bending compounds* are more expensive than sand or resin but they may be used over and over, indefinitely, if not overheated. They melt at a temperature of 150° F., and should be melted in a ladle or in boiling water. The tubing should be preheated before the melted compound is poured in.

Resin is used in much the same manner as the special bending alloys. The resin is melted in a ladle and poured into the plugged and preheated tube. If the tube is not preheated the resin will harden before it reaches the plugged end of the tube. Allow the resin filler to cool to room temperature before you bend the tubing.

Caution: When you are ready to melt out a filler material, apply the heat first to the open end of the tube, and then move the heat along the tube as the melting filler runs out. If you forget to start at the end there is danger of an explosion, especially if resin is the filler material.

One way to straighten a piece of tubing that is dented or flattened is to connect the tubing to a compressed air line and tap it into shape with a mallet. Be sure to anneal the tubing for best results. To straighten a length of tubing, just roll it on the bench top with your hands.

Tubing that is subject to vibration should be annealed again after bending, as the bending process hardens the metal. Annealed tubing will absorb vibration shocks.

FLARING

Flaring (or *belling*, as it is sometimes called) is the stretching of the end of the tubing into a funnel shape so it can be held by a fitting. Figure 10 shows how a flared end is held by two different types of fittings.

Before you flare both ends of a tube, be sure you have all the necessary fittings on the tube. If you do not, fittings will not slip over the flare.

There are several types of tools for forming flares. The *ball type* flaring tool (Fig. 11) is recommended for thin wall, soft copper or aluminum tubing not over ¾ inch (1.905 cm.) in diameter.

The *hammer type* flaring tool shown in Fig. 12 is a good one for general use, because it is a combination outfit that will

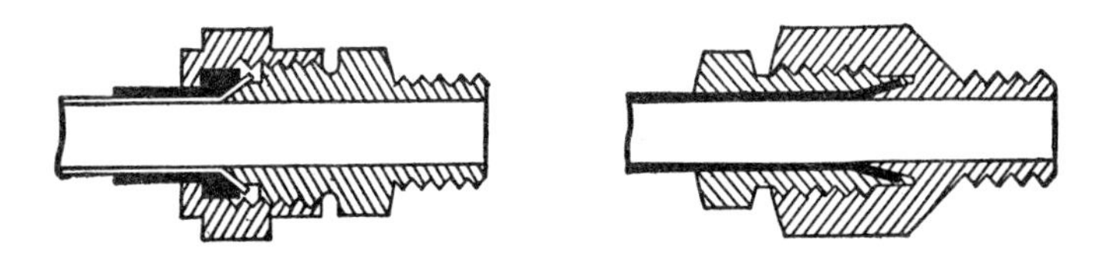

Fig. 10. Triple and standard fittings.

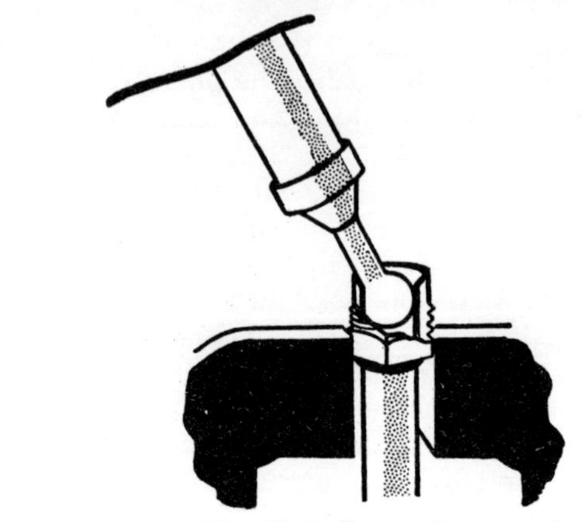

Fig. 11. Ball type flaring tool.

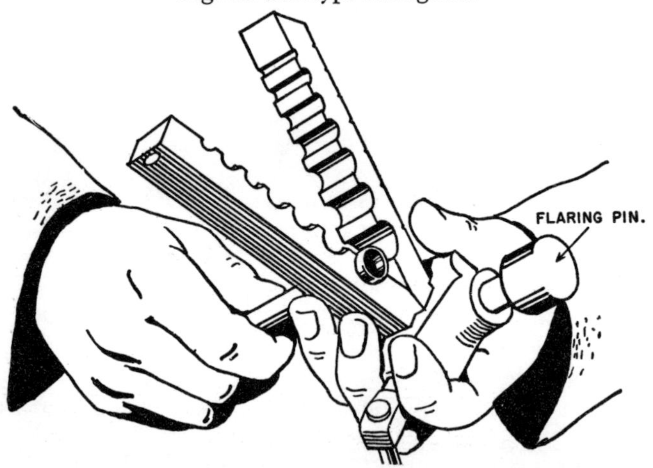

Fig. 12. Combination flaring tool.

flare several sizes of tubing. Another type is similar to the hammer type except that it is not hammered. Its cone-shaped anvil is forced against the end of the tubing by a threaded rod with a T-handle which screws through the clamp head.

The *standard* and *triple hammer types* of tools are designed for sizes up to 2 inches (5.08 cm.) in diameter. The standard type is used with a two-piece fitting; the triple hammer type on either two- or three-piece fittings. Each size of tubing requires a separate tool. (*See* Fig. 13.)

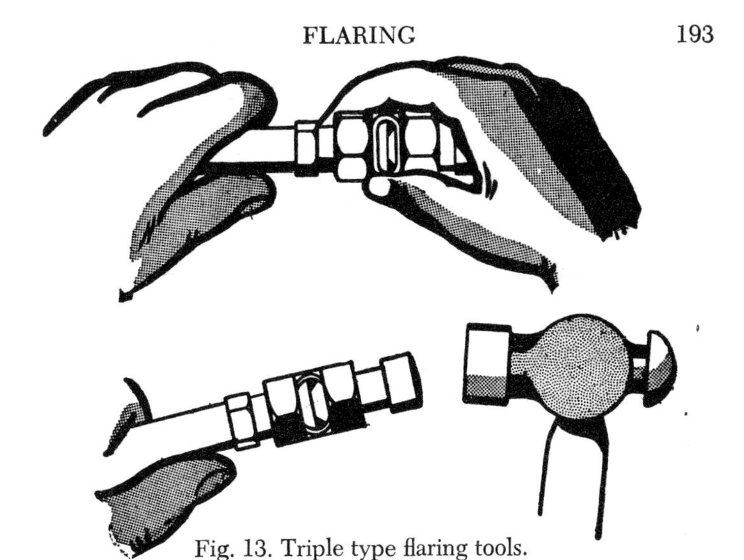

Fig. 13. Triple type flaring tools.

The important points to keep in mind are the *length* of the flare, the *squareness* of the end, and the *fit* of the flare against the fitting. In addition, you must avoid cracks, dents, pocks, and scratches on the flared surface.

The tubing flare and the fitting must form a joint that is tight and strong. A, Figure 14 illustrates a correct flare. Both views at B, Fig. 14 show how the angle and radius of the flare should match the contour of the fitting seat.

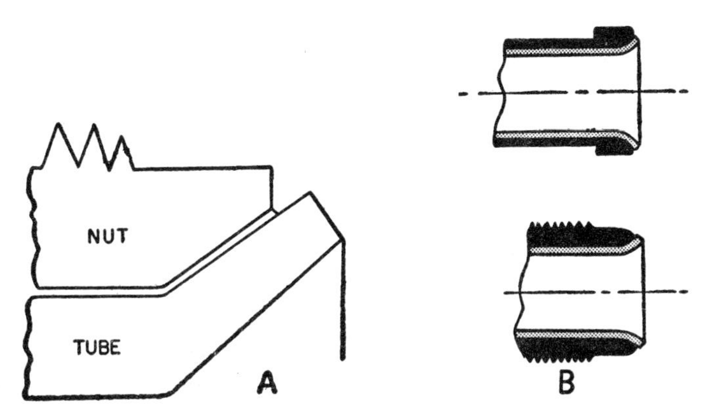

Fig. 14. Flares.

If a flare is too short, the full clamping area of the fitting is not utilized. Because of the small area of the tube that is clamped, the flare may be squeezed thin and weakened. Such joints do not provide maximum security against pullout strains, leakage, or vibration damage.

Tubes flared too long will stick and jam on the threads during assembly, and prevent the parts of the fitting from seating properly. This condition is often the cause of leakage.

Uneven flares are usually the result of carelessness. Either the tube end has not been cut squarely, or the flaring tool has been used incorrectly. Such a flare will be forced into shape by the parts of the fitting. This forcing strain weakens the metal, and may cause eventual breakage or leakage.

SOLDERED FITTINGS

Soldered fittings involve the use of a *union tail* (or sleeve) (Fig. 15) which is soldered over the end of the tubing. No flare is required. *Silver solder*, or another hard solder, is used so the joint will be strong. (*See* Chap. 17.)

TUBING ASSEMBLY

All joints of tubing should be tested under pressure before they are installed. Repair shops are equipped to make these tests under conditions similar to those under which the installation will eventually be used.

Before you install or replace a piece of tubing, make sure that its interior is clean. All foreign material (chips, dust, and

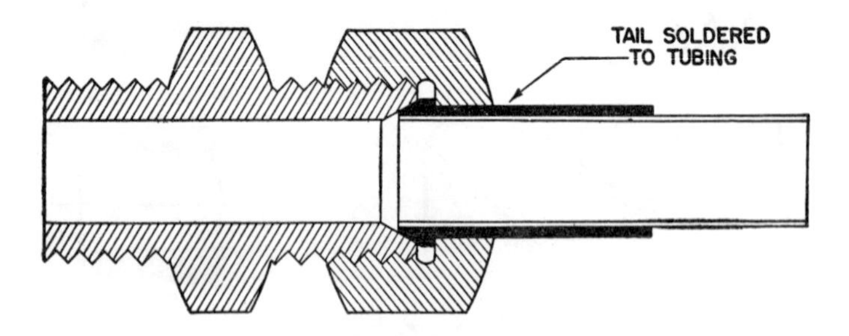

Fig. 15. Soldered fittings.

scale) must be removed. This is particularly true of the tubing used for *hydraulic* lines of equipment. Those lines must be clean, because they contain delicate and sensitive valves.

Cleaning directions for copper tubing and pipe for hydraulic lines are as follows.

1. Treat machined faces or threads of end fittings with hot paraffin wax.
2. Dip in an acid bath solution (and remove immediately). The solution should be two parts sulphuric acid, one part nitric acid, and four parts fresh water.
3. Wash in fresh water.
4. Immerse for one minute in a neutralizing bath of 1½ pounds of Magnus No. 2 per half gallon of fresh water.
5. Soak in boiling water for ten minutes.
6. Pass a frayed wire rope through the pipe.
7. Wash with stream of fresh cold water at high pressure.
8. Dry thoroughly, being careful not to leave threads of rags, toweling, or waste in the pipe or tube.
9. Seal both ends until ready to install.

Never spring or force tubing into position. Tubing must be bent to fit so that the threads of the fittings are perfectly alined. If a cement is to be used on a joint, start the threads about three turns before applying the cementing material. The tinning of the outside threads of a hydraulic line fitting produces a superior seal.

Avoid leaving any dirt or other foreign matter in a line. Plug the open ends of tubes and pipes if any drilling, filing, welding, or soldering is done near them.

To sum it all up, tubing and pipe joints must fit, and the line must be kept clean, tight, and strong.

Chapter 4

Plumbing Fixture Arrangement

The location of plumbing fixtures to achieve convenient use and economical installation is a very important design item. It is too often given only casual consideration in the design of dwellings. As a result the plumbing system must bear the burden of unnecessary complexities. Once located, the fixtures must be connected according to standards or regulations, regardless of the quantity of materials or labor involved. Forethought can reduce these costs a considerable amount.

Logical arrangement of rooms containing plumbing fixtures and orientation of fixtures within the rooms will require minimum lengths of pipe and will provide an economical installation with probable operating and maintenance economies.

Plan fixture location, where possible, to group all the fixtures on each floor and orientate the fixtures in each group, so that the shortest length of drain pipe can be run to a central stack. This will automatically reduce to a minimum the length of required water piping.

Where fixtures are on two or more floors, other conditions permitting, group the fixtures on each floor and locate the groups approximately over each other so that all fixtures can be connected to the same drain and vent stacks.

ROOM ORIENTATION

Plumbing Services

Locate the rooms containing fixtures near the point of entrance of the water and sewer services. This is particularly applicable in the design of long, narrow houses such as ramblers, or those houses with the short dimension of the building toward the utilities service entrance, and those which are oriented to take advantage of site or weather conditions.

The length of the house sewer and the water service is affected by orientation of the house. Some plumbing economy could be achieved in house B shown in Fig. 1, if the kitchen and the bath could have been designed to be at the front of the house, so that the plumbing stack could have been located at X, Fig. 1, which shows same house plan with kitchen located under bathroom.

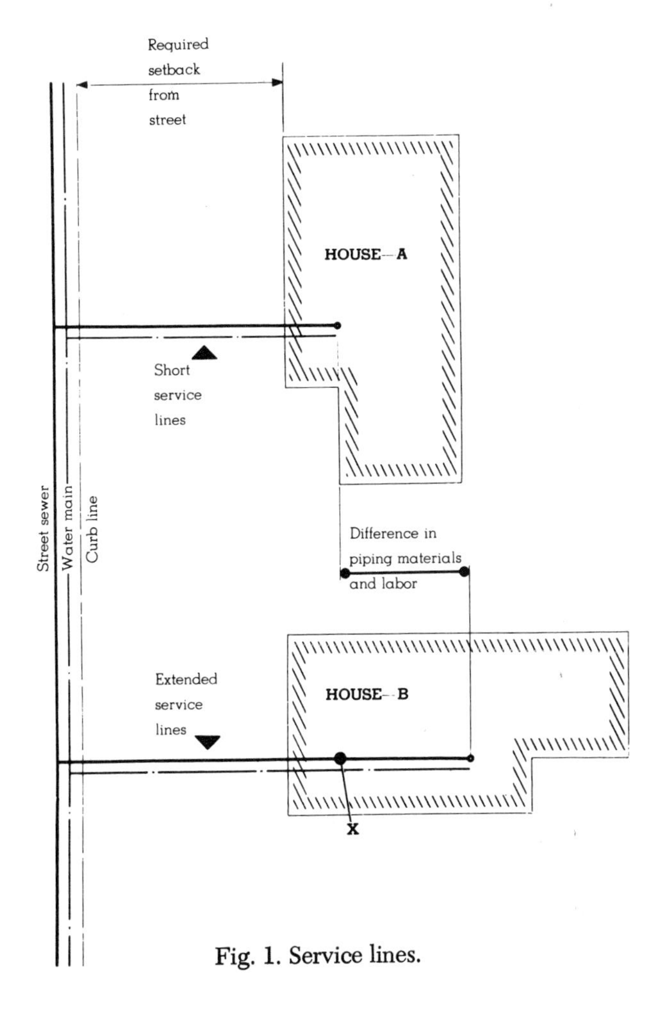

Fig. 1. Service lines.

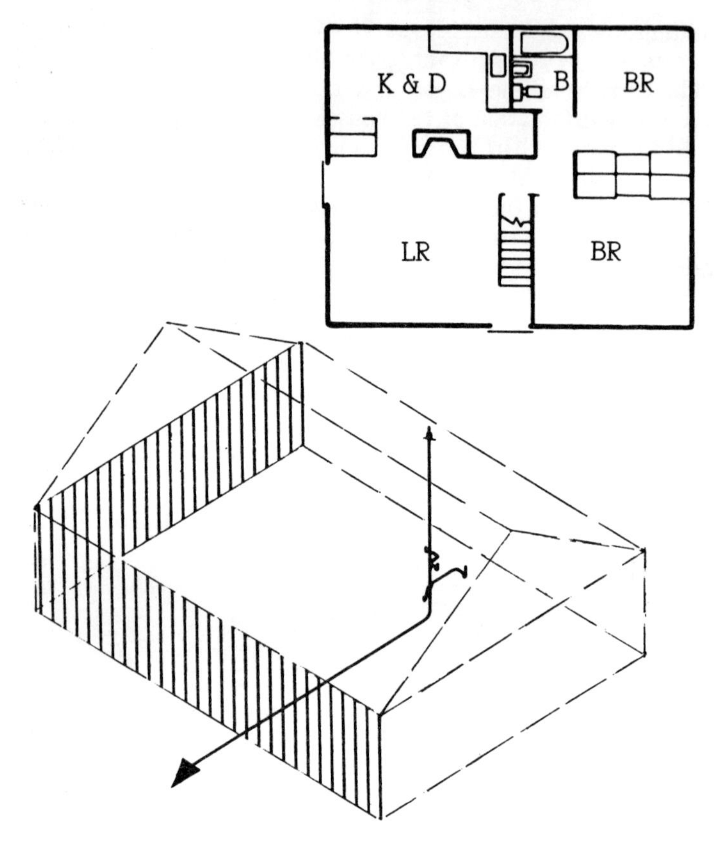

Fig. 2. Kitchen and bath (back-to-back).

Backing-up of Rooms

In single-story houses, locate the bathroom, the kitchen and the utility room so that a common wall may be used to contain as much of the piping as possible.

Figure 2 shows the principle of backing-up the kitchen and the bath as contrasted with Fig. 3 wherein these rooms are not backed-up. The piping diagrams in Figs. 2 and 3 show the necessary drainage and vent piping for each condition in a single-story house with no basement. The difference in the cost of materials and labor for installing fixtures where rooms are backed-up and where they are not can be visualized

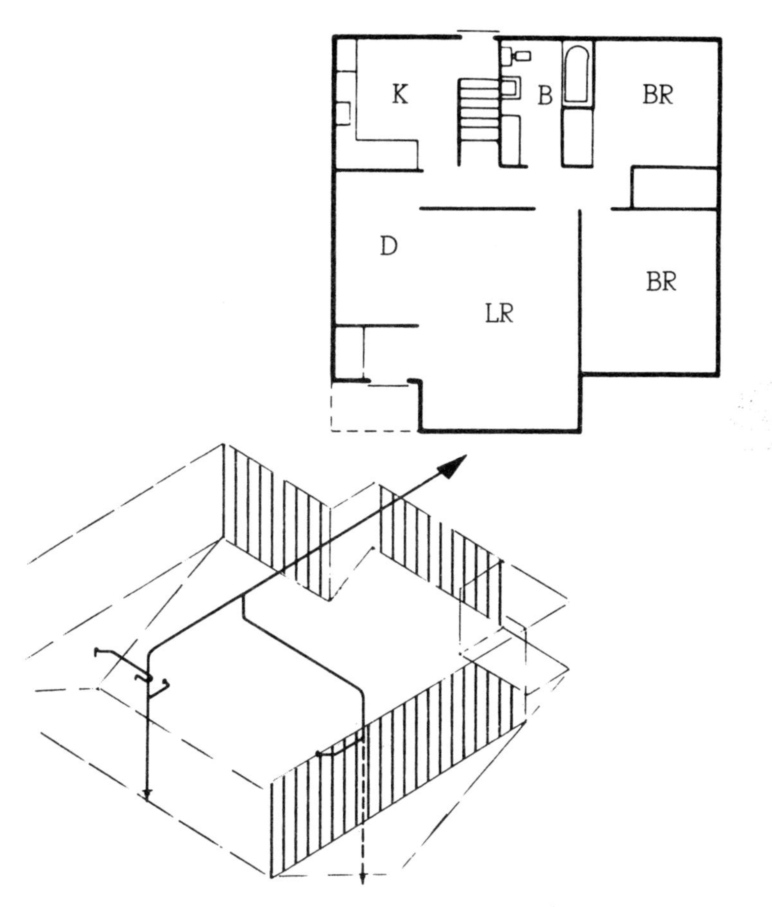

Fig. 3. Kitchen and bath (remote).

by examining the piping diagrams. Although not shown in Figs. 2 and 3, additional water piping is also required in cases where the kitchen and the bath are not backed-up.

Figure 4 shows the principle of backing-up rooms containing plumbing fixtures as compared with Fig. 5 wherein the kitchen, the bath, and the utility rooms are separated. The piping diagrams in Figs. 4 and 5 show the necessary drainage and vent piping for each condition in a single-story house with no basement. The difference in the cost of labor and materials for installing fixtures in each case can be visualized by examining the piping diagrams. Where the kitchen, the bath and the utility rooms are separated, additional water

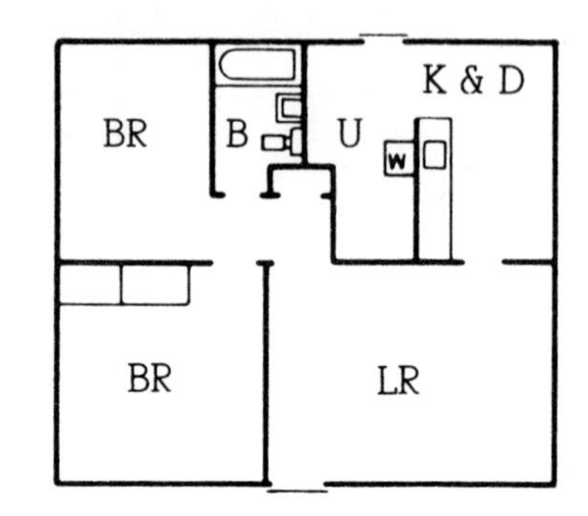

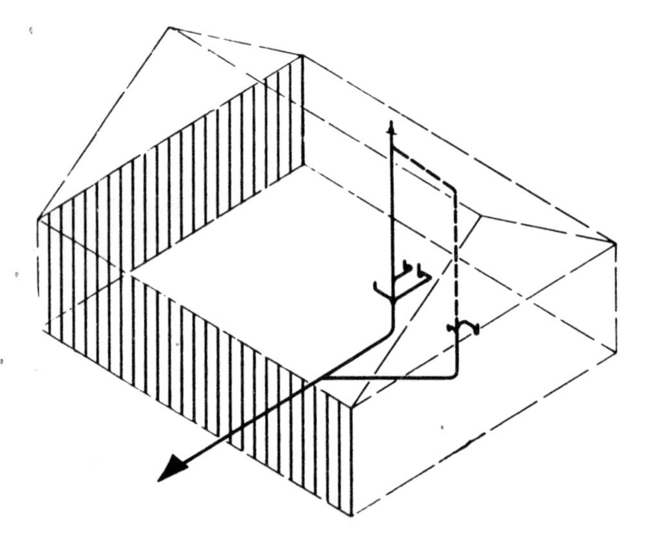

Fig. 4. Kitchen, bath, and utility room (adjacent).

piping is also necessary although it is not shown in Figs. 4 and 5.

Vertical Alignment

In *two-story houses* locate the bathroom, or the bathrooms, on the second floor over the kitchen or powder room so that all fixtures may discharge to a single stack.

Figure 6 shows the simple layout required for a bath located

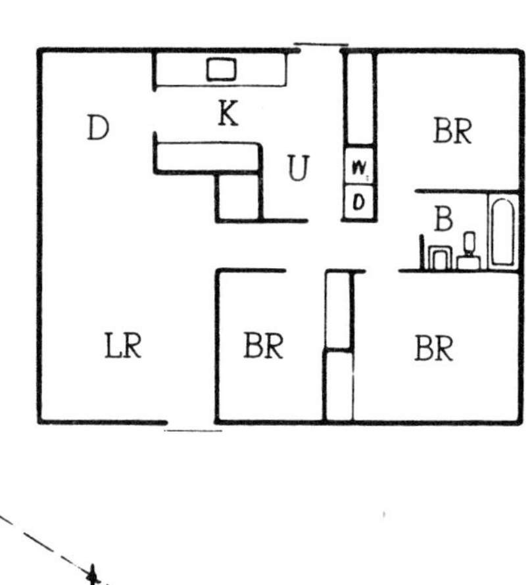

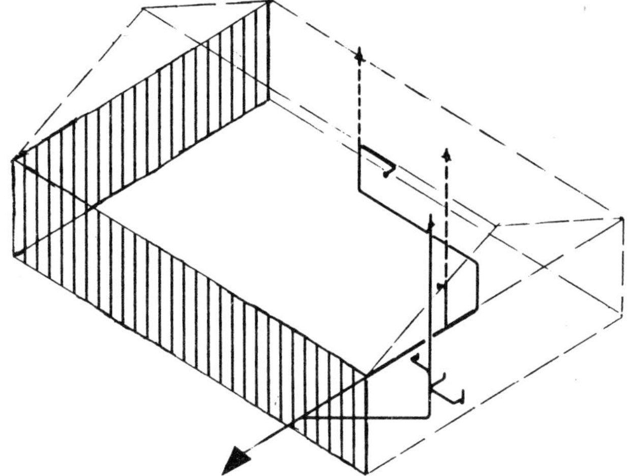

Fig. 5. Kitchen, bath, and utility room (remote).

over a kitchen or laundry. The illustration also indicates the additional material required where the sink or the laundry tray is over code limits, usually five to eight feet from the stack.

Figure 7 shows the layout required for the arrangement of two baths backed-up and aligned over a kitchen and a powder room backed-up. The additional necessary material, where the kitchen and the powder room are not located under the backed-up baths, is indicated in Fig. 7.

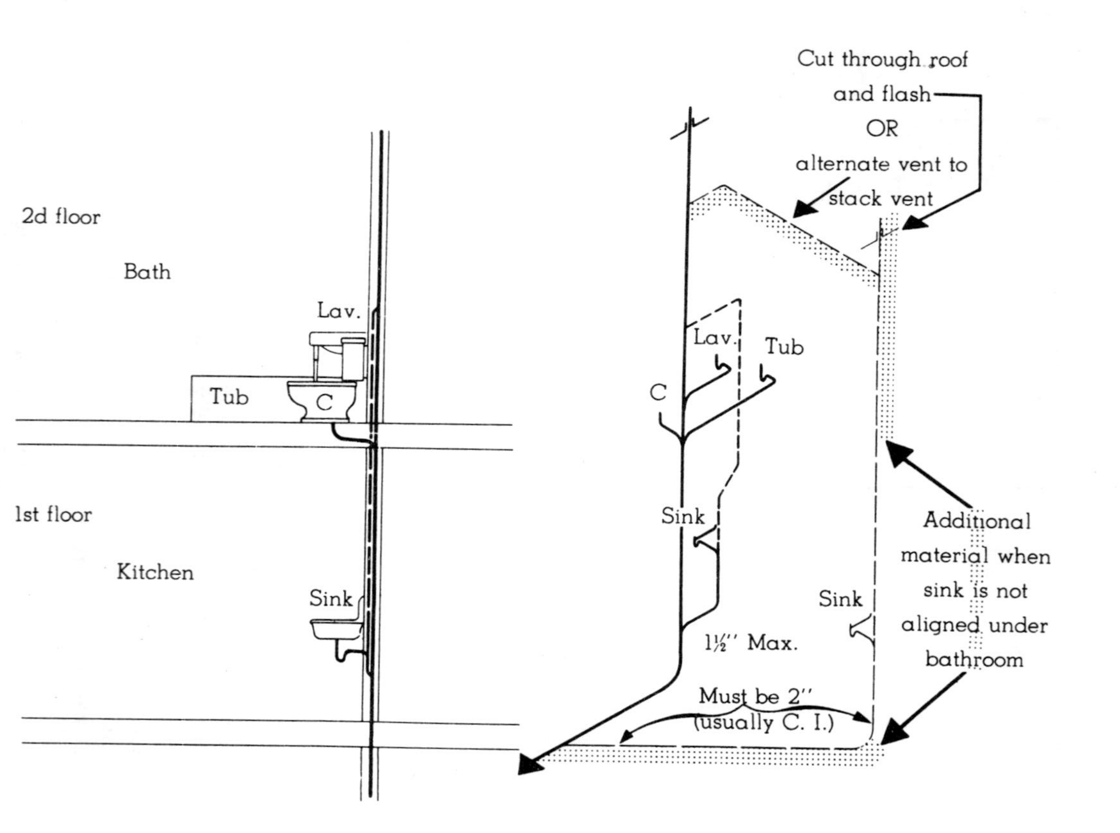

Fig. 6. Bathroom aligned over kitchen.

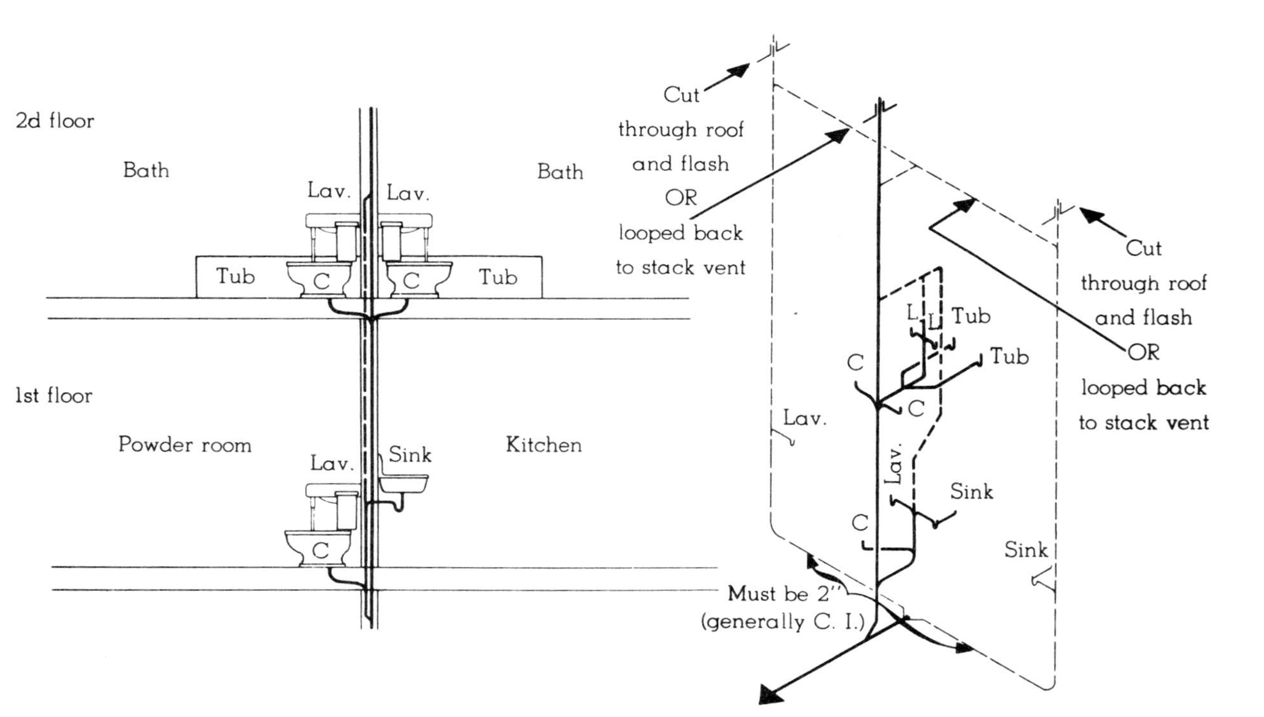

Fig. 7. Two bathrooms aligned over a kitchen and over a powder room.

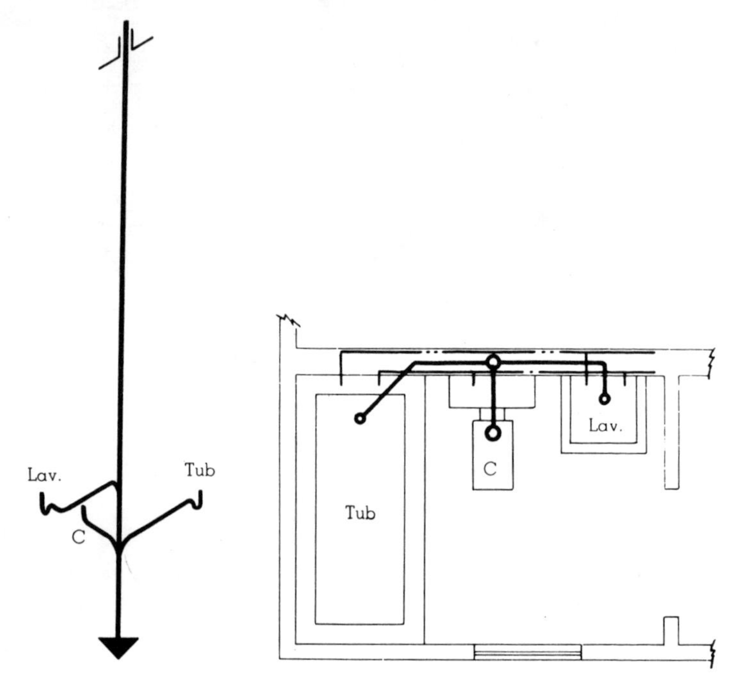

Fig. 8. Compact fixture layout and required drainage piping.

FIXTURE ARRANGEMENTS

Fixtures in Line

The fixtures in a single room, where feasible, should be placed against, and discharge toward, the same wall.

In Fig. 8 the layout is compact and all the fixture traps may be connected directly to the stack without additional vent piping. In Fig. 9 the fixtures are extended and some of the fixture traps need separate venting to protect their water seals. The added piping due to poor layout is obvious.

Vertical Proximity

Basement laundry trays, lavatories, and the like should be placed directly under the kitchen or the first-floor bathroom.

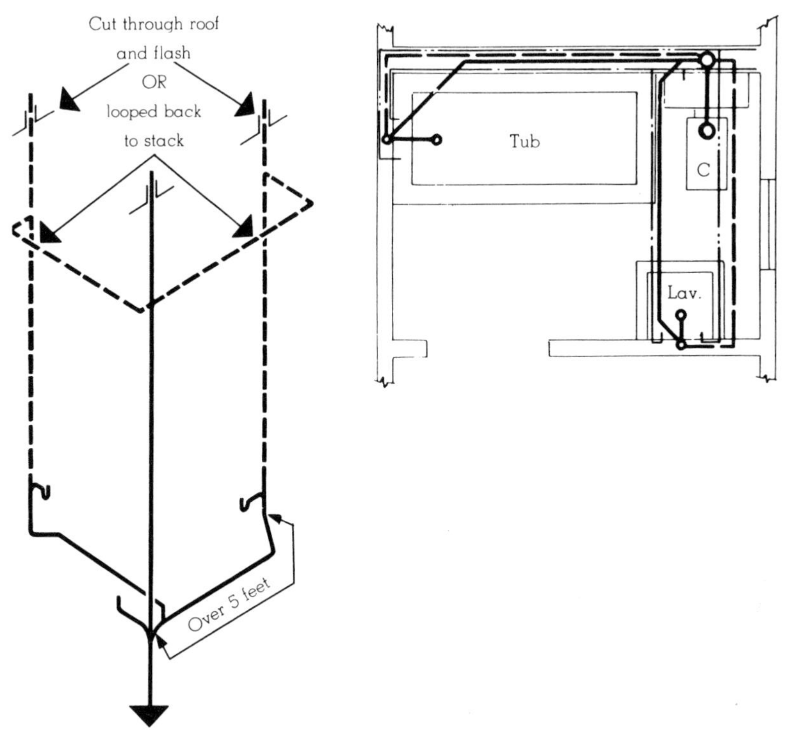

Fig. 9. Extended figure layout with required drainage piping.

Figure 10 shows the extra material needed when the laundry trays are away from the kitchen or the first-floor bathroom stack.

APPARATUS LOCATION

Water Heaters

Hot water heaters should be located as close as possible to those fixtures using the most hot water. Consideration must be given to the flue or the chimney location.

Figure 11 shows a plan wherein consideration has been given to the water heater location and also to the location of the chimney. The short runs of the water piping are economical regarding material and labor, and result in a minimum loss

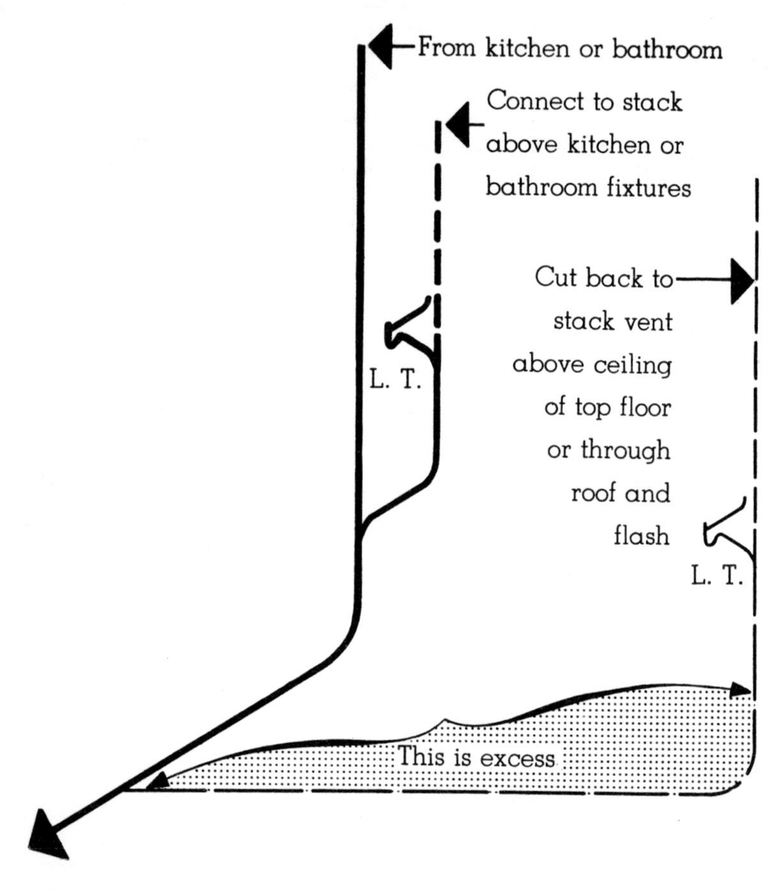

Fig. 10. Basement laundry trays
located under kitchen or first floor bath.

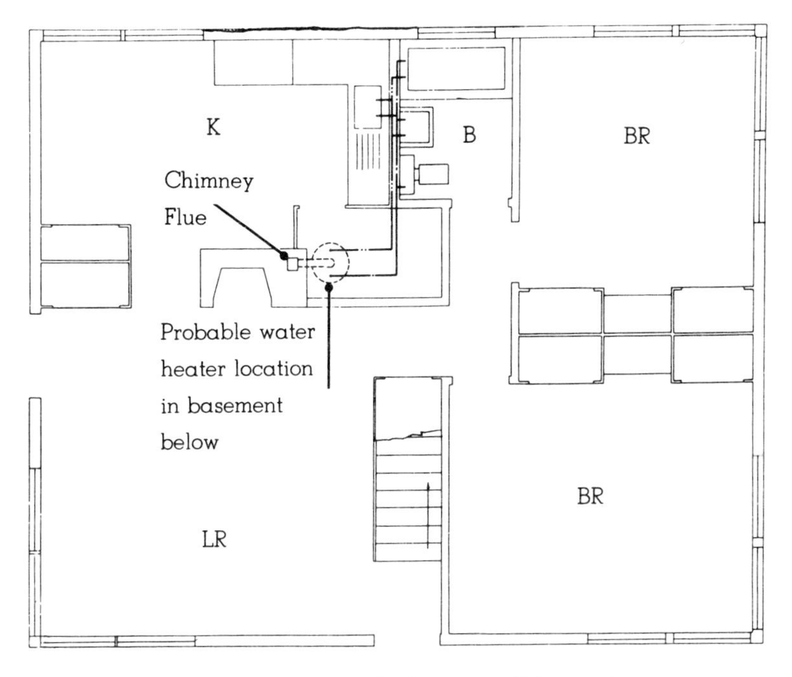

Fig. 11. Water heater and chimney well located for economical hot water piping.

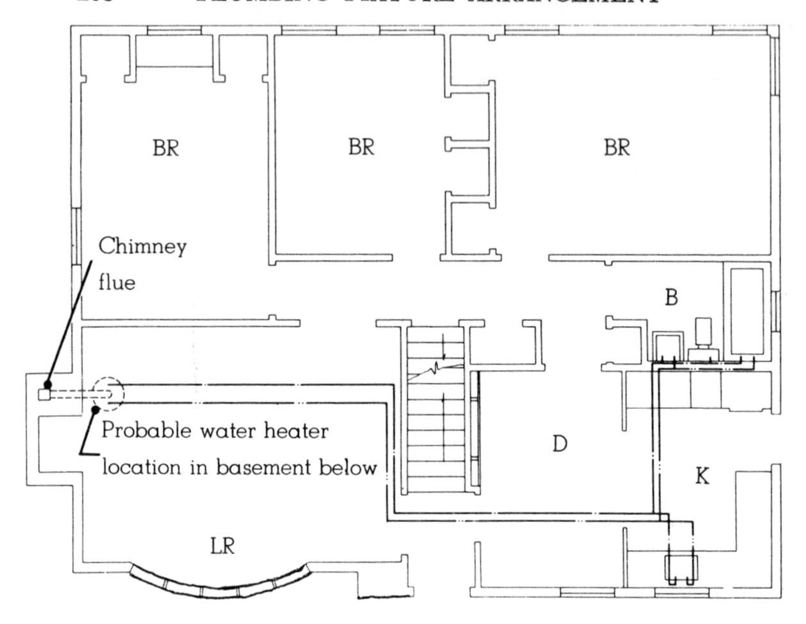

Fig. 12. Extended hot water lines.

of heated water during standby periods and in warming-up.

Figure 12 shows a plan wherein little attention has been given to the hot water system. The chimney flue is far from the hot water fixtures—consequently the heater must be set at a distance. Although the kitchen and the bath are logically grouped for simple drainage piping, the water piping is extended. The result is a more expensive installation with a higher operational cost, due to heat lost through lengthy piping.

Clothes Washers

If possible, plan space for the *fixed clothes washer* with regard to the water supply and the drainage facilities, especially in the small, more compact houses. Long, uneconomical piping should be avoided.

For the *movable clothes washer*, try to provide space adjacent to a laundry tray for easy access to hot water and drainage, or to some other hot water supply and drain facility. In either case, the hot water supply for the washer

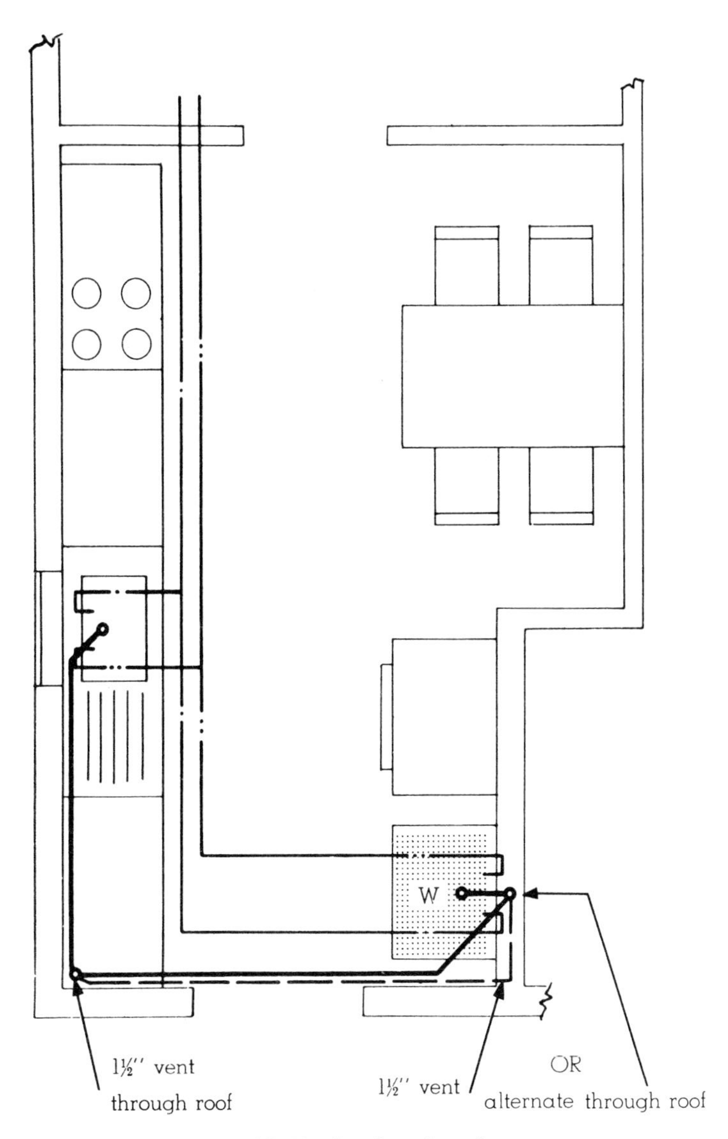

Fig. 13. Kitchen laundry plan.

should be fitted with a hose connection for filling the washer. If the washer is to be stored in another location after being used, then plan convenient storage space for minimum moving effort.

Figure 13 shows a *kitchen-laundry plan* in which the clothes washer requires extended water lines to be run from the

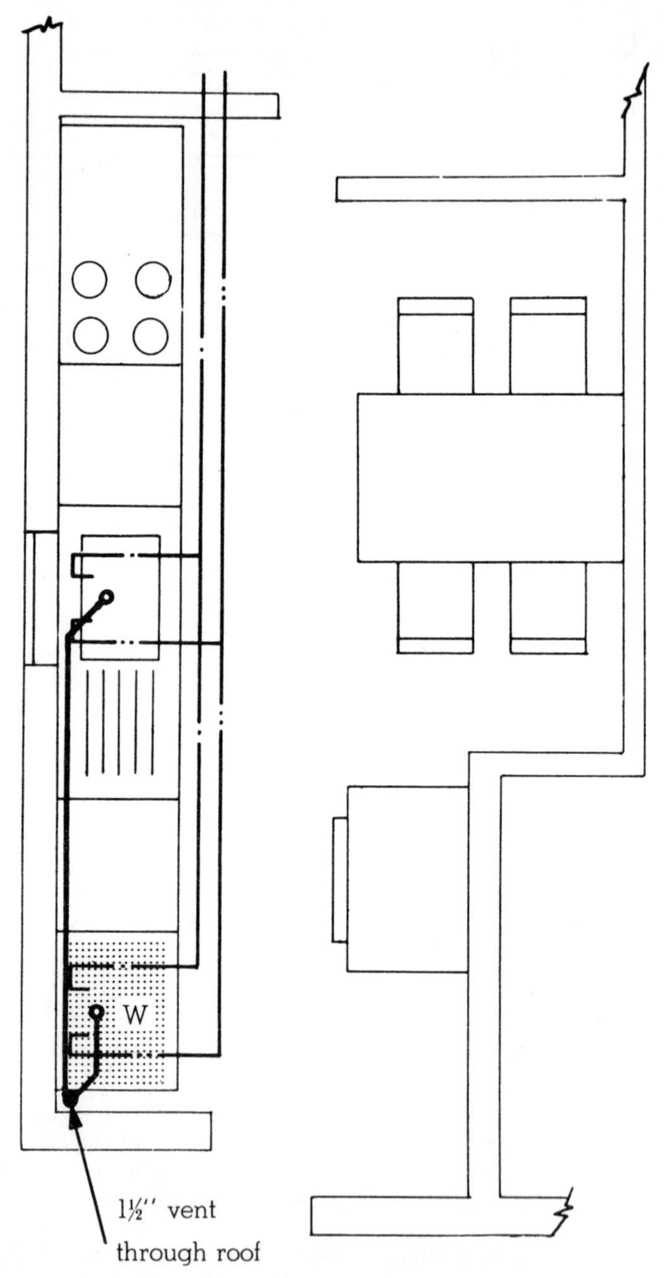

Fig. 14. Compact clothes washer and sink arrangement.

supplies serving the kitchen sink, as well as an additional vent stack through the roof to serve the waste connection. Rearrangement of the equipment to place the clothes washer adjacent to the sink would have saved this expense at no apparent sacrifice in operating service or convenience, as shown in Fig. 14.

Dishwashers

Domestic type dishwashing machines, when installed as a separate piece of equipment should be located adjacent to the sink to take advantage of the shortest possible piping and drainage connections and to increase their convenience.

OTHER CONSIDERATIONS

The utility of the conventional three-fixture bathroom may be increased by placing the water closet, or the water closet and the lavatory, in a partitioned compartment, separated from the tub or shower. This arrangement permits simultaneous private use of the fixtures.

Figure 15 shows a plan for locating the water closet in a separate compartment without incurring excess plumbing costs due to complicated piping connections.

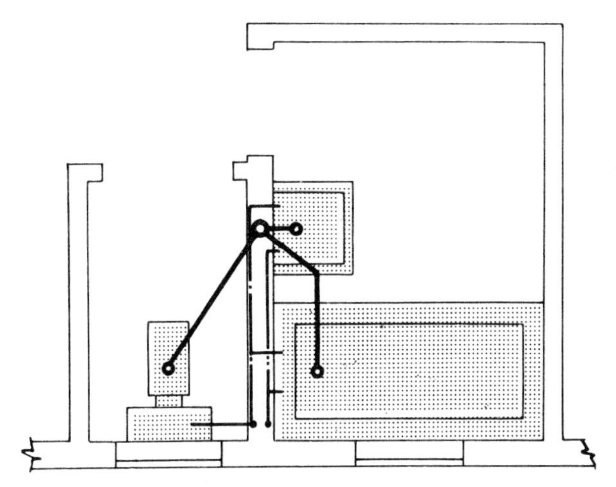

Fig. 15. Water closet in separate compartment.

Figure 16 illustrates an arrangement providing separation of the bathtub from the other fixtures while retaining the advantage of compact piping connections. This type of separation is usually considered to be more desirable than that shown in Fig. 15.

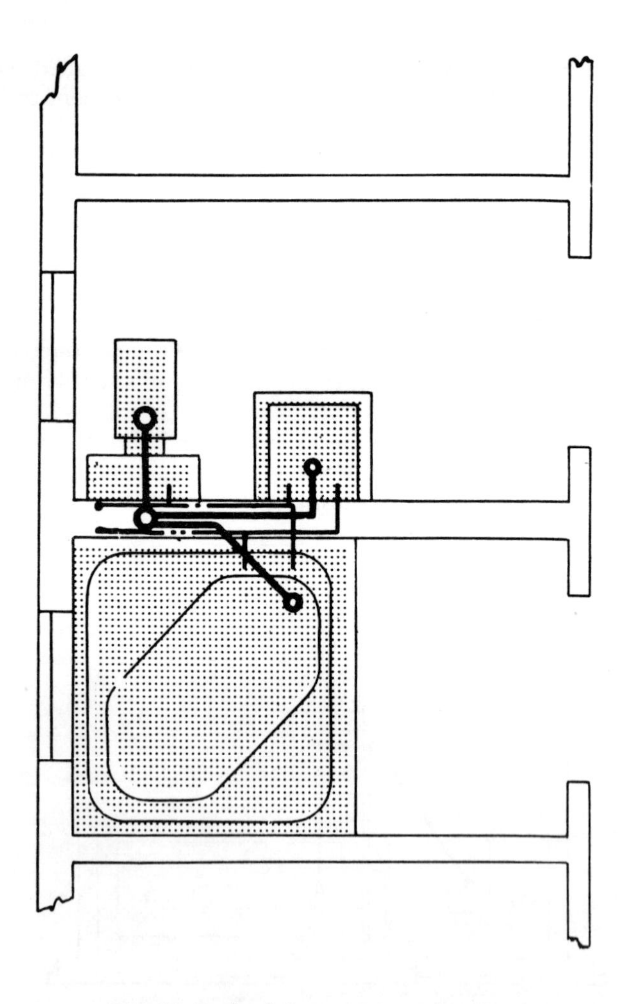

Fig. 16. Bathtub in separate compartment.

SUMMARY

Piping Materials and Supplies

Careful consideration and use of the principles of fixture orientation and location described in this chapter will generally result in appreciable savings of the piping materials and supplies. It is conservatively estimated that, for the average small home having one bathroom, savings should average about 10 feet of 3- or 4-inch cast iron and pipe; 5 feet of galvanized steel drainage pipe; 10 feet of galvanized steel vent pipe; and 15 feet of water piping over the more extended piping systems commonly used when fixture locations are not logically planned. These savings amount to about 84 pounds of cast iron, 5 pounds of lead (for cast iron pipe joints), 48 pounds of steel and 5½ pounds of copper, per dwelling.

A considerable saving will be returned to the homeowner in the form of reduced costs and operating expenses resulting from fuel saved by the use of efficient hot water piping installations.

CODES AND STANDARDS

Most *plumbing codes* and *standards* will permit the simplified installations suggested in this chapter, provided that the fixtures are positioned so as to reduce the length of horizontal drains within their particular limitations. Cooperation among the planner, the designer of the plumbing system and the homeowner during the initial stages of the design of the house, or when remodeling appears to be the most reasonable method for obtaining a simple and economical plumbing installation.

Chapter 5

Septic Tank—Soil Absorption Systems for Your Home

Population movement within the United States continues to be from rural to metropolitan areas. Because of the difficulty of providing adequate sewerage systems for this new growth, individual septic tank—soil absorption systems continue to be an important method of sewage disposal where they are acceptable. Roughly one-fourth of the new homes in the United States are being constructed with these systems.

The decision on the suitability of *septic-tank* installations must be based on many factors outside of those covered in this chapter. It is emphasized, however, that connection to an adequate public sewerage system is the most satisfactory method of disposing of sewage. Every effort should be made, therefore, to secure public-sewer extensions. Where connection to a public sewer is not feasible, and when a considerable number of residences are to be served, consideration should be given next to the construction of a community sewerage system and treatment plant. Specific information on this matter should be obtained from the local authority having jurisdiction.

Individuals proposing to construct individual sewage-disposal systems should consult the officials in charge of such installations in their area. A number of states and localities have developed requirements which have been incorporated in their official regulations, in many cases soundly based on conditions peculiar to those areas and adequately representing good practice there. The recommendations contained in this chapter should be considered as supplemental to such local requirements. Homeowners and builders and others interested in septic-tank systems should seek guidance from the local authorities prior to land acquisition, in order to have the benefit of their experience as well as their approval of plans and construction.

Safe disposal of all human and domestic wastes is necessary to protect the health of the individual family and the community and to prevent the occurrence of nuisances. To accomplish satisfactory results, such wastes must be disposed of so that—

1. They will not contaminate any drinking water supply.
2. They will not give rise to a public health hazard by being accessible to insects, rodents, or other possible carriers which may come into contact with food or drinking water.
3. They will not give rise to a public health hazard by being accessible to children.
4. They will not violate laws or regulations governing water pollution or sewage disposal.
5. They will not pollute or contaminate the waters of any bathing beach, shellfish breeding ground, or stream used for public or domestic water supply purposes, or for recreational purposes.
6. They will not give rise to a nuisance due to odor or unsightly appearance.

These criteria can best be met by the discharge of domestic sewage to an adequate public or community sewerage system. Where the installation of an individual household sewage disposal system is necessary, the basic principles outlined in this chapter on design, construction, installation, and maintenance should be followed. When these criteria are met, and where soil and site conditions are favorable, the septic tank system can be expected to give satisfactory service. Experience has shown that adequate supervision, inspection and maintenance of all features of the system are required to insure compliance in this respect. Underground portions of the system should be inspected before being covered, so necessary corrections can be made.

SOIL

The first step in the design of subsurface sewage disposal systems is to determine whether the *soil* is suitable for the absorption of septic tank effluent and, if so, how much area is required. The soil must have an acceptable percolation rate, without interference from ground water or impervious strata below the level of the absorption system. In general, the following two conditions must be met.

1. The percolation time should be within the range of those specified in Table 2.

2. The maximum seasonal elevation of the ground water table should be at least four feet below the bottom of the trench or seepage pit. Rock formulations or other impervious strata should be at a depth greater than four feet below the bottom of the trench or the seepage pit.

Unless these conditions can be satisfied, the site is unsuitable for a conventional subsurface sewage disposal system.

TABLE 2
ABSORPTION-AREA REQUIREMENTS FOR INDIVIDUAL RESIDENCES (A).

[Provides for garbage grinder and automatic clothes washing machines]

Percolation rate (time required for water to fall one inch, in minutes)	Required absorption area, in sq. ft. per bedroom (b), standard trench (c), seepage beds (c), and seepage pits (d)	Percolation rate (time required for water to fall one inch, in minutes)	Required absorption area in sq. ft. per bedroom (b), standard trench (c), and seepage beds (c), and seepage pits (d)
1 or less	70	10	165
2	85	15	190
3	100	30 (e)	250
4	115	45 (e)	300
5	125	60 (e), (f)	330

(a) It is desirable to provide sufficient land area for entire new absorption system if needed in future.

(b) In every case sufficient land area should be provided for the number of bedrooms (minimum of 2) that can be reasonably anticipated, including the unfinished space available for conversion as additional bedrooms.

(c) Absorption area is figured as trench-bottom area and includes a statistical allowance for vertical side wall area.

(d) Absorption area for seepage pits is figured as effective side wall area beneath the inlet.

(e) Unsuitable for seepage pits if over thirty.

(f) Unsuitable for absorption systems if over sixty.

PERCOLATION TESTS

Subsurface explorations are necessary to determine subsurface formations in a given area. An auger with an extension handle, as shown in Fig. 1, is often used for making the investigation. In some cases, an examination of road cuts,

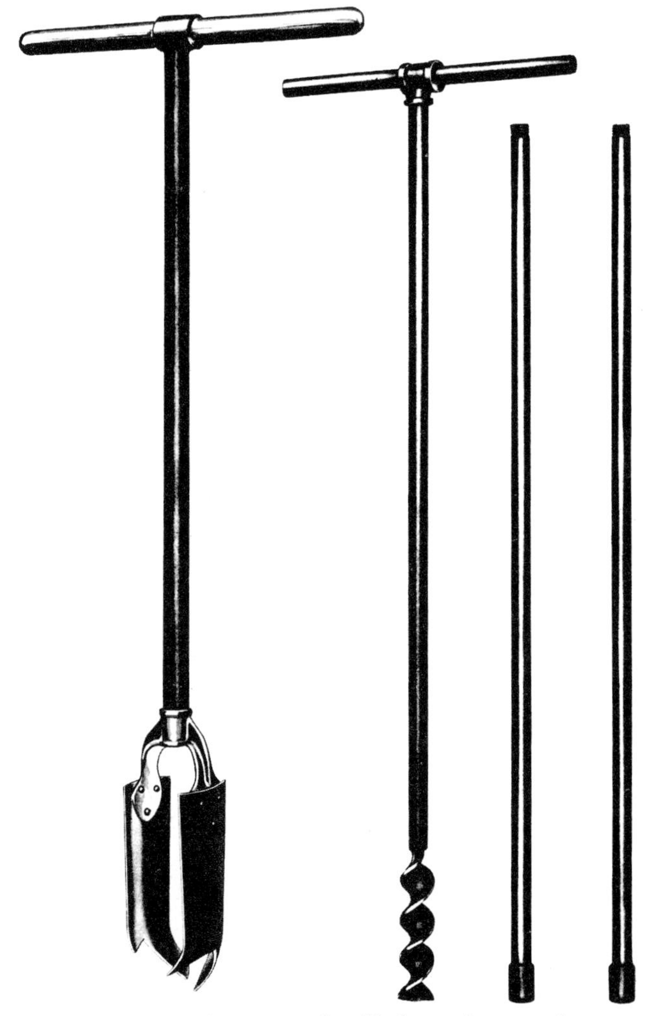

Fig. 1. Auger and extension handle for making test borings.

stream embankments, or building excavations will give useful information. Wells and well drillers' logs can also be used to obtain information on ground water and subsurface conditions. In some areas, subsoil strata vary widely within short distances, and borings must be made at the site of the system. If the subsoil appears suitable, percolation tests should be made at points and elevations selected as typical of the area in which the disposal field will be located.

The percolation tests help to determine the acceptability of the site and establish the design size of the subsurface disposal system. The length of time required for percolation tests will vary in different types of soil. The safest method is to make tests in holes which have been kept filled with water for at least four hours, preferably overnight. This is particularly desirable if the tests are to be made by an inexperienced person, and in some soils it is necessary even if the individual has had considerable experience (as in soils which swell upon wetting). Percolation rates should be figured on the basis of the test data obtained after the soil has had the opportunity to become wetted or saturated and then swell for at least 24 hours. Enough tests should be made in separate holes to assure that the results are valid.

The use of the percolation test is particularly recommended when knowledge of soil types and soil structure is limited. When previous experience and information on soil characteristics are available, some persons prefer other percolation test procedures.

PROCEDURE FOR PERCOLATION TESTS

1. *Number and location of tests.* Six or more tests should be made in separate test holes spaced uniformly over the proposed absorption field site.

2. *Type of test hole.* Dig or bore a hole with horizontal dimensions of from 4 to 12 inches and vertical sides to the depth of the proposed absorption trench. In order to save time, labor, and volume of water required per test, the holes can be bored with a 4-inch auger (Fig. 1). (*See* Fig. 2.)

3. *Preparation of test hole.* Carefully scratch the bottom and sides of the hole with a knife blade or sharp-pointed instrument, in order to remove any smeared soil surfaces and to

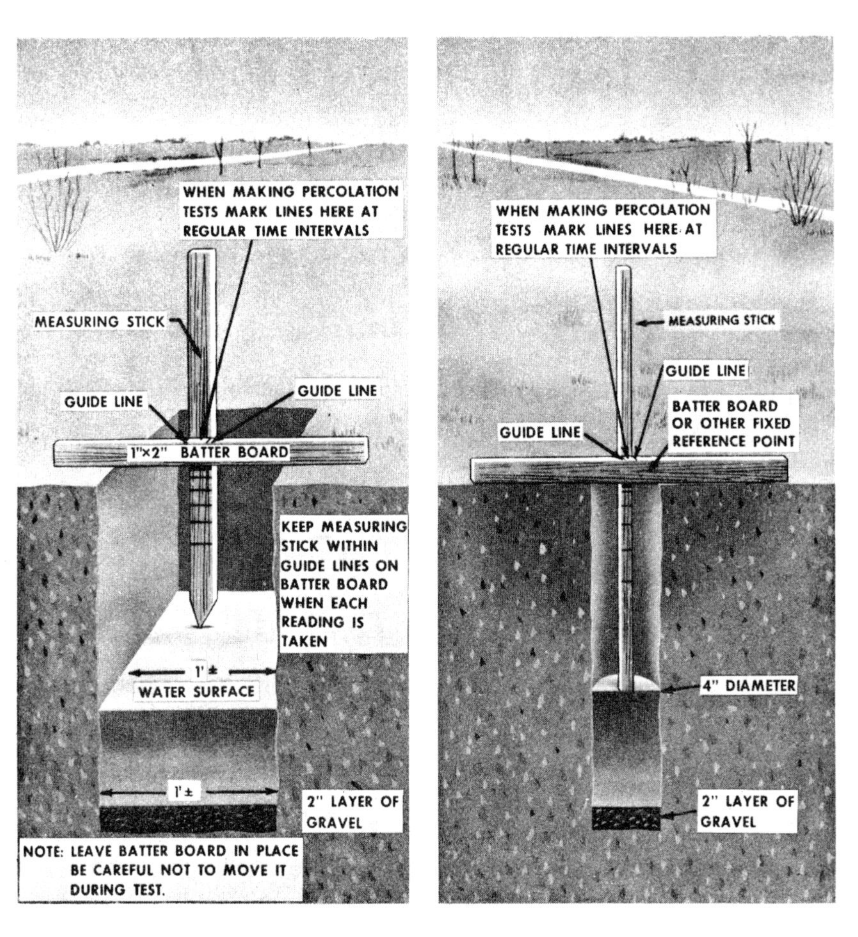

Fig. 2. Methods of making percolation tests.

provide a natural soil interface into which water may percolate. Remove all loose material from the hole. Add two inches of coarse sand or fine gravel to protect the bottom from scouring and sediment.

4. *Saturation and swelling of the soil.* It is important to distinguish between saturation and swelling. Saturation means that the void spaces between soil particles are full of water. This can be accomplished in a short period of time. Swelling is caused by intrusion of water into the individual soil particle. This is a slow process, especially in clay-type soil, and is the reason for requiring a prolonged soaking period.

In the conduct of the test, carefully fill the hole with clear water to a minimum depth of 12 inches over the gravel. In most soils, it is necessary to refill the hole by supplying a surplus reservoir of water, possibly by means of an automatic syphon, to keep water in the hole for at least four hours and preferably overnight. Determine the percolation rate 24 hours after water is first added to the hole. This procedure is to insure that the soil is given ample opportunity to swell and to approach the condition it will be in during the wettest season of the year. The test will give comparable results in the same soil, whether made in a dry or in a wet season. In sandy soils containing little or no clay, the swelling procedure is not essential, and the test may be made as described in section 5(C), after the water from one filling of the hole has completely seeped away.

5. *Percolation-rate measurement.* With the exception of sandy soils, percolation-rate measurements should be made on the day following the procedure described in section 4.

(A) If water remains in the test hole after the overnight swelling period, adjust the depth to approximately six inches over the gravel. From a fixed reference point, measure the drop in water level over a 30-minute period. This drop is used to calculate the percolation rate.

(B) If no water remains in the hole after the overnight swelling period, add clear water to bring the depth of water in the hole to approximately six inches over the gravel. From a fixed reference point, measure the drop in water level at approximately 30 minute intervals for four hours, refilling six inches over the gravel as necessary. The drop that occurs during the final 30 minute period is used to calculate the percolation rate. The drops during prior periods provide information for possible modification of the procedure to suit local circumstances.

(C) In sandy soils (or other soils in which the first six inches of water seeps away in less than 30 minutes after the overnight swelling period), the time interval between measurements should be taken as ten minutes and the test run for one hour. The drop that occurs during the final ten minutes is used to calculate the percolation rate.

SOIL ABSORPTION SYSTEM

For areas where the percolation rates and soil characteristics are good, the next step after making the percolation tests is to determine the required absorption area from Table 2 or from Fig. 3, and to select the soil absorption system that will

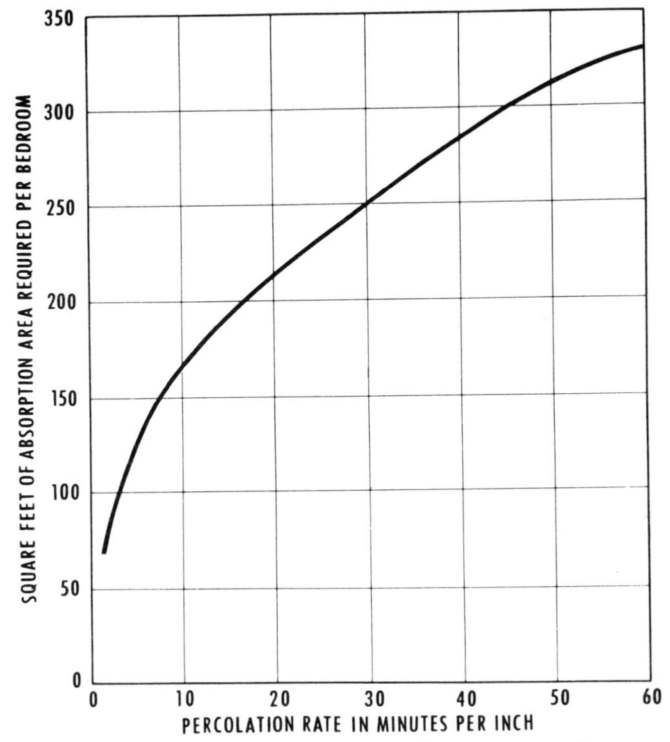

Fig. 3. Absorption area requirements for private residences.

be satisfactory for the area in question. Note in Table 2, soil in which the percolation rate is slower than one inch (2.54 cm.) in 30 minutes is unsuitable for seepage pits, and that slower than one inch (2.54 cm.) in 60 minutes is unsuitable for any type of soil absorption system.

When a soil absorption system is determined to be usable, three types of design may be considered: Absorption trenches, seepage beds, and seepage pits. A modification of the standard absorption trench is discussed later in this chapter giving credit for more than the standard 12 inches (30.48 cm.) of gravel depth in the trench.

The *selection* of the *absorption system* will be dependent to some extent on the location of the system in the area under consideration. A safe distance should be maintained between the site and any source of water supply. Since the distance that pollution will travel underground depends upon numerous factors, including the characteristics of the subsoil formations and the quantity of sewage discharged, no specified distance would be absolutely safe in all localities. Ordinarily, the greater the distance, the greater will be the safety provided. In general, location of components of sewage disposal systems should be as shown in Table 3.

Seepage pits should not be used in areas where domestic water supplies are obtained from shallow wells, or where there are limestone formation and sinkholes with connection to underground channels through which pollution may travel to water sources.

Details pertaining to local water wells, such as depth, type of construction, vertical zone of influence, and the like, together with data on the geological formations and porosity of subsoil strata should be considered in determining the safe allowable distance between wells and subsurface disposal systems.

Absorption Trenches

A soil absorption field consists of a field of 12-inch (30.48-cm.) lengths of 4-inch (10.16-cm.) agricultural drain tile, 2- to 3-foot (60.96- to 91.44-cm.) lengths of vitrified clay sewer pipe, or perforated, nonmetallic pipe. In areas having unusual soil or water characteristics, local experience should be reviewed before selecting piping materials. The individual

TABLE 3
MINIMUM DISTANCE BETWEEN COMPONENTS OF SEWAGE DISPOSAL SYSTEM.

Component of System	Horizontal Distance (feet)				
	Well or suction line	Water supply line (pressure)	Stream	Dwelling	Property line
Building sewer	50	10 (a)	50	....	....
Septic tank	50	10	50	5	10
Disposal field and Seepage Bed	100	25	50	20	5
Seepage Pit	100	50	50	20	10
Cesspool (b)	150	50	50	20	15

(a) Where the water supply line must cross the sewer line, the bottom of the water service within 10 feet of the point of crossing, shall be at least 12 inches above the top of the sewer line. The sewer line shall be of cast iron with leaded or mechanical joints at least 10 feet on either side of the crossing.

(b) Not recommended as a substitute for a septic tank. To be used only when found necessary and approved by the health authority.

laterals preferably should not be over 100 feet long, and the trench bottom and tile distribution lines should be level. Use of more and shorter laterals is preferred because if something should happen to disturb one line, most of the field will still be serviceable. From a theoretical moisture flow viewpoint, a spacing of twice the depth of gravel would prevent taxing the percolative capacity of the adjacent soil.

Many different designs may be used in laying out subsurface disposal fields. The choice may depend on the size and shape of the available disposal area, the capacity required, and the topography of the disposal area.

Typical layouts of absorption trenches are shown in Figs. 4 and 6.

To provide the minimum required gravel depth and earth cover, the depth of the absorption trenches should be at least 24 inches. Additional depth may be needed for contour ad-

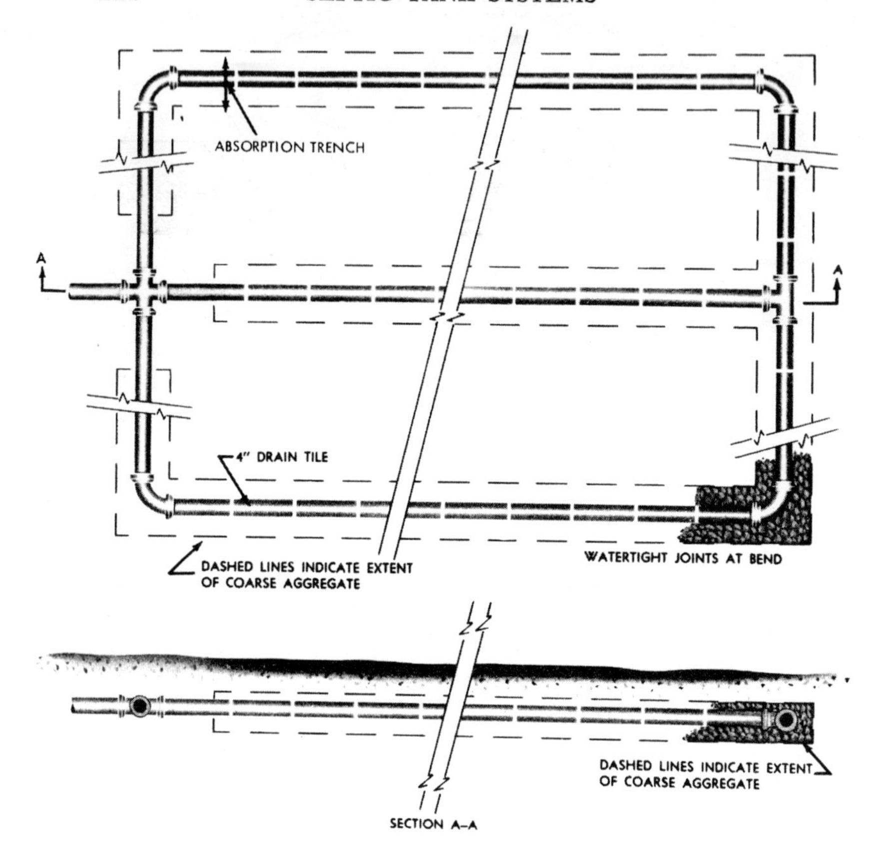

Fig. 4. Typical layout of absorption trench.

justment, extra aggregate under the tile, or other design purposes. The maintenance of a 4-foot separation between the bottom of the trench and the water table is required to minimize ground water contamination. In considering the depth of the absorption field trenches, the possibility of the lines freezing during a prolonged cold period is raised. Freezing rarely occurs in a carefully constructed system kept in continuous operation. It is important during construction to assure that the tile lines are surrounded by gravel. Pipe under driveways or other surfaces which are usually cleared of snow should be insulated.

The required absorption area is predicated on the results of

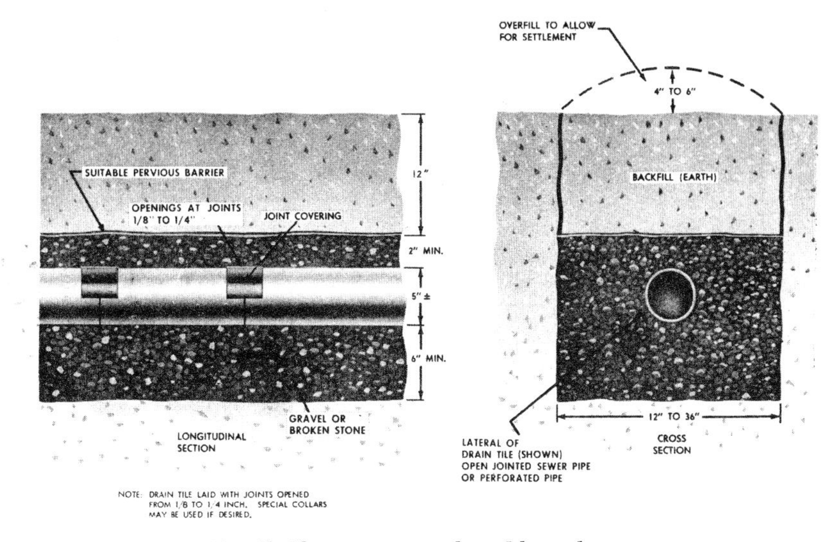

Fig. 5. Absorption trench and lateral.

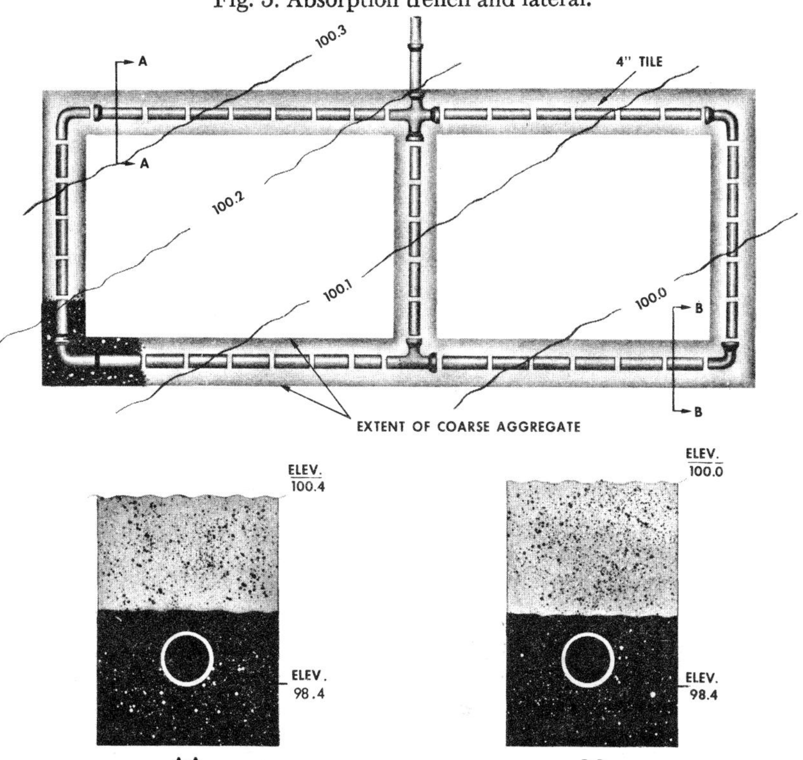

Fig. 6. Absorption-field system for level ground.

the soil percolation test, and may be obtained from Table 2, columns 2 or 4 or from Fig. 3. Note especially that the area requirements are per bedroom. The area of the lot on which the house is to be built should be large enough to allow room for an additional system if the first one fails. Therefore, for a three-bedroom house on a lot where the minimum percolation rate was one inch in 15 minutes, the necessary absorption area will be three bedrooms × 190 square feet per bedroom, or 570 square feet. For trenches two feet wide with six inches of gravel below the drain pipe, the required total length of trench would be 570 ÷ 2, or 285 feet. If this were divided into five portions (i.e., 5 laterals), the length of each line would be 285 ÷ 5, or 57 feet. The spacing of trenches is generally governed by practical construction considerations dependent on the type of equipment, safety, and the like. For serial distribution on sloping ground, trenches should be separated by six feet to prevent short circuiting. Table 3 gives the various distances the system has to be kept away from wells, dwellings, and the like.

In the example just described, trenches are 2 feet wide × 5 trenches = 10 feet plus 6 feet between trenches × 4 spaces = 24 feet. The total width of 34 feet × 57 feet in length = 1,938 square feet, plus additional land required to keep the field away from wells, property lines, and the like.

Construction Considerations. Careful construction is important in obtaining a satisfactory soil absorption system. Attention should be given to the protection of the natural absorption properties of the soil. Care must be taken to prevent sealing of the surface on the bottom and sides of the trench. Trenches should not be excavated when the soil is wet enough to smear or compact easily. Soil moisture is right for safe working only when a handful will mold with considerable pressure. Open trenches should be protected from surface runoff to prevent the entrance of silt and debris. If it is necessary to walk in the trench, a temporary board laid on the bottom will reduce the damage. Some smearing and damage is bound to occur. All smeared or compacted surfaces should be raked to a depth of 1 inch (2.54 cm.) and loose material removed, before the gravel is placed in the trench.

The pipe, laid in a trench of sufficient width and depth, should be surrounded by clean, graded gravel or rock, broken hard burned clay brick, or similar aggregate. The material may range in size from ½ inch to 2½ inches (1.27 to 6.35 cm.). Cinders, broken shale, and similar material are not recommended, because they are usually too fine and may lead to premature clogging. The material should extend from at least 2 inches (5.08 cm.) above the top of the pipe to at least 6 inches (15.24 cm.) below the bottom of the pipe. If tile is used, the upper half of the joint openings should be covered as shown in Fig. 5. The top of the stone should be covered with untreated building paper, a 2-inch (5.08-cm.) layer of hay or straw, or similar pervious material to prevent the stone from becoming clogged by the earth backfill. An impervious covering should not be used, as this interferes with evapotranspiration at the surface. Although generally not figured in the calculations, evapotranspiration is often an important factor in the operation of horizontal absorption systems.

Drain tile connectors, collars, clips, or other spacers with covers for the upper half of the joints are of value in obtaining uniform spacing, proper alinement, and protection of tile joints, but use of such aides is optional. They have been made of galvanized iron, copper, and plastic.

It has been found that root problems may be prevented best by using a liberal amount of gravel or stone around the tile. Clogging due to roots has occurred mostly in lines with insufficient gravel under the tile. Furthermore, roots seek the location where moisture conditions are most favorable for growth and, in the small percentage of cases where they become troublesome in well-designed installations, there is usually some explanation involving the moisture conditions. At a residence which is used only during the summer for example, roots are most likely to penetrate when the house is uninhabited, or when moisture immediately below or around the gravel becomes less plentiful than during the period when the system is in use. In general, trenches constructed with ten feet of large trees or dense shrubbery should have at least 12 inches of gravel or crushed stone beneath the tile.

If trees are near the sewage disposal system, difficulty with roots entering poorly joined sewer lines can be anticipated. Lead-calked cast iron pipe, a sulfur base or bituminous pipe joint compound, mechanical clay pipe joints, copper rings over joints, and lump copper sulfate in pipe trenches have been found effective in resisting the entrance of roots into pipe joints. Roots will penetrate into the gravel in tile field trenches rather than into the pipe. About two or three pounds of copper sulfate crystals flushed down the toilet bowl once a year will destroy roots the solution comes in contact with, but will not prevent new roots from entering. The application of the chemical should be done at a time, such as late in the evening when the maximum contact time can be obtained before dilution. Copper sulfate will corrode chrome, iron, and brass; hence it should not be allowed to come into contact with these metals. Cast iron is not affected to any appreciable extent. Some time must elapse before the roots are killed and broken off. Copper sulfate in the recommended dosage will not interfere with the operation of the septic tank.

The top of a new absorption trench should be hand-tamped and should be overfilled with about four to six inches of earth. Unless this is done, the top of the trench may settle to a point lower than the surface of the adjacent ground. This will cause the collection of storm water in the trench, which can lead to premature saturation of the absorption field and possibly to complete washout of the trench. Machine tamping or hydraulic backfilling of the trench should be prohibited.

Where sloping ground is used for the disposal area, it is usually necessary to construct a small temporary dike or surface water diversion ditch above the field, to prevent the disposal area from being washed out by rain. The dike should be maintained or the ditch kept free of obstructions until the field becomes well covered with vegetation.

A heavy vehicle would readily crush the tile in a shallow absorption field. For this reason, heavy machinery should be excluded from the disposal area unless special provision is made to support the weight. All machine grading should be completed before the field is laid.

The use of the field area must be restricted to activities which will not contribute to the compaction of the soil with the consequent reduction in soil aeration.

Seepage Beds

Common design practice for soil absorption systems for private residences provides for trench widths up to 36 inches (91.44 cm.). Variations of design utilizing increased width are being used in many areas. Absorption systems having trenches wider than three feet are referred to as seepage beds. The design of trenches is based on an empirical relationship between the percolation test and the bottom area of the trenches. The use of seepage beds has been limited by the lack of experience with their performance and the absence of design criteria comparable to that for trenches.

Studies sponsored by the Federal Housing Administration have demonstrated that the seepage bed is a satisfactory device for disposing of effluent in soils that are acceptable for soil absorption systems. The studies have further demonstrated that the empirical relationship between the percolation test and bottom area required for trenches is applicable for seepage beds.

There are three main elements of a *seepage bed:* Absorption surface, rockfill or packing material, and the distribution system. The design of the seepage bed should be such that the total intended absorption area is preserved, sufficient packing material is provided in the proper place to allow for further treatment and storage of excess liquid, and a means for distributing the effluent is protected against siltation of earth backfill and mechanical damage. Construction details for a conventional seepage bed are outlined in the following material in such a way that these principal design elements are incorporated. Tabulation of construction details for the conventional seepage bed is not intended to preclude other designs which may provide the essential features in a more economical or otherwise desirable manner. Specifically, there may be equally acceptable or even superior methods developed for distributing the liquid than by tile or perforated pipe covered with gravel.

The use of seepage beds results in the following advantages.

1. A wide bed makes more efficient use of land available for absorption systems than a series of long narrow trenches with wasted land between the trenches.

2. Efficient use may be made of a variety of modern earth moving equipment employed at housing projects for other purposes such as basement excavation and landscaping, resulting in savings on the cost of the system.

Construction Considerations. When seepage beds are used, the following design and construction procedures providing for rockfill or packing material, an adequate distribution system, and protection of the absorption area, should be observed.

1. The amount of bottom absorption area required should be the same as shown in Table 2.

2. Percolation tests should be conducted in accordance with section on Percolation Tests previously described.

3. The bed should have a minimum depth of 24 inches below natural ground level to provide a minimum earth backfill cover of 12 inches.

4. The bed should have a minimum depth of 12 inches of rockfill or packing material extending at least two inches above and six inches below the distribution pipe.

5. The bottom of the bed and distribution tile or perforated pipe should be level.

6. Lines for distributing effluent should be spaced not greater than six feet apart and not greater than three feet from the bed sidewall.

7. When more than one bed is used: (a) there should be a minimum of six feet of undisturbed earth between adjacent beds; and, (b) the beds should be connected in series in accordance with the section concerning serial distribution.

8. Applicable construction considerations for standard trenches previously described should also be followed.

Distribution Boxes

1. *Distribution boxes* can be eliminated from septic tank-soil absorption systems in favor of some other method of distribution without inducing increased failure of disposal fields. In fact, evidence indicates that distribution boxes as presently used may be harmful to the system.

2. Data indicates that on level ground, equal distribution without inducing increased failure of disposal fields. In fact, evidence indicates that distribution boxes as presently used may be harmful to the system.

3. On sloping ground a method of distribution is needed to prevent excessive build-up of head and failure of any one trench before the capacity of the entire system is utilized. It is doubtful that distribution boxes as presently used give equal distribution. Rather, they probably act as diversion devices sending most of the liquid to part of the system.

For these reasons it is recommended that distribution boxes not be used for individual sewage disposal systems.

Serial Distribution

Serial distribution is achieved by arranging individual trenches of the absorption system so that each trench is forced to pond to the full depth of the gravel fill before liquid flows into the succeeding trench.

Serial distribution has the following advantages.

1. Serial distribution minimizes the importance of variable absorption rates by forcing each trench to absorb effluent until its ultimate capacity is utilized. The variability of soils even in the small area of an individual absorption field raises doubt of the desirability of uniform distribution. Any one or a combination of factors may lead to non-uniform absorptive capacity of the several trenches in a system. Varying physical and chemical characteristics of soil, construction damage such as soil interface smearing or excessive compaction, poor surface drainage, and variation in depth of trenches are some of the factors involved.

2. Serial distribution causes successive trenches in the absorption system to be used to full capacity. Serial distribution has a distinct advantage on sloping terrain. With imperfect division of flow in a parallel system, one trench could become overloaded, resulting in a surcharged condition. If the slope of the ground and elevation of the distribution box were such that a surcharged trench continued to receive more effluent than it could absorb, local failure would occur before the full capacity of the system was utilized.

3. The cost of the distribution box is eliminated in serial distribution. Also, long runs of closed pipe connecting the box to each trench are unnecessary.

Fields in Flat Areas. Where the slope of the ground surface does not exceed six inches in any direction within the area utilized for the absorption field, the septic tank effluent may be applied to the absorption field through a system of interconnected tile lines and trenches in a continuous system. The following specific criteria should be followed.

1. A minimum of 12 inches of earth cover is provided over the gravel fill in all trenches of the system.

2. The bottom of the trenches and the distribution lines should be level.

3. One type of satisfactory absorption system layout for level ground is shown in Fig. 6.

4. Construction considerations for standard trenches, as described previously in the section on Absorption Trenches, should be followed.

Fields in Sloping Ground. Serial distribution may be used in all situations where a soil absorption system is permitted and should be used where the fall of the ground surface exceeds approximately six inches in any direction within the area utilized for the absorption field. The maximum ground slope suitable for serial distribution systems should be governed by local factors affecting the erosion of the ground used for the absorption field. Excessive slopes which are not protected from surface water runoff or do not have adequate vegetation cover to prevent erosion should be avoided. Generally, ground having a slope greater than one vertical to two horizontal should be investigated carefully to determine if satisfactory from the erosion standpoint. Also, the horizontal distance from the side of the trench to the ground surface should be adequate to prevent lateral flow of effluent and breakout on the surface and in no case should be less than two feet.

In serial distribution, each adjacent trench (or pair of trenches) is connected to the next by a closed pipe line laid on an undisturbed section of ground, as shown in Fig. 7. The arrangement is such that all effluent is discharged to the first trench until it is filled. Excess liquid is then carried by means of a closed line to the next succeeding or lower trench. In that manner, each portion of the subsurface system is used in succession. When serial distribution is used, the following design and construction procedures should be followed.

1. The bottom of each trench and its distribution line should be level.

2. There should be a minimum of 12 inches of ground cover over the gravel fill in the trenches.

3. The absorption trenches should follow approximately the ground surface contours so that variations in trench depth will be minimized.

4. There should be a minimum of six feet of undisturbed earth between adjacent trenches and between the septic tank and the nearest trench.

5. Adjacent trenches may be connected with the relief line or a drop box arrangement (Fig. 7) in such a manner that each trench is completely filled with septic tank effluent to

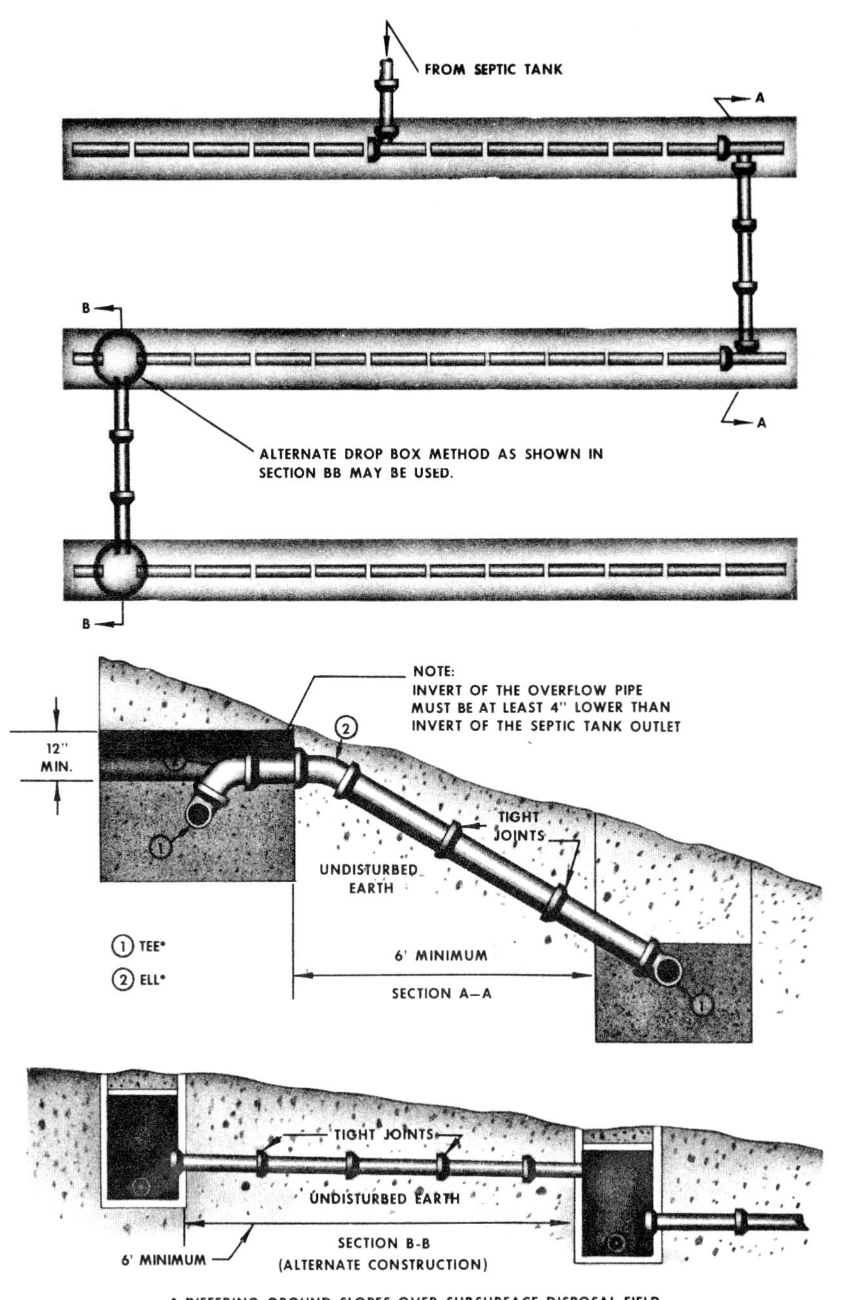

Fig. 7. A relief line arrangement for serial distribution.

the full depth of the gravel before effluent flows to succeeding trenches. (Figure 7 does not preclude the use of other arrangements to provide serial distribution.)

(A) Trench connecting lines should be four-inch, tight-joint sewers with direct connections to the distribution lines in adjacent trenches or to a drop box arrangement.

(B) Care must be exercised in constructing relief lines to insure an undisturbed block of earth between trenches. The trench for the relief pipe, where it connects with the preceding absorption trench, should be dug no deeper than the top of the gravel. The relief line should rest on undisturbed earth and backfill should be carefully tamped.

(C) The relief lines connecting individual trenches should be as far from each other as practicable in order to prevent short circuiting.

6. Invert of the overflow pipe in the first relief line must be at least four inches lower than the invert of the septic tank outlet (Fig. 7).

7. All other construction features of the disposal field are the same as recommended previously in section on Absorption Trenches.

Deep Absorption Trenches and Seepage Beds

In cases where the depth of filter material below the tile exceeds the standard six-inch depth, credit may be given for the added absorption area provided in deeper trenches with a resultant decrease in length of trench. Such credit should be given in accordance with Table 4 which gives the percentage of length of standard absorption trench (as computed from Table 2), based on six-inch increments of increase in depth of filter material.

To use Table 4 consider the example previously given in section on Absorption Trenches. Using a trench two feet wide with six inches of gravel under tile, 285 feet are required. If the depth of gravel is increased to 18 inches, keeping trench width at two feet, only 66 percent of 285 feet is required, or 188 feet. If four laterals are used, the length would be 188 divided by 4 = 47 feet.

The space between lines for serial distribution on sloping ground is six feet × 3 spaces = 18 feet, plus 4 lines × 2 feet = 8 feet. Total land required is 26 feet in width × 47 feet in length = 1222 square feet, plus additional arcs required to keep the field away from wells, property lines, and the like.

Seepage Pits

Seepage pits as with all soil absorption systems, should never be used where there is a likelihood of contaminating underground waters, nor where adequate seepage beds or trenches can be provided. When seepage pits are to be used, the pit excavation should terminate four feet above the ground water table.

In some states, seepage pits are permitted as an alternative when absorption fields are impracticable, and where the top three or four feet of soil is underlaid with porous sand or fine gravel and the subsurface conditions are otherwise suitable for pit installations. Where circumstances permit, seepage pits may be either supplemental or alternative to the more shallow

TABLE 4

PERCENTAGE OF LENGTH OF STANDARD TRENCH.[1]

Depth of Gravel Below Pipe in Inches[2]	Trench width 12″	Trench width 18″	Trench width 24″	Trench width 36″	Trench width 48″	Trench width 60″
12	75	78	80	83	86	87
18	60	64	66	71	75	78
24	50	54	57	62	66	70
30	43	47	50	55	60	64
36	37	41	44	50	54	58
42	33	37	40	45	50	54

[1] The standard absorption trench is one in which the filter material extends two inches above and six inches below the pipe.

[2] For trenches or beds having width not shown in Table 3, the percent of length of standard absorption trench may be computed as follows:

$$\text{Percent of length standard trench} = \frac{w + 2}{w + 1 + 2d} \times 100$$

Where w = width of trench in feet
d = depth of gravel below pipe in feet

Fig. 8. Deep percolation test for seepage pit.

absorption fields. When seepage pits are used in combination with absorption fields, the absorption areas in each system should be prorated, or based upon the weighted average of the results of the percolation tests.

It is important that the capacity of a seepage pit be computed on the basis of percolation tests made in each vertical stratum penetrated. The weighted average of the results should be computed to obtain a design figure. Soil strata in which the percolation rates are in excess of 30 minutes per inch should not be included in computing the absorption area. As will be apparent from Fig. 8, adequate tests for deep pits are somewhat difficult to make, time consuming, and expensive. Although few data have been collected comparing percolation test results with deep pit performance, nevertheless the results of such percolation tests, while of limited value, combined with competent engineering judgment based on experience, are the best means of arriving at design data for seepage pits.

TABLE 5
VERTICAL WALL AREAS OF CIRCULAR SEEPAGE PITS.
[In Square Feet]

Diameter of seepage pit (feet)	Effective strata depth below flow line (below inlet)									
	1 foot	2 feet	3 feet	4 feet	5 feet	6 feet	7 feet	8 feet	9 feet	10 feet
3	9.4	19	28	38	47	57	66	75	85	94
4	12.6	25	38	50	63	75	88	101	113	126
5	15.7	31	47	63	79	194	110	126	141	157
6	18.8	38	57	75	94	113	132	151	170	188
7	22.0	44	66	88	110	132	154	176	198	220
8	25.1	50	75	101	126	151	176	201	226	251
9	28.3	57	85	113	141	170	198	226	254	283
10	31.4	63	94	126	157	188	220	251	283	314
11	34.6	69	104	138	173	207	242	276	311	346
12	37.7	75	113	151	188	226	264	302	339	377

Example: A pit of 5 foot diameter and 6 foot depth below the inlet has an effective area of 94 square feet. A pit of 5 foot diameter and 16 foot depth has an area of 94 + 157, or 251 square feet.

Example: A pit of five foot diameter and six foot depth below the inlet has an effective area of 94 square feet. A pit of five foot diameter and 16 foot depth has an area of 94 + 157, or 251 square feet.

Table 2 or Fig. 3 gives the absorption area requirements per bedroom for the percolation rate obtained. The effective area of the seepage pit is the vertical wall area (based on dug diameter) of the pervious strata below the inlet. No allowance should be made for impervious strata or bottom area. With this in mind, Table 5 may be used for determining the effective side-wall area of circular or cylindrical seepage pits.

Example: Assume that a seepage pit absorption system is

to be designed for a 3-bedroom home on a lot where the minimum percolation rate of one inch in 15 minutes prevails. According to Table 2, 3×190 (or 570) square feet of absorption area would be needed. Assume also that the water table does not rise above 27 feet below the ground surface, that seepage pits with effective depth of 20 feet can be provided, and that the house is in a locality where it is common practice to install seepage pits five feet in diameter (i.e., four feet to the outside walls, which are surrounded by about six inches of gravel). Design of the system is as follows:

Let d = depth of pit in feet; D = pit diameter in feet:

$$\pi Dd = 570 \text{ square feet}$$

$$3.14 \times 5 \times d = 570 \text{ square feet}$$

Solving for d = depth of pit = 36 feet (approx.)

In other words, one five foot diameter pit 36 feet deep would be needed, but since the maximum effective depth is 20 feet in this particular location, it will be necessary to increase both of these. This is illustrated in the following example.

(a) Design for 2 pits with a 10 foot diameter; d = depth of each pit.

$$2 \times 3.14 \times 10 \times d = 570 \text{ square feet}$$

$$d = 9.1 \text{ feet deep}$$

Use 2 pits 10 feet in diameter and 9.1 feet deep.

(b) Design for 2 pits with a 5 foot diameter; d = depth of each pit.

$$2 \times 3.14 \times 5 \times d = 570 \text{ square feet}$$

$$d = 18 \text{ feet (approx.)}$$

Use 2 pits 5 feet in diameter and 18 feet deep.

Experience has shown that seepage pits should be separated by a distance equal to 3 times the diameter of the largest pit. For pits over 20 feet in depth, the minimum space between pits should be 20 feet. (*See* Fig. 9.) The area of the lot on which the house is to be built should be large enough to maintain this distance between the pits while still allowing room for additional pits if the first ones should fail. If this can be done, such an absorption system may be approved, if not, other suitable sewerage facilities should be required.

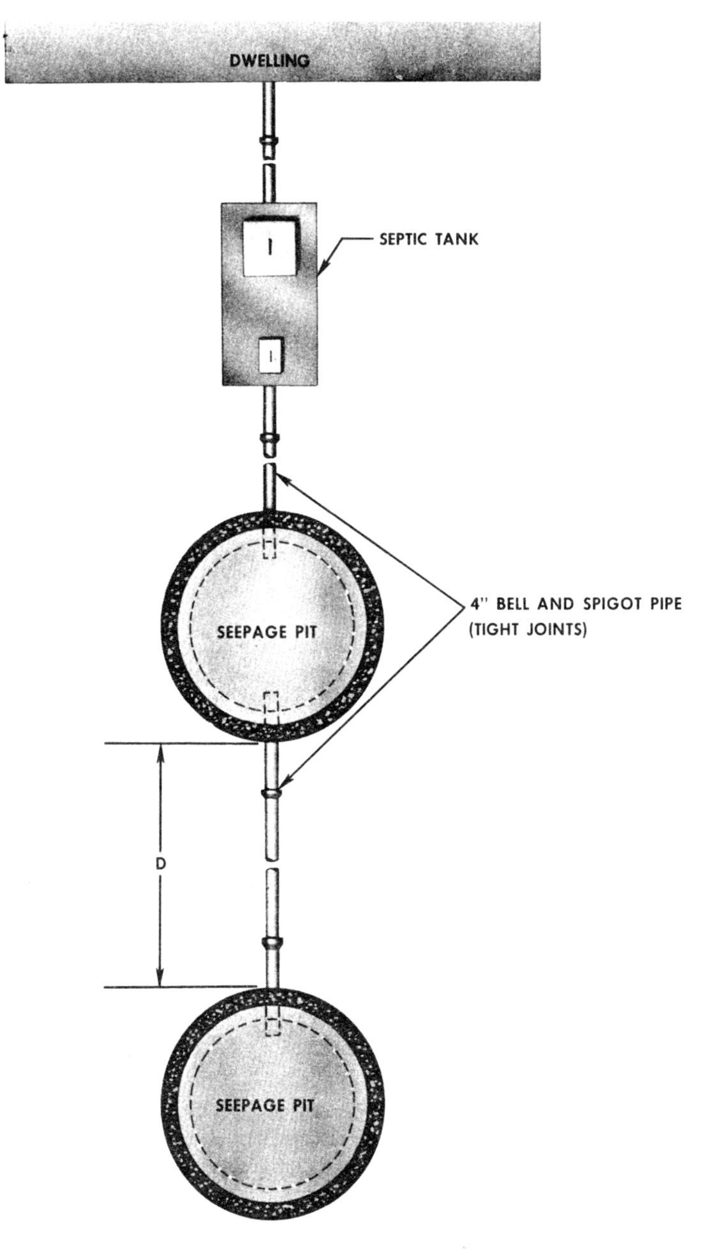

Fig. 9. Disposal system using two seepage pits.

Construction Considerations

Soil is susceptible to damage during excavation. Digging in wet soils should be avoided as much as possible. Cutting teeth on mechanical equipment should be kept sharp. Bucket augered pits should be reamed to a larger diameter than the bucket. All loose material should be removed from the excavation.

Pits should be backfilled with clean gravel to a depth of one foot above the pit bottom or one foot above the reamed ledge to provide a sound foundation for the lining. Preferred lining materials are clay or concrete brick, block, or rings. Rings should have weep holes or notches to provide for seepage. Brick and block should be laid dry with staggered joints. Standard brick should be laid flat to form a four-inch wall. The outside diameter of the lining should be at least six inches less than the least excavation diameter. The annular space formed should be filled with clean, coarse gravel to the top of the lining as shown in Fig. 10.

Either brick dome or flat concrete covers are satisfactory. They should be based on undisturbed earth and extend at least 12 inches beyond the excavation and should not bear on the lining for structural support. Bricks should be either laid in cement mortar or have a two-inch covering of concrete. If flat covers are used, a prefabricated type is preferred, and they should be reinforced to be the equivalent in strength of an approved septic tank cover. A nine-inch capped opening in the pit cover is convenient for pit inspection. All concrete surfaces should be located with a protective bitumastic or similar compound to minimize corrosion.

Connecting lines should be of a sound, durable material the same as used for the house to septic tank connection. All connecting lines should be laid on a firm bed of undisturbed soil throughout their length. The grade of a connecting line should be at least two percent. The pit inlet pipe should extend horizontally at least one foot into the pit with a tee or ell to divert flow downward to prevent washing and eroding of the sidewalls. If multiple pits are used, or in the event repair pits are added to an existing system, they should be connected in series.

Abandoned seepage pits should be filled with earth or rock.

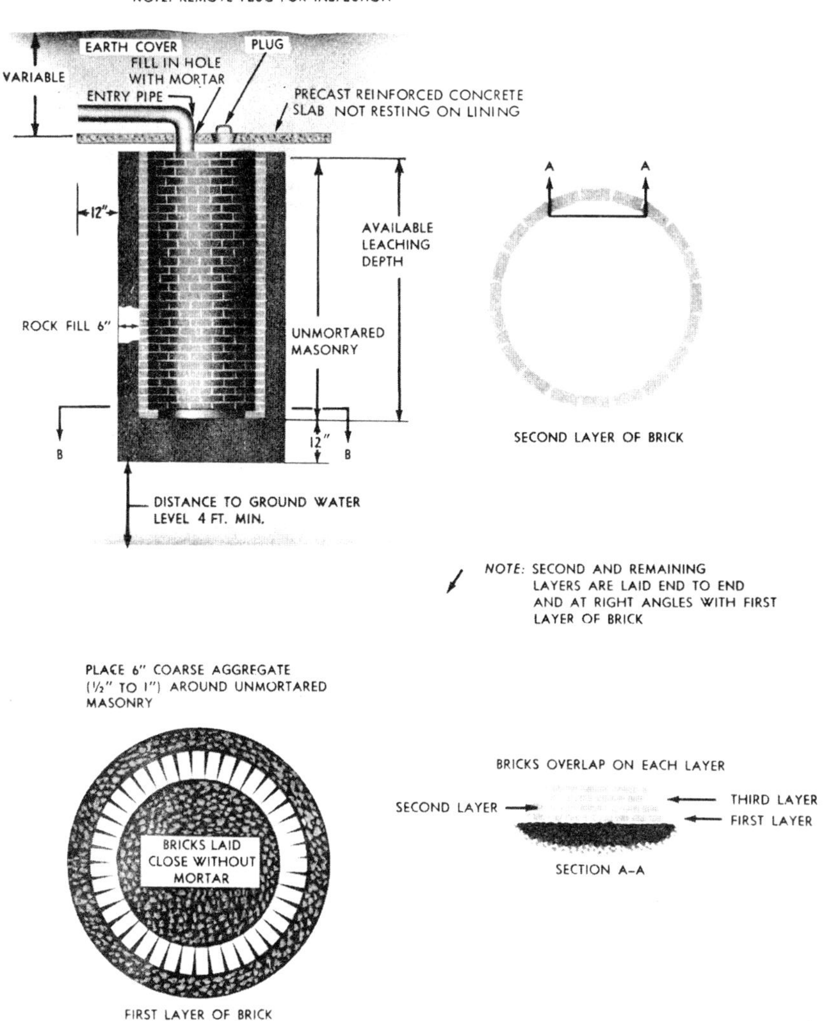

Fig. 10. Seepage pit.

SELECTION OF A SEPTIC TANK

Assuming that the lot will be large enough to accommodate one of the types of absorption systems, and that construction of the system is permitted by local authority, the next step will be selection of a suitable septic tank.

Functions of Septic Tanks

Untreated liquid household waste (sewage) will quickly clog all but the most porous gravel formations. The tank conditions sewage so that it may be more readily percolated into the subsoil of the ground. Therefore, the most important function of a septic tank is to provide protection for the absorption ability of the subsoil. The three functions that take place within the tank to provide this protection are as follows.

1. *Removal of solids.* Clogging of soil with tank effluent varies directly with the amount of suspended solids in the liquid. As sewage from a building sewer enters a septic tank, its rate of flow is reduced so that larger solids sink to the bottom or rise to the surface. These solids are retained in the tank, and the clarified effluent is discharged.

2. *Biological treatment.* Solids and liquid in the tank are subjected to decomposition by bacterial and natural processes. Bacteria present are of a variety called anaerobic which thrive in the absence of free oxygen. The decomposition or treatment of sewage under anaerobic conditions is termed "septic," hence the name of the tank. Sewage which has been subjected to such treatment causes less clogging than untreated sewage containing the same amount of suspended solids.

3. *Sludge and scum storage. Sludge* is an accumulation of solids at the bottom of the tank, while *scum* is a partially submerged mat of floating solids that may form at the surface of the fluid in the tank. Sludge and scum to a lesser degree, will be digested and compacted into a smaller volume. However, no matter how efficient the process is, a residual of inert solid material will remain. Space must be provided in the tank to store this residue during the interval between cleanings, otherwise sludge and scum will eventually be scored from the tank and may clog the disposal field.

If adequately designed, constructed, maintained, and operated septic tanks are effective in accomplishing their purpose.

The relative position of a septic tank in a typical subsurface disposal system is shown in Fig. 11. The liquid contents of the house sewer (A, Fig. 11) are discharged first into the septic tank (B, Fig. 11), and finally into the subsurface absorption field (C, Fig. 11).

The heavier sewage solids settle to the bottom of the tank, forming a blanket of sludge. The lighter solids, including fats

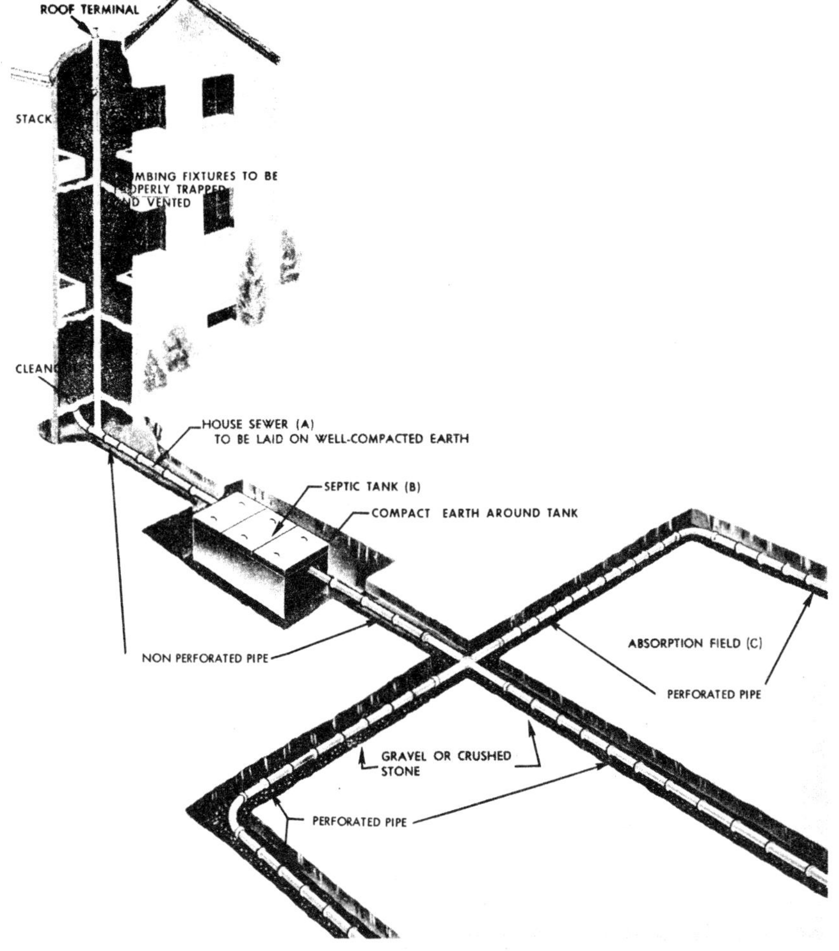

Fig. 11. Septic-tank sewage-disposal system.

and greases, rise to the surface and form a layer of scum. A considerable portion of the sludge and scum are liquefied through decomposition or digestion. During this process, gas is liberated from the sludge, carrying a portion of the solids to the surface, where they accumulate with the scum. Ordinarily, they undergo further digestion in the scum layer, and a portion settles again to the sludge blanket on the bottom. This action is retarded if there is much grease in the scum layer. The settling is also retarded because of gasification in the sludge blanket. Furthermore, there are relatively wider fluctuations of flow in small tanks than in the large units. This effect has been recognized in Table 6 which shows the recommended minimum liquid capacities of household septic tanks.

TABLE 6
LIQUID CAPACITY OF TANK (GALLONS).

[Provides for use of garbage grinders, automatic clothes washers, and other household appliances]

Number of bedrooms	Recommended minimum tank capacity	Equivalent capacity per bedroom
2 or less	750	375
3	900	300
4 [1]	1,000	250

[1] For each additional bedroom, add 250 gallons.

Location. Septic tanks should be located where they cannot cause contamination of any well, spring, or other source of water supply. Underground contamination may travel in any direction and for considerable distances, unless filtered effectively. Underground pollution usually moves in the same general direction as the normal movement of the ground water in the locality. Ground water moves in the direction of the slope or gradient of the water table, i.e., from the area of higher water table to areas of lower water table. In general, the water table follows the general contour of the ground surface. For this reason, septic tanks should be located downhill from wells or springs. Sewage from disposal systems occasionally contaminate wells having higher surface elevations. Obviously, the elevations of disposal systems are almost always higher than the level of water in such wells as may be located nearby; hence, pollution from a disposal system on a lower surface elevation may still travel downward to the water bearing stratum as shown in Fig. 12. It is necessary, therefore, to rely upon horizontal as well as vertical distances for protection. Tanks should never be closer than 50 feet from any source of water supply. Greater distances are preferred where possible.

Fig. 12. Pollution of well from sources with lower surface elevations.

The septic tank should not be located within five feet of any building, as structural damage may result during construction or seepage may enter the basement. The tank should not be located in swampy areas, nor in areas subject to flooding. In general, the tank should be located where the largest possible area will be available for the disposal field. Consideration should also be given to the location from the standpoint of cleaning and maintenance. Where public sewers may be installed at a future date, provision should be made in the household plumbing system for connection to such a sewer.

Effluent. Contrary to popular belief, septic tanks do not accomplish a high degree of bacteria removal. Although the sewage undergoes treatment in passing through the tank, this does not mean that infectious agents will be removed. Therefore, septic tank effluents cannot be considered safe. The liquid that is discharged from a tank is, in some respects, more objectionable than that which goes in; it is septic and malodorous (ill-smelling). This, however, does not detract from the value of the tank. As previously described, its primary purpose is to condition the sewage so that it will cause less clogging of the disposal field.

Further treatment of the effluent, including the removal of pathogens, is effected by percolation through the soil. Disease producing bacteria will, in time, die out in the unfavorable environment afforded by soil. In addition, bacteria are also removed by certain physical forces during filtration. This combination of factors results in the eventual purification of the sewage effluent.

Capacity. Capacity is one of the most important considerations in septic tank design. Studies have proved that liberal tank capacity is not only important from a functional standpoint, but is also good economy. The liquid capacities recommended in Table 6 allow for the use of all household appliances, including garbage grinders.

Specifications for Septic Tanks

Materials. Septic tanks should be watertight and constructed of materials not subject to excessive corrosion or decay, such as concrete, coated metal, vitrified clay, heavyweight concrete blocks, or hard burned bricks. Properly cured precast and cast-in-place reinforced concrete tanks are believed to be acceptable everywhere. Steel tanks meeting Commercial Standard 177-62 of the U.S. Department of Commerce are generally acceptable. Special attention should be given to job-built tanks to insure water tightness. Heavyweight concrete block should be laid on a solid foundation and mortar joints should be well filled. The interior of the tank should be surfaced with two ¼-inch (.635-cm.) thick coats of portland cement sand plaster. Some typical septic tanks are shown in Fig. 13. Suggested specifications for watertight concrete are given in Appendix IV.

Precast tanks should have a minimum wall thickness of three inches, and should be adequately reinforced to facilitate handling. When *precast slabs* are used as covers, they should be watertight, adequately reinforced, and have a thickness of at least three inches. All concrete surfaces should be coated with a bitumastic or similar compound to minimize corrosion.

General. Backfill around septic tanks should be made in thin layers thoroughly tamped in a manner that will not produce undue strain on the tank. Settlement of backfill may be done with the use of water, provided the material is

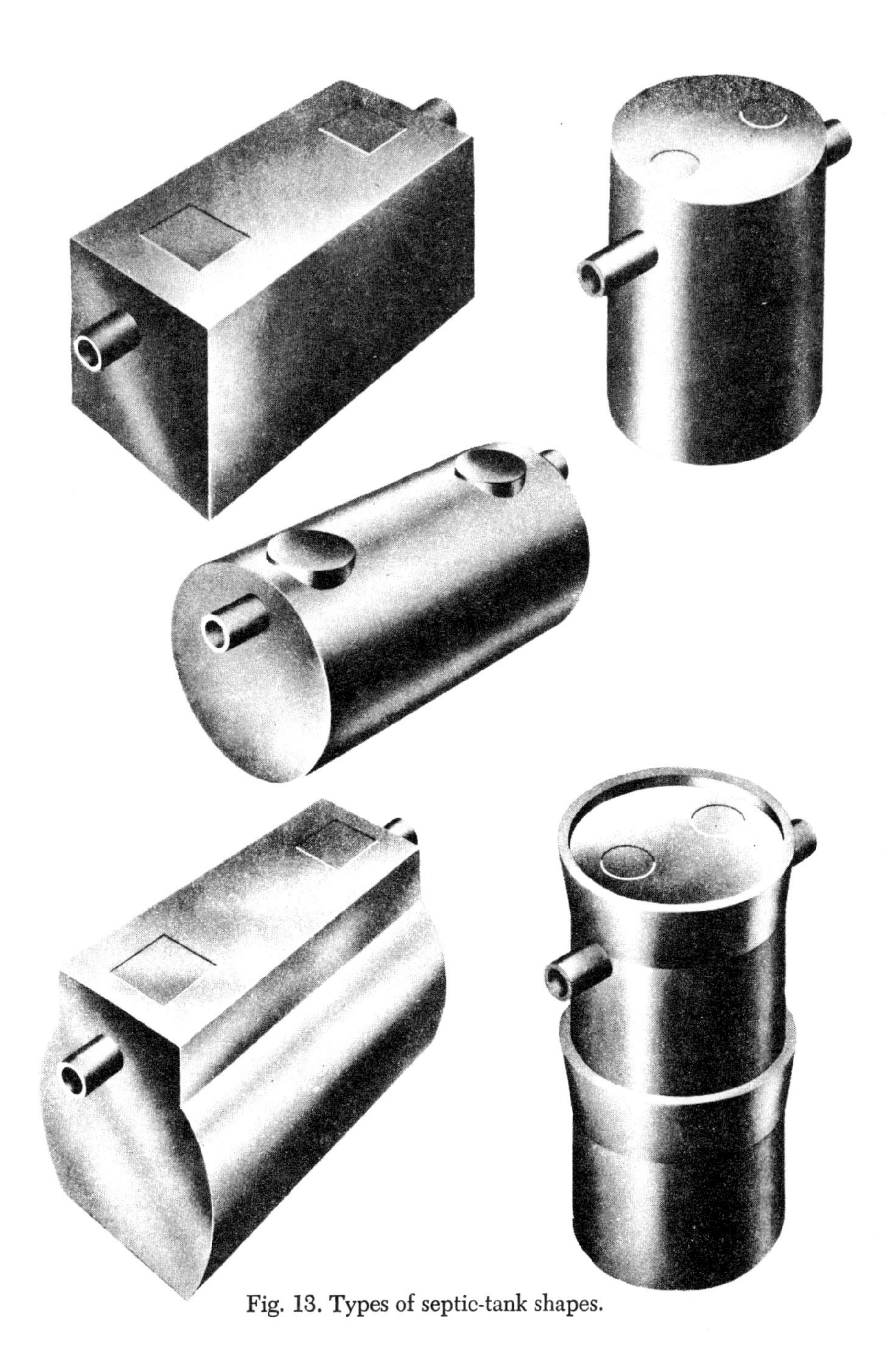

Fig. 13. Types of septic-tank shapes.

thoroughly wetted from the bottom upwards and the tank is first filled with water to prevent floating.

Adequate access must be provided to each compartment of the tank for inspection and cleaning. Both the inlet and outlet devices should be accessible. Access should be provided to each compartment by means of either a removable cover or a 20-inch manhole in least dimension. Where the top of the tank is located more than 18 inches below the finished grade, manholes and inspection holes should extend to approximately eight inches below the finished grade (Fig. 14), or can be extended to finished grade if a seal is provided to keep odors

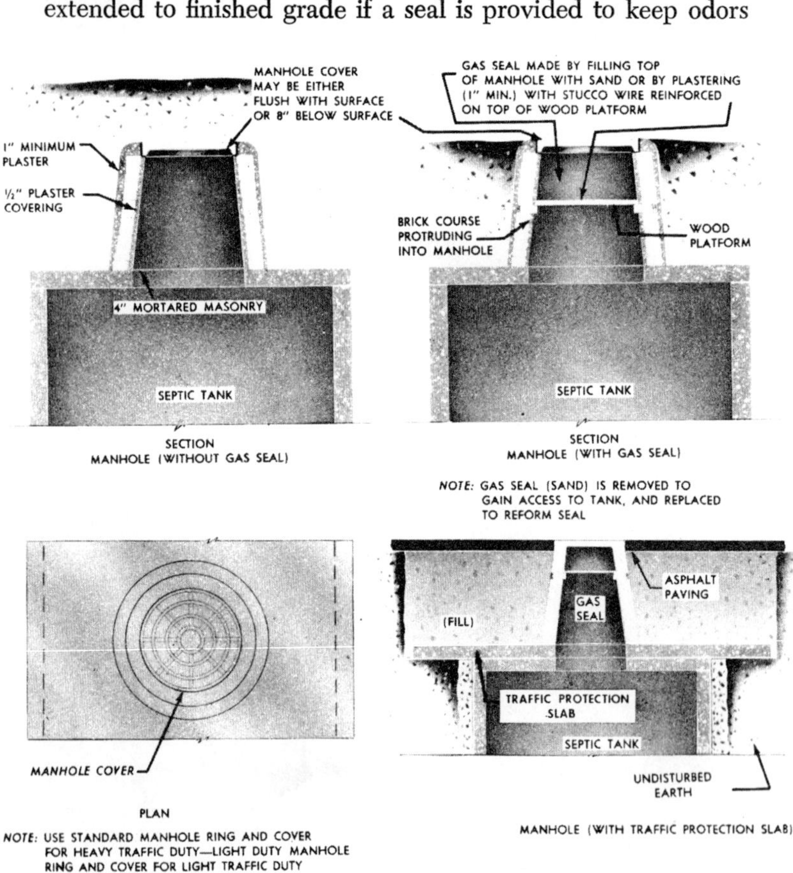

Fig. 14. Design of manholes.

from escaping. In most instances, the extension can be made using clay or concrete pipe, but proper attention must be given to the accident hazard involved when manholes are extended close to the ground surface. Typical single and double compartment tanks are shown in Figs. 15 and 17.

Inlet. The inlet invert should enter the tank at least three inches above the liquid level in the tank, to allow for momentary rise in liquid level during discharges to the tank.

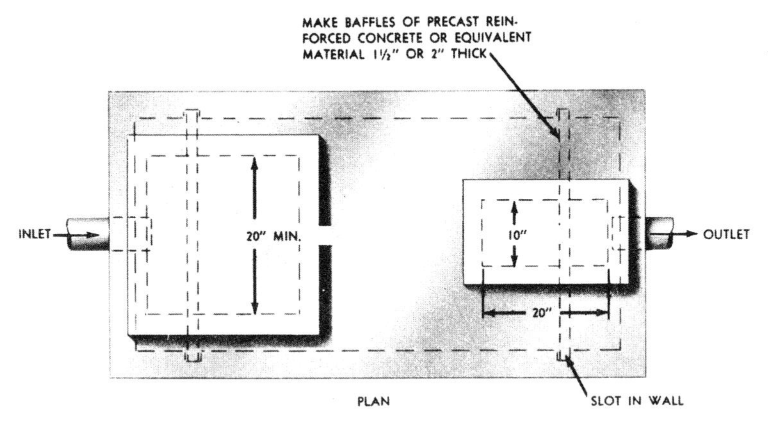

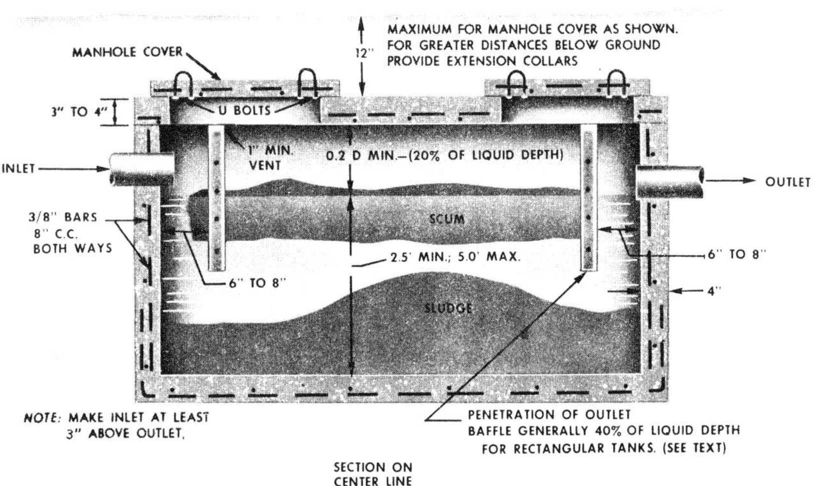

Fig. 15. Household septic tank.

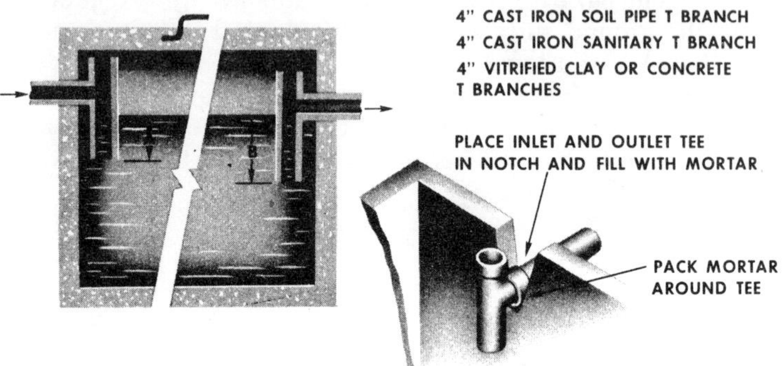

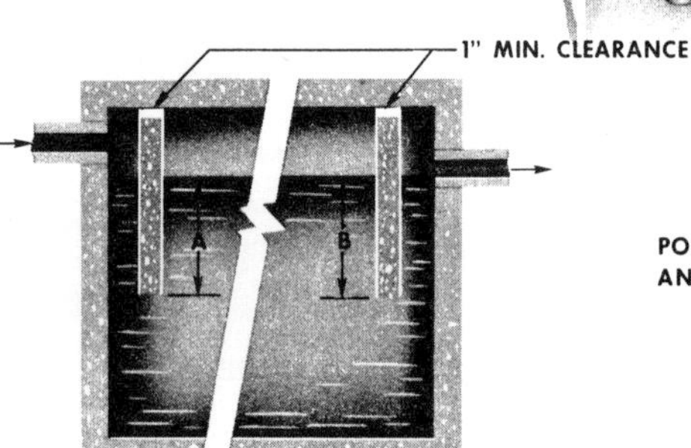

SEMI-CIRCULAR BAFFLE

PREFABRICATED CONCRETE , CLAY
TILE OR STEEL

D
D
A
B

NOTE: "A" SHOULD BE NO LESS THAN 6"
AND NO GREATER THAN B.

"B" PENETRATION OF OUTLET DEVICE
GENERALLY 40% OF LIQUID DEPTH FOR
TANKS WITH VERTICAL SIDES & 35% FOR
HORIZONTAL CYLINDERS TANKS.

SECTION D-D

Fig. 16. Types of inlet and outlet devices.

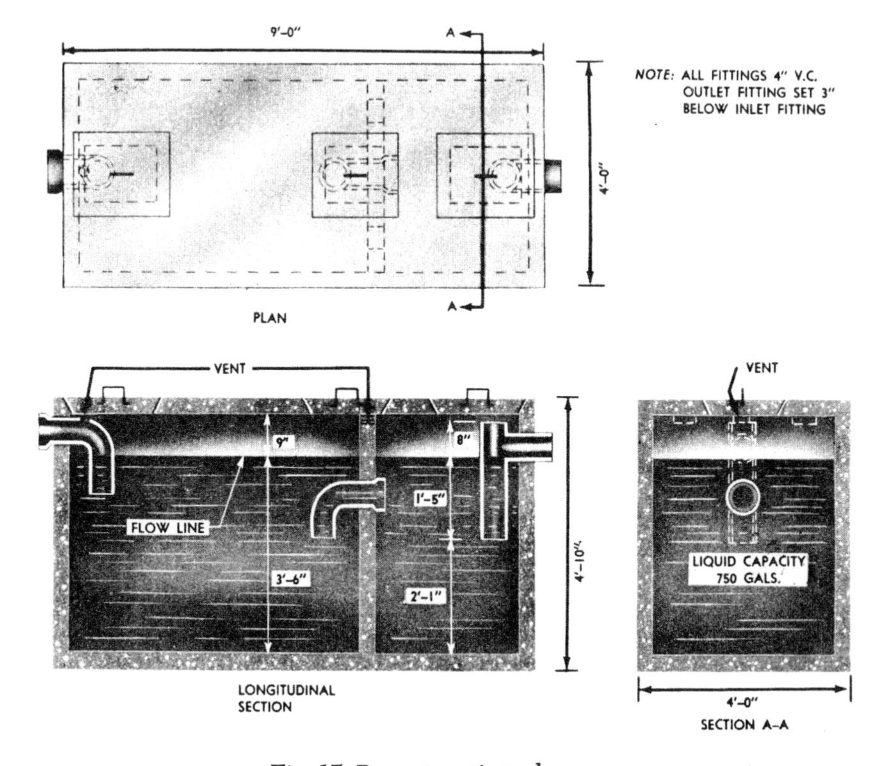

Fig. 17. Precast septic tank.

This free drop prevents backwater and stranding of solid material in the house sewer leading to the tank.

A *vented inlet tee* or *baffle* should be provided to divert the incoming sewage downward. It should penetrate at least six inches below the liquid level, but in no case should the penetration be greater than that allowed for the outlet device. A number of arrangements commonly used for inlet and outlet devices are shown in Fig. 16.

Outlet. It is important that the outlet device penetrate just far enough below the liquid level of the septic tank to provide a balance between sludge and scum storage volume; otherwise, part of the advantage of capacity is lost. A vertical section of a properly operating tank would show it divided into three distinct layers—scum at the top, a middle zone free of solids (called "clear space"), and a bottom layer of sludge. The outlet device retains scum in the tank, but at the same time, it limits the amount of sludge that can be accommodated without scouring, which results in sludge discharging in the effluent from the tank. Studies indicate that the outlet device should generally extend to a distance below the surface equal to 40 percent of the liquid depth. For horizontal, cylindrical tanks, this should be reduced to 35 percent. For example, in a horizontal cylindrical tank having a liquid depth of 42 inches, the outlet device should penetrate $42 \times .35 = 14.7$ inches below the liquid level.

The outlet device should extend above the liquid line to approximately one inch from the top of the tank. The space between the top of the tank and the baffle allows gas to pass off through the tank into the house vent.

Tank Proportions. The available data indicate that for tanks of a given capacity, shallow tanks function as well as deep ones. Also, for tanks of a given capacity and depth, the shape of a septic tank is unimportant. However, it is recommended that the smallest plan dimension be at least two feet. Liquid depth may range between 30 and 60 inches.

Storage Above Liquid Level. Capacity is required above the liquid line to provide for that portion of the scum which floats above the liquid. Although some variation is to be expected, on the average about 30 percent of the total scum will accumulate above the liquid line. In addition to the provision for scum storage, one inch is usually provided at the top of the tank to permit free passage of gas back to the inlet and house vent pipe.

For tanks having straight, vertical sides, the distances between the top of the tank and the liquid line should be equal to approximately 20 percent of the liquid depth. In horizontal, cylindrical tanks, area equal to approximately 15 percent of the total circle should be provided above the liquid level. This condition is met if the liquid depth (distance from outlet invert to bottom of tank) is equal to 79 percent of the diameter of the tank.

Use of Compartments. Although a number of arrangements are possible, compartments, as used here, refer to a number of units in series. These can be either separate units linked together, or sections enclosed in one continuous shell as shown in Fig. 17, with watertight partitions separating the individual compartments.

A single compartment tank will give acceptable performance. The available research data indicate, however, that a two-compartment tank, with the first compartment equal to one-half or two-thirds of the total volume, provides better suspended solids removal which may be especially valuable for protection of the soil absorption system. Tanks with three or more equal compartments give at least as good performance as single-compartment tanks of the same total capacity. Each compartment should have a minimum plan dimension of two feet with a liquid depth ranging from 30 to 60 inches.

An access manhole should be provided to each compartment. Venting between compartments should be provided to allow free passage of gas. Inlet and outlet fittings in the compartmented tank should be proportioned as for a single tank. (*See* Fig. 16.) The same allowance should be made for storage above the liquid line as in a single tank.

MAINTENANCE OF A SEPTIC TANK

Cleaning

Septic tanks should be cleaned before too much sludge or scum is allowed to accumulate. If either the sludge or scum approaches too closely to the bottom of the outlet device, particles will be scored into the disposal field and will clog the system. Eventually, when this happens, liquid may break through to the ground surface, and the sewage may back up in the plumbing fixtures. When a disposal field is clogged in this manner, it is not only necessary to clean the tank, but it also may be necessary to construct a new disposal field.

The tank capacities given in Table 6 will give a reasonable period of good operation before cleaning becomes necessary. There are wide differences in the rate that sludge and scum will accumulate from one tank to the next. For example, in one case out of twenty, the tank will reach the danger point, and should be cleaned, in less than three years. Tanks should be inspected at least once a year and cleaned when necessary.

Although it is difficult for most homeowners, actual inspection of sludge and scum accumulations is the only way to determine definitely when a given tank needs to be pumped. When a tank is inspected, the depth of sludge and scum

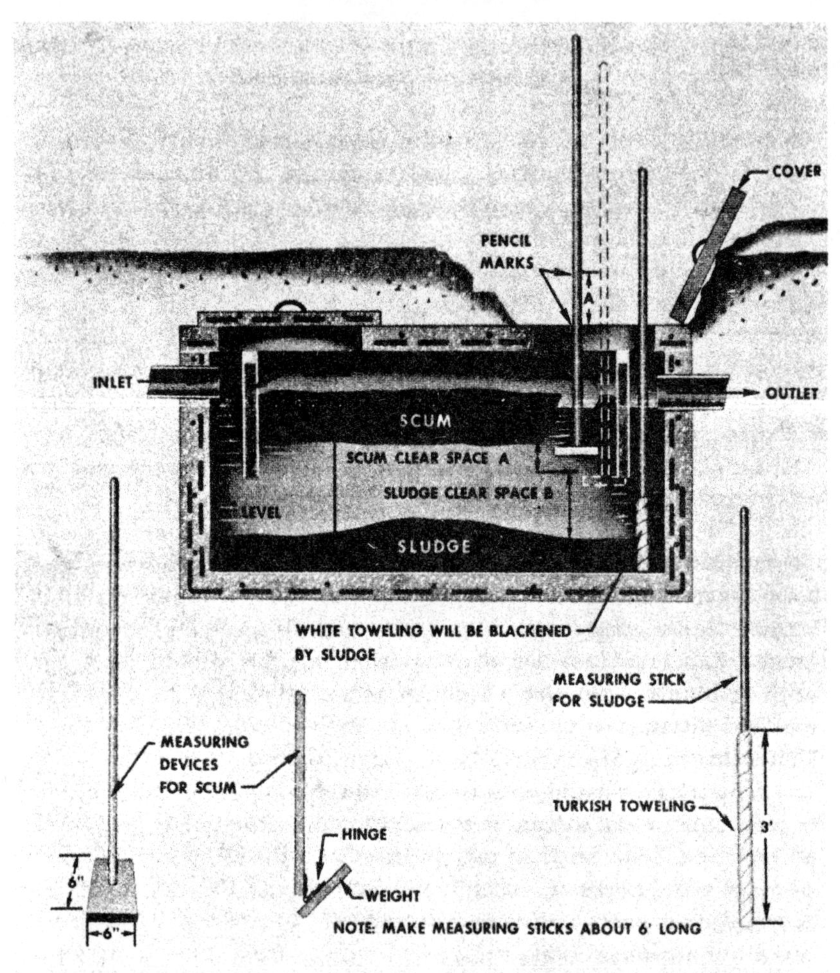

Fig. 18. Devices for measuring sludge and scum.

should be cleaned if either: (1) The bottom of the scum mat is within approximately three inches of the bottom of the outlet device; or (2) sludge comes within the limits specified in Table 7. (*See* Fig. 18.)

TABLE 7
ALLOWABLE SLUDGE ACCUMULATION.

Liquid capacity of tank, gallons [a]	Liquid depth			
	2½ feet	3 feet	4 feet	5 feet
	Distance from bottom of outlet device to top of sludge, inches			
750	5	6	10	13
900	4	4	7	10
1,000	4	4	6	8

[a] Tanks smaller than the capacities listed will require more frequent cleaning.

Scum can be measured with a stick to which a weighted flap has been hinged, or with any device that can be used to feel out the bottom of the scum mat. The stick is forced through the mat, the hinged flap falls into a horizontal position, and the stick is raised until resistance from the bottom of the scum is felt. With the same tool, the distance to the bottom of the outlet device can be found (Fig. 18).

A long stick wrapped with rough, white toweling and lowered to the bottom of the tank will show the depth of sludge and the liquid depth of the tank. The stick should be lowered behind the outlet device to avoid scum particles. After several minutes, if the stick is carefully removed, the sludge line can be distinguished by sludge particles clinging to the toweling.

In most communities where septic tanks are used, there are firms which conduct a business of cleaning septic tanks. The local Health Department can make suggestions on how to obtain this service. Cleaning is usually accomplished by pumping the contents of the tank into a tank truck. Tanks should not be washed or disinfected after pumping. A small residual of sludge should be left in the tank for seeding purposes. The material removed may be buried in uninhabited places or, with permission of the proper authority, emptied into a sanitary sewer system. It should never be emptied into storm drains or discharged directly into any stream or watercourse. Methods of disposal should be approved by the health authorities.

When a large septic tank is being cleaned, care should be taken not to enter the tank until it has been thoroughly ventilated and gases have been removed to prevent explosion hazards or asphyxiation of the workers. Anyone entering the tank should have one end of a stout rope tied around his waist, with the other end held above ground by another person strong enough to pull him out if he should be overcome by any gas remaining in the tank.

Grease interceptors. Grease interceptors (grease traps) are not ordinarily considered necessary on household sewage disposal systems. The discharge from a garbage grinder should never be passed through them. The septic tank capacities recommended in this chapter are sufficient to receive the grease normally discharged from a home.

Chemicals

The functional operation of septic tanks is not improved by the addition of disinfectants or other chemicals. In general, the addition of chemicals to a septic tank is not recommended. Some proprietary products which are claimed to "clean" septic tanks contain sodium hydroxide or potassium hydroxide as the active agent. Such compounds may result in sludge bulking and a large increase in alkalinity, and may interfere with digestion. The resulting effluent may severely damage soil structure and cause accelerated clogging, even though some temporary relief may be experienced immediately after application of the product.

Frequently, however, the harmful effects of ordinary household chemicals are overemphasized. Small amounts of chlorine bleaches added ahead of the tank may be used for odor control and will have no adverse effects. Small quantities of lye or caustics, normally used in the home, added to plumbing fixtures is not objectionable as far as operation of the tank is concerned. If the septic tanks are as large as herein recommended, dilution of the lye or caustics in the tank will be enough to overcome any harmful effects that might otherwise occur.

Some 1200 products, many containing enzymes, have been placed on the market for use in septic tanks, and extravagant claims have been made for some of them. As far as is known, however, none has been proved of advantage in properly controlled tests.

Soaps, detergents, bleaches, drain cleaners, or other material, as normally used in the household, will have no appreciable adverse effect on the system. However, as both the soil and essential organisms might be susceptible to large doses of chemicals and disinfectants, moderation should be the rule. Advice of responsible officials should be sought before chemicals arising from a hobby or home industry are discharged in the systems.

Miscellaneous

It is generally advisable to have all sanitary wastes from a household discharge to a single septic tank and disposal system. For household installations, it is usually more economical to provide a single disposal system than two or more with the same total capacity. Normal household waste, including that from the laundry, bath, and kitchen, should pass into a single system.

Roof drains, foundation drains and drainage from other sources producing large intermittent or constant volumes of clear water should not be piped into the septic tank or absorption area. Such large volumes of water will stir up the contents of the tank and carry some of the solids into the outlet line; the disposal system following the tank will likewise become flooded or clogged, and may fail. Drainage from garage floors or other sources of oily waste should also be excluded from the tank.

Toilet paper substitutes should not be flushed into a septic tant to the homemaker in the kitchen than in any other room sticks may not decompose in the tank, and are likely to lead to clogging of the plumbing and disposal system.

Waste brines from household water softener units have no adverse effect on the action of the septic tank, but may cause a slight shortening of the life of a disposal field installed in a structured clay type soil.

Adequate venting is obtained through the building plumbing if the tank and the plumbing are designed and installed properly. A separate vent on a septic tank is not necessary.

A chart showing the location of the septic tank and disposal system should be placed at a suitable location in dwellings served by such a system. Whether furnished by the builder, septic tank installer, or the local health department, the charts should contain brief instructions as to the inspection and maintenance required. The charts should assist in apprising homeowners of the necessary maintenance which septic tanks require, thus forestalling failures by assuring satisfactory operation. The extension of the manholes or inspection holes of the septic tank to within eight inches of the ground surface will simplify maintenance and cleaning.

Abandoned septic tanks should be filled with earth or rock.

INSPECTION

After a soil absorption system has been installed and before it is used, the entire system should be tested and inspected. The septic tank should be filled with water and allowed to stand overnight to check for leaks. If leaks occur, they should be repaired. The soil absorption system should be promptly inspected before it is covered to be sure that the disposal system is installed properly. Prompt inspection before backfilling should be required by local regulations, even where approval of plans for the subsurface sewage disposal system has been required before issuance of a building permit. Backfill material should be free of large stones and other deleterious material and should be overfilled a few inches to allow for settling.

Chapter 6

What You Should Know About Plumbing

The fundamentals of good plumbing may be summed up in the word quality—quality in fixtures manufactured by firms of national reputation and quality in installation by experienced and reputable plumbing contractors. Trouble-free operation of the plumbing system is the test of quality; a beautiful, convenient, and admired kitchen or bathroom is the result.

In planning the plumbing for a new home, or for remodeling an old one, the choice of a plumbing contractor is one of the most important decisions which the prospective homeowner must make. It is wise to settle on who is to do the plumbing while plans are still in the dream stage. It is more satisfactory to deal directly with the master plumber and heating contractor, who will be able to offer helpful advice, than to have the plumbing and heating in the general contract.

The *master plumber* should be chosen as carefully as the family doctor, because good plumbing protects the health of you and your family.

The prospective homeowner, in planning the plumbing for a new home, should keep in mind that there are two parts to his or her plumbing system—the fixtures and the piping.

Plumbing *fixtures* are used in several rooms in the house such as the bathroom, the powder room or first-floor washroom, the kitchen, and the laundry or utility room. Sometimes a corner of a bedroom is furnished with a lavatory, providing a bedroom dressing-table or shaving unit; or a shower cabinet is placed outside the bathroom at some convenient place in the house.

The *piping* consists of all pipes leading to and away from the fixtures. Most of this piping (nearly 300 feet in the average home) is out of sight behind the walls and floors.

The *fittings*, such as faucets, showerheads, drains, valves, and the like, are the parts of the plumbing system which connect the piping and the fixtures.

The basis of all plumbing is *piping*. The piping system of a building may be compared to the veins and arteries of the human body. It is invisible, it does its work quietly and efficiently, yet it is absolutely vital to the physical well-being of the occupants of a house. (*See* Fig. 1.)

The *piping system* is divided into the supply piping and the waste piping. As indicated by the name, the *supply piping* brings the pure water to the fixtures and the *waste piping* removes the impure water.

It is important that all piping be adequate in diameter, durable, leakproof, that all joints be tight, and that the piping be pitched or slanted so that it can be drained. The experienced plumber will design a plumbing system with piping which not only meets all these specifications but whose capacity will

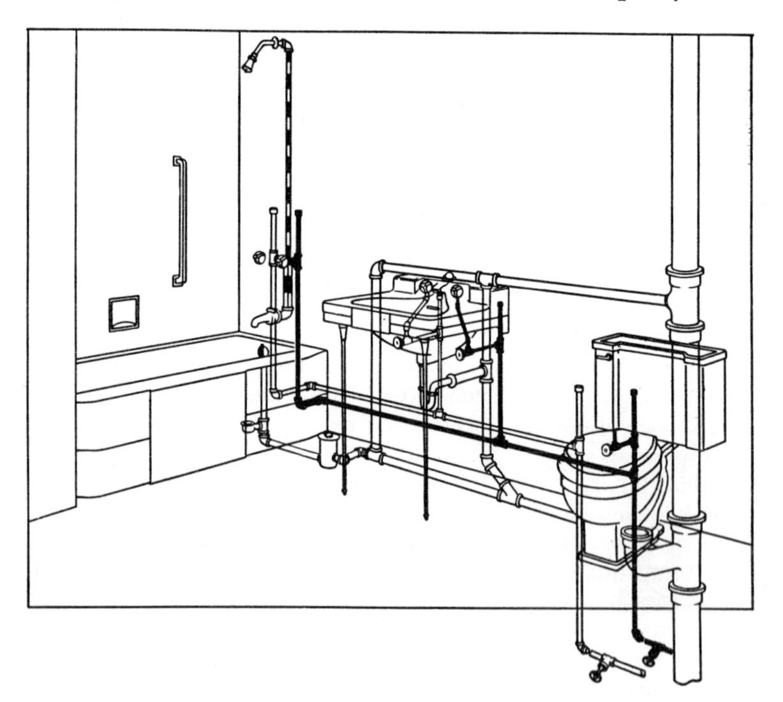

Fig. 1. Roughing-in sheet of bathroom installations.

be sufficient, at the very least, to permit the simultaneous use of those fixtures which experience indicates are likely to be used at the same time. Inadequate or poorly designed piping may cause vacuums in supply pipes, which are dangerous because they can create suction. Under certain conditions this may cause back-siphonage, that is, pollution of the drinking water with impure or waste water.

In planning to build a home on a city lot, the prospective homeowner should find out about the water supply of the city, and the sewage disposal, about charges for connecting the home plumbing system to both the city water supply and sewers, whether or not the city has installed municipal water softening equipment, and something about local plumbing and building codes. The plumbing contract, incidentally, should include a statement to the effect that the plumbing contractor guarantees all work to be done in accordance with local building codes.

THE BATHROOM AND POWDER ROOM

One of the important developments in homes of the future will be the ratio of more baths per room. In smaller homes, the convenience of a bath with every bedroom will be achieved by ingenious arrangement of plumbing fixtures to allow for multiple use of the bathroom.

In some homes there will be a compact bathroom and a separate powder room. In other homes, the same result may be obtained by putting the bathtub in a separate room and placing the lavatory and water closet in an adjoining room.

Another popular small-home plumbing arrangement has three dimensions—a bathtub and overhead shower in one room with a powder or washroom on either side. This layout is the equivalent of two bathrooms.

Privacy may also be obtained in the family bathroom by recessing some fixtures. For instance, the bathtub may be set in a recess and the panel at one end used to form a compartment, with or without door, for the water closet. The panel may be of opaque glass, inexpensive wallboard tile, or other material.

The bathroom with twin lavatories will add to living comfort and subtract from family friction (Fig. 2). Planning a

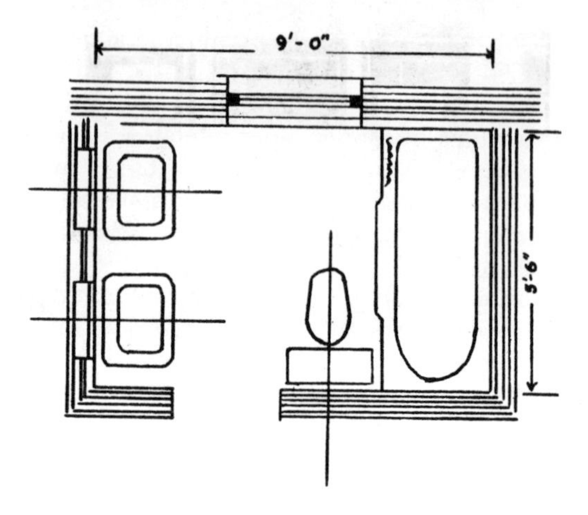

Fig. 2. Twin lavatories.

bathroom with two water closets, too, is a thoughtful provision. The extra one may be in a compartment while the other is in the bathroom proper.

Individual family schedules and preference should be consulted to select the multiple use family bathroom plan which best solves each family's problems. (*See* Figs. 3, 4, 5, 6, and 7.)

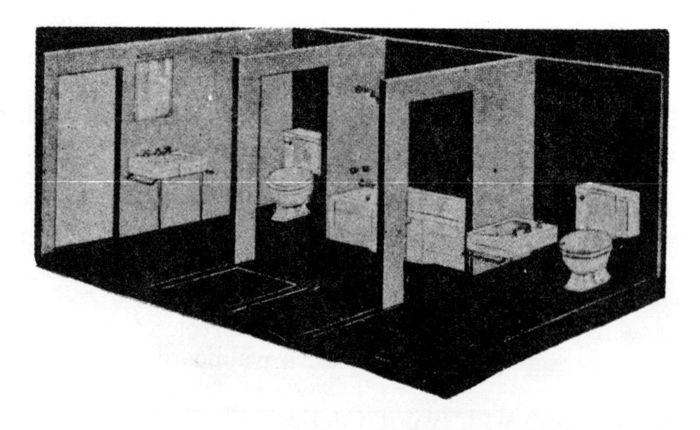

Fig. 3. Double duty bath.

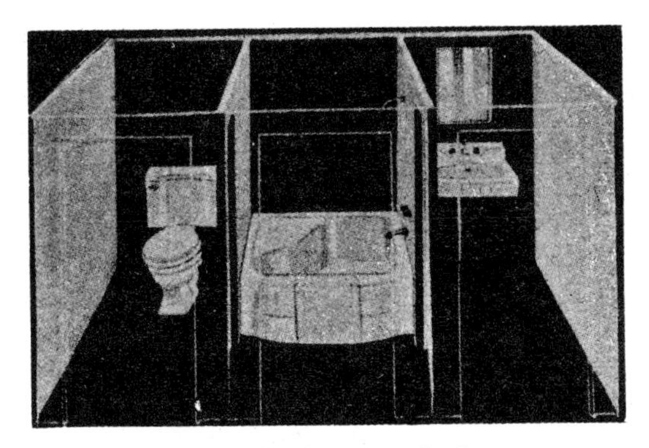

Fig. 4. The three-way bath.

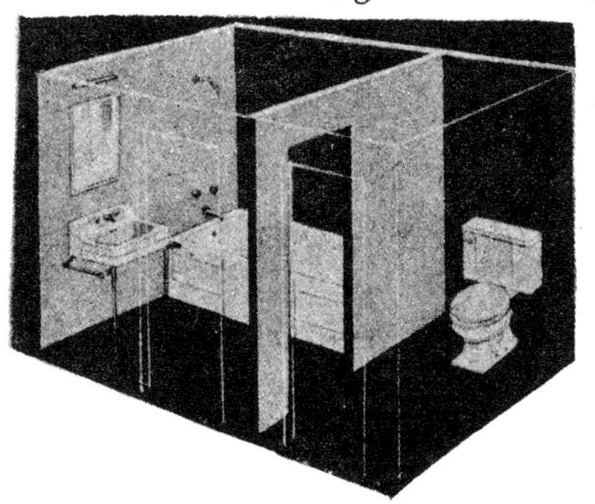

Fig. 5. The toilet compartment.

Fig. 6. The corner lavatory.

Fig. 7. The bedroom lavatory.

Things to remember in planning the multiple-use bathroom include: Backing up of some fixtures to economize on behind-the-wall piping, placing the tub in an alcove to avoid drafts, placing the window over the radiator or convector, using extra wall space for built-in towel and medicine storage cabinets or a dressing table, placing a shower cabinet in a compartment or separate room. Cabinets of composition board may be built on either side of the medicine cabinet as well as either side of the lavatory. Above the water closet is another good place for a storage cabinet. There should be enough towel bars to serve all those using the family multiple bathroom.

Selection of the fixtures for the bathroom and powder room should be done carefully because plumbing fixtures are relatively permanent pieces of "furniture" and cannot be moved around at will after the family moves in.

It is wise to go "fixture-shopping," visiting showrooms of contractors, wholesalers, and manufacturers, and model homes, before making the final selection of fixtures.

Materials from which fixtures are made include porcelain enamel on cast iron, vitreous china, earthenware (vitreous glazed), and sheet steel.

Bathtub

All of the better tubs are now made with some type of built-in seat. There is a choice between a seat integral with the rim or built-in across the end of corners. Tubs are approximately five feet long, or four feet square.

The bathtub should be selected with due regard to its dual function, namely as a fixture for bathing and as a receptor for a shower. Strength and rigidity are foremost considerations. However, other important factors in the selection of bathtubs are durability and ease of cleaning.

The better tubs are equipped with a raised edge on the side toward the wall and ends, which tends to prevent seepage of water. A bathtub hanger, an iron device built into the wall, will prevent the danger of seepage, too, and also prevent a bathtub from settling with shrinkage of the house.

Safety features are important in the selection of a bathtub. Preference should be given to the low bathtub with steep sides and ends. It is easier to get in and out of the bathtub with low sides. Important, too, from the standpoint of safety are grab-bars which may be attached to the wall above or to the bathtub. A convenient type is an **L**-shaped bar. The same effect may be achieved by installing two bars—one vertical and one horizontal. Shower compartments should be provided with vertical grab-bars. (*See* Figs. 8 and 9.)

Shower

No bathroom is complete without a shower, whether it be over the tub, in a separate compartment, or both. The shower compartment may be built in or a complete pre-fabricated leakproof metal cabinet unit. These shower cabinets may be installed any place in the home—in the bathroom, in an adjoining room, in the basement, attic, or garage. Shower cabi-

Fig. 8. Properly designed, adequately positioned, and substantially installed assistance accessories.

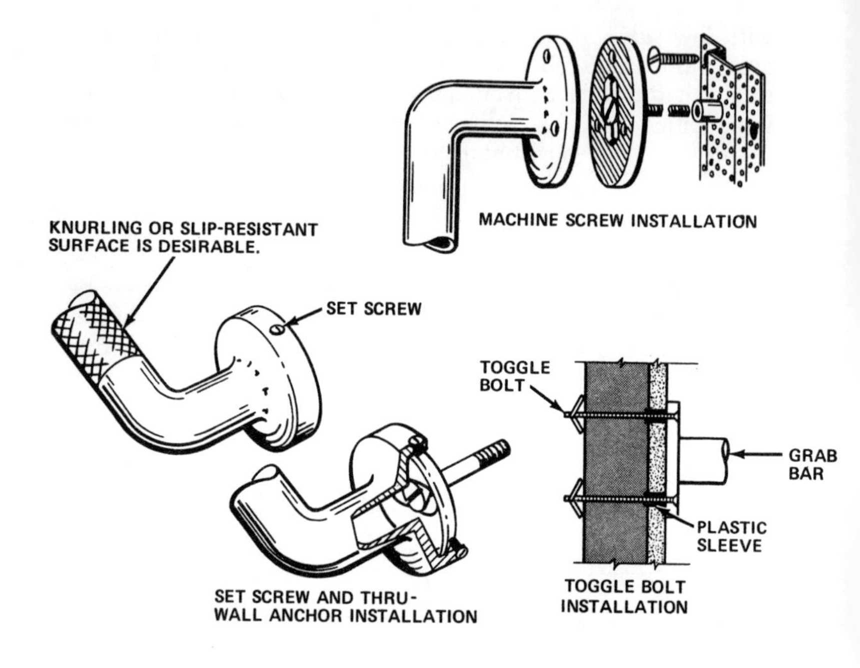

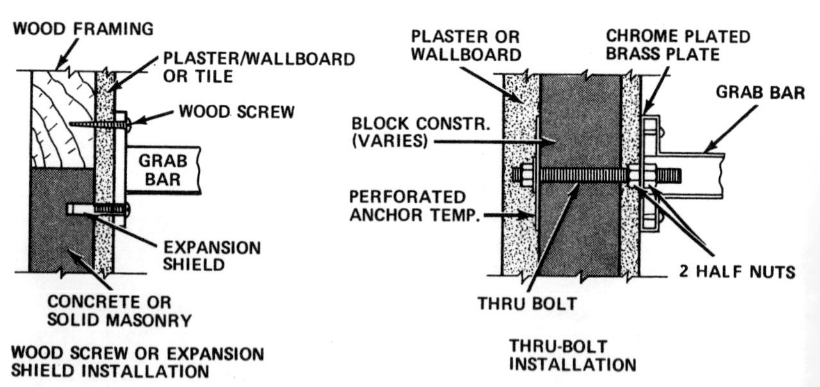

Fig. 9. Typical grab bar installations.

nets may have galvanized steel walls, enameled steel walls, glass walls with doors of metal or glass, or with curtains.

It is customary nowadays to install a shower over the tub in all bathrooms, particularly for family or multiple use.

Lavatory

Lavatories, sometimes called *washbasins*, are made in a variety of styles and sizes. They may be wall hung, or with chromium or glass legs. They may have towel bars integral with the legs, or attached to the sides. They may have shelves at the back or just wide slab surfaces. The position of the faucets will vary, as well as the shape of the basin itself.

Faucets in some lavatories are set at an angle in the shelf-back, others are in a recess. Some spouts are made integral with the lavatory while others are chromium like the faucets. The mixing valve faucet, which has handles for hot and cold water, and a single spout to allow tempering of water before using, is the preferred choice today.

To prevent back-siphonage, the spout on the lavatory is set well above the water level of the fixture. All bathtub spouts in modern tubs are placed in the wall above the tub for the same reason. Fittings for tub and lavatory should be selected in matching design.

The shape of lavatory basins has tended toward the oblong in recent years because this has been found to be more convenient than a round basin. Some lavatories have D-shaped basins. This shape is determined by the position of the arms when using a lavatory. Many lavatories are made with non-splash rims across the front.

Water Closet

The *water closet* should be selected with the utmost care because of its hygienic importance. Factors to be taken into consideration are quiet and efficient action, ease in cleaning, and pleasing design.

Sanitation and trouble-free operation are provided by a large passageway in the trap of the fixture, by rim holes of adequate size, and by rugged and dependable closet tank mechanism.

Functionally, there are three types of closets: siphon jet, reverse trap, and the washdown. The best is the *siphon jet* which has all of the desirable features of a good water closet. The next best is the *reverse trap*. The *washdown* is the lowest in cost.

Water closets are also divided as to means of flushing. By

water in a tank controlled by a *float valve*, or by a *flush valve* which may or may not be concealed in the wall.

An important recent innovation in exterior water closet design is the one-piece unit with low tank. This style and the close-coupled model may be placed away from the wall, which facilitates cleaning and prevents condensation on walls. Other styles of tank closets are placed against the wall. There is a new type of flush valve water closet recently developed for use in the small house. The valve mechanism is concealed by a streamlined housing with a molded rubber bumper on the center which, by pressure from the raised seat cover, operates the valves.

Faucets and Fittings

Faucets, valves, showerheads, and drain controls are the operating parts of plumbing fixtures. They are subjected to constant wear and should be selected for durability and service first, and design second. Better valves and faucets are made with renewable seats, which can easily be refitted.

There are three types of water controls for showers. The thermostatically controlled mixing valve, which provides water at the desired temperature and protects against danger from violent fluctuations in water; the mixing valve, which tempers the water; and the two-valve system, where the bather by turning the hot and cold faucets tempers the water.

Lighting and Accessories

To provide adequate illumination in the bathroom, there should be lighting fixtures around the medicine cabinet, over the tub, in the shower compartment, and a central light which should be controlled by a switch convenient to each door. A small night light is a convenience for children and also in case of illness.

One of the newer developments in medicine cabinets is the illuminated wing cabinet. Indirect lighting is cleverly concealed behind each wing so as to shine through frosted portions of the wing mirrors. Perfect illumination is given to the face from all angles. Another type of medicine cabinet has tubular lights installed in the cabinet on either side of the mirror door. Another interesting development is the round seamless cabinet without dirt-catching corners, and with a handsome round mirror.

Medicine cabinets should be as large and roomy as the site of the bathroom will permit. The better cabinets are made of seamless porcelain or seamless baked enamel trimmed with chromium-plated brass with a mirrored door.

It is a good idea to provide a separate cabinet well out of the reach of the children where harmful medicines may be kept. This cabinet may be a regular medicine cabinet, or it may be a built-in cupboard with door. Wall cabinets of wood or composition board may also be used to solve this problem.

With the perfection for home use of fluorescent lighting, it will find a wider use in homes because fluorescent light is cool, soft and diffused.

Mirrors in the bathroom may be on the medicine cabinet, bathroom door, or shower door, or on a convenient wall space. At least one mirror should be low enough for children. A wall mirror over the lavatory, extending high enough for the grown-ups and low enough to reach the edge of the lavatory at the wall, will not only be convenient for the youngsters, but will also protect the wall at the back of the lavatory from splashing.

THE KITCHEN

Time-saving equipment and arrangement are more important to the homemaker in the kitchen than in any other room of the house. That is because her daily schedule includes a majority of tasks involving the use of the kitchen.

In planning the kitchen for tomorrow's home, or for remodeling an old one, the first step is to decide on the basic type. Will it be a compact kitchen with separate dining-room or dinette, a larger kitchen to include a dinette or lunch bar, or a combination dining-room and kitchenette?

There are many interesting arrangements for kitchens. Perhaps one of the great questions of the day for homemakers with new kitchens in the making is whether to have the old-fashioned activities-center kitchen with modern equipment or a scientific type of kitchen workshop and laboratory. Either may be charming and attractive. Either may follow the same basic arrangement.

There are three general kitchen plans: Straight wall or

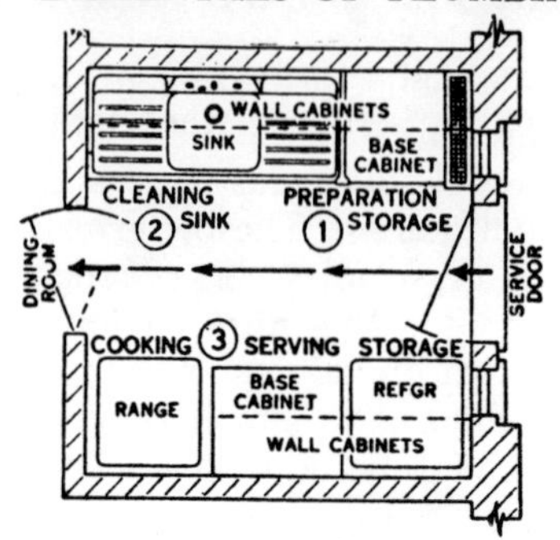

Fig. 10. The "corridor" type kitchen.

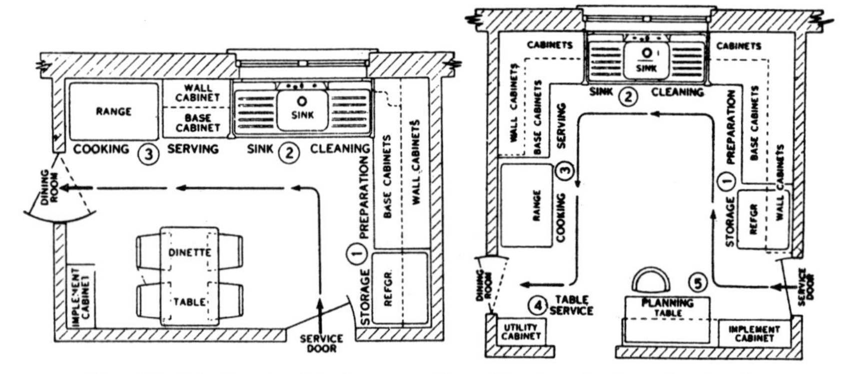

Fig. 11. The L-plan kitchen. Fig. 12. The ideal U-plan kitchen.

corridor type (Fig. 10); L-shaped (Fig. 11); and U-shaped (Fig. 12). The three major pieces of equipment in the kitchen are the sink, range, and refrigerator, and it is the arrangement of these in relation to the wall space that decides the basic plan.

More light in the kitchen is an outstanding trend. Corner windows of plain or structural glass provides more light and offer interesting possibilities for kitchen planners. The sink should be placed near or under the window.

Sink

The *sink* is one of the three most important pieces of equipment in the kitchen. The first consideration, therefore, in planning an efficient, time-saving kitchen is the selection of the sink. Both food preparation and the cleaning away processes involve the use of the sink, and this means more than half of the homemaker's time in the kitchen is spent at this fixture.

Preference is for the sink with two compartments, one deeper than the other. Choice may also be made between a sink with single or double drainboards, or none, and with some kind of shelfback. The sink faucet should have a swinging spout and a disappearing rinsing hose. Sink faucets are made of long wearing chromium-plated brass.

Sinks may be made of durable cast iron enameled with a beautiful glossy finish in white or color, pressed steel, or of ceramic materials with a hard glossy finish.

An extra sink with single basin may be specified in the kitchen when the laundry room is in the basement.

In building a new kitchen, it is important to plan to have the kitchen sink, cabinets, lunch bar, built-in dinette, stove, and refrigerator specified or built at height and widths which are most convenient for those who will use the equipment. Cabinets and sinks are built to fit together.

THE UTILITY OR LAUNDRY ROOM

While some laundry rooms have moved upstairs next to the kitchen, most prospective builders or some homeowners will want them in the basement.

The *first floor laundry room* is usually referred to as the *utility room* and contains not only the laundry equipment but the domestic hot water and house heating equipment. The utility room is usually found in small basementless houses. Modern developments in compact hot water heating boilers and radiators have made possible the installation of heating equipment on the first floor.

Partitioning the basement so that there is a laundry room separate from the general basement space is a good idea. Because of the improvements in heating which have reduced and streamlined the size of central heating equipment, most home builders will be able to have a basement recreation room. Therefore a separate laundry room is desirable. With the laundry and recreation room in the basement, it is well to plan for a basement washroom containing lavatory and closet.

The arrangement of the equipment in the home laundry, or

utility room, as in the kitchen, should be in accordance with the work flow. Sorting space should be near the laundry trays. The washer, stove, and drier are arranged next in order of use.

For the laundry room of the small home, two compartment laundry trays may be in one piece, or two separate trays. Laundry tubs of the right height and right depth with hard, durable, easily cleaned surfaces, rounded corners, and handy swinging faucets of chromium-plated brass are the answer to how-to-make-washday-pleasant.

Whether the laundry room is in the basement or part of the first floor utility room, it offers a logical place for the hot water heater. Because hot water is indispensable to the health and physical well-being of every member of the family, it is important to select water heating equipment that will be adequate to provide a plentiful supply of hot water whenever it is required.

Hot Water Heater

Equipment differs with respect to the fuel used and the method used. Fuel may be gas, oil, or coal.

Modern hot water boilers are equipped with built-in coils called indirect heaters for the heating of domestic water. There are three types of gas-fired water heaters: automatic, semi-automatic, or manually operated. Small coal heaters, called bucket-a-day heaters, may also be used to heat the domestic hot water. Storage tanks are necessary with indirect, coal, or gas-fired heaters.

Devices to regulate pressure and temperature should be installed on hot water or range boiler storage tanks. A temperature of 140 degrees is ample for all domestic uses.

The use of a tempering tank between the cold water service pipe and the water heater is recommended. The tempering tank, exposed to room or basement temperatures, will increase temperature of the water 10 to 15 degrees before it enters the heater, thus affording a saving in fuel consumption.

Special Equipment

In planning a house for the utmost livability and convenience, it may be desirable to include special equipment such as devices to keep the basement dry, to reduce water hammer, to prevent grease from entering waste pipes, to ventilate the kitchen, to soften the water, and to temper hot water.

CARE AND MAINTENANCE

Every homeowner should know how to take good care of his or her plumbing system. The first requisite is to have all plumbing work done under the supervision of a reputable plumbing contractor. This includes not only the original installation but all repair and maintenance work. Good workmanship is less expensive, safer, and more enduring than any other.

THE HOME BEYOND THE WATER MAINS

The prospective homeowner planning to build a home in the country or just beyond the city limits will want the same convenience of running water that is available in the city. Today this is possible by installing an *electric water system* and a *septic tank.*

The modern electric water system is entirely automatic and should require only an occasional inspection to check mechanical conditions and state of lubrication. In such systems, the pumps are automatically started and stopped, automatically oiled and even the proper air supply in the storage tank is automatically controlled. A modern electric water system with all these automatic features provides a water supply that is as dependable as that received from any large municipal water plant.

It is particularly important in the case of rural or suburban homes with their own water systems and septic tanks to use *grease interceptors* because grease interferes with the action of bacteria in septic tanks. Otherwise sinks should be drained into a separate catch basin and not into the septic tank.

IMPORTANCE OF PLANNING AHEAD

So many things must be taken into consideration in planning the plumbing for a new house or modernizing the plumbing in an old one, that it is wise to start making plans with the architect and plumbing contractor well in advance.

While the magic of color makes the difference between a house and a home, color belongs to the interior decoration scheme and may be changed but quality in design and installation must be built into the home to provide that background of durability which insures a satisfactory dwelling.

Plumbing equipment is evolutionary, not revolutionary, and changes in design come slowly. Plumbing fixtures of quality maintain a high standard of beauty, smartness, and sanitation for many years. Quality fixtures insure low repair and maintenance costs. Homes that are well-planned will be livable and up-to-date for the lifetime of the house.

Chapter 7

Planning Your Bathroom

You can have bathroom areas in your new or remodeled home that provide maximum family convenience and give satisfactory service for many years. It is all possible if you—

1. Plan carefully.
2. Insist on good workmanship.
3. Use the best materials you can afford.

The right kind of bath areas add greatly to the livability of a home. Prospective homeowners give high priority to conveniently located, nicely equipped, and attractively decorated bathrooms. Adding a bathroom or converting existing space into a bath area is a remodeling project undertaken by many families.

Whether you are building or remodeling, it is a good idea to plan the decorating scheme for your bath areas early, before you order bathroom fixtures. This is particularly true if you are using colored fixtures.

Bathroom fixtures, counters, cabinets, and floor coverings are costly. You want these permanent furnishings to be as pleasing in 5, 10, or even 20 years, as they are now. So take plenty of time and shop widely before you make your final selections. You can vary the decorative effect of a bathroom inexpensively from time to time by changing wall color, curtains, and accessories.

Begin your overall planning by considering all the ways a bath area will be used. The family bathroom, in particular, deserves careful study.

When you are installing a new bathroom or remodeling an old one be sure to—

1. Comply with plumbing codes, regulations, and guides that will insure a safe and satisfactory installation.
2. Choose an experienced person to install your bathroom.

3. Have an agreement in writing with whoever is installing your plumbing fixtures. This agreement should include price, general descriptions of fixtures and materials to be furnished, and a statement that places the liability for an unsatisfactory installation or damaged fixtures on the installer.

MODERN BATHROOM PLANS

Modern plumbing fixtures and colorful accessories may make the bathroom beautiful, but experience has shown that to continue in favor a bathroom must also be efficient in use. Therefore a bathroom must be well-planned and the fixtures properly installed under the supervision of a qualified plumber.

The bathroom plan should be suitable to the house and to the family who will use the bathroom and the fixtures should conform to these requirements. Quality fixtures are available from plumbing contractors or dealers in various price ranges.

The plans in this chapter have been developed to serve a three-fold purpose for the prospective homeowner or modernizer: (1) Suggest good plans for family bathrooms; (2) show economical combinations of rooms or fixtures; (3) indicate how it is possible to obtain bathroom efficiency in minimum areas.

Bathroom planners should always remember that it is advisable to have the bathroom as large as the family budget and the type of house will permit.

While the plans shown in this chapter have been carefully checked by architects and master plumbers, home planners should discuss the plans for their bathrooms and powder rooms with their own plumbing contractors and architects to see that the plans fit the home they are building or modernizing.

The scale of the plans shown is ¼ inch (.635 cm.) to one foot (30.48 cm.).

FAMILY BATHROOMS

Recommended size for family bathroom (Fig. 1) is 6 feet by 8 feet. All fixtures are on one wall, and window is over the radiator.

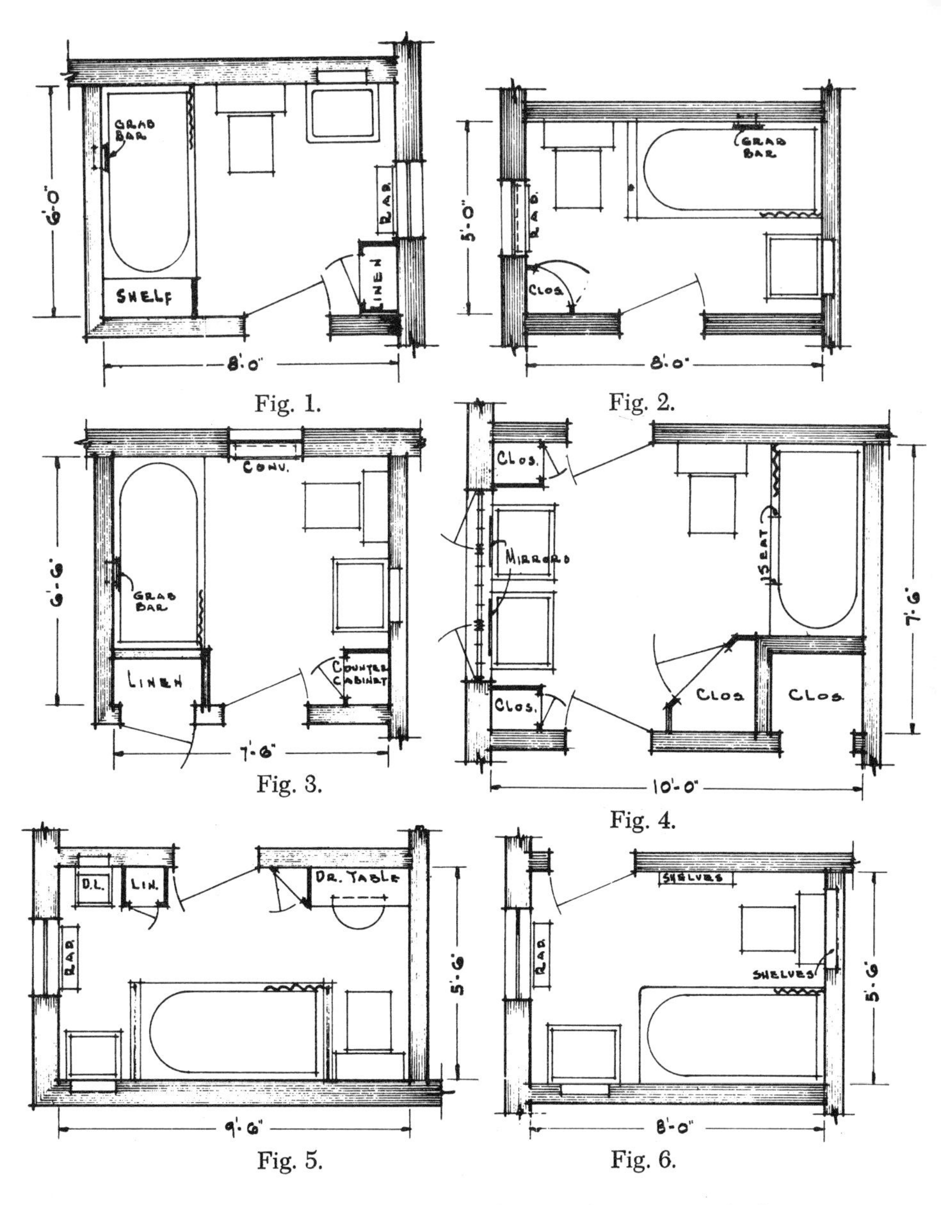

Fig. 1.

Fig. 2.

Fig. 3.

Fig. 4.

Fig. 5.

Fig. 6.

Figures 2 and 3 show popular fixture placements which use less space, and more behind-wall plumbing.

Twin lavatories and twin entrances make a fine double duty family bath (Fig. 4).

For efficiency, recess one or more fixtures, use the dental lavatory, and have shelves over the water closet (Figs. 5 and 6).

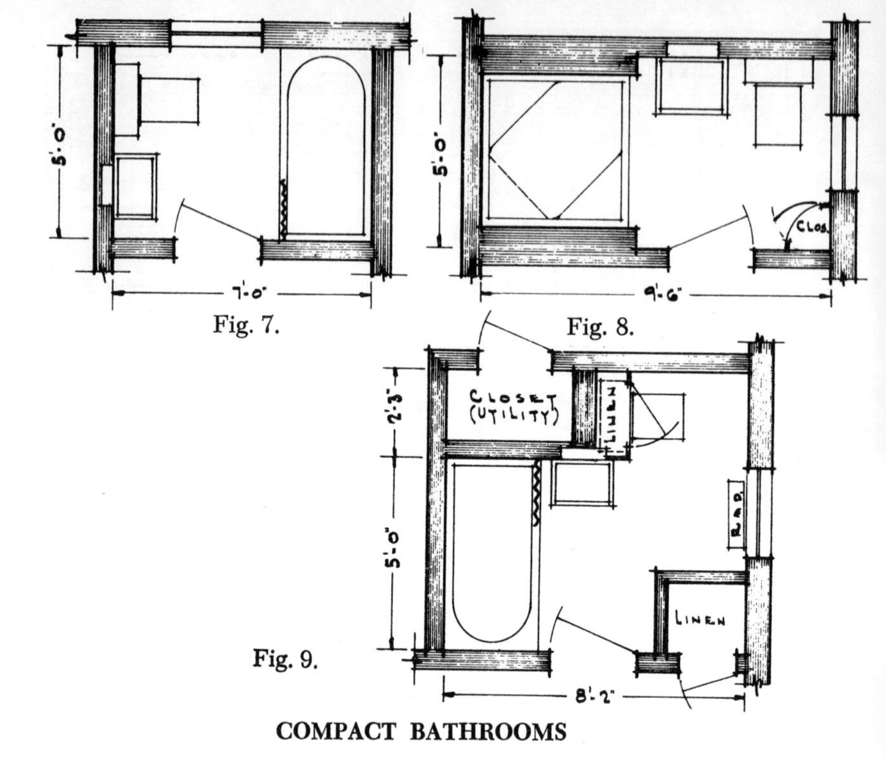

Fig. 7.

Fig. 8.

Fig. 9.

COMPACT BATHROOMS

Space-saving fixtures, such as the lavatory shown in Fig. 7, allow sufficient room for a compact family bath.

Square tub is practical for a long narrow room (Fig. 8). Figure 9 shows a recessed water closet under shelves in an odd space.

MINIMUM BATHROOMS

Figures 10, 11, and 12 show minimum bathroom sizes that are inadequate for family baths, but illustrations show what is possible in a small space.

Recommendation of architects is 28-inch doors which open into bath. Doors in these plans are 24 inches.

Space between fixtures should be six inches. For economy, as shown, less space may be allowed.

Small receptor bathtub, two possible small size lavatories shown in Fig. 12 are space-saving fixtures, useful in a minimum area.

Another space-saving fixture is the corner shower cabinet (Fig. 13).

The minimum plans shown in Figs. 13, 14, and 15 will be helpful in planning modernization of unused or odd spaces

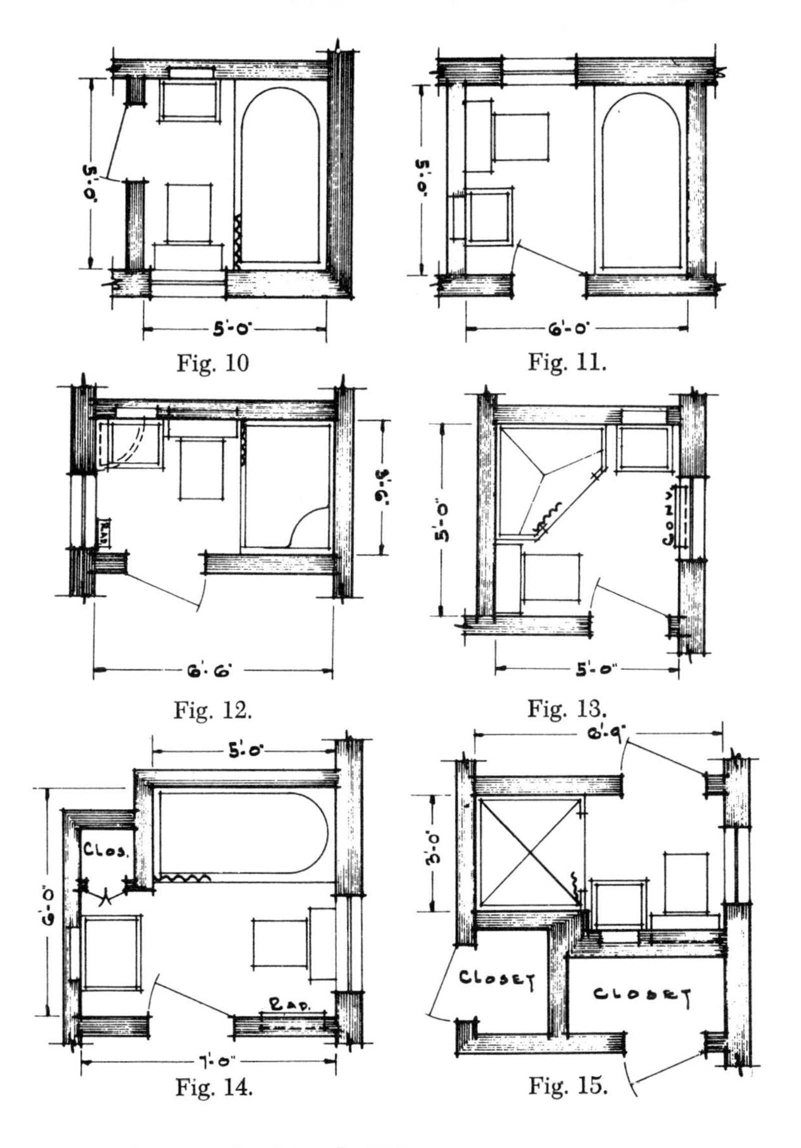

Fig. 10 Fig. 11. Fig. 12. Fig. 13. Fig. 14. Fig. 15.

to provide extra plumbing facilities.

Partially extend dimensions for clever space use as shown in Fig. 14.

Figure 15 shows how closets can be used as insulation.

COMPARATIVE ARRANGEMENTS

For convenience in use, bathrooms must be well-planned and the fixtures installed under the direction of a qualified master plumber. Good planning involves the proper arrangement of the fixtures in the allotted space and for the family who will use the room. (*See* Figs. 16 and 17.) Figure 16 illustrates a poor plan from the point of view of the designer of plumbing systems and the user: A fixture is installed on each wall which means more behind-the-wall piping and increased cost; the window is over the tub which is drafty for the bather and inconvenient when curtains are to be hung, windows raised or washed, or screens put up; requires water resistant curtains; steam may cause woodwork to swell; lavatory is compact when space-saving is not necessary; door opens out of room; view through door should be of lavatory or tub, not water closet. Note how Fig. 18 remedies all these faults, and adds convenience: Dressing table, shelf, grab-bar, and radiator under the window.

Compare Figs. 17 and 19. Figure 19 illustrates a better use of the space for powder room: Front door does not provide view into the room and more room for use of fixtures is allowed.

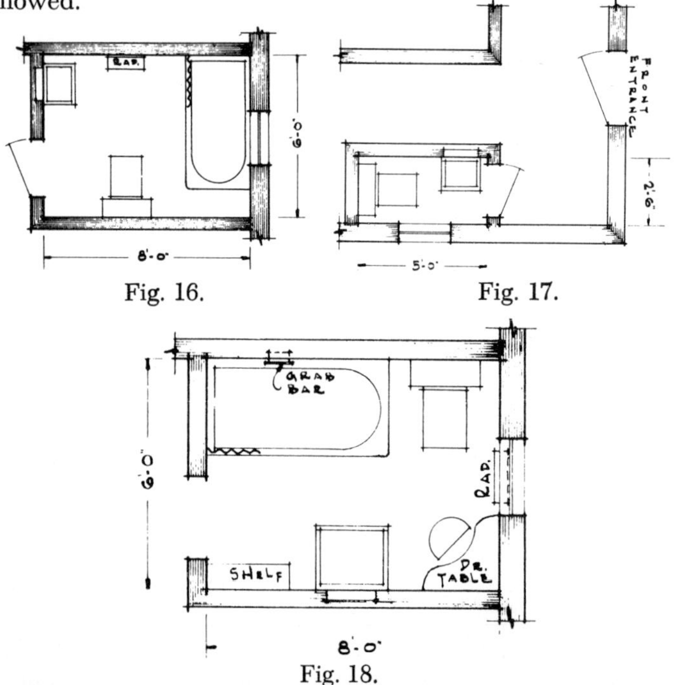

Fig. 16.

Fig. 17.

Fig. 18.

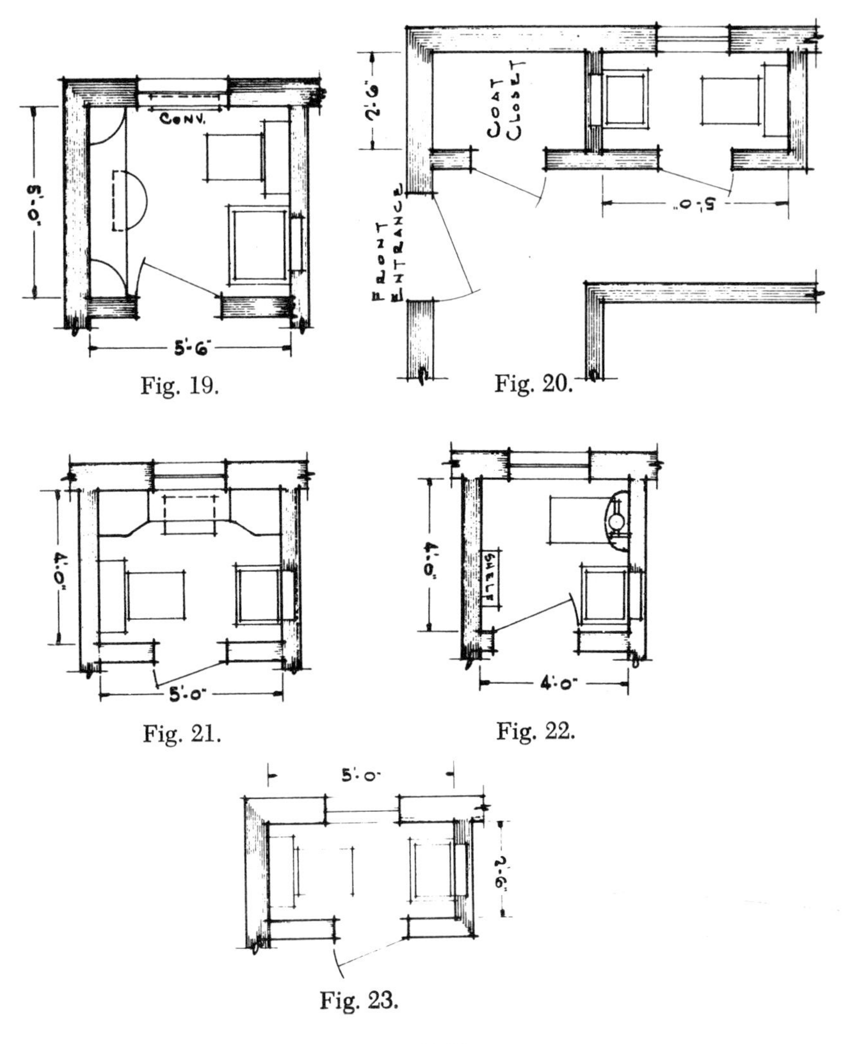

Fig. 19. Fig. 20. Fig. 21. Fig. 22. Fig. 23.

POWDER ROOMS

A powder room saves many steps. In a two-story home a first-floor washroom will help prevent stairway accidents. It is possible to have the convenience of a powder room in a small area, but it is better to allow more space. Guests like a dressing table, and mothers with children like enough room for two. (*See* Figs. 20, 21, 22 and 23.)

Figure 22 shows a space-saving small home flush valve water closet.

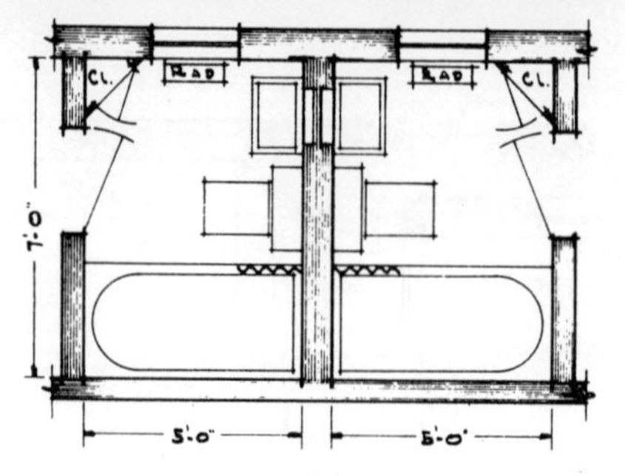

Fig. 24.

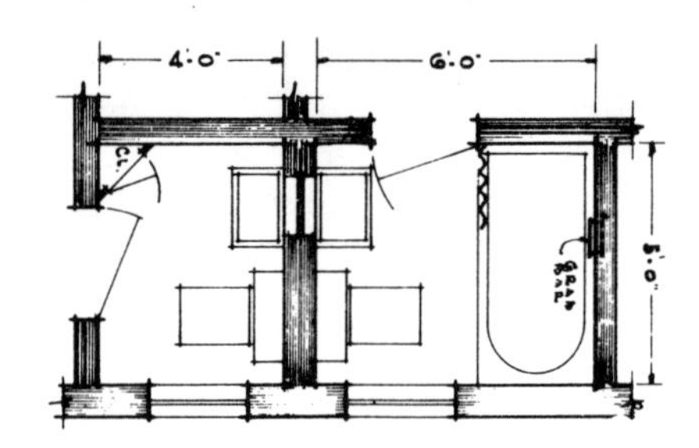

Fig. 25.

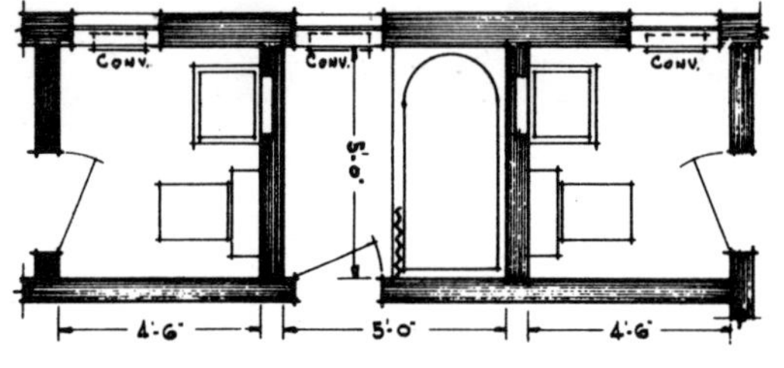

Fig. 26.

COMBINATIONS

Planning the home so that bathrooms or bathroom and powder room are next to each other enables plumbers to back up the fixtures and requires less behind-the-wall piping. Backing lavatory to lavatory, water closet to water closet is recommended, but an 8-inch wall partition is necessary in order to have medicine cabinets over each lavatory and provide for the stack or soil pipe. If the wall partition is narrow, it is advisable to back water closet to lavatory. (*See* Figs. 24 and 25.)

An in-a-line arrangement creates the same efficiency as three bathrooms (Fig. 26). It is ideal for large families, also assures privacy for guests.

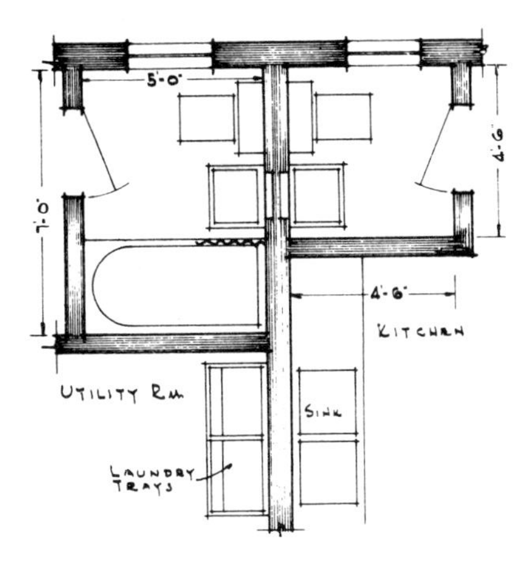

Fig. 27.

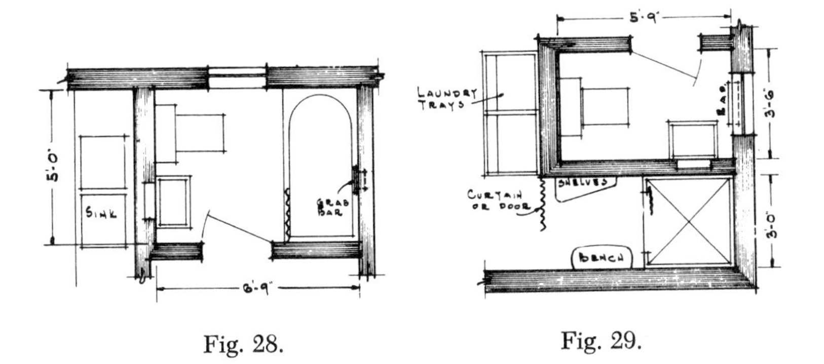

Fig. 28.

Fig. 29.

Figure 27 suggests how all plumbing may be centrally located for economical installation.

Backing up of the bathroom and the kitchen plumbing fixtures is an economical arrangement for a one-story house, or where there is a first-floor powder room or bathroom (Fig. 28). If the sink is placed under a window, a popular location, keep the sink as near the bathroom piping as possible for economy of installation. In a two-story house, the bathroom should be located above the kitchen sink for piping economy.

Plumbing facilities are a convenience in the basement. They should be placed near the laundry trays. Basement shower facilities are a wise investment for families who have amateur mechanics or mechanics, gardeners, or budding athletes (Fig. 29). This plan is easily adapted to other locations in the home for modernization.

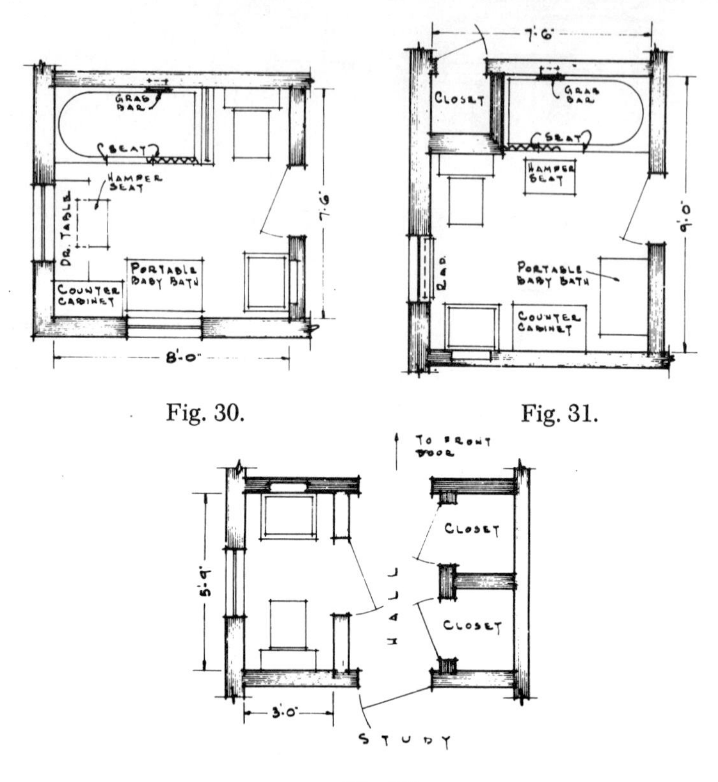

Fig. 30.

Fig. 31.

Fig. 32.

Few bathrooms are adaptable to growing families. No bathroom was ever too big—so, whenever possible, plan to have as large a bathroom as the budget will allow. Extra spaces to accommodate baby baths should be provided. Have counter-high cabinets near this space. Have a tub with seat ledge and a hamper seat, for this is useful when dressing small children. Have a safety grab-bar for the child to use in getting out of the tub. Have the radiator recessed under the window for safety. Have towel bars on lavatory, convenient for all. (*See* Figs. 30 and 31.)

A powder room near the front door and adjacent to the study, the den, or the library will prove a godsend in the case of visiting relatives, sickness or overnight guests (Fig. 32). This unit arrangement makes possible an extra bedroom (studio couch or daybed converted to bed) with its own half-bath, providing privacy. It is an ideal plan for use in modernizing older homes or for the expandable home. The tub may be added later in the closet area.

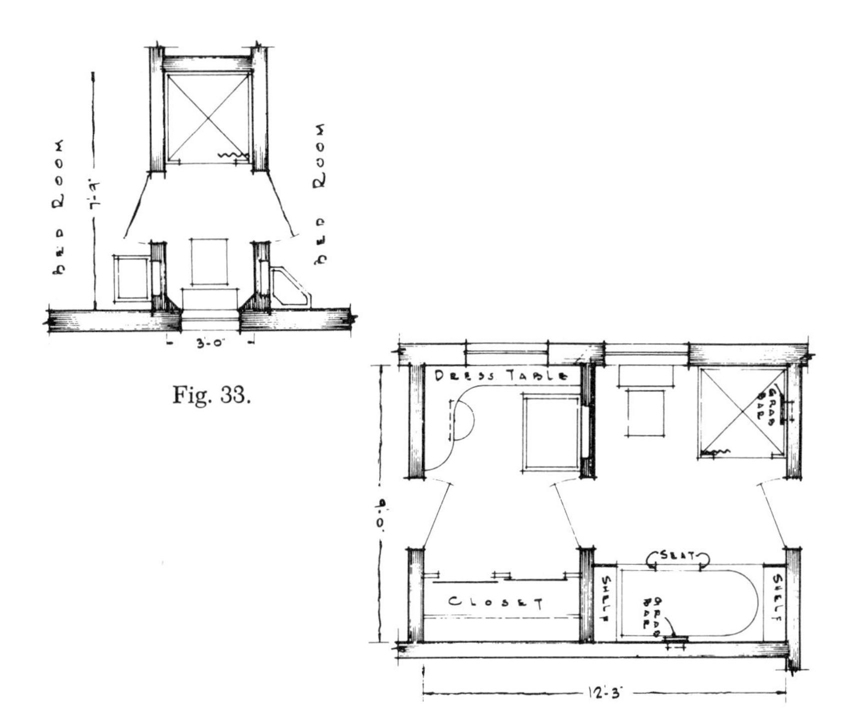

Fig. 33.

Fig. 34.

Lavatories in bedrooms complete this bathroom-between-bedroom arrangement (Fig. 33). This plan is efficient when modernizing for more plumbing facilities; it is also good for teen-agers' bathrooms.

Provision of a dressing room next to the bedroom is as acceptable in moderate homes as in the more luxurious type because this is an efficient use of space. Smaller bedrooms are possible because the dressing room contains much of the closet and storage area. It is easy to keep the bathroom and dressing room warm although bedroom windows may be open. Bathers need no longer be confined to a steamy bathroom while dressing, also they need no longer hold up bathroom traffic. (*See* Fig. 34.)

Variations of this bath-dressing room include: Placing the lavatory in the bathroom in place of the shower stall and having over-the-tub shower; having the lavatory in the bathroom proper and a second lavatory set in a counter-top dressing table. Sliding doors between the bath-dressing room are advisable. The dressing room may be smaller or larger, depending upon the number of persons who share the closet and storage area.

LOCATION OF BATH AREAS

Once you decide on the kind and the number of bath areas you need, the next step is to consider the best possible location for each.

Figure 35 illustrates a good location of a single bathroom in a one-story house that has no other toilet facilities. The bathroom can be reached from the back door without going through the work area of the kitchen and from the kitchen without going through the living room. It is located next to the utility room for a compact, economical plumbing arrangement that requires a short run of supply and waste pipes. The bathroom is accessible from all rooms through the hall. Another desirable feature is that the bathroom door is not visible from the living room or the front entrance.

Usually in a 1½- or 2-story house the bathroom is located on the second floor. But you might consider locating it on the first floor near the stairway for daytime convenience if there is no other wash-up area on the first floor.

For safety, avoid placing an upstairs bathroom at the head of the stairs or next to the stairs. If, however, this is the only possible location for the bathroom, install night lighting on the stairway, or a gate at the top of the stairs.

When more than one complete bathroom is planned for a

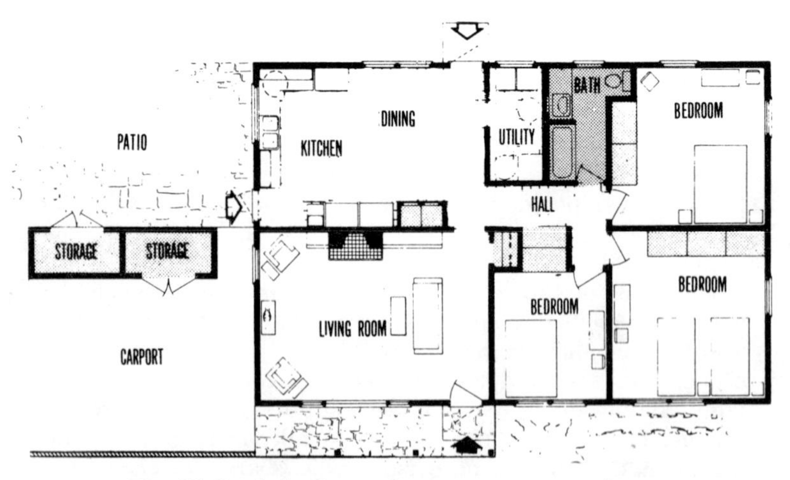

Fig. 35. Single bathroom location in one-story house.

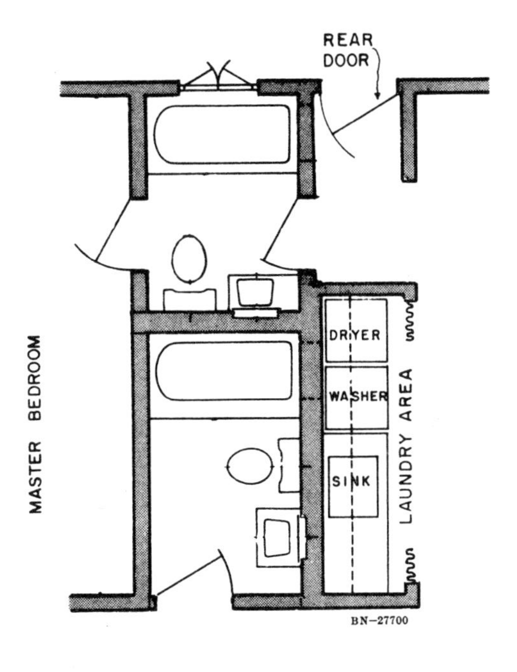

Fig. 36. Master bedroom bath arrangement.

home, the second frequently opens from the master bedroom. Such a bathroom can be located to serve a dual purpose. Figure 36 shows an arrangement in which the master bathroom, located conveniently near the rear entrance, is also the wash-up area. Note how the family bathroom, master bathroom, and laundry area are grouped together for an economical installation of plumbing.

Compartmented baths are popular with families with growing children. The addition of one or two fixtures and the multiple use of others add convenience and flexibility. In remodeling, a compartmented bath often makes the best use of space, particularly if a large area is being converted into a bathroom.

A single lavatory installed in a bedroom is one way to add convenience at a nominal cost. The lavatory can be enclosed or shielded by a screen. (*See* Fig. 37.)

Fig. 37. Extra lavatory in bedroom.

CHOICE OF FIXTURES

Bathroom fixtures are available at different price levels. The price depends on the material which the fixture is made of, and the size, color, and styling of the fixture.

Vitreous china is always used for toilets, and may be used for lavatories. *Porcelain enameled cast iron* and *pressed steel* are used for tubs and lavatories. All white and colored china and porcelained enamel fixtures now on the market are acid resisting.

In recent years, good quality *fiberglass* fixtures have become available. The gloss and color of the gel coat finish is similar in appearance to that of enameled or china fixtures and is resistant to ordinary household chemicals. Tub-shower units (Fig. 38) and shower stalls including the surrounding wall area are of leakproof one-piece construction.

Porcelain enameled cast iron and steel tubs are heavy. If remodeling, have joists checked by an experienced builder to

Fig. 38. Fiberglass tub-shower unit for one-piece construction.

make sure they will support the proposed installation. Fiberglass fixtures are comparatively lightweight and therefore lend themselves to remodeling where the structure may not support the heavier fixtures.

Built-in tubs of cast iron and enameled steel must be partially supported by the studs to prevent their pulling away from the wall. A 2- by 4-inch support secured to the studs or special hangers are used. Fiberglass tubs and shower stalls have nailing flanges and are nailed directly to enclosing stud partitions.

Wall-hung lavatories are supported by special brackets or hangers. China or metal legs can be added to some designs. Be sure the legs can be adjusted to fit the desired height of your lavatory.

Whatever type of lavatory you choose, be sure to install it at a comfortable height for the adults of the family. A height of 33 to 36 inches from the floor suits most adults.

Lavatory cabinet combinations usually come in two heights —31 inches and 34 inches. If you want the counter surface on a lavatory cabinet to be higher, you can increase the height of the toe space.

Toilets are classified according to the water action used. The three types most commonly installed in homes are siphon jet, reverse trap, and washdown.

The *siphon jet* is the most expensive, and has the quietest action of any of the three types. The trapway (E, Fig. 38a), located at the rear of the bowl, and the water surface are extra large for maximum cleanliness (A, Fig. 38a). A deep water seal (B, Fig. 38a) gives maximum protection from sewer gases. It has top supply and integral flushing rim and jet (F, Fig. 38a). It is available in plain or elongated rim (C, Fig. 38a), and in floor and wall type.

The *reverse trap* has the same water action as the siphon jet (F, Fig. 38b), but a smaller trapway (E, Fig. 38b), less water surface (A, Fig. 38b), and not as deep a water seal (B, Fig. 38b).

The *washdown*, the least expensive of the three types, has the trapway (E, Fig. 38c) at the front of the bowl. The flushing action is noisier, the water surface smaller (A, Fig. 38c), and the water seal (B, Fig. 38c) not as deep as in the other two types. This type is not suitable for elongated rim (C, Fig. 38c), and is available in floor type only.

Details of size, style, and color of bathtubs, lavatories, and toilets need to be checked with your local dealer.

Showers and Fittings

The most economical way to provide a shower is to add a shower head over the tub. If the shower fittings are installed

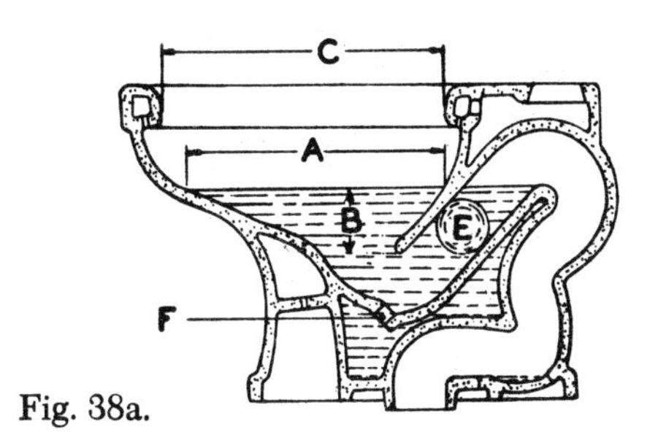

Fig. 38a.

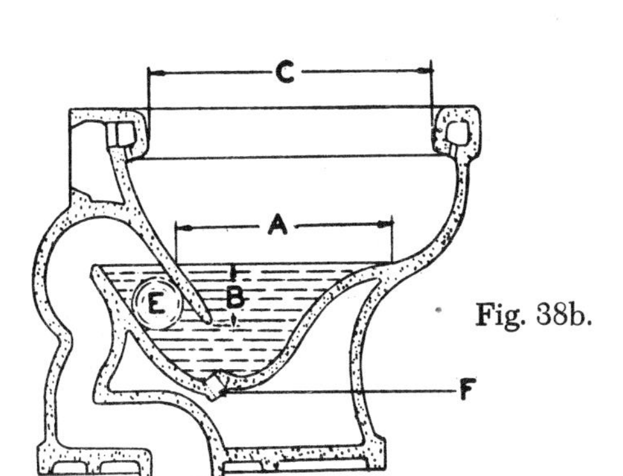

Fig. 38b.

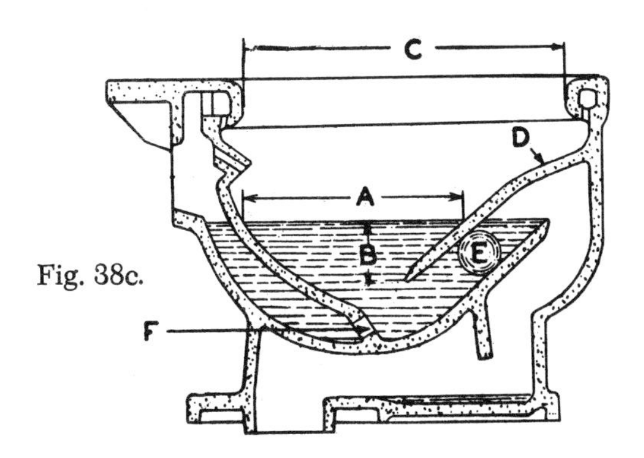

Fig. 38c.

at the time the bathroom is built, the pipes for the shower can be concealed in the wall. Shower fixtures with exposed pipes are available.

Shower heads are usually made of chrome-plated brass, and have swivel joints for directing the spray. Some models also have volume regulators, or both volume and spray regulators. A fitting that diverts the water from the tub faucets to the shower head is combined with the tub, but separate faucets or mixer valves for the shower can be used.

To insure head clearance for adults, the shower head should be installed at least 6 feet 2 inches from the floor. Tubs installed with shower heads can be enclosed with permanent rigid enclosures, or with shower curtains. Install the rod for the shower curtain at a height of 6 feet 6 inches.

For separate shower facilities, build a shower stall of masonry or tile, or buy a pre-fabricated enclosure. Pre-fabricated stalls come in porcelain enameled steel and fiberglass, and range in floor size from 30 by 30 inches to 36 by 36 inches to 34 by 48 inches. Height ranges from 74 to 80 inches. Prefabricated bases or receptors range in sizes from 32 by 32 inches to 32 by 48 inches.

BATHTUBS

Tubs for recess (fit flush between two walls) or for corner installation are 4, 4½, 5, or 5½ feet long. The 5-foot tub is the most used length. Tubs with widened rims are usually 32 or 33 inches wider; tubs with straight fronts, 30 or 31 inches wide. (*See* Figs. 39, 39a and 40.)

Square tubs are about 4 feet by 3½ or 4 feet, and are available for either recess or corner installation. Some styles have one built-in seat, others two. A square tub is heavier than a rectangular tub and may require additional framing for support. (*See* Fig. 41.)

Receptor tubs (Figs. 42 and 42a) are approximately 36 to 38 inches long, 39 to 42 inches wide, and 12 inches high. They are most suitable for shower installations, but, because of lower height, are also convenient for bathing children and others who need assistance.

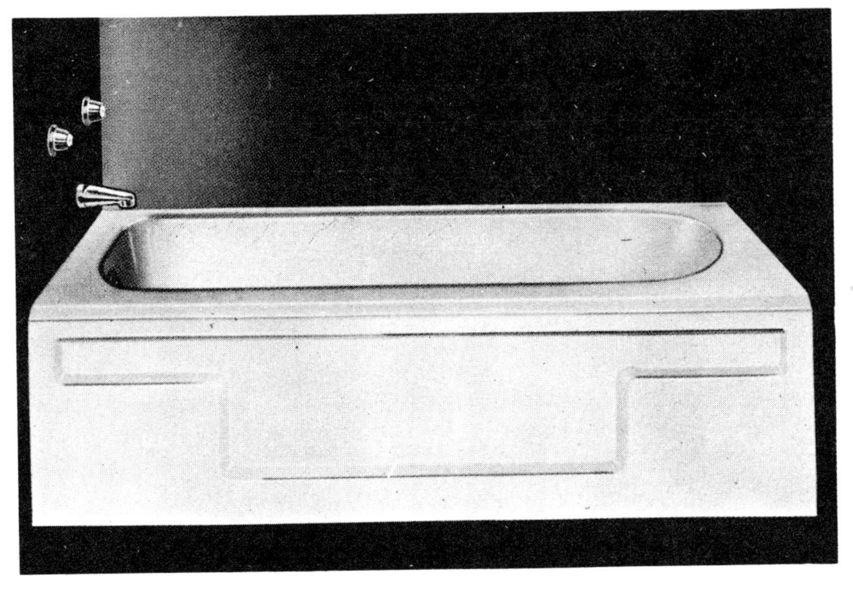

Fig. 39. Recess tub.

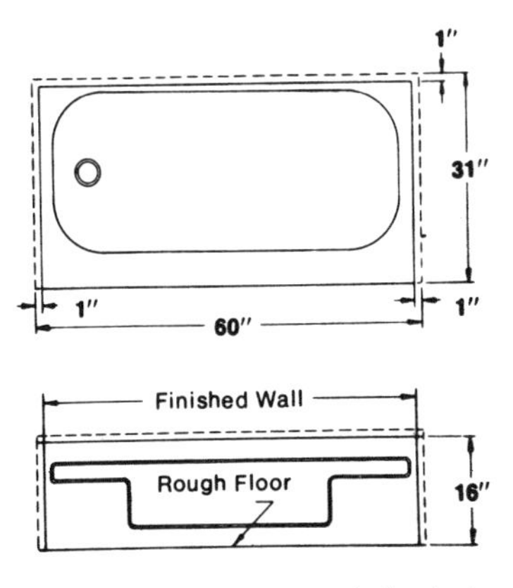

Fig. 39a.

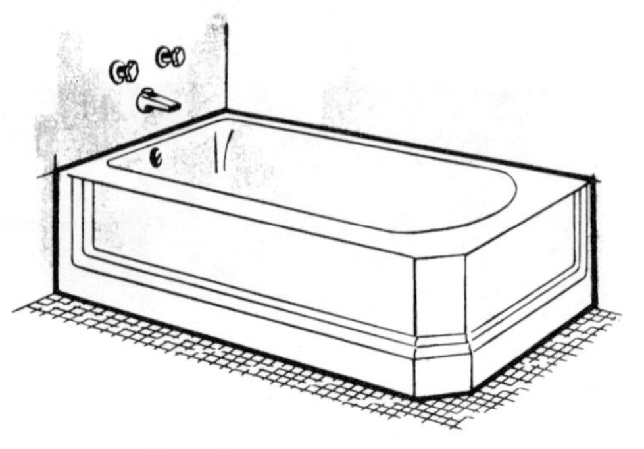

Fig. 40. Corner tub.

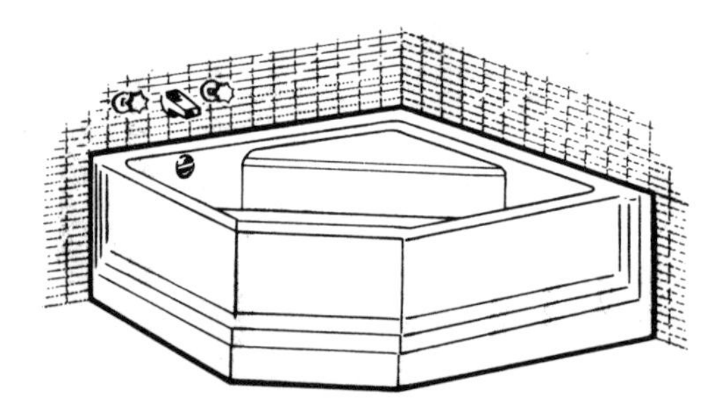

Fig. 41. Corner square tub.

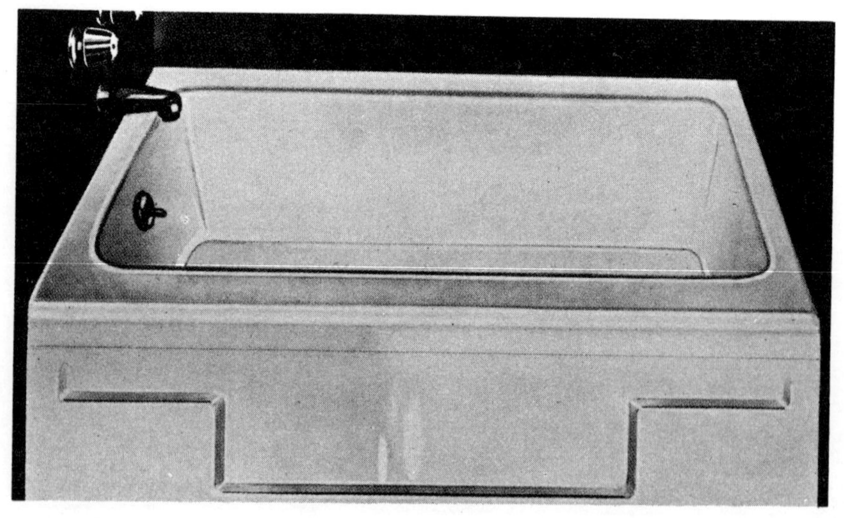

Fig. 42. Receptor tub.

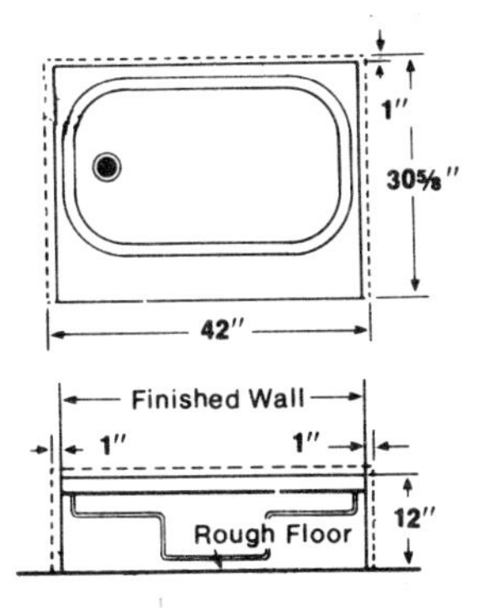

Fig. 42a. Dimensions shown are nominal; actual dimensions may vary (±) ½-inch.

LAVATORIES

Lavatories are available in a variety of types, sizes, and colors. (*See* Figs. 43, 44, 45, 46, 47, 48, 49, 50 and 50a.)

The decorator-styled *vanity cabinet* shown in Fig. 50 is constructed of sturdy ⅝-inch (1.5875-cm.) flakeboard and is covered with a tough sueded vinyl finish that looks and feels like a natural wood grain surface. The finish is highly resistant to surface damage and stains from most medicines and household liquids.

Topping the vanity cabinet is a custom, acid resistant, vitreous china lavatory. This contemporary rectangular lavatory mounts easily on the vanity and features two soap depressions and integral rear splash panel for wall protection.

Vanity cabinets are available in various sizes and also colors

Fig. 43. Ledge back lavatory.

Fig. 44. Splash back lavatory.

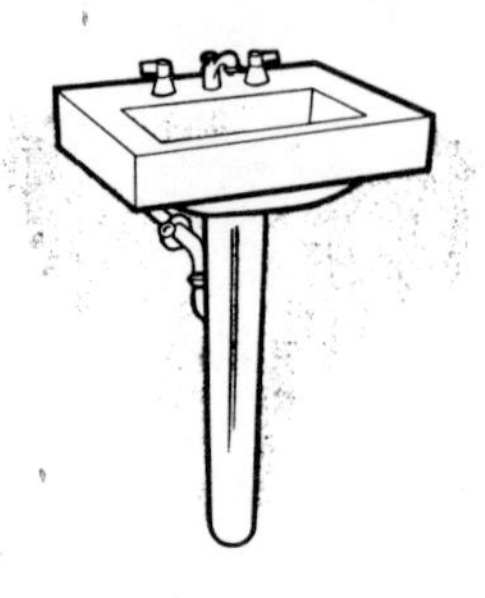

Fig. 45. Slab with china leg lavatory.

Fig. 46. Shelf back lavatory.

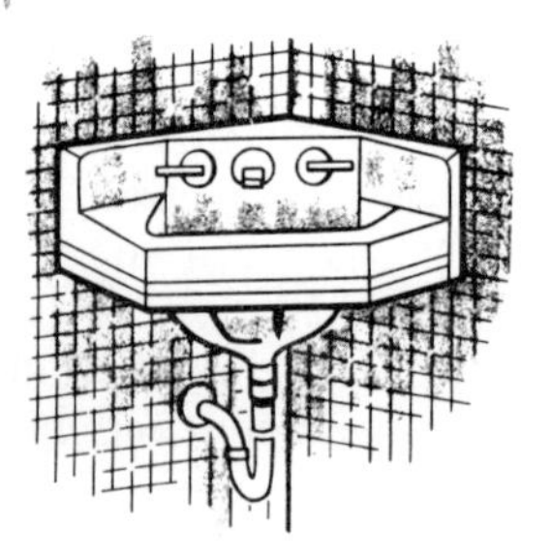

Fig. 47. A typical size for a corner lavatory is—length along wall, 17 inches; extension from wall, 19½ inches.

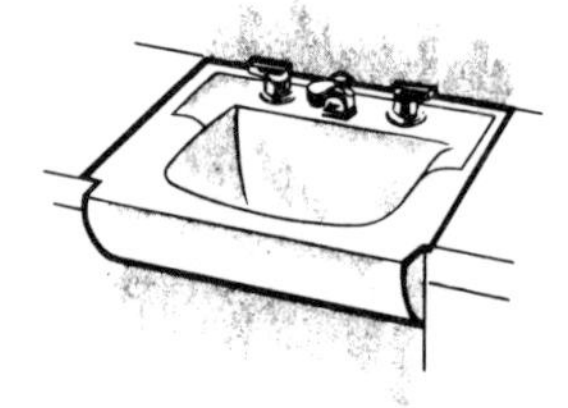

Fig. 48. Lavatory-counter top with rolled rim basin.

Fig. 49. Lavatory-counter top set on a cabinet.

Fig. 50. Flat rim lavatory set into cabinet.

and materials. The precision-made components are easily handled and assembled in a few minutes. A special slip-together system assures a solid custom fit and appearance. Doors have self-closing hinges.

A, Figure 50 shows an oval self-rimming vitreous china lavatory with front overflow and splash lip around the front edge. It has an all-brass supply and indirect lift waste fitting with aerator. This modern lavatory includes a soap recess that can be reached without dripping on counter surfaces. This is achieved by a soap receptacle concealed under the front rim. The soap receptacle includes combination drain holes and overflow. (See B, Fig. 50.)

Fig. 50a. Modern design self-rimming lavatory.

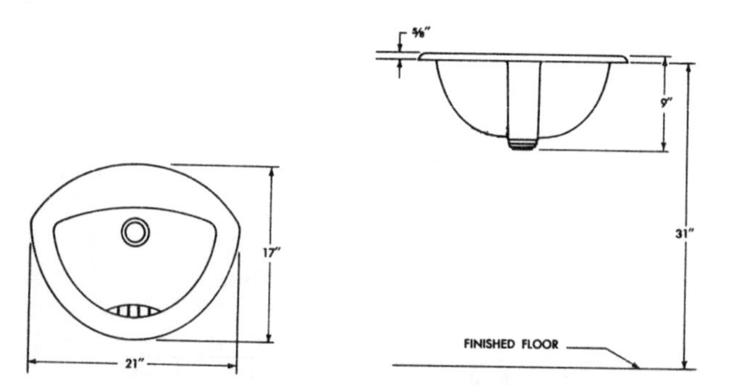

Fig. 50b.

TABLE 8
TYPICAL SIZES OF LAVATORIES.

	Width (inches)	Front to back (inches)
Wall-hung:		
Ledge back	19	17
	24	20
Splash back	19	17
	20	18
	24	20
Slab	20	18
	24	20
	27	22
Shelf back	19	17
	20	14
	22	18 or 19
Set in or on cabinets:		
Rolled rim	20	18
	21	17
	27	20
Flat rim	20	18
	24	20
	19½	15¼
Lavatory on cabinet	27	21
	21½	17¾
	36	18

TOILETS

One-piece toilets (Fig. 51) are neat in appearance and easily cleaned, but are more expensive than two-piece models.

Close-coupled tank and bowl—the tank is a separate unit and is attached to the bowl (Fig. 52).

A *two-piece toilet* with a *wall-hung tank* is shown in Fig. 53.

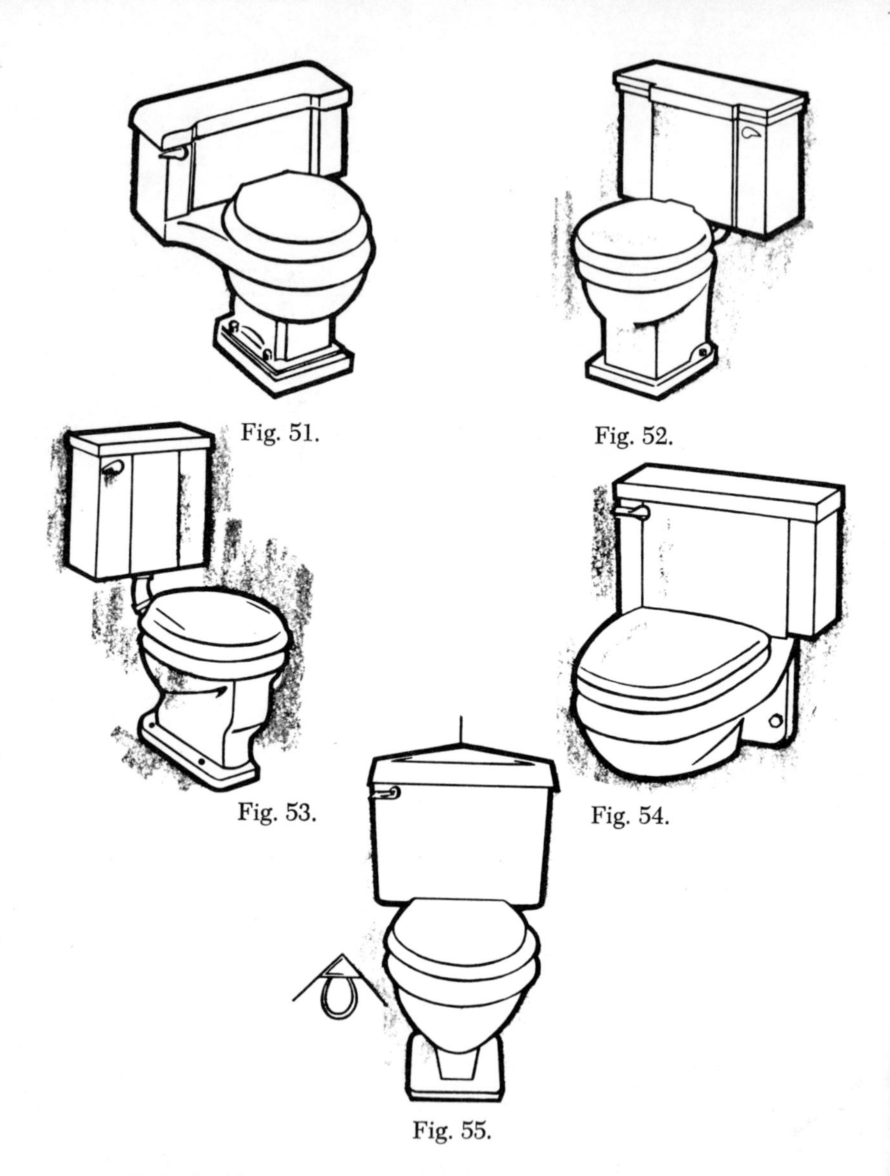

Fig. 51.

Fig. 52.

Fig. 53.

Fig. 54.

Fig. 55.

Completely wall-hung toilets and *tanks* make it possible to clean the floor under and around the toilet (Fig. 54).

The *corner toilet* (Fig. 55) is a space saver. Note the triangular tank. (*See* Table 9.)

Figure 55a shows a siphon jet, quiet whirlpool action, elongated rim bowl with integral wrap-around tank, tank lid, and bolt caps.

Fig. 55a.

TABLE 9
APPROXIMATE DIMENSIONS FOR TOILETS.

	Tank		Extension of fixture into room (*inches*)
	Height (*inches*)	Width (*inches*)	
One-piece toilet	18½ to 25	26¾ to 29¼	26¾ to 29¼.
Close coupled tank and bowl	28½ to 30⅞	20⅝ to 22¼	27½ to 31⅜.
Wall-hung toilet	27 to 29½	21 to 22¼	26 to 27½ (concealed tank 22).
Wall-hung tank	32 to 38	17¾ to 22	26½ to 29½.
Corner toilet	28¾	19¼	31.

Bathroom fixtures that get proper care before, during, and after installation usually give satisfactory service for the lifetime of the house.

During Finishing and Installation

A careful workman protects bathroom fixtures from blows, scratches, falls, and other damage during delivery, room finishing, and installation. He sees that fixtures are well covered with suitable materials and that plaster, paints, and acids do not get on them.

The damaged surface of a porcelain fixture cannot be restored. Special repair materials of the same chemical composition as the basic material are available for fiberglass fixtures. When competently applied, the damaged areas can be restored to new condition.

Here are some tips on fixture care during finishing and installation.

1. Uncrate fixtures carefully. Leave protective wrappings on fixtures.

2. If fixtures are delivered uncovered, cover them with several layers of strong wrapping paper held in place with tape; or cover them with corrugated board, or with special coverings available from plumbing supply dealers, or the special coatings that can be sprayed or brushed on.

3. Do not use newspaper or dyed paper next to enamel, as they may leave permanent stains. Newspapers can be used

for added protection if fixture is first covered with unprinted paper or plastic.

4. Avoid using paste made with flour to attach covering. Do not use sawdust as a protective filler. Flour paste and sawdust ferment when wet and produce an acid which etches the enamel.

5. Keep fixtures clear of tools, scrap lumber, wet paper or burlap, and other debris.

6. Remove carefully any plaster or cement on a fixture with water or a nongritty cleaning compound.

7. Soften paint drips with the recommended solvent and remove carefully.

After Installation

Harsh gritty cleansers soon scratch the surface of a fixture regardless of the material of which it is made. To test the abrasiveness of a cleanser, put a small amount between two pieces of glass and rub them together. If the glass is scratched, the cleanser is too harsh to use on the fixtures. Liquid detergents are recommended for fiberglass fixtures because of their somewhat softer finish.

Other precautions to observe in fixture care are as follows.

1. Do not use bathtubs or lavatories for washing venetian blinds or sharp-edged articles. If it is necessary to stand in the bathtub or to place a stepladder in it when washing walls and windows, cover the bottom of the tub with a rug or mat with a nonskid backing.

2. Do not drop bottles or other heavy objects in enameled fixtures. The surface can be chipped. Fiberglass fixtures are more resilient and not as subject to this type of damage.

3. Do not allow strong solutions including household and hair bleaches to stand in fixtures. Even acid resisting enamel will be damaged by continued contact with acid. Stains from iodine and burning cigarettes are the most difficult to remove from fiberglass fixtures. Take the precaution of rinsing the lavatory after using cosmetic lotions, hair tints, and medicines.

4. Do not allow faucets to drip constantly—the minerals in some water discolor and stain enameled surfaces.

5. Do not leave wet non-slip mats in tub. Some of them make permanent stains. Hang them to dry after each use before replacing in the tub.

STORAGE AND ACCESSORIES

Well-appointed bathrooms have convenient storage and functional accessories.

In planning any bath area, add storage units, either built-in or free-standing, whenever possible.

Toiletry Cabinets

Toiletries, such as toothpaste and shaving supplies, are conveniently stored in a cabinet above or within reach of the lavatory.

The toiletry cabinet is frequently called the medicine cabinet, but it is not wise to combine the storage of medicine and cosmetics. Preferably, medicine should be stored in a special place by itself so there is no danger of confusing it with other supplies.

In households where there are small children, provide a separate cabinet—one that can be locked—for medicines. Install it out of reach of the persons you wish to protect.

Toiletry cabinets can be wall hung (the least expensive type) or recessed. Recessed cabinets can be purchased ready for installation, or made on the job. Readymade cabinets usually have mirror doors.

Adjustable shelves permit the best use of cabinet space. Shelving should be made of plastic, glass, or enameled metal that is not damaged by moisture or spilled cosmetics.

Place the toiletry cabinet at a convenient height for family needs. The top of the mirror is usually placed 69 to 74 inches from the floor. If you measure from the bottom of the mirror, a distance of 48 to 54 inches from the floor is satisfactory for the person of medium height.

Towel Cabinets

You can save steps by storing some bath linens in the bathroom.

Regular-sized bath towels folded in thirds lengthwise fit on a shelf that is 12 inches deep; folded in half they fit on a shelf 16 inches deep.

The only available space for a towel cabinet in a minimum-sized bath may be above the toilet. If you put a cabinet for servicing the tank, the cabinet can be built into the stud space to provide additional depth if the location of the soil stack permits.

Metal-pole-supported shelves are easily installed over a water closet and provide some shelf storage for a nominal cost.

Utility Cabinets

Plan some storage space in the bathroom for reserve supplies of tissues and soap and for cleaning tools and cleaners.

In a limited size bathroom make use of the space under the lavatory for storage.

Small Accessories

Bathroom accessories include soap dishes, grab bars, towel bars or rings, tissue dispensers, and hot and cold water controls for the tub, lavatory, and shower. Any of these accessories that are improperly designed and installed can become safety hazards.

Towel Rod. Each family member needs rod space for a towel and washcloth. In addition, you will want some extra space for guest use. Towel rods on the sides of the lavatory are a convenient height for small children. A towel pole provides for extra towels in a minimum of space.

Paper Holder. Paper holders of china or metal can be recessed in the wall or fastened to the wall. Place the paper holder so that its bar is about 30 inches from the floor, and if on a sidewall, about 6 to 8 inches beyond the front edge of the toilet.

Grab Bar. Grab bars are available in many configurations; installation methods are nearly as numerous. Grab bars can be used to help maintain balance and ensure footing in assuming the normal standing, stooping, and sitting postures during the bathing process. These bars become useful additions to the overall safety of the bathroom. Grab bars (and towel racks which may also serve as grab bars) should be designed without pinch points, sharp edges, and pointed ends. The materials selected for use in the fabrication of grab bars and grab bar installation components should be shatterproof and should possess sufficient strength and durability to withstand their maximum intended use over the planned life of the dwelling. (*See* Fig. 56.)

Fig. 56. Angle grab bar for tub and shower.

Soap Holders and Clothes Hooks. Soap holders for the tub and shower are usually recessed. Vitreous china and metal are commonly used materials. For tub use, place the soap holder at about the middle of the wall beside the tub and within easy reach from a sitting position in the tub.

In the shower stall, the soap holder is usually placed about shoulder height, and far enough forward so the shower spray does not reach it. Or, if you prefer, you can install a corner shelf in the shower stall for soaps, shampoos, and rinses.

Nonrusting hooks for hanging bathrobes and other clothing add convenience. Place the hooks from 5 feet 5 inches to 6 feet from the floor. They should be above eye level for safety.

Toothbrush and Tumbler Holder. These accessories are often combined, but can be bought separately. Those made of vitreous china are set into the tile wall. Metal holders may be recessed or wall mounted. Some of these accessories are stationary, others revolve and close flush with the wall. Revolving combination units hold soap, tumbler, and toothbrushes.

Drying Lines and Racks. If clothes—especially those made of drip dry and wash-and-wear fabrics—are to be dried in the bathroom, it is best to make special provision for the job, rather than to depend on towel rods for hanging space.

Here are suggested ways to provide bathroom drying.

1. Place hooks in the walls at each end of the built-in tub for attaching clotheslines across the tub when needed.

2. Put a telescope rod with rubber suction cups over a recessed tub. This rod may be left in place permanently or stored after each use.

3. Mount a drying rack on which to hang hangers on the wall at one end or on the side of the tub. The rack will fold flat against the wall when not in use.

4. Install a clothesline reel with retractable plastic line over the bathtub. Line is hooked to opposite wall for use.

VENTILATION, LIGHTING, AND HEATING

Ventilation

Every bathroom or wash-up area should be ventilated either by a window or an exhaust fan. Natural or forced ventilation is necessary to comply with local building codes and to meet requirements of various agencies.

If your bathroom is ventilated by a window, avoid, if possible, locating the tub under the window. If there is no other location for the tub, a window that opens with a crank is easier to operate than a double-hung window.

To help prevent excessive humidity in the house, exhaust fans vented to the outside can be installed in all bathrooms whether or not they have windows. Fans are particularly necessary in humid climates. Exhaust fans in combination with lights and heater are good choices for small bathrooms. Lights and exhaust fans can be installed with one wall switch, but separate switches are preferred if such an installation is permitted by codes and ordinances.

Lighting

The well-lighted bathroom has good, glare-free, general illumination and properly placed area lights at the lavatory or dressing counters. The lights at the lavatory or dressing counter should be located so the light shines on the face, not on the mirror.

If proper fixtures are used in the small bathroom, the lights at the lavatory generally give enough illumination for the entire area. To provide good lighting for grooming at the mirror over the lavatory, place one light in the ceiling and one light on each side of the mirror.

Place the lavatory side lights 30 inches apart with the center of the light bulb 60 inches above the floor. Center the ceiling light above the front edge of the lavatory.

In a large bathroom general illumination will be needed in addition to area light. You may need extra light in your shower. Select a vapor-proof fixture.

Because it is easy to touch water and metal while switching on lights in the bathroom, make certain that lights are controlled by wall switches out of reach of anyone in the bathtub or shower, or anyone using a water faucet. Defective wiring and frayed cords on electrical equipment can result in severe electrical shock. Locate a grounded convenience outlet near the lavatory at a comfortable height for electrical appliances used in the bathroom.

Heating

Remember to plan for heat in your bathroom. If you do not have a central heating system, you will need to install either gas or electric wall space heaters. Plan the location of these carefully. Place the wall heater where there is no possibility of a person being burned on it or of towels or curtains catching fire from it.

Make certain that an electric heater is properly grounded and equipped with a thermostat, and that a gas heater is vented and has safety pilot shut-off features.

Portable heaters are not recommended as the general source of heat for the bathroom. For small areas, ceiling radiant heaters combined with a light or a fan or both are often used for general or for auxiliary heat.

WALL FINISHES

The varied materials used to finish bathroom walls today are pleasing to the eye, remarkably practical, and easily cleaned. Some of these decorative wall materials will last many years, others will need to be renewed from time to time.

You have a choice of paint, ceramic or plastic tile, plastic-coated hardboards, plastic laminates, wallpapers, or fabric-backed wall coverings. The kind of wall finish you select will depend on how much money you want to spend, your personal taste, and the way the bath area is used.

If you decide to *paint* the walls, choose a paint that is recommended for bathroom use—one that withstands moisture, is resistant to mildew, and is easy to clean. Gloss or semi-gloss enamel is usually recommended. Follow application directions carefully. Painted surfaces are not recommended for the interior of shower stalls because they do not withstand the constant wetting (for long periods) and are subject to wrinkling, blistering, and discoloration.

You may want to consider rigid wall coverings, such as *plastic-coated hardboards.* These are available in a nice assortment of colors, may have a plain finish or be scored to resemble tile. Rigid *plastic laminates,* familiar as counter coverings, are increasingly popular as bathroom wall coverings. *Sheet vinyl* with a moisture-resistant backing can also be used for bathroom walls and counters.

Washable wallpaper is practical for the bathroom and, if applied with a moisture-resistant or waterproof adhesive as recommended by the manufacturer, can be used successfully even on the wall around the tub. A satisfactory job of hanging paper can usually be done by the home workman. However, it is wise to test a sample of the paper to make certain that the colors are fast and that it can be cleaned satisfactorily.

Coated fabric wall coverings are well suited to bathrooms. They are colorful and easy to apply. One type is made of paper stock bonded to rugged woven cloth, coated with a vinyl resin, and printed in various patterns and colors. Still others are fabrics to which pure vinyl has been applied by heat and pressure, or several coats of enamel have been baked on.

Ceramic tile and plastic tile are in wide demand as bathroom wall coverings. Ceramic tile is made from clay that has been fired—it comes glazed and unglazed. Glazed tile, the type commonly used for walls, has a white body with a vitreous glaze of the desired color on the face. Unglazed tile has a dense vitreous body and is the same color throughout.

Ceramic wall tiles come in a wide variety of colors and a

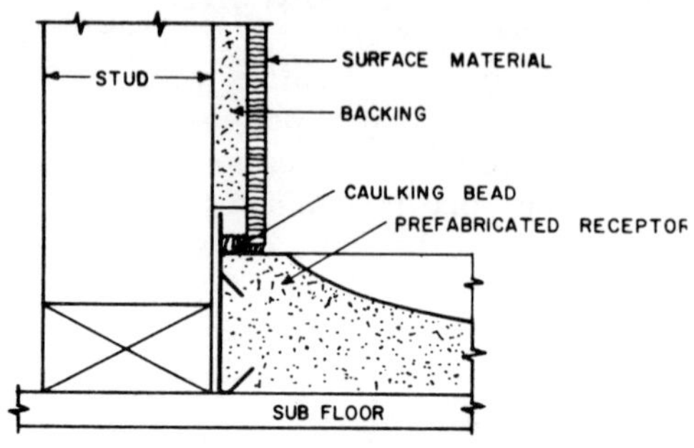

Fig. 57. Construction of a shower stall joint using a precast receptor.

number of sizes; a commonly used size for bathroom walls is approximately 4¼ by 4¼ inches (10.795 by 10.795 cm.). They can be ordered from the factory assembled in blocks on mesh or paper sheets. Tiles assembled in blocks can be installed in less time than it takes to install individual tiles.

Plastic wall tile is inexpensive, and comparatively easy for a home workman to install. Like ceramic tile, plastic tile is available in multiple colors that can be coordinated nicely into a decorating scheme.

The performance of any wall finish depends on the care with which it is installed and maintained. Always follow the manufacturer's recommendations exactly for type of adhesive and backing material, and for the method of installation. Backing material around tubs and showers should be thoroughly sealed with waterproofing materials prior to application of the wall finish. For recommended construction of base joints around showers and tubs, *see* Figs. 57 and 58.

After installation, protect the beauty and durability of wall finishes by cleaning only with mild detergent solutions and nonabrasive cleaners. With periodic care—wiping with a damp cloth—all finishes can be kept in acceptable condition

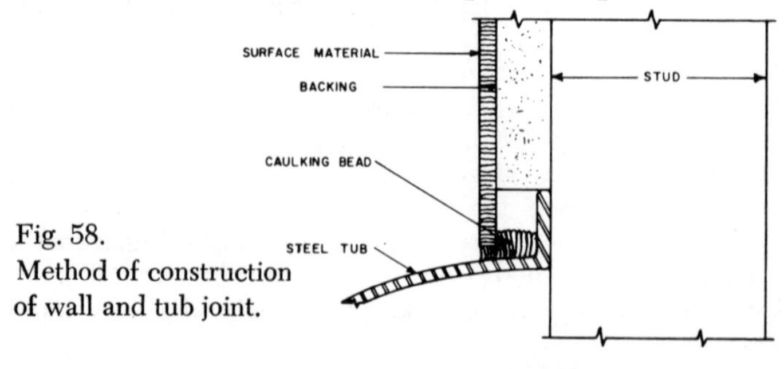

Fig. 58.
Method of construction of wall and tub joint.

without the use of harsh cleaners. Abrasive cleaners cause color fading and loss of gloss, particularly on plastic materials. Grout lines between tiles tend to darken with age, but can be cleaned with a small brush and a slightly abrasive cleanser.

FLOOR FINISHES

Today's bathroom floor finishes are of two main types: (1) Nonresilient floor finishes, such as ceramic tile and concrete, and (2) smooth-surface resilient floorings, such as linoleum, asphalt, rubber, and vinyl. Wood floors are rarely seen in bathrooms now. They will give satisfactory service, however, if they are refinished periodically with a water-resistant seal or varnish.

No one floor finish has all the properties desirable in a bathroom flooring. It is up to you to decide what properties you want most and choose accordingly. Necessarily, installation requirements and cost will help determine your choice. Other considerations include durability, appearance, ease of installation and upkeep, resistance to soil, moisture, and indentation, dimensional stability, and quietness.

Some flooring materials, including linoleum and cork, cannot be used on grade or below grade in a basement; others such as asphalt and vinyl asbestos can be used. Get the flooring and adhesive recommended for such installations.

If you plan to install the floor finish yourself, your choice of materials may be limited by the skill required for a satisfactory job. Many flooring problems can be traced to faulty installation. It is extremely important to follow the manufacturer's installation instructions and recommendations in working with any type of floor finish. If a specific adhesive is recommended, be sure to use it; do not substitute.

Nonresilient Floor Finishes

Concrete. Concrete floors can be used for bathrooms and wash-up areas on or below grade, and are satisfactory if the concrete surface is hard, dense, and smooth. They can be made more attractive by the addition of color to the concrete or by painting with one of the special concrete-floor paints.

Ceramic Tile. This widely used and popular finish for bathroom floors comes glazed or unglazed—with a bright or dull finish—and in multiple colors and shapes. Most ceramic tile sold today is factory assembled on paper or mesh. The traditional method of setting ceramic tile is in cement mortar. However, organic adhesives are extensively used today.

Ceramic tile floors are easy to keep clean. Washing with mild soap, powdered cleanser, or a synthetic detergent solution is usually sufficient. In areas where the water is hard, soap is less satisfactory than synthetic detergent or cleaning powder because of the insoluble film that forms from the reaction of the soap with salts in the water. If necessary, scouring powder can be used on heavily soiled areas. Ceramic floors should not be waxed.

Resilient Floor Coverings

Smooth-surface resilient floor coverings suitable for bathrooms include asphalt, homogeneous vinyl, and vinyl asbestos tiles; linoleum, backed vinyl, and rubber, in either sheet or tile form. Some companies also offer homogeneous vinyl in sheet form.

Inexpensive enameled or printed floor coverings are also available in sheet form. Some of these floorings now have a top layer of vinyl. The wear life of most printed and enameled floorings is limited. For this reason they are not recommended as a permanent installation in heavily used areas.

The home workman can usually do a more acceptable job of floor installation with tiles than with sheet goods. Sheet material has the advantage of fewer seams.

The 9- by 9-inch (22.86 by 22.86-cm.) square is the most commonly used size of resilient tile, but tiles 6 by 6 inches (15.24 by 15.24 cm.) and 12 by 12 inches (30.48 by 30.48-cm.) are available in some materials. Oblong and diagonal tiles are also made by some manufacturers.

Backed vinyl or linoleum in sheet form is usually 6 feet wide; rubber, 36 or 45 inches wide; and homogeneous vinyl, 45 inches wide.

The thickness or gage of flooring materials varies. Linoleum is usually 1⁄16 (.15875 cm.)- or 1⁄8-inch (.3175-cm.) thick. Asphalt and rubber tiles are 1⁄8-inch (.3175-cm.) thick; vinyl floorings vary from 1⁄16 (.15875 cm.)- to 1⁄8-inch (.3175-cm.) thick. Feature strips, insets, and moldings are available for all these floorings.

Chapter 8

Planning Your Kitchen and Workroom

Kitchen safety is influenced by kitchen efficiency, which results from careful, deliberate plans by the architect, designer, and homeower. Once a house has been built and the equipment installed, the correction of errors or elimination of conditions becomes costly and, in some instances, impossible without major structural renovations. It is, therefore, extremely important that proper planning for the efficient use of the kitchen and the safety of the occupants be given priority in the early stages of planning and design.

Safe kitchens result from choosing suitable space standards, from planning the kitchen's location and arrangement in relationship to the entire house, and from planning the type and location of fixtures and appliances for the most efficient and effective operation. To accomplish these ends successfully requires a realistic determination of the proper amount of space to allow for storage, appliances, counters, and traffic or other activity. It also requires establishing the relationship of the kitchen parameters to the anticipated traffic flows and living habits of the occupants.

A safe, well-planned kitchen is one designed so that only a minimum number of steps is required between the most frequently used work centers, such as the sink, oven/range unit, refrigerator, and other work areas. However, this area of heaviest traffic must not be so small as to be crowded, thereby increasing the probability of accidents. Although it may not be feasible to establish a fixed design that is optimum from a safety viewpoint, it is desirable to specify alternate configurations and dimensional ranges within which the placement of fixtures and appliances have proven to be conducive to a safer kitchen design.

The kitchen equipment (sink, cabinets, range, refrigerator, and the like) should be grouped and located so as to place each unit immediately available to the homemaker in the fewest steps, the very minimum of retraced steps and with the least amount of bending and reaching.

Improvements and features which should be looked for in a modern kitchen sink are depressed drainboards, acid resisting porcelain enamel, low back, cup-strainer, safety ledge along the back for placing stemware, eight-inch deep basin, swinging spout, control panel for housing supply fittings to allow full and unobstructed use of basin area, squared corners to permit continuous counter top and flush front cabinet installation. The best kitchen sink is one of one piece construction and integral drainboards which is easier to clean and eliminates unsanitary cracks or crevices. Single or double compartment sinks are obtainable with the double basin offering many advantages for ease and speed in dishwashing. (*See* Fig. 1.)

No sink is complete without a baked enamel or wood cabinet with the innumerable step- and labor-saving devices it provides. Facilities for storing the many everyday working tools, kitchen equipment and supplies within easy reach are necessary for efficient functioning. The same reasoning prompts cabinets to be placed wherever possible over the working area of the kitchen. Besides the cabinet underneath the sink, additional cabinets are obtainable to make a counter type installation and are called base cabinets. These are produced in various sizes and types—the cupboard type with drawers and door, the drawer type, the combination type, and the corner type. All may be combined with the sink to form a flush front, counter installation 36 inches high, the height of the sink, which was determined as the most practical and comfortable for the average person. Some of these cabinets are designed with black recessed sub-bases for toe room to provide comfort in standing and prevent scuffing of shoes and cabinet. Some sink cabinets have a recessed front to provide knee room while working at the sink. The counter tops of these base cabinets are made of a solid base of plywood on which is cemented a durable linoleum.

To further utilize precious space above the sinks and counters, other cabinets are available in the wall type, sometimes termed wall cupboards. They consist of various sizes and heights in the straight and the corner type to make a continu-

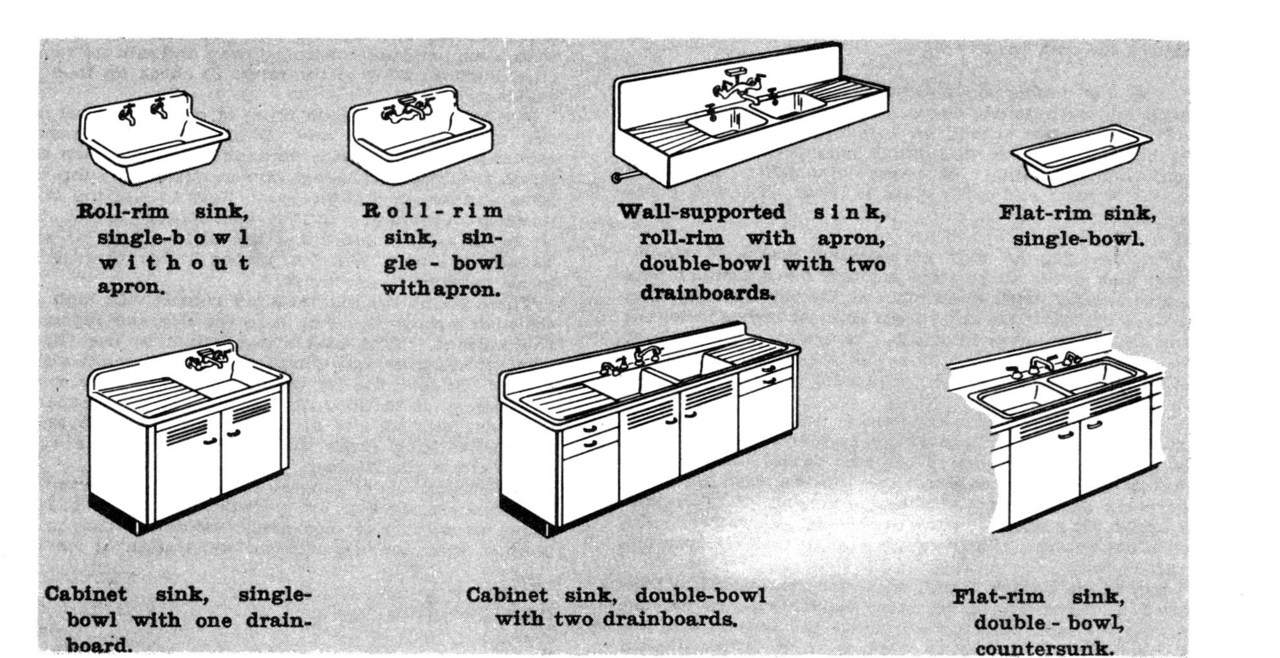

Fig. 1. Types of sinks.

ous wall cabinet effect. To serve a very necessary kitchen need, implement and utility cabinets for brooms and mops up to 90 inches high complete the all purpose service of these convenient storage cabinets.

With the variety and types available, the plumbing contractor or homeowner is able to design and install a complete cabinet kitchen that will fit any of the various kitchen plans and which will make for the utmost in efficiency.

Besides space for general storage, some of the cabinets may be fitted with other accessories to extend their usefulness, such as a wooden breadboard and a sliding sectionalized tray for forks and spoons in the top drawer of the base cabinet, garbage receptacle on door of the sink cabinet, ventilated vegetable storing bin, sliding dish towel rack, flour, sugar, and meal bins for bulk storage, a wire baking pan file, and other refinements.

All steel cabinets, however, are of sanitary and easily cleaned, acid-resisting and baked-on enamel and the doors and drawer fronts are insulated to increase the normal rigidity and eliminate any "tinny" sound. In some cabinets the drawers are on roller bearings to provide practically effortless opening and closing.

To form a neat joint between sink and counter in making continuous work surface, the sink edge must be flat and a special joint molding strip should be used. Sinks are made so they may be adapted to this type of installation and a special patented joint molding strip of polished metal is available to join the sink and counter top in a watertight joint. A plastic cement is also used to complete this junction.

PLANNING THE SINK CENTER

The reason the position of the sink in the new kitchen is decided first is that it is the most important work center. Two thirds of all kitchen work is carried on here. The location of the sink and the position of the range, the refrigerator, and the storage equipment in relation to the sink determine the efficiency of a kitchen.

From the standpoint of economical building or remodeling costs, the location of the sink is important also. In a new house, if the same main water supply and drain pipes can serve the sink and other plumbing connections, costs will be less. When remodeling an old kitchen, it is practical to place the new cabinet sink so that existing plumbing connections can be used. There may be times, however, when it will be advisable to change the sink's location completely.

The nature of the sink, as well as its position, is an important factor in planning.

Also available are the electric sink and the automatic dishwasher. Many families, realizing that they are certain to want a dishwasher for complete modernization within a few years, believe that it is wiser to get it now and save themselves the cost of conversion when the time comes. And, as you will see, it is also possible to start with a cabinet sink but plan on the addition of a dishwasher later, so that installation will be easier.

An extremely practical idea that is a delight to homemakers, and the family as well, is a ventilating fan above the stove which draws out kitchen air and prevents cooking odors from reaching the rest of the house and provides fresh air as well.

The *home laundry* should be planned as carefully as any room in the house to make for convenience and step saving in a normally disagreeable task. Location of laundry tubs, washing machine, drying equipment and mangle are important. Although some laundry rooms are on the main floor, most are separate rooms or partitioned parts of the basement. In selecting location of the room, thought should be given to a convenient airy place that does not require unnecessary lugging of laundry baskets or does not necessitate too long a walk to drying areas either in winter or summer. Good equipment should be selected and washing machines that are of large enough capacity installed.

Laundry tubs should be easily cleaned, sanitary, smooth and safe for the finest silks and nylons. The correct height is important to prevent washday backaches; 36 inches is considered the most universally desirable height.

A one-piece two-compartment tub is the most popular and has many advantages because of the elimination of crevices between tubs. An integrally molded washboard in one tub is a great convenience and extremely practical. An extended shelf where washing powders and cleaning materials can be laid is of much assistance. A single swinging spout faucet providing tempered water to either tub should be a "must." The

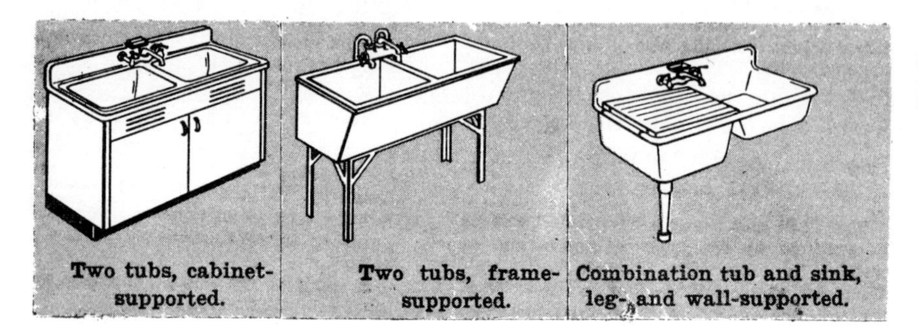

Fig. 2. Types of laundry tubs.

preferred material should be acid-resisting to counteract those disinfectants or bleaches that might be used. (*See* Fig. 2.)

Where more than two compartments are needed or desired, these are available as single units or as interconnected individual trays.

By carefully considering equipment arrangement and locations, a labor saving washroom can eliminate much of the unpleasantness of washday.

WORK TRIANGLE

The *work triangle* is one means of measuring the efficiency of a kitchen. It is defined as a triangle whose sides are formed by lines connecting the center front of the sink, range, and refrigerator. This triangle encompasses the area of major activity in the kitchen. Time and steps are saved and fatigue is avoided if the configuration is designed so that the sum of the sides of the work triangle is relatively small, but yet is large enough so that occupants are not crowded. With all other design factors considered to be satisfactory, the sum of the sides of the work triangle should not exceed 23 feet. Large kitchens are not necessarily more efficient or safer.

Figure 3 shows a well-planned kitchen layout with ample task space provided. Convenient traffic flow provides easy access to and from adjacent areas without crossing the work triangle.

Fig. 3. A well-planned kitchen layout.

TRAFFIC FLOW

Anticipated *traffic flow* patterns of the entire home environment must be taken into consideration when designing safe, efficient kitchens. In today's home, the patio and/or other outdoor recreational areas have become significant factors contributing to the occupant's daily life style. As such, these areas should be located adjacent to, or easily accessible from, the kitchen. The normal meal-serving areas, including dining room and breakfast room, should be located as convenient to the kitchen activities as practicable. In no case, however, should there be an interference between the work triangle and normal traffic flow patterns for these areas. The presence of hot dishes, appliances, and utensils in the kitchen, combined with a traffic flow that traverses the work triangle, creates an accident-prone condition. All traffic flow patterns that originate outside the home should be convenient to the kitchen but should not pass through the work triangle.

Another traffic flow factor of safe, efficient kitchen design which must be considered during the planning stage is adequate clearance between fixtures that are opposite each other. There should be enough space for one person to safely pass another who is using one of the fixtures, without crowding or physically contacting that person. A minimum of 48 inches provides adequate space between most kitchen fixtures (including oven, refrigerator, and cabinet doors) for normal traffic to pass, without undue interference to the person functioning in the kitchen work area.

The following recommendations for kitchen fixture placement provides for a more efficient and safer kitchen.

1. Kitchen should be designed so that no traffic flows through the work triangle.

2. The clear space between cabinets and appliances opposite each other should be a minimum of 48 inches. When such fixtures are at right angles to each other and separated by a passageway, they should be spaced a minimum of 30 inches apart.

3. In L-shaped or U-shaped kitchens, the minimum edge distance between the appliance and adjacent corner should be 9 inches from the edge of the sink, 16 inches from the refrigerator, and 14 inches from the center of the nearest range burner.

Figure 4 shows a well-planned kitchen which provides ample working space, an adequately proportioned work triangle, and a thorough traffic plan which does not interfere with kitchen activities.

CABINETS

As a major component of major activity area of the home,

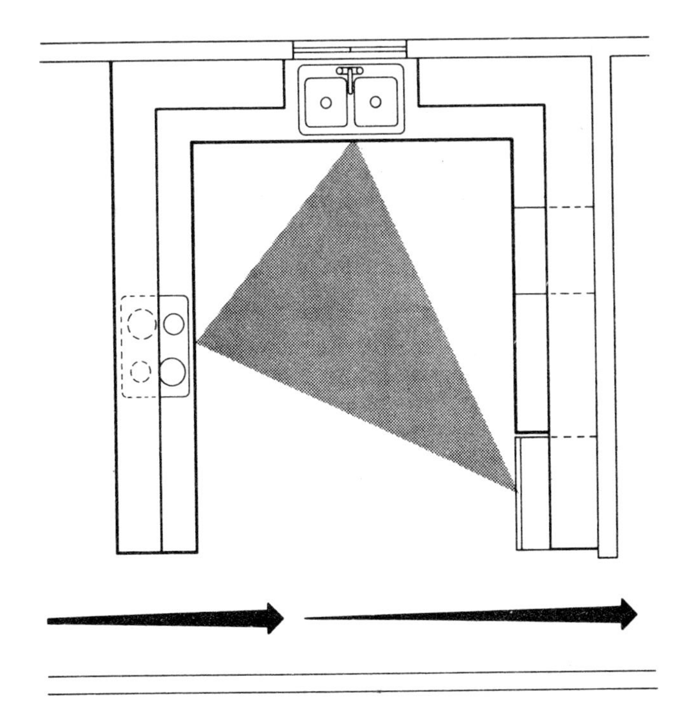

Fig. 4.

cabinets deserve appropriate attention during the planning, design, and construction stages. The size, number, and the location of the cabinets should be carefully determined while the kitchen is being designed. A high degree of craftsmanship is required during the construction and installation of cabinets to ensure easy and safe operation. The presence of base cabinets exceeding 24 inches in depth makes it difficult for the occupant to reach objects in the rear of the wall cabinets. This inconvenience may force the occupant to use chairs, stools, or other makeshift methods of reaching the uppermost cabinets. Top shelves of wall cabinets should not be placed more than 72 inches above the floor.

Cabinet counter levels should be flush with the installed appliances to provide a smooth, clear work area. Physiological studies show that counter tops that are 36 inches above the floor provide a convenient, efficient level for the type of activities normally performed in the kitchen. Manufacturers of kitchen appliances and fixtures usually design their products for this level of operation.

Base cabinets and counter tops should not be more than 24 inches in depth if there are wall cabinets located above them. Otherwise, the width of the base cabinet presents an obstacle to the individual using the wall cabinet.

For safer usage of kitchen cabinets and counters, the following recommended dimensions are given.

1. Counter top depth with wall cabinet above—24 inches maximum.
2. Counter top depth without wall cabinet above—30 inches maximum.
3. Counter top height above floor—36 ± ¾ inches.
4. Base cabinet depth—24 inches maximum.
5. Wall cabinet depth—12 inches maximum.
6. Wall cabinet top shelf height above floor—72 inches maximum.
7. Trim used on counter top edges should be free of all sharp edges, burrs, and points.

DOORS

The greatest potential hazard associated with kitchen doors is the conflicting relationship of *door swing* to fixtures or appliances. Door swings which conflict with or restrict the use of appliances should be avoided. Doors should be installed to swing against the side or end of a cabinet or out of the kitchen. The placement of standard sliding, pocket, or folding doors should be considered in locations where standard hinged doors would present a hazardous condition.

The following recommendations should be considered when designing for door placement in the kitchen.

1. Hinged doors should not be placed in such a way that they conflict with or restrict use of cabinets or appliances.
2. Use of sliding, pocket, or folding doors.

APPLIANCES

Kitchen appliances, regardless of safety precautions observed in design, manufacture, and installation, are inherently hazardous by reason of high temperatures, cutting edges, shearing or mashing actions, and other physical or conditional factors which are necessary in the conduct of their intended function. Exceptional care must be employed in planning the location of these necessary modern tools in the home, and special attention must be given to their proper and safe installation.

Refrigerator location may be hazardous unless the refrigerator is placed so that door swing does not interfere, to an unreasonable degree, with adjacent cabinet doors or the traffic flow. Refrigerators should also be located so that with doors open the interior is easily accessible from the work triangle. (*See* Fig. 5.)

Ranges and *ovens* are hazardous appliances because of the high temperatures associated with them. Convenient and adequate counter space adjacent to these units is a necessity

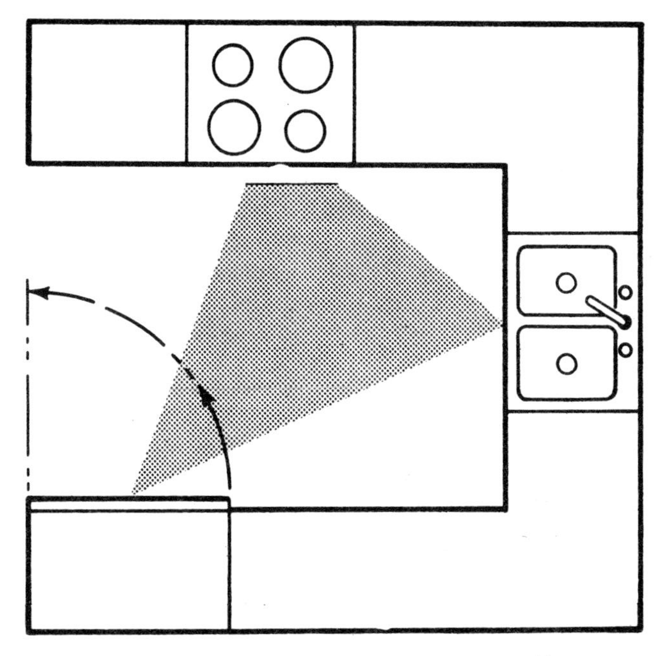

Fig. 5. Well-planned refrigerator door-swing, allowing ample passage in the opened position.

to minimize the distances required to transport hot utensils and to provide space for utensil handles. A minimum of 18 inches on each side of ranges and surface of separate oven units is considered adequate for these purposes.

Figure 6 shows dimensional limits for the safe installation of standard size *range hoods.* Care in planning a range hood can prevent or diminish accidents related to range hood design and installation.

In installations utilizing a *wall oven,* or ovens, and separate range surface, counter top accessibility must be considered to accommodate these two cooking appliances. Additionally, the design and placement of the range hood over the range surface requires adequate safety consideration. Range hoods are intended primarily to remove heat, smoke, moisture, and

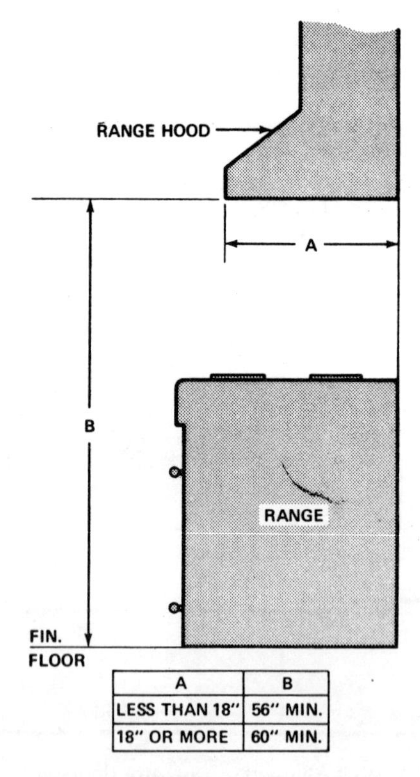

A	B
LESS THAN 18"	56" MIN.
18" OR MORE	60" MIN.

Fig. 6.

odors from the kitchen, as well as to provide a light source immediately above the cooking surface. Range hoods should be designed without sharp corners and sharp edges. Rounded corners and rolled edges will eliminate or substantially reduce this hazard. The height of the lower edge of the range hood above the floor should be such that it does not restrict the view of the interior of utensils on the rear heating elements. Studies show that the optimum height-range for hood installation is 56 to 60 inches above the floor. The depth of the hood determines which limit of the dimensional range should be used for a particular hood. A hood 17 inches or less in depth should be placed no less than 56 inches above the floor, whereas one 18 inches or more in depth should be installed no less than 60 inches above the floor.

Exhaust systems incorporated within a hood design should mechanically direct their flow to the outside atmosphere and not into the attic or other overhead, unused space in the house.

Two types of *garbage disposers* are in general use in homes today, the batch-feed type and the continuous-feed type. *Batch feed disposers* are loaded to capacity and are activated when the cover is positioned over the opening. *Continuous-feed* disposers permit continuous loading during operation. Operation is activated by a remote switch nearby. (*See* Fig. 7.)

Both types of disposers have inherent safety hazards, since the heart of the disposer function is a grinding action. Injuries occur when a person attempts to extract, or otherwise touch, an item placed in the disposer unit while it is in operation. In batch-feed installations, inserting the hand through the opening may activate the disposer, which is normally deactivated with the cover removed. An inexpensive method of eliminating or reducing this type of accident is the use of a momentary contact switch located a minimum of six linear feet from the disposer opening.

To provide safety considerations in the use of major kitchen appliances, the following recommendations are made.

1. Refrigerator door swing should be away from the "work triangle." For double-door refrigerator/freezer combinations, the refrigerator door should swing away from the "work triangle."

2. Provide minimum counter space of 18 inches on each

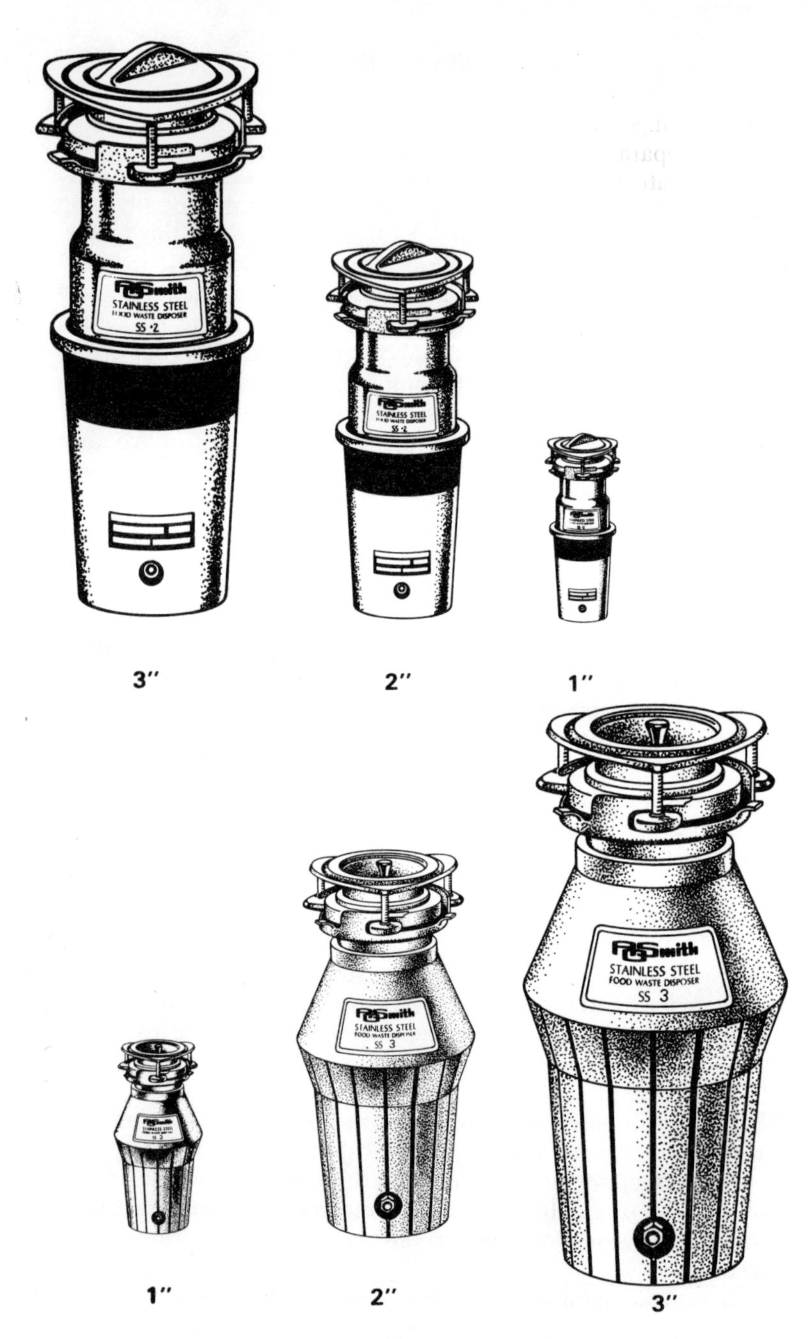

Fig. 7. Garbage disposers.

side of ranges and surface units and 24 inches on at least one side of separate oven units.

3. Locate the controls for ranges and surface units to the front or side.

4. Eliminate storage above the range and surface units.

5. Range hoods should have rounded corners and rolled edges.

6. Range hoods 17 inches or less in depth, should be placed no more than 56 inches above the floor. Those 18 inches or more in depth should be placed no more than 60 inches above the floor.

7. Range hood exhaust should not be to attic or other unused spaces.

8. The switch for the garbage disposer should be of the momentary contact type and should be located a minimum of six linear feet from the opening of disposal. (*See* Fig. 8.)

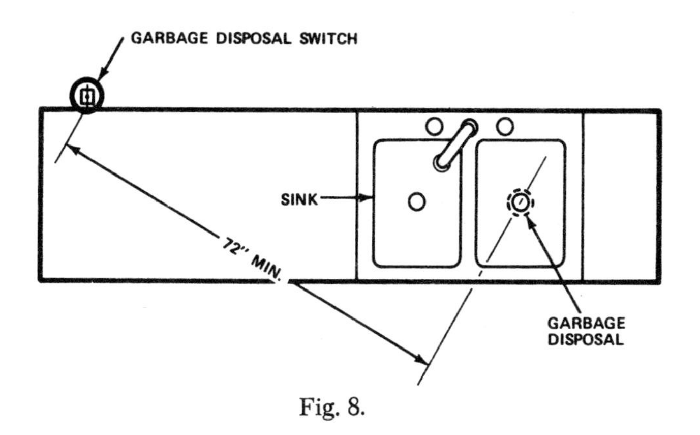

Fig. 8.

LIGHTING

Minimum lighting requirement recommendations have been revised extensively in recent years, with a corresponding increase in efficiency and a decrease in eye-strain associated with inadequate lighting. *Lighting* is used for utility and for creating decorative accents. Since there are no established lighting system requirements for most areas within a home, the decorative aspect often supersedes the utility aspect of home lighting.

Lighting for the kitchen area should consist of two basic types: Task area illumination and general area illumination. *Task area lighting* provides illumination for a small work area or surface, whereas *general*, or *environmental illumination*, provides a source for the entire area. Neither of these lighting types can adequately substitute for the other. Both must be considered as necessary factors in the properly lighted kitchen, from the standpoint of aesthetics and safety.

The kitchen is an area of multi-activities and one where many potentially dangerous appliances and items of equipment are used. Therefore, such factors as good lighting cannot be disregarded from a safety standpoint.

Certain design considerations must be analyzed for each specific task when selecting the proper lighting system for any particular application. These include the task that is to be performed, the quantity of light required to perform the task safely, the type of light that best provides this quantity, the location of the task, and special considerations relative to the task.

The following light values are recommended by the Illuminating Engineering Society (IES) for efficient and safe lighting in the kitchen:

Task Area	*Footcandles*
Kitchen sinks	70
Range cooking surface	50
Kitchen counters	50
Ironing boards and machines	50
Kitchen desks	30
Laundry trays	50
Washing machines	30
General lighting	30

Lighting in the kitchen should be planned for comparatively bright environmental coverage. Individual task areas should be illuminated with task lighting which creates a minimum of shadows directly within the task area or on the task surface. (*See* Fig. 9.)

KITCHEN PLANS

The ideal plan is as readily adaptable to the kitchen of the new home as it is to a remodeled one.

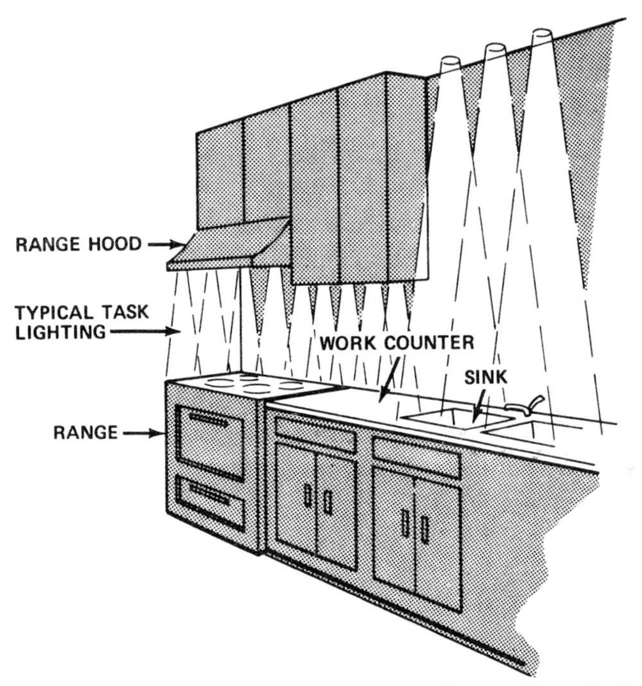

Fig. 9. Dimensional limits for the safe installation of standard sized range hoods.

When a kitchen is designed for a new house built to owner specifications, the doors, windows, and wall space can be planned at the outset to permit the arrangement of equipment best suited to the family needs. The sink area is planned first, and other major equipment, including cabinets for storage space and work surface, is placed as near the sink as possible to obtain the most functional arrangement.

FLOOR PLANS

Kitchen floor plans fall into the following four basic patterns U-shape, L-shape, one-wall, and parallel-wall. There may be variations of all these types. Then there is the irregular kitchen in which wall or floor space is broken up into odd lengths. No matter which of these types of kitchens is designed for the new house, the same basic plan may be followed to advantage.

In the *U-shaped kitchen*, the major appliances and storage and work cabinets are arranged on three sides of the room (Fig. 10). In the *L-shaped kitchen*, the equipment is located on two walls (Figs. 11 and 12). Figure 11 shows a partially completed kitchen with space left for additional equipment as the budget permits. Corner wall and base cabinets are often used in these kitchens so that corner space, which otherwise might be wasted, becomes usable. In the one-wall kitchen, all equipment is lined up along one wall just as the name implies. Use of an electric sink gives this small kitchen big-kitchen efficiency. (*See* Fig. 13.) This type of installation is used most often in small, minimum-cost homes or apartments. The parallel-wall arrangement, as shown in Fig. 14, is particularly well adapted to a long, narrow room with doors at both ends.

When the sink area is installed as the pivot around which the rest of the equipment is grouped, any of the four types of kitchens will be comfortable and convenient rooms in which to work.

If wall-to-wall cabinets are desired, the floor plan should be made to have walls of a length that can be evenly divided by three. However, it is not necessary for cabinets and other major equipment to form an unbroken line around the room. A space may be reserved for a specific purpose, such as a built-in ironing board, a small planning desk, or a small shelf for cookbooks and a few decorative items.

CHOICE OF EQUIPMENT

The size and arrangement of the kitchen and the budget are the chief factors in determining what equipment should be installed. But an adequate two-bowl modern sink and ample work and storage space pay big dividends in convenience.

Many homemakers now know how worthwhile it is to plan on a dishwasher. The electric sink and automatic dishwasher completely modernize dishwashing, and the savings in time and work are thought by many homemakers to more than justify the cost.

When the family already owns a range and refrigerator in usable condition, certainly they should be installed in the new kitchen. If possible, enough space should be left between

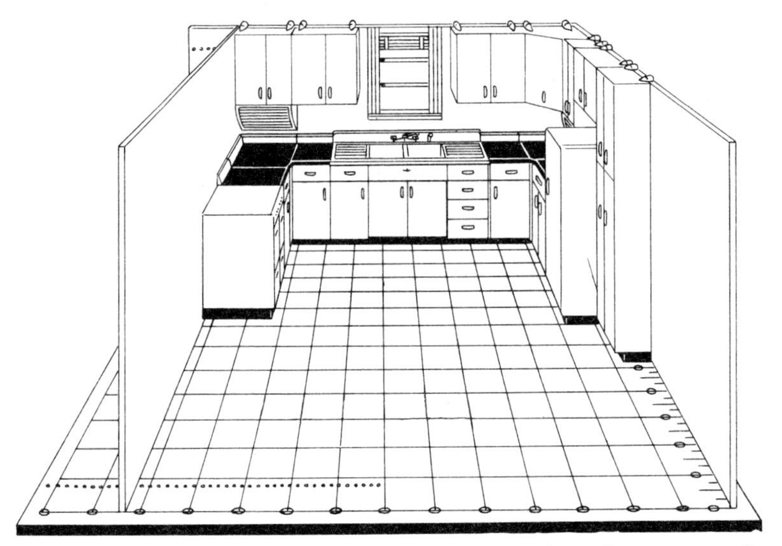

Fig. 10. U-shaped kitchen with the cabinet sink in the center and other equipment as near it as possible to give adequate storage space.

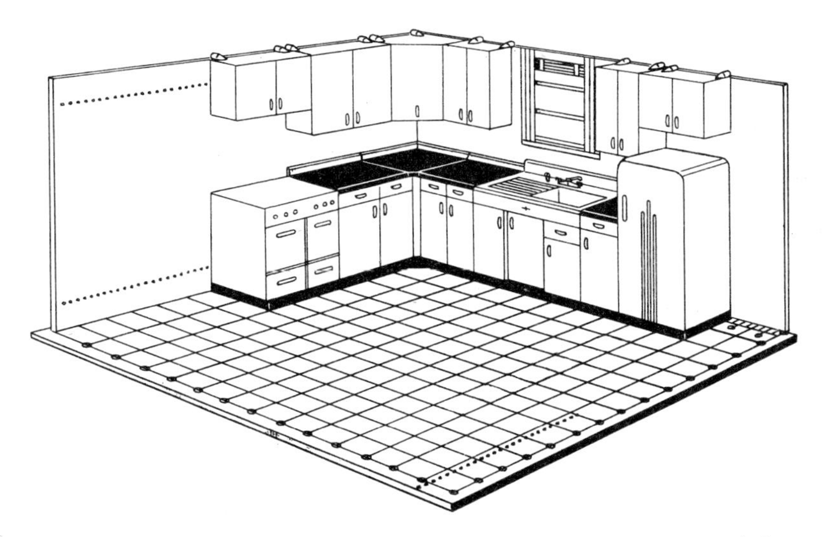

Fig. 11. An L-shaped kitchen, partially completed, with space left for additional equipment as the budget permits.

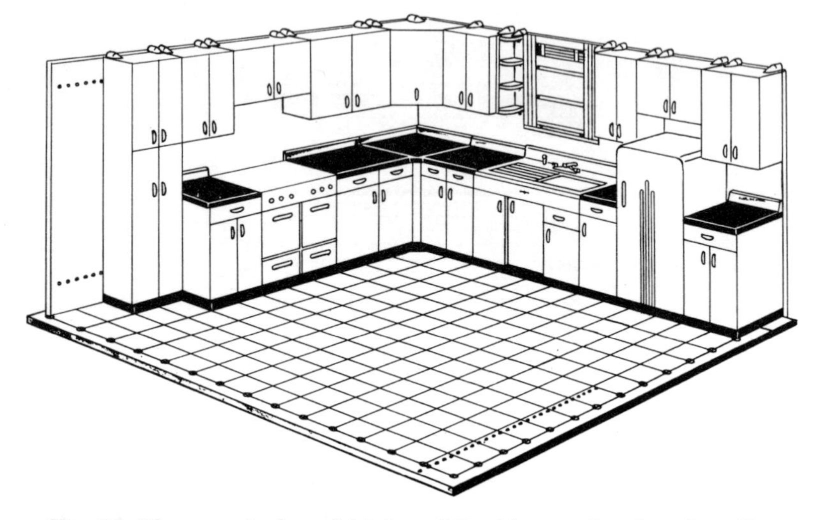

Fig. 12. The same L-shaped kitchen (Fig. 3), completed, with wall and base cabinets, what-not shelves, and a utility cabinet added.

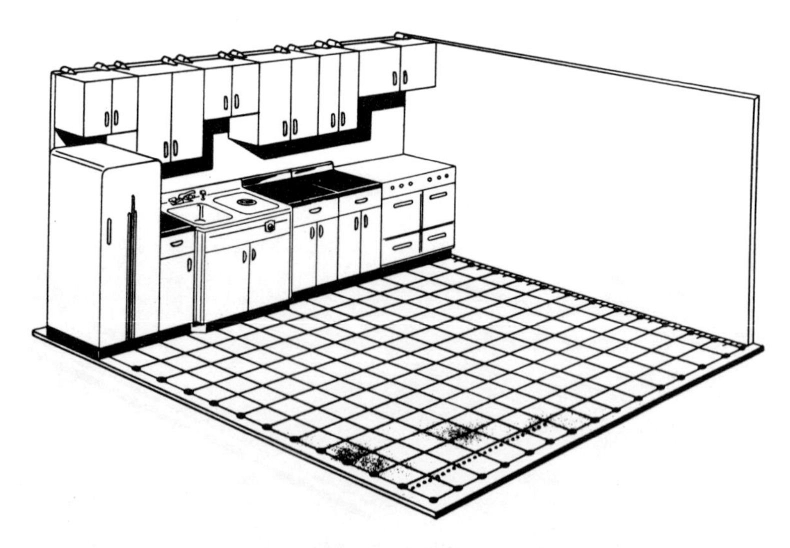

Fig. 13. A one-wall installation placed on an inside wall. Use of an electric sink gives this small kitchen big-kitchen efficiency.

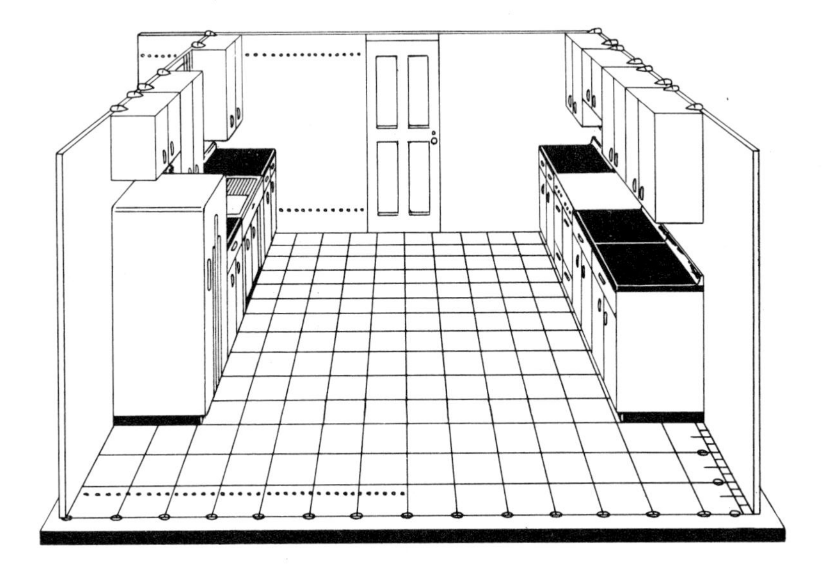

Fig. 14. This parallel-wall arrangement is most suitable for a long, narrow room with doors at both ends.

these two appliances and adjacent cabinets to permit easy cleaning and servicing of the appliances, or replacement of either of them at a future date.

ESTIMATE OF COST

Before embarking upon any kitchen building plan, it is only good business practice to get an idea of how much it will cost. This planning and estimating ahead eliminates many costly mistakes. Too often, quick, "economical" purchases become the largest expense, for they are never used. Such buying defeats the purpose of the entire project. Thinking through the selection of each item and buying the best that the budget will allow will always mean greater satisfaction for everyone.

Once the decision has been made to remodel a kitchen or to furnish a new one, careful consideration must be given to the selection of the new sink. The sink should be chosen with special attention to family habits and preferences, in order to ease and simplify work.

The range must be considered either when planning a new kitchen or remodeling an old one. Where the range is to be placed in the kitchen is of basic importance. For the greatest efficiency in meal preparation and service, locate the range near the sink work center, and at the same time have it as convenient to the dining area as possible.

If an old kitchen is to be remodeled, it may not be necessary to change the location of the range. On the other hand, if a new range is to be installed, one of the first decisions to be made is which type of heat will be used, and it must be determined how gas pipes or electric installation can best be located.

The need of the family will determine the size and style of the range. If the purchase of a new and larger model is contemplated at a later date, it is a good idea to leave enough space between the present range and adjoining cabinets. Work surfaces at either end of the range give added convenience in preparing and serving meals.

The refrigerator and the sink are very closely related in an efficient, modern kitchen. Many needless steps will be avoided if the two are placed close together.

If a new refrigerator is to be purchased, it is wise to choose one of sufficient capacity to provide for those special occasions when there is more than the usual amount of food to be refrigerated. On the other hand, if the present refrigerator is to be used with the idea of eventually replacing it, sufficient space should be allowed to accommodate a larger model. Experience shows that families tend to buy larger refrigerators when making replacements.

It is also practical to allow space around the refrigerator so that it can be moved easily for cleaning or for servicing. Mechanical refrigerators should be placed at least three inches from the back wall to provide the circulation of air necessary for efficient operation.

The location of the refrigerator in the kitchen determines whether its door should open from the left or from the right side. The door should open so that it is easy to move from refrigerator to the sink and the counter where most food preparation is done. Base cabinets next to the refrigerator are important to provide both work and storage space at this area. The extra counter space makes a handy place to set dishes or food before placing it in or removing it from the refrigerator.

Cabinets placed above the refrigerator are excellent for storage of seldom-used household items.

There is a growing tendency to include space in modern kitchens for serving breakfasts or other informal meals. In a large kitchen, table and chairs placed at one side of the room offer comfortable eating facilities.

Sometimes a low partition between the kitchen and a utility room may also serve as a counter for a snack bar. A drop-leaf counter built on a wall where it would be inconvenient to have a counter permanently set up is another practical solution. A separate dining alcove adjoining the kitchen is particularly desirable, if there is room.

Any of these types of dining areas may be as spacious and well appointed as the available space and family finances will allow. The purpose of such facilities is to save steps and lighten the homemaker's work.

Just as much thought should be given to the decoration of the kitchen as is given to planning other rooms in the house. Colors should be selected that are pleasing and harmonious. The small items on the open shelves can add variety or may offer contrast in the color scheme. Curtains can be tailored or ruffled, depending upon individual preference. Cabinet-top material comes in basic colors to ease your color-planning job and make a bright, pleasant scheme easy to create.

Some kitchens have wallpaper at the dining area, which adds design and further interest to the room. So much time is spent in the kitchen that the modern homemaker is wise to plan decorative colors that are easy to work with and still easy to keep clean.

A utility area is just what the name implies—a place where many of the miscellaneous homemaking activities can be accomplished. It usually includes laundry facilities. It may also afford storage and work space for sewing, for a children's playroom or a hobby center. If such an area is part of the kitchen, it helps the homemaker dovetail many of her regular jobs with resultant savings in time and energy.

Many homemakers prefer to have their laundry equipment in the kitchen. When this is the case, the deep bowl of the double-bowl sink will serve the purpose of both sink and laundry tub. Should an automatic washer be used, it is practical to install it as near the main plumbing line as possible. A conventional washer may be rolled to the sink for filling, using and emptying. This type of washer can be stored when not in use. In either case, cabinet space provided at the sink for washing supplies will streamline this very necessary homemaking task.

The same tall utility cabinet which houses mops and brooms will also accommodate the average portable ironing board, along with other ironing equipment. Or there may be a narrow wall space where a built-in ironing board will fit. If an ironing machine is used, space must be allowed in which to store the ironer.

If the kitchen is to include a utility area, special care must be given when planning. Work and storage space must not be sacrificed.

The type of floor covering is determined by individual preference, by the nature of the basic flooring, and by costs. Coverings such as linoleum, rubber tile or asphalt tile are easy to clean. They also offer a certain amount of resilience which makes them comfortable surfaces on which to stand. Whatever type of floor covering is selected, a durable, easy-to-care-for covering is a good investment.

If the flooring and the floor covering in an old kitchen are in good condition, there is no need to change them when remodeling. If a major change is to be made, it pays to do it before installing the equipment, unless you wish to investigate cove installation, which prevents hard-to-clean cracks and crevices. When this method is used, covings are curved to fit snugly against walls or bases of cabinets so that sharp corners and crevices between the floor and quarter round are eliminated.

The color of the floor covering is usually the basis for the color scheme of the kitchen. Plain colors are smart and modern, but figured or mottled patterns are more practical. The patterned type requires less attention since footprints and other marks are less conspicuous than on plain-colored floors.

Good lighting by both day and night is a most important consideration. If windows give adequate daytime lighting at the main preparation area, the situation is ideal. Sink and work counters are not necessarily placed directly under a window. However, they should be arranged so that a person using this area does not work in a shadow.

There are many types of modern lighting fixtures to give well-distributed artificial illumination. In addition to good over-all lighting, separate fixtures in each work area add to the comfort and efficiency of the kitchen.

Kitchen odors are usually appetizing, but unless some means is provided to air the kitchen, these odors become stale and undesirable. Cross ventilation provided by an outside door and windows is a natural means of ventilation.

A kitchen ventilating fan is an efficient all-weather device for carrying off odors. Such a fan, installed in a window, an outside wall, or a flue, will circulate fresh air and draw stale air out of the kitchen, thus preventing an accumulation of undesirable odors.

A portable fan placed a foot or two from an open window with the blades directed toward the window will also carry off kitchen odors.

Chapter 9

The Water Works and Hot Water Heating Systems

The ideal home plumbing system has an adequate supply of clean, incoming cold water, plus a water heating unit that provides a constant supply of hot water. The system will have water at your command in almost every part of the house—easily accessible through fittings designed for convenient on, off and flow control.

Chances are your water company puts cold water into your home's 300 or so feet of plumbing at about 40-150 pounds per square inch. If the pressure is about 80 pounds your plumbing may rattle loudly because the pressure is too high. You can have this corrected by having your plumber install a pressure reducing device just ahead of the water meter. It is a worthwhile investment, because excessive pressure is damaging to appliance valves and to plumbing system joints and valves.

Your water supply lines should run far enough apart so the hot water is not cooled by the neighboring cold water lines. Wrap insulation on long runs of hot water line to maintain heat.

Be sure you and your family know where to locate water shutoff valves in event of an emergency, such as an unexpected leak.

Cold weather can wreck unguarded plumbing. Drain outdoor faucets each fall after first closing the shutoff valves inside the house. Your plumbing drain system has a water-holding trap installed in every drain line to block odors. If you must leave a house unheated in winter, add liquid anti-freeze to these traps and to toilet bowls.

If you live in a hard water area, install a water softener to prevent damaging water deposits and fixture corrosion.

WATER HEATERS

Hot water is one of civilized man's most highly prized comforts. Perhaps this is why we have developed modern water heaters that are among the most efficient and trouble-free of household appliances. There are a few tips, however, that will help you get the best from your water heater at minimum cost and trouble.

A single-family dwelling requires a heater of from 30- to 60-gallon capacity in gas-fired units and 80-gallon in electric-fired units. It is better to buy a little more capacity than you currently need. Most popular are glass-lined, gas or electric-fired heaters.

The most useful thing to know about hot water is the temperature required for household use. The ideal minimum temperature is 140°F. While this may seem a little warm for hand-washing, it is the required level for really effective performance from your dishwasher and washing machine. A quick way to check the water temperature is to hold a thermometer under a hot water faucet. Check temperature under a faucet that does not have an aerator.

Your water heater is equipped with a drain valve to enable you to drain off sediment which accumulates at the bottom of the water heater. Just open the drain faucet located at the bottom of the heater every couple of months, and run out a few gallons. This will pull off the sediment that would otherwise continue to build and appear in your water supply during a time of heavy demand.

If hot water takes some time to appear at your faucets, there are probably long runs of bare hot water pipe in the system. Having your plumber insulate these pipes saves water, fuel, and your time at the faucet. If you install a new water heater in your home, locate it as close as possible to the bathrooms and the kitchen, as this reduces the hot water travel distance and holds down water heating cost.

Other features of the water heater include a thermostat, actuated by the heater's water temperature that holds temperature at the desired level, and if it is a gas-fired water heater, a draft diverter or vent hood to keep chimney downdrafts from knocking out the pilot flame. If you have frequent trouble with a pilot going out and the heater shutting down, check to see if a draft diverter or vent hood has been omitted.

The tank safety relief valve is a pressure relief designed to automatically let off some water and steam should all other controls fail and your heater superheat the stored water. Some water heaters have a valve which may be manually tripped occasionally to test its free operating condition.

WASH TUBS

Laundry room wash tubs have been made of many materials. Most popular today are *fiberglass tubs* that are light and easy to clean.

Laundry tubs have so many good uses that every home should have a side-by-side pair if space permits.

Ideal water supply calls for a hot and cold water mixing faucet with a swing spout, so both tubs can be served from one faucet. You may find a spray nozzle handy for cleaning tubs after bleaching or dyeing operations.

WASHING MACHINES

The *washing machine* should be installed with hoses direct from hot and cold water supply lines, including a shutoff valve for each. Shut water off with these valves when the washer is not running to relieve pressure on the machine's hoses and water inlet valves. Screens in hoses should be checked and cleaned periodically.

The most frequent cause of washer breakdown is water inlet solenoid failure. If hot or cold water will not enter the machine, check these first. They are easy to replace and inexpensive.

If water will not drain, check the drain system from the bottom up. If drain hose is clear, check the spin solenoid. If, when you manually activate this solenoid, the machine drains and spins okay, the solenoid needs replacement.

Another source of drainage problems is the impeller-type pump used in most washers. The cover must be removed and checked for heavy lint clogging. This is generally a job for a qualified appliance serviceman.

SHOWER STALL

Pinpoint control is the secret behind comfortable showers. Ideal shower plumbing and fittings will include an automatic mixing valve to give you the right blend of hot and cold water. A ball joint shower head permits spray angle adjustment.

Completely *plumbed shower stalls*, economically priced, can be quickly installed in a basement or laundry room. This is a good solution to peak time bathroom crowding and to scrub up the family's football players, gardeners, and beach fans without routing their dirt through the house.

Shower stalls should be placed to take advantage of a short run to the drain system. There should be adequate lighting in the stall area, since you cannot safely have a light in the stall; for comfort, it should be in a section of the basement or laundry room that can be heated to normal room temperature.

WATER METER

Generally, a shutoff valve will be in the incoming water supply pipe just ahead of the meter. In case of a leak or other emergency, this is the valve you close to cut off the water supply.

If you feel your water bills are higher than they should be, flip open the metal register cover and take a reading. Match your readings (cubic feet of consumption) to the time periods shown on the water bill. Sometimes a combination of visible and hidden water leaks can eat up several hundred gallons of water per month.

DISHWASHER

Dishwashers thrive on good, hot water.

Water to your dishwasher should be at least 140 degrees. If your household hot water is not hot enough to suit your dishwashing mood, a plumber can equip the hot water heater with a dual temperature valve. This will furnish hotter water to your dishwasher than to the rest of the house.

Most modern machines do a good job of cleaning food from dishes. Nevertheless many housewives make good use of plastic bristled vegetable brushes or plastic netted sponges to give dishes a quick advance scrub. Besides loosening heavy food, this rinsing softens all the food on the dishes.

Do not locate your dishwasher next to the refrigerator or deepfreeze. The heat it generates will make nearby freezing units work harder.

The most frequent cause of dishwasher failure is aging of the hot water inlet solenoid. Have your serviceman check it and replace it if needed.

KITCHEN SINK

Stainless steel sinks have pretty much replaced enameled ones these days in kitchens. And hot and cold single lever mixing valves with handy swing spouts have become vastly popular. A hose and spray nozzle is a useful rinsing accessory.

A disposal unit installed under your kitchen sink saves considerable time and labor. Run lots of cold water when operating a disposal. This facilitates grinding and reduces wear on moving parts. The disposal will have its own trap for blocking odors, it should also have a reset button somewhere on its frame for those times when it stalls.

BATHROOM SINK

The single-lever, washerless faucets with their convenience of controls have revolutionized bathroom sinks. Now, with a judge of the hand or wrist, you can get the right flow at the right temperature easily. The washerless faucets in single lever or two-handle models are available in matching sets for lavatory, bath, and shower.

If you find the water from the faucet is discoloring porcelain in the sink or tub, consult your plumber. Often he can eliminate the mineral deposits that cause these stains by installing a filter.

It is a good idea to clean the bathroom sink's removable drain plug every few weeks. Hair wads removed from here help prevent blockages further down the drain system.

If your bathroom sink clogs, it may well be in the trap just beneath. Disconnect the trap to check it after placing a bucket or pan to catch spill water.

Be sure all the bathroom plumbing is leakfree. In addition to wasting water, bathroom leaks will rot the bathroom floor in time, forcing a complete and expensive reconstruction job.

Keep the bathroom well ventilated in summer. This helps reduce dampness caused by condensation on flush tanks.

TOILET

The major parts of the toilet are the tank and bowl. The tank is for flush water, while the bowl is for entrapment of gases. The tank has an inlet valve and a flush valve. The function of these valves is to empty the tank to flush the bowl, and to refill the tank and the trap in the bowl. Main source of trouble, which results in water leakage, is misalignment of the flush valve. If you hear water running in the tank, it is probably due to this problem. This can be corrected by replacing with a new part (preferably original equipment).

SHOWER/TUB

The shower/tub provides great bathing convenience, particularly if equipped with a mixing valve which cycles you safely through the cold and hot zones.

If you are selecting shower/tub fixtures, choose models that are sturdily built and of simple, easy-to-clean design.

Safety is important in the tub. Strip the bottom of your tub with sections of tight-gripping adhesive tape to improve the footing. Tape kits may be purchased at a local hardware store.

Additional no-slip safety can be provided with grab bars, both on the tub/shower wall and just outside the tub.

Never touch an electrical appliance when in a tub or shower because of the danger of receiving a severe shock which could prove fatal.

Water is a most versatile commodity, but can be a pest, too. If tiles come loose around the tub area, get all loose ones removed and properly replaced immediately. A prepackaged grout material is available for easy sealing around bath crevices. There are many good quality tile adhesives available for tile replacement. If left unchecked, a breach in the tile will let shower and tub water get at walls, floors and substructure and ruin them.

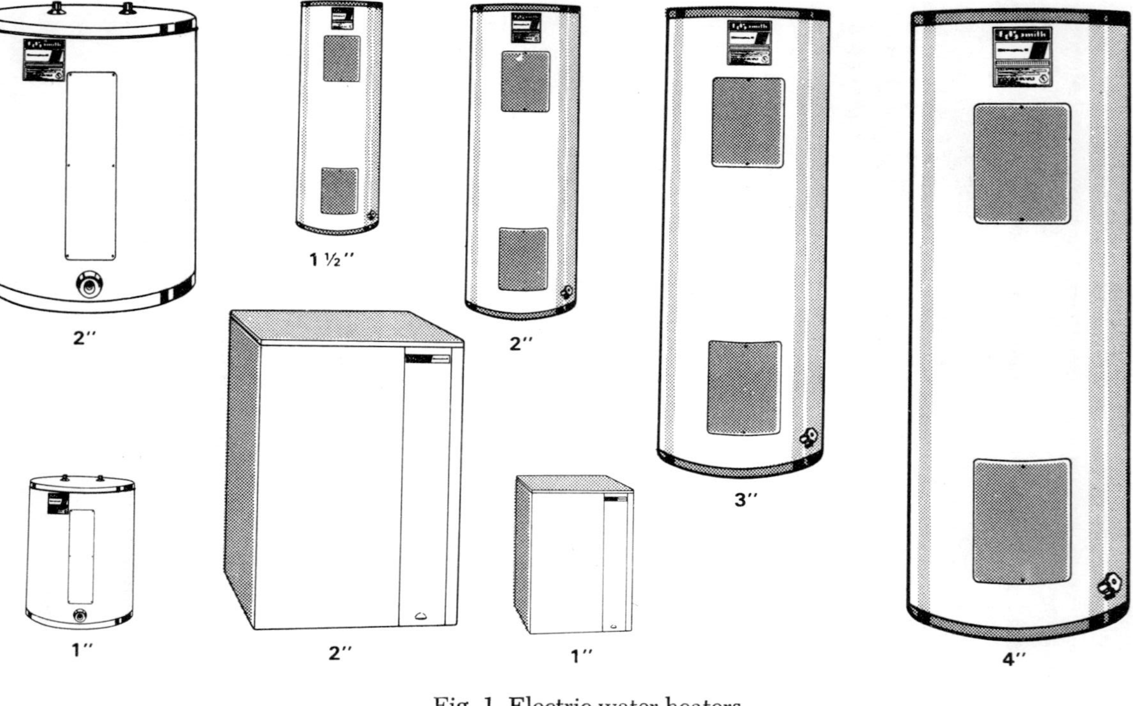

Fig. 1. Electric water heaters.

ELECTRIC WATER HEATER

All heaters must be installed in accordance with local codes and utility company requirements. In the absence of local codes or utility company requirements, the heater must be installed in accordance with the National Electrical Code. (*See* Fig. 1.)

Installation

The wiring diagrams (Fig. 2) show factory installed internal wiring to meet local utility requirements. Make the ground connection to the green screw provided in the junction box.

The heater should be located in an area where leakage of the tank or connections will not result in damage to the area adjacent to the heater or to lower floors of the structure.

When such locations cannot be avoided, a suitable drain pan should be installed under the heater. Such pans should be at least two inches deep, have a minimum length and width of at least two inches greater than the diameter of the heater and should be piped to an adequate drain.

Relief Valve

Caution: For protection against excessive pressures and temperatures in this water heater, install temperature-and-pressure protective equipment required by local codes, but not less than a combination temperature-and-pressure-relief valve certified as meeting the requirements in the *Listing Requirements for Relief Valves and Automatic Gas Shutoff Devices for Hot Water Supply Systems*, by a nationally recognized testing laboratory that maintains periodic inspection of production of listed equipment or materials. The valve should be so oriented or provided with a drain line so that any discharge will not contact any live electrical part. The valve or drain line orientation must be directed toward a suitable drain or location so the discharge of water will not cause damage. Do not thread, plug, or cap the end of the drain line. Omission of improper installation of the temperature-and-pressure relief valve voids the manufacturer's warranty and liability. (*See* Figs. 3, 4, 5 and 6.)

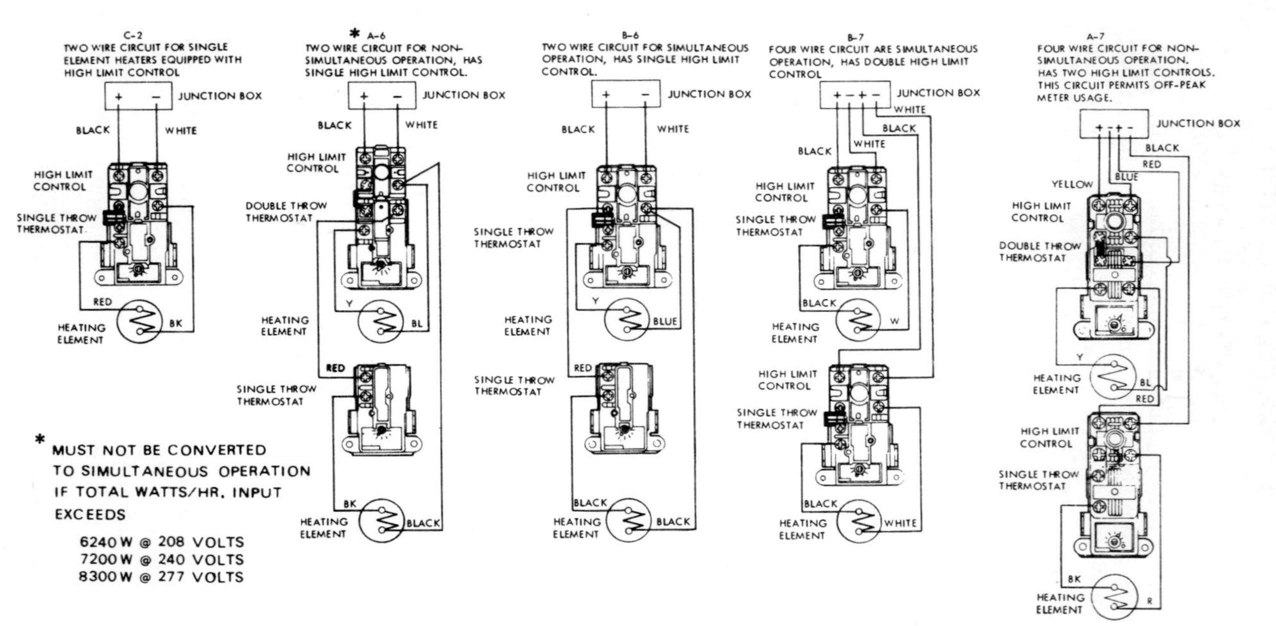

Fig. 2. Wiring diagrams.

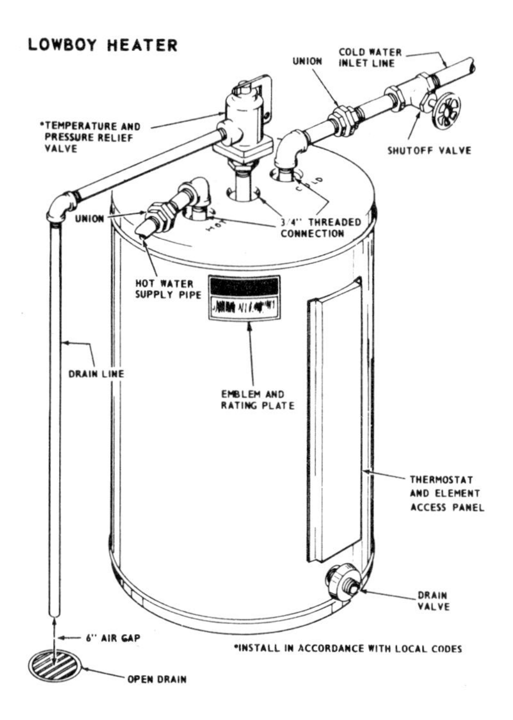

Fig. 3. Lowboy heater.

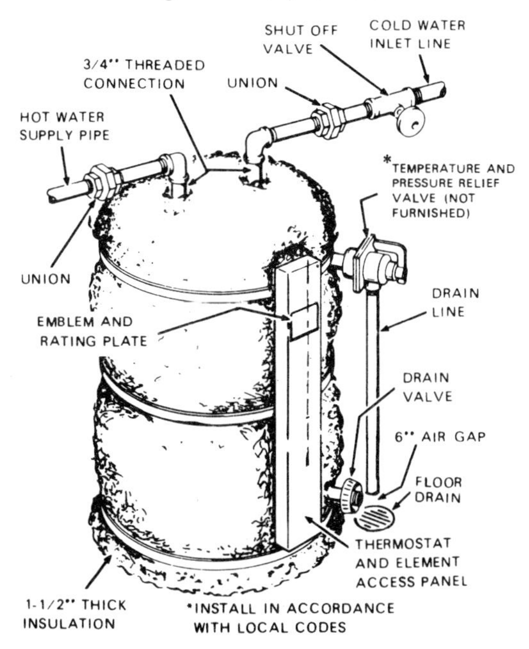

Fig. 4. Unjacketed heater.

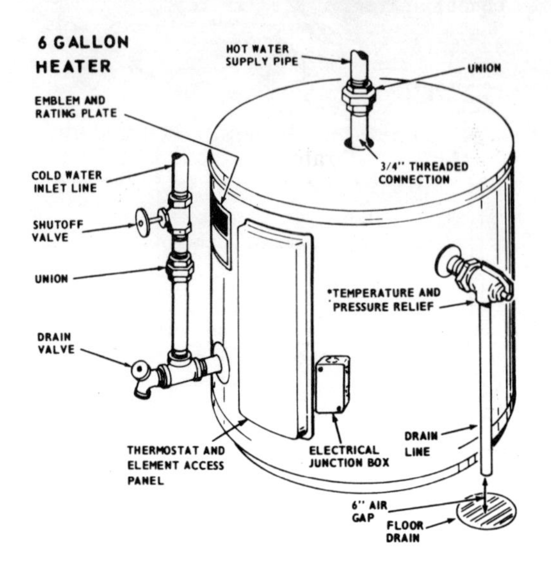

Fig. 5. Six-gallon heater.

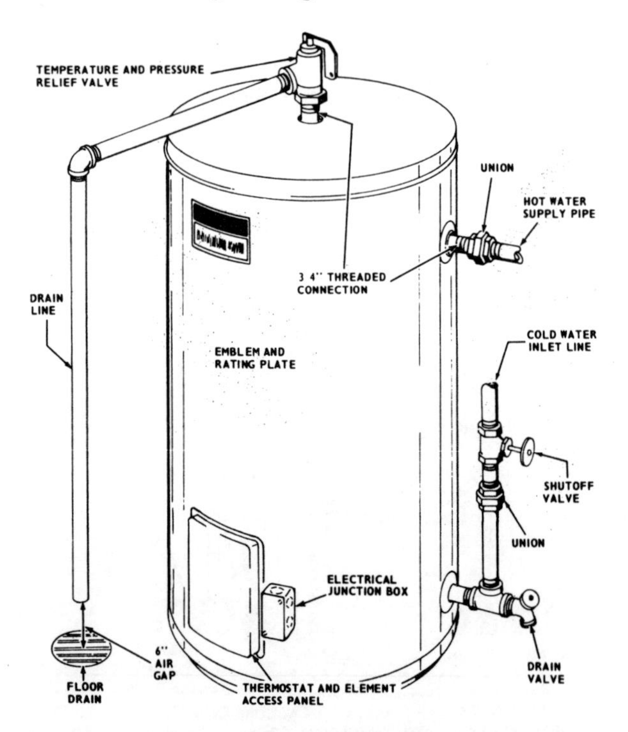

Fig. 6. Fifteen-gallon heater.

Operation

Caution: Do not operate the heater without installing an approved temperature and pressure relief valve in the opening provided in the tank. Ground the heater to guard against electric shock from the heater or water system. Never operate the heater without filling with water per the filling instructions. Failure to do so will damage internal parts.

Filling the Water Heater

1. Close the water heater drain valve by turning handwheel to right (clockwise).
2. Open a nearby hot water faucet to permit the air in the system to escape.
3. Fully open the cold water inlet valve allowing the heater and piping to be filled.
4. Close the hot water faucet as water starts to flow.
5. Turn on the electrical switch to the water heater.

Temperature Regulation

Each heating element operates under the control of a thermostat which is factory set at 150° F. This setting has proven to be most satisfactory from the standpoint of operational costs and household needs.

The thermostat(s) can be adjusted to provide a warmer or cooler water temperature. Contact your dealer or utility company if adjustment is necessary.

For safety a non-adjustable high temperature limit control will break the electrical circuit when excessive water temperatures are reached. All heaters are equipped with a high limit control which can be manually reset should it operate. This limit is shown in the wiring diagrams (Fig. 2) and has the reset button. The limit only operates when abnormally high water temperatures are present, so it is very important your dealer be contacted to determine the reason for operation.

Draining

If the heater is to be shut down and exposed to freezing temperatures, it must be drained. Water, if left in the tank and allowed to freeze, will expand and damage the heater.

1. Turn off the electrical switch and cold water inlet valve. (A hose must be connected to the drain valve to carry the water away.)

2. Open a nearby hot water faucet and the heater drain valve. (Be careful to grasp the drain valve handle in such a way that the hand will not be exposed to hot water.)

3. The drain valve must be left open during the shutdown period. (To restart heater, refer to the foregoing filling instructions.)

Chemical Vapor Corrosion

Water heater corrosion and component failure can be caused by the heating and breakdown of airborne chemical vapors. Spray can propellants, cleaning solvents, refrigerator and air conditioning refrigerants, swimming pool chemicals, calcium and sodium chloride, waxes, and process chemicals are typical compounds which are potentially corrosive. These materials are corrosive at very low concentration levels with little or no odor to reveal their presence.

Products of this sort should not be stored near the heater. Also, air which is brought in contact with the water heater should not contain any of these chemicals. If necessary, uncontaminated air should be obtained from remote or outside sources.

Maintenance

Electric water heater maintenance consists of cleaning the tank and removing lime (or scale) from the heating elements in hard water areas. Your dealer should be contacted for element cleaning. In some instances a hissing sound may be heard as the scale builds up. This noise is normal.

To assure long life and efficiency, the water heater tank must have a small amount of water drained periodically.

From time to time the drain valve should be opened and the water allowed to run until it flows clean. This will help to prevent sediment buildup in the tank bottom. Periodically the temperature and pressure relief valve should be checked to insure that it is in operating condition. Lift the lever at the top of the valve several times until the valve seats properly and operates freely.

CUSTOMER'S COPY

PACKING SLIP

IMPORTANT – THE BOOKS IN THIS SHIPMENT HAVE BEEN COUNTED AND THE COUNT VERIFIED AS BEING CORRECT. KINDLY CHECK YOUR COUNT WITH OURS BEFORE MATCHING TITLES AGAINST YOUR INVOICE. IN THE EVENT OF A CLAIM – THIS FORM MUST BE RETURNED WITH PARTICULARS.

B&T THE BAKER & TAYLOR COMPANY
A DIVISION OF W.R. GRACE CO.
GLADIOLA AVENUE, MOMENCE, ILLINOIS 60954

Ⓤ FORM #37

DATE: 5-15 ACCOUNT NO. 5-102814

PROCESSED	
PROCESSED - W/O JACKET	
KIT LOOSE & JACKET	
JACKET ONLY	
NAP	
KIT ONLY	
KIT WITH BOOK	
CARDS LOOSE	
TOTAL BOOKS	

PACKER 250 CHECKER AK

CARTONS	PACKAGES	BOOKS PACKED
	1	1

Caution: The water passing through the valve during this checking operation may be extremely hot.

Checklist

Before contacting your dealer, check the water heater to see if the apparent malfunction is caused by some external fault. Consulting this checklist may eliminate the need for a repair call and restore hot water service.

If there is not enough or no hot water check the following.

1. Be sure that the water heater electrical switch is turned to ON position. (In some areas an additional special meter, controlled by a timer, is used to govern the periods electricity is available. If the heater operates on a timed electrical circuit, recovery will be limited to certain hours.)

2. Check for loose or blown fuses in the water heater circuit.

3. If the water has been excessively hot and is now cold, the high temperature limit control may have operated. To restore service, contact your dealer or utility company. (*See* section on Temperature Regulation.)

4. The storage capacity of the heater may have been exceeded by large demands for hot water.

5. If the heater were installed when incoming water temperatures were warm, colder incoming temperatures will create the effect of less hot water.

6. Look for leaking or open hot water faucets.

Water is too hot. (*See* section on Temperature Regulation.)

Water heater makes sounds. Lime or scale has accumulated on the heating element(s) causing a hissing sound. (The noise is normal although the element[s] should be cleaned. Your dealer should be consulted for this service.)

When water leakage is suspected or encountered:

1. Check to see if the heater drain valve is tightly closed.

2. The apparent leakage may be condensation which forms on cool surfaces of the heater and piping.

3. If the outlet of the relief valve is leaking it may represent:

(a) Excessive water pressure.
(b) Excessive water temperature.
(c) Faulty relief valve.

Excessive water pressure is the most common cause of relief valve leakage. It is often caused by a "closed system." A check valve in the inlet system will not permit the expanded hot water volume to equalize pressure with the main. A relief valve must release this water or the water heater or plumbing system will be damaged.

4. When such a condition is encountered, local codes or inspection agency should be consulted to determine which system is acceptable in your area. These may consist of the following:

(a) Installation of a second relief valve with lower setting than the primary safety relief valve.

(b) An expansion tank of suitable pressure and provision to avoid water logging.

(c) Removal of the check valve.

If you cannot identify or correct the source of the malfunction:

1. Place the water heater electrical switch in the OFF position.
2. Close the cold water inlet valve to the heater.
3. Contact your dealer.

GAS WATER HEATER

Like all water heaters, this heater must be installed in accordance with local codes and utility requirements. In the absence of local codes, it is suggested that the recommended practices set forth in the ANSI booklet "Installation of Gas Appliances and Gas Piping" be followed.

The heater must be connected to a chimney. The vent pipe from the heater to the chimney must be the same diameter as the outlet of the draft diverter on the heater and should slope upward to the chimney at least ¼ inch per foot (.635 per 30.48 cm.)

The heater must be installed with a minimum of 2 inches (5.08 cm.) clearance from combustible surfaces to prevent a possible fire hazard condition.

Flammable items, pressurized containers or any other potential fire hazardous articles must never be placed on or adjacent to the water heater. Open containers of flammable material should not be stored or used in the same room with the water heater.

An accumulation of lint or other foreign material that

restricts or blocks the air openings to the heater or burner will, by reducing the amount of air necessary for combustion, create a hazardous condition.

Gas Connections

Refer to recommended installation as shown in Fig. 7. Before attaching the gas line to the water heater, be sure that all gas pipe used is clean on the inside. If dirt and scale from the gas pipe get into the thermostat, the gas valve in the thermostat will fail to close properly, causing the water heater to overheat. Care must be taken not to apply too much pressure when attaching gas supply pipe to thermostat gas inlet, in order to prevent damage to the thermostat.

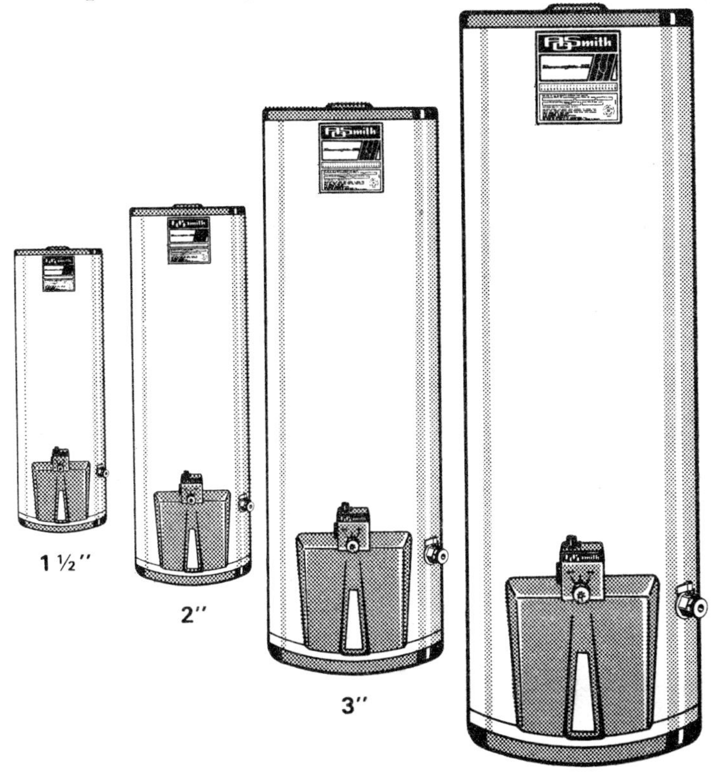

Fig. 7. Gas water heaters.

Air Adjustment

All except 65-, 70-, 80-, and 100-gallon models are designed with a fixed primary air setting that requires no adjustment. All 65-, 70-, 80-, and 100-gallon models should have the main burner air shutter adjusted to obtain a blue flame. To adjust proceed as follows. Loosen air shutter locknut, operate main burner, rotate air shutter to obtain blue flame. Tighten locknut.

Relief Valve

An unplugged relief valve opening is provided on top of the water heater for installing a temperature and pressure relief valve. An approved (A.S.M.E. or A.G.A.) ¾-inch by ¾-inch (1.905 by 1.905-cm.) temperature and pressure relief valve must be installed in the opening (Fig. 8). The drain line at this valve should terminate near a suitable drain. Do not thread, plug, or cap the end of this drain line. Pressure rating of relief valve must not exceed the working pressure shown on the rating plate of this heater. (*See* Fig. 8.)

Operation

Never operate the heater without first being certain it is filled with water and a temperature and pressure relief valve is installed in the relief valve opening on top of the heater. Do not attempt to operate heater with cold water inlet valve closed.

Filling

1. Close the water heater drain valve by turning the handle clockwise. (Place the protective cap furnished with the drain valve over the valve opening.)
2. Fully open the cold water inlet pipe valve allowing the heater and piping to be filled.
3. Open a nearby hot water faucet to permit the air in the system to escape.
4. Close the hot water faucet as water starts to flow.
5. The heater is ready to be lighted.

Lighting

Lighting and operating instructions are on a plate attached

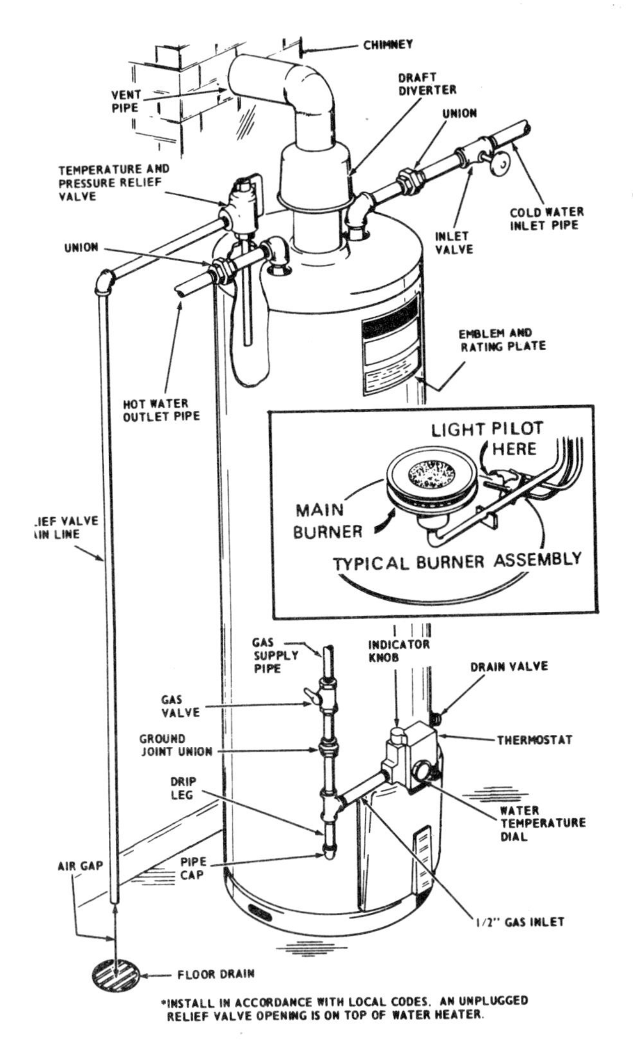

Fig. 8. Detailed gas water heater.

to the water heater. For convenience, these instructions are repeated here.

1. Fill heater with water. (*See* Filling Instructions.)

2. Turn thermostat indicator knob to OFF position. (*See* Fig. 8.)

(a) This shuts off gas supply to heater.

(b) Wait five minutes for any gas that might be in the combustion chamber to clear the heater.

(c) If closed, open gas valve in gas supply pipe.

(d) Turn indicator knob to *pilot* position.

3. Depress indicator knob and light pilot. (Continue holding indicator knob for one minute after pilot is lit. Pilot flame should remain on when knob is released.)

4. Turn indicator knob to ON position.

(a) Main burner will ignite.

(b) Adjust main burner air shutter to obtain a blue flame. (*See* Air Adjustment section previously described.)

5. Set water temperature dial to desired temperature. (*See* following section on Temperature Regulation.)

6. Repeat these instructions if it is necessary to relight heater.

Temperature Regulation

The water heater is equipped with a thermostat and with an automatic temperature control. (*See* Fig. 8.)

The water temperature dial is factory set for *warm* water temperature but may be adjusted as desired.

The water temperatures produced in the tank for each setting are approximately:

120°—WARM
160°—HOT

For general household usage, the midpoint setting will be satisfactory. However, the knob may be set to any position which satisfies temperature requirements.

Condensation

Whenever the water heater is filled with cold water there will usually be a certain amount of condensation formed on the cooler tank surfaces.

Drops of water may fall on the floor or the hot burner surfaces producing a sizzling sound. This is normal and will disappear as the water in the tank becomes heated.

Condensation appearing in the vent pipe (water dripping from draft diverter) during heater operation is evidence of poor vent action. Possible causes are too long a vent pipe or improper chimney operation.

To Shut Off Gas

The gas supply to the heater may be shut off by—

1. Turning the thermostat indicator knob on the thermostat to the OFF position (Fig. 8).
2. By closing the gas valve in the gas supply pipe (Fig. 8).

Draining

If the heater is to be shut off and exposed to freezing temperatures, it must be drained. Water, if left in the tank and allowed to freeze, will expand and damage the heater.

1. Turn off the gas and cold water inlet valves at the water heater.
2. Open a nearby hot water faucet.
3. Remove the protective cap from the drain valve opening. (Connect a hose to the drain valve to carry away the hot water.)
4. Open the drain valve by slowly turning the drain valve handle counterclockwise. (*Caution:* Grasp handle so that hand or body is not exposed to the flow of hot water. The drain valve has a high hot water discharge rate.)
5. The drain valve should be left open and the protective cap left off the valve during the shutdown period.
6. Refer to the section on Filling when restarting the water heater.

MAINTENANCE

To assure long life and efficiency, the water heater tank must have a small amount of water drained periodically.

Once a month the drain valve should be opened and the water allowed to run until it flows clean. This will help to prevent sediment buildup in the tank bottom. Once a month the temperature pressure relief valve should be checked to insure that it is in operating condition. Lift the lever at the top of the valve several times until the valve seats properly and operates freely.

Caution: The water passing out of the valve during this checking operation may be extremely hot.

CHECKLIST

Before contacting your dealer, check the water heater to see if the apparent malfunction is caused by some external fault. Consulting this checklist may eliminate the need for a repair call and restore hot water service.

High Temperature Limit Switch (Energy Cutoff)

All models have a factory installed non-adjustable limit switch that guards against excessive water temperatures.

If the high limit switch should open, it would reclose automatically when the tank water temperature drops. However, the heater must be relit as the main and pilot burners have been extinguished as a safety measure. Follow the lighting instructions in this chapter or on the side of the heater. Adjust the water temperature dial to a lower setting.

Continued pilot outage preceded by higher than usual water temperatures is evidence of high limit switch operation. It is important that your dealer be contacted to determine the reason for operation.

Not Enough or No Hot Water

1. Check to see if the pilot flame is lit.

(a) To relight the pilot, follow the instructions on the heater or in this chapter.

(b) Check to see if the gas valve in the gas supply pipe is partially closed.

(c) The thermostat may be set too low. Refer to the section on Temperature Regulation for details of water temperature adjustment.

(d) Check to see if thermostat indicator knob is in ON position (Fig. 8).

2. Look for leaking or open hot water faucets. Check for excessive usage.

3. Your gas company can check the gas input to the heater to see that it is correct. An underfired heater will not produce hot water at its normal recovery rate.

4. If the heater were installed when incoming water temperatures were warm, colder incoming temperatures will create the effect of less hot water.

5. If you cannot determine the cause of the problem, contact your dealer.

Water Temperature Is Too Hot

1. The water temperature dial may be set too high. Refer to section on Temperature Regulation for adjustment details.

2. If lower control settings do not reduce the water temperature, contact your dealer.

Gas Smell at the Heater

1. Close the gas valve in the gas supply pipe near the heater (Fig. 8). The thermostat includes a gas valve (indicator knob which can be closed).

2. Call your gas company.

Water Leakage Is Suspected

1. Check to see if the heater drain valve is tightly closed.

2. The apparent leakage might be condensation. In warm or humid locations condensation can accumulate and run from within the heater or its piping.

Note: When a water heater is first installed and filled, the bottom head of the tank might condense water. The water accumulation, if excessive, can drip into the floor shield. During normal operation there may be occasions when large quantities of water are drawn, chilling the tank bottom. This too can result in condensation.

3. If the leakage is from the temperature and pressure relief valve or its drain line it may represent a normal condition. In some installations a "closed system" may be created by a check valve or a water meter containing a check valve. In these installations the relief valve will periodically operate. Continuous relief valve leakage should be investigated by your dealer.

4. If you cannot identify or correct the source of water leakage proceed as follows.

(a) Shut off the gas valve in the gas supply pipe at the heater.

(b) Close the cold water inlet valve to the heater.

(c) Contact your dealer.

Water Heater Makes Sounds

1. Occasional excessive condensation, as explained under Leakage, can cause a sizzling sound as the moisture is vaporized by the gas flame. This is a normal sound and may be disregarded.

2. A sediment accumulation may be causing a rumbling noise. Contact your dealer for details of flushing the heater.

3. If you cannot identify or remedy the condition, contact your dealer.

SAFETY PROCEDURES FOR OIL FIRED WATER HEATERS

Be sure to turn off power when working on or near the electrical system of the heater. Never touch electrical components with wet hands or when standing in water. When replacing fuses always use the correct size for the circuit.

INSTALLATION

General

The heater (Fig. 9) must be installed in accordance with local codes and utility requirements. In the absence of local codes, it is suggested that the recommended practices set forth in the following publications be followed.

Standard For The Installation of Oil Burning Equipment, NFPA No. 31. Available from National Fire Protection Association, 60 Battery-March Street, Boston, MA 02110,

Code For the Installation of Heat Producing Appliances. Available from American Insurance Association, 85 John Street, New York, NY 10038,

The National Electrical, NFPA No. 70. Availability same as NFPA No. 31.

Fig. 9. Tank-type oil-fired water heater.

Flue Gas Disposal

The chimney connector diameter should be the same size as the heater flue outlet. A minimum rise of ¼ inch per foot (.635 per 30.48-cm.) of horizontal connector length must be maintained between the heater and chimney opening. The connector length should be kept as short as possible.

A draft regulator must be installed in the same room as the heater. Locate the regulator as close as possible to the heater and at least 18 inches (45.72 cm.) from a combustible ceiling or wall. A manually operated damper should not be placed in the chimney connector.

Clearances

The heater must be installed with a minimum of 6 inches (15.24-cm.) clearance from combustible surfaces (sides and top) to prevent a possible fire hazard condition.

At least 24 inches (60.96-cm.) should be provided at the front of the unit for proper servicing.

Some models must be installed on non-combustible floors.

Flammable items, pressurized containers or any other potential fire hazardous articles must never be placed on or adjacent to the water heater. Open containers of flammable material should not be stored or used in the same room with the water heater.

FUEL SYSTEMS

The standard for the installation of oil burning equipment (NFPA No. 31), local codes, and these instructions must be adhered to when installing the tank and piping. In addition, an oil pump installation sheet and oil burner certificate are packed with the burner for use and completion by the installer.

The manual and the completed oil burner certificate are to be left with the user for future reference.

Figure 10 shows a typical single stage one- or two-line fuel system. When two or more tanks are connected to one burner, the supply line from each tank should run to a header fitted with an approved three-way valve. Normally only one tank may be drawn at a time unless local codes permit simultaneous feeding of two tanks on gravity type installations.

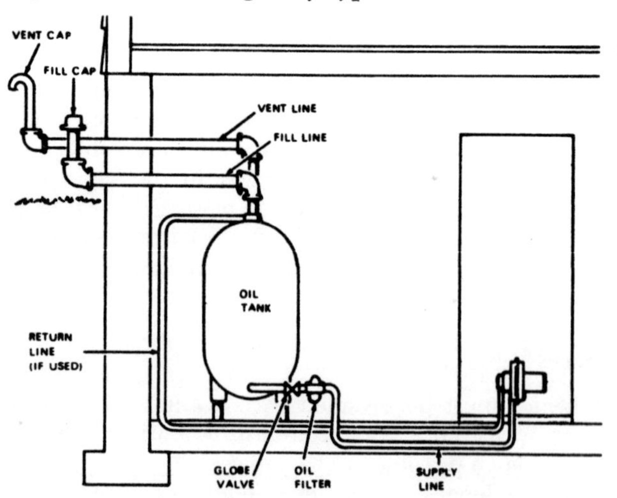

Fig. 10. Typical single stage one or two line full system.

System Types

Single Stage, One Line. The bottom of the oil storage tank must be above the level of the fuel unit (Fig. 11). The fuel oil will flow by gravity to the burner. A single pipe is run between the tank and fuel unit. The heater shown in Fig. 10 is for this type of service as shipped (the by-pass plug is not installed).

Single Stage, Two Line. This type of system is self-priming (Fig. 12). The heater shown in Fig. 17 with by-pass plug field installed is for this type of service.

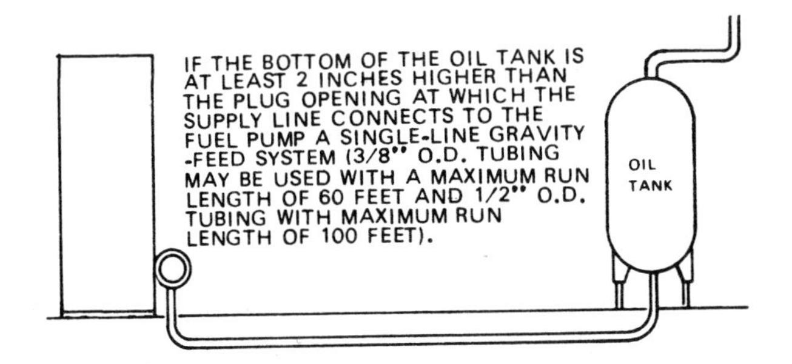

Fig. 11. Single stage, one line oil storage tank.

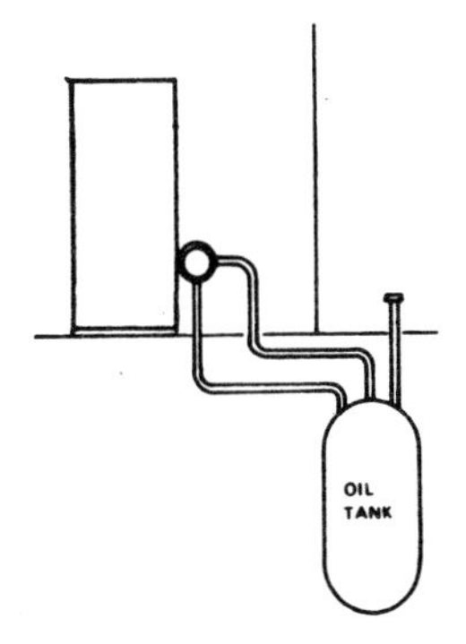

Fig. 12. Single stage, two line oil storage tank.

TABLE 10
SINGLE STAGE PUMP.

Distance Tank Bottom Below Pump Plug (Feet)	Max. Run Length (Ft.)		Distance Tank Bottom Below Pump Plug (Feet)	Max. Run Length (Ft.)	
	3/8" O. D. Tubing	1/2" O. D. Tubing		3/8" O. D. Tubing	1/2" O. D. Tubing
1	66	100	6	36	100
2	55	100	7	31	100
3	50	100	8	26	100
4	45	100	9	21	83
5	40	100	10	16	64

Two Stage, Two Line. This system is required when long lines and high lifts (requiring up to 20 inches of vacuum—10 feet vertical lift) are encountered (Fig. 13). A two-stage pump may be used on one-line, gravity feed installations.

TABLE 11
TWO STAGE PUMP.

Distance Tank Bottom Below Pump Plug (Feet)	Max. Run Length (Ft.)		Distance Tank Bottom Below Pump Plug (Feet)	Max. Run Length (Ft.)	
	3/8" O. D. Tubing	1/2" O. D. Tubing		3/8" O. D. Tubing	1/2" O. D. Tubing
1	74	100	9	51	100
2	71	100	10	48	100
3	69	100	11	45	100
4	66	100	12	42	100
5	63	100	13	39	100
6	60	100	14	37	100
7	57	100	15	34	100
8	54	100			

Multiple Heater Fuel Lines

Where two or more heaters form a water heating system, each burner shall have an entirely separate oil supply line run from the tank to the burner.

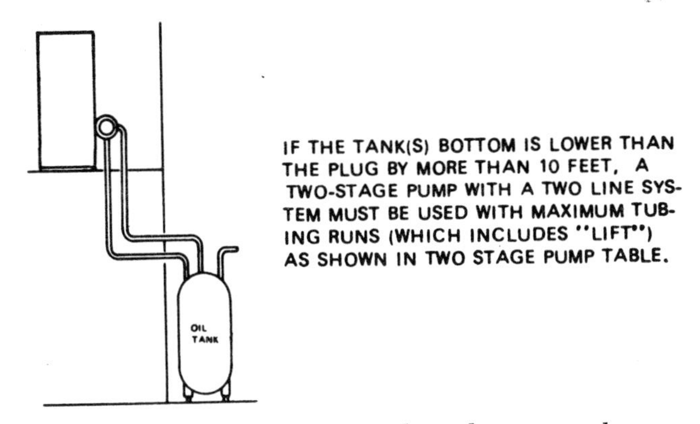

Fig. 13. Two stage, two line oil storage tank.

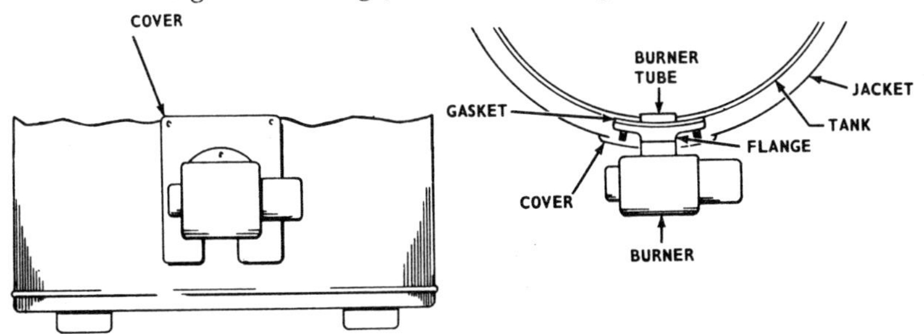

Fig. 14. Connecting oil line(s) and electrical wires to burner.

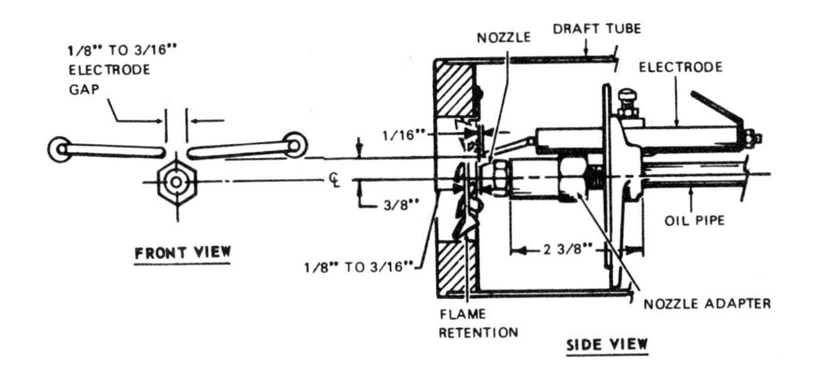

Fig. 15. Firing assembly dimensions.

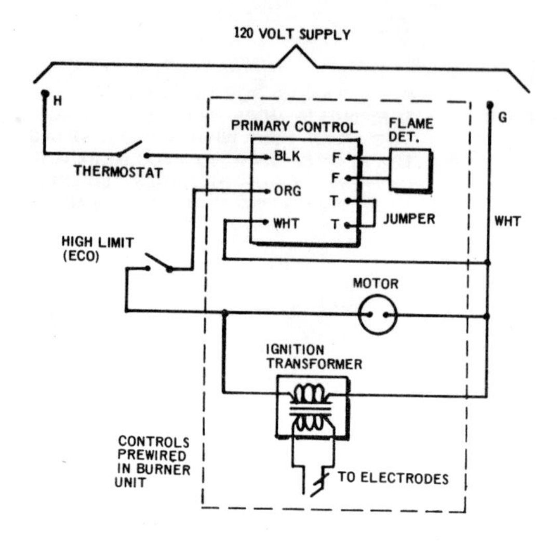

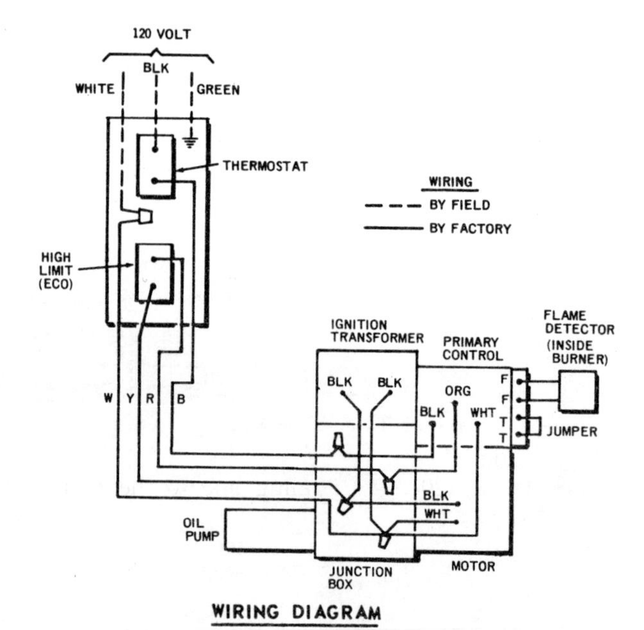

Fig. 16. Schematic and wiring diagrams.

Relief Valve

An unplugged ¾-inch relief valve opening is provided on top of the water heater for installing a listed temperature and pressure relief valve.

Install temperature and pressure protective equipment required by local codes, but not less than a combination temperature and pressure relief valve certified as meeting requirements by a nationally recognized testing laboratory that maintains periodic inspection of production of listed equipment or materials. The drain line of this valve should terminate near a suitable drain. Do not thread, plug, or cap the end of this drain line.

The pressure setting of the relief valve should not exceed the pressure capacity of any component in the system. The temperature setting of the relief valve should not exceed 210°F.

Burner

The burner model number includes a code which identifies the major features of the oil burner. The burners are to be used with fuel oil not heavier than No. 2.

The oil burner is connected to the heater at the factory. The installer is only required to hook up the oil line(s) and make the necessary electrical connections.

Connect the oil line(s) and electrical wires to the burner as follows.

1. The oil pump manufacturer's instructions should be checked for connection and bleeding information. (The burner is approved for use with fuel oil not heavier than No. 2.) (*See* Figs. 14 and 15.)

2. An approved, separately fused circuit with disconnect switch should be available for the oil burner. Using the wiring diagram as a guide,

(a) Connect the 120-volt incoming line to the thermostat as shown in Fig. 16.

(b) Ground the heater in accordance with the NEC to guard against electrical shock from the heater or water system.

3. All burners have "intermittent ignition" as defined by UL. (Ignition is on during the time the burner is on and off when the burner is off.)

4. Do not "test fire" the heater to complete the Oil Burner Certificate until the tank is filled with water. (*See* the following section on Operation.)

An oil burner certificate is furnished with the heater. After the heater has been installed, perform all necessary tests and record results on the certificate. The certificate, instructions for the heater, and the guarantee must be left with the user for future reference.

Operation

Never operate the heater without first being certain it is filled with water and a temperature and pressure relief valve is installed in the relief valve opening on top of the heater.

Filling

1. Turn off the oil burner electrical disconnect switch.
2. Close the heater drain valve.
3. Open a nearby hot water faucet to allow the air in the system to escape.
4. Fully open the cold water inlet valve, filling the heater and piping.
5. Close the hot water faucet as water starts to flow from the opening. The heater is now ready for start up and temperature regulation if being placed in operation for the first time.

Start Up

The following checks should be made by the installer when the heater is placed into operation for the first time.

1. Check all factory and field made water, oil, and electrical connections for tightness. Also check flue gas disposal provisions on top of the heater. (Repair any water and oil leaks. Tighten electrical flue connections as necessary.)

Be sure the oil burner, related piping, valves, and controls are in place, adjusted and ready for operation before turning on the electricity.

Temperature Regulation

Set thermostat for desired water temperature. (It is suggested the thermostat be set to the lowest setting which satisfies the hot water requirements of the system. This helps minimize scale formation in the heater.)

Air Adjustment

The air adjustment is located next to the opening in accordance with the instructions in the Combustion Test Specifications section. Figure 17 shows two types of air adjustment in use.

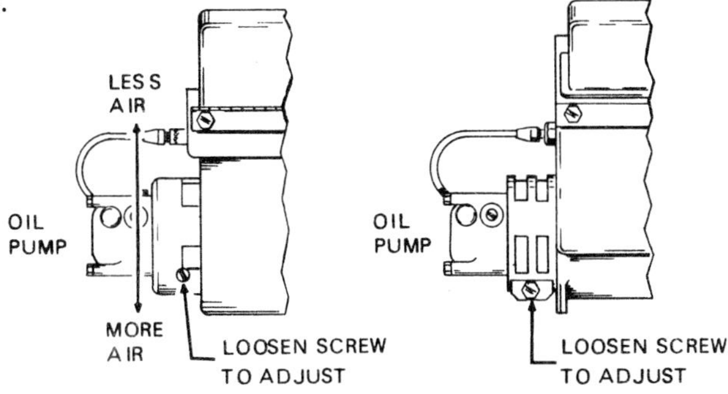

Fig. 17.

Burner Certificate (Combustion Test)

The Commercial Standard CS75-56 Oil Burner Certificate form must be filled in and posted in the vicinity of the water heater.

Instructions for filling in are on the back of the certificate. This must be done by the installer at the time the heater is first operated.

(*See* the water heater Combustion Test Specifications later in this chapter. The heater must be filled with water prior to burner operation.)

Draining

The water heater must be drained if it is to be shut down and exposed to freezing temperatures. Maintenance and service procedures may also require draining the heater.

1. Turn off the oil burner electrical disconnect switch. (If required by the reason for draining the heater, turn off the oil line supply valve.)

2. Close the cold water inlet valve to heater.
3. Open a nearby hot water faucet to vent the system.
4. Open the heater drain valve.
5. If the heater is being drained for an extended shutdown, it is suggested the drain valve be left open during this period.

Follow filling instructions when restoring hot water service.

MAINTENANCE

Periodically the temperature pressure relief valve should be checked to insure that it is in operating condition. Lift the lever at the top of the valve several times until the valve seats properly and operates freely.

Caution: The water passing out of the valve during this checking operation may be extremely hot.

Water heater maintenance includes periodic tank flushing and cleaning, and removal of lime scale. The oil burner should be inspected and adjusted to maintain proper combustion. When used, the water heating system circulating pump should be oiled.

Initially, maintenance should be performed on schedule until experience indicates the interval for a given operation should be changed—to a shorter or longer interval. Oil burner inspection and adjustment should be performed by a competent technician. (*See* Fig. 18.) For performing some of the recommended maintenance proceed as follows.

Flushing

1. Turn off the oil burner electrical disconnect switch.
2. Open the drain valve and allow water to flow until it runs clean.
3. Close the drain valve when finished flushing.
4. Turn on the oil burner electrical disconnect switch.

Sediment Removal

Water borne impurities consist of fine particles of soil and sand which settle out and form a layer or sediment on the bottom of the tank.

For convenience, sediment removal and lime scale removal should be performed at the same time.

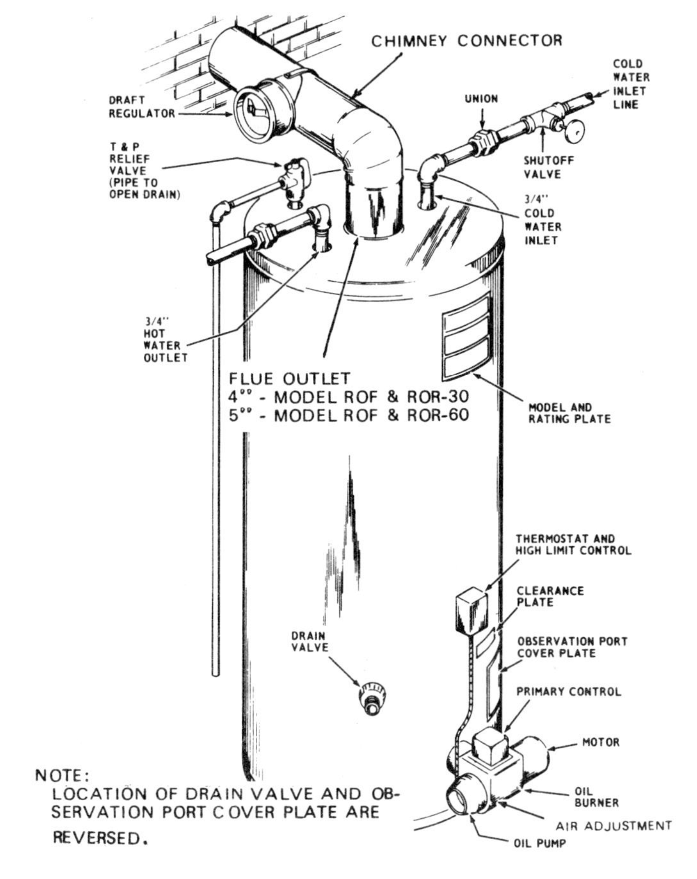

Fig. 18. Detailed oil fired water heater.

Lime Scale Removal

It is normal for lime scale to accumuate in the heater tank. Factors which affect the amount of this formation are as follows.

1. Amount of hot water used. As the volume of water heated increases, more scale results.
2. Water temperature. As the temperature of the water is increased, more scale is deposited.
3. Characteristics of water supply. Regardless of water treatment, the tank bottom should be examined regularly.

Sediment and lime scale accumulations reduce the ability of the heater to heat water and may cause noises to occur during operation.

Sediment and lime scale removal may be accomplished through the ¾-inch (1.905-cm.) drain valve opening. The heater must be drained. (*See* Draining section, before removing drain valve.)

Circulating Pump

The water heater or water heating system may include a *circulating pump*. Where used, it should be lubricated once every four months with SAE No. 20 non-detergent motor oil or as directed by the manufacturer. (Place two to three teaspoonsful in the bearing oil cup and 10 to 12 drops in the motor oil cups.)

Field installed replacement circulating pumps should be of all bronze construction.

Oil Burner Motor

The *oil burner motor* has two oil cups. Annually lubricate the motor with SAE No. 20 non-detergent motor oil.

CHECKLIST

Before calling for service, check the following points to see if the cause of trouble can be identified and corrected. Reviewing this checklist may eliminate the need of a service call and quickly restore hot water service.

Be sure to turn off the electricity when checking equipment.

TABLE 12
MAINTENANCE SCHEDULE.

Component	Operation	Interval	Required
Tank	Flushing	Monthly	
	Sediment Removal	Semi-Annually	
	Lime Scale Removal	Semi-Annually	BorCoil Delimer
Circulating Pump	Oiling	Four Months	SAE No. 20 non-detergent motor oil.
Oil Burner Motor	Oiling	Semi-Annually	SAE No. 20 non-detergent motor oil.
Oil Burner	Inspection and Adjustment	Semi-Annually	Combustion test kit & test specifications
	Nozzle Replacement	Semi-Annually	New Nozzle

Not Enough or No Hot Water

1. Be certain the oil burner electrical disconnect switch serving the water heater is in the ON position.
2. Check the fuses. (The oil burner electrical disconnect switch usually contains fuses.)
3. The capacity of the heater may have been exceeded by a large demand for hot water. (Large demands require a recovery period to restore water temperature.)
4. Cold incoming water temperature will lengthen the time required to heat water to the desired temperature. (If the heater were installed when incoming water temperature was warm, colder water creates the effect of less hot water.)
5. Look for hot water wastage and leaking or open hot water faucets.
6. Sediment or lime scale may be affecting water heater operation. Refer to the section on Maintenance for details.
7. Heater or burner may be dirty.
   (a) Clean all flue passages and the smoke pipe.
   (b) Have burner properly cleaned and readjusted.

8. Burner may not be firing at proper rate.
   (a) Check nozzle size.
   (b) Check oil pump pressure setting.

9. Burner may be short cycling. Short cycling (too frequent off and on) of burner will cause sooting. If unit or burner become dirty at frequent intervals, after correcting the "dirt condition" also correct the control settings (or other cause of the short cycling).

10. Oil burner fan wheel may be dirty. (Clean fan wheel with a stiff brush.)

11. Draft regulator may be stuck. (Check to see if vane swings freely. Clean if vane is stuck.)

12. Primary control safety reset is open. (Reset safety switch on burner mounted primary control.)

13. Burner motor safety reset is open. (Reset safety switch on motor.)

Burner Starts—Will Not Operate

1. No oil in tank.
2. Oil line valve closed.
3. Loose connection in primary control. (Check and tighten all wire connections.)
4. Electrodes out of adjustment. (*See* Fig. 15.) (Clean firing head and readjust electrodes.)
5. Clogged burner nozzle. (Replace with new nozzle.)
6. Dirty filter. (Replace element in oil filter.)

Water Is Too Hot

1. (*See* section on Temperature Regulation.)

Water Heater Makes Sounds

1. Sediment or lime scale accumulations cause noises when the heater is operating. (The sounds are normal; however, the tank bottom should be cleaned. *See* section on Maintenance for details.)

2. Some of the electrical components of the water heater make sounds which are normal.
   (a) Contacts "click" or snap as the heater starts and stops.
   (b) Transformers often hum.

Water Leakage Is Suspected

1. Check to see if the water heater drain valve is tightly closed.
2. The apparent leakage may be condensation which forms on cool surfaces of the heater and piping.
3. If the outlet of the relief valve is leaking it may represent the following.
   (a) Excessive water pressure.
   (b) Excessive water temperature.
   (c) Faulty relief valve.

Excessive water pressure is the most common cause of relief valve leakage. It is often caused by a "closed system." A check valve in the inlet system will not permit the expanded hot water volume to equalize pressure with the main. A relief valve must release this water or the water heater or plumbing system will be damaged.

When such a condition is encountered, local codes or inspection agency should be consulted to determine which system is acceptable in your area. These may consist of the following.

1. Installation of a second relief valve with lower setting than the primary relief valve.
2. An expansion tank of suitable pressure and provision to avoid water logging.

If You Cannot Identify or Correct the Source of Malfunction

1. Place the oil burner electrical disconnect switch in the OFF position.
2. Close the cold water inlet valve to the heater.
3. Contact your dealer.

COMBUSTION TEST SPECIFICATIONS

General

A *combustion test kit* capable of testing CO_2 content, stack temperature, draft, and smoke must be available to aid in adjusting the unit and filling out the Oil Burner Certificate. A pressure gage is needed to measure and adjust oil pump pressure.

Procedure

1. Check nozzle size.
2. Open air band about halfway and, being certain heater is filled with water, start burner.
3. Check oil pump pressure with pressure gage. Pump pressure should be 100 psig. Adjust setting as necessary.
4. Allow burner to operate for 15 minutes before proceeding with test.
5. After 15 minutes' operation, check the heater outlet draft (between the heater and draft regulator) and adjust the draft regulator until the correct reading is obtained.

HEATER OUTLET DRAFT	.02 to .04 Inches of Water

6. Decrease air supply by closing air band until flame has smoky tips. Immediately increase air supply until the smoky tips just disappear.
7. Using combustion test kit, check smoke density and CO_2 in the chimney connector. (Adjust air supply with an acceptable smoke density reading. Test and readjust as necessary.)

SMOKE DENSITY	Preferably No. 1 spot (Not over No. 2 spot)
CO_2	10 to 12%

8. Check the stack temperature in the chimney connector about halfway between the heater and the draft regulator. (If stack temperature is too high, check for a soot accumulation in heater or excessive oil pump pressure.)

STACK TEMPERATURE	Min. 400°F Max. 650°F

9. Recheck combustion efficiency against specifications when final adjustments have been made.

HYDRONICS

Hydronics signifies modern forced circulation hot water heating (also known as *baseboard* and *radiant heating*). It is the method of using fluids in a sealed system to provide heating and/or cooling for indoor comfort. Although steam is used for hydronic heating in some large projects, the most common method is to circulate heated water in the winter, and chilled water in the summer, using the same distribution system and room outlets. The equipment can be used concurrently to provide domestic hot water to the house or apartment faucets and to the laundry, as well as swimming pool heating and for snow-melting in driveways where needed.

FOR HEATING

Hydronic boilers may be installed almost anywhere—in the basement, in a closet, or in a corner of the kitchen. Advanced engineering provides top efficiency, giving you more heat for your money with any fuel—gas, oil, or electricity. A small pump in the boiler circulates the hot water quietly through finger-size tubing to room heating units, usually baseboard units, convectors or modern recessed radiators.

The water is heated in a boiler fired by natural fuels such as gas or oil, or by electricity where the rates are sufficiently low. The boiler is delivered to the house as a complete package with all operating accessories in place.

Where there is heating without air conditioning the most common terminal units in the rooms are ½-inch (1.27-cm.) or ¾-inch (1.905-cm.) baseboard panels, and each house or apartment is provided with a *thermostat* to control its own heat through a zone valve or circulator. The baseboard heater is made of copper tubing with metal fins, enclosed in a low protective panel and installed at the floor level along the outside wall of the room.

A circulator, mounted adjacent to the boiler, sends hot water through the tubing mains. When the thermostat calls for heat, its zone valve opens permitting positive circulation through the room baseboard. The water gives off some of its heat to the room and then flows back to the boiler. Whenever the boiler water requires additional temperature the burner comes on automatically, activated by a *built-in control.*

After the initial filling of the system the equipment is energized by turning on the burner switch, and no further work is required except for semi-annual oiling of the motor on the circulator and the oil burner.

An *expansion tank* mounted above the boiler provides an internal air cushion to permit the heated water to expand, and the water can be heated to high temperatures without boiling, as it is a sealed system. The tank also serves to eliminate any air which may have entered the system during the initial fill.

Careful balancing of flow to each house or apartment is not required because the thermostats automatically adjust the zone valves to provide suitable heat as needed by the occupants. If the zone control is not provided, you can control the heat by adjusting the *fingertip damper*, provided as standard on each baseboard heater.

The cost of installing the hydronic system is closely comparable with any other modern system.

FOR AIR CONDITIONING AND HEATING

The hydronic heating system is compatible with any air conditioning system, the choice of the latter being dependent on local custom and competitive conditions. One familiar method is the use of *individual through-the-wall* air conditioners, mounted slightly above the hot water baseboard, and is applicable where internal and external noise of the unit does not create a problem. A variation of this in some northern areas is to install wall sleeves.

In some multifamily houses it is found more feasible in comfort and economy to install an independent cooling system with ducts, the registers being mounted high on an inside wall for best results. The hot water baseboard is installed along the outside wall and operates separately, thereby eliminating the ON-OFF cycling of blowing air during winter operation.

Another method of cooling independently is the use of *valance units*, a finned element concealed in a troffer at the ceiling line along the outside wall, using circulating chilled water to reduce room temperature and humidity.

Energy Conservation

A new method now in use is the combination of hydronic heating and direct-expansion cooling, incorporated into one console installed under the window. The piping and controls are arranged to provide alternate heating or cooling to various sides of the house, particularly during interim seasons, depending on sun and wind loads.

This type of equipment and other sophisticated systems can be arranged for *energy conservation*, in that the heat gain is absorbed from the warm areas by the air conditioning, and transmitted by the flowing water to other areas where that heat in the water is utilized. Conversely, when the warm water has given up some of its heat on the cool side of the house, it can then be directed to aid in cooling the other side, thereby reducing energy costs.

Fan-Coil Operation

The most common method of providing a unified hydronic system for both heating and cooling is the *fan-coil unit*, consisting of a small centrifugal fan and a heat-transfer coil mounted together in a casing. A *two-pipe connection* from the tubing main provides chilled water to the coil in summer and hot water in winter.

Such units can be provided in various forms—as an *exposed cabinet* installed under the window, or a compact, *sheet-metal unit* concealed in a soffit or in a vertical chase. A separate fan-coil unit may be provided for each major room, or one concealed unit can serve an entire house or apartment by running small flexible ducts to each room.

The main water tubing for the system comes from the equipment room where there is a boiler and a water chiller. At seasonal changes twice a year the chiller or the boiler is alternately valved off, manually or automatically.

For combined heating and cooling, the fan-coil system provides many practical benefits, and primary among these is the reduced operating and maintenance cost, as well as simplified, low-cost installation.

Cost Reductions

Unitary air conditioners such as the through-the-wall types, can provide from five to ten B.T.U.s of cooling per watt of electrical input, while central equipment normally provides 13 to 15 B.T.U.s per watt. As the equipment will cycle off automatically when satisfied, this can result in major savings. Maintenance of central equipment provides additional savings in that the machinery is built to industrial standards, rather than a residential quality. Centralized service, being limited primarily to the equipment room, also reduces the need for maintenance personnel to enter the house or apartments.

Another economic consideration is the diversity factor which is based on the fact that all apartments will not require cooling simultaneously. For this reason it is good engineering practice to install central equipment with a cooling capacity substantially below the total calculated heat gain of the building, usually from 40 percent to 65 percent less than the expected total load. However, if individual units are installed, each one must be large enough to fully cool its own area, so the advantage of diversity is lost.

Another important savings in initial cost results from the elimination of large ductwork. The use of tubing in sizes from ½ inch to 2 inches (1.27 to 5.08 cm.) instead of bulky sheet metal ducts reduces space requirements and installation costs, and eliminates the need to insulate ducts.

HYDRONIC CONTROL CENTER

The hydronic control center as shown in Fig. 19 is located in the center hall, and includes (left to right) humidistat, air cleaning control, and thermostat for heating and cooling. This thermostat controls the heating in the bedroom area. A second heating zone covers the family living area—living room-dining room, family room and kitchen—and its thermostat is located in the family room.

Fig. 19.

Fig. 20.

Fig. 21.

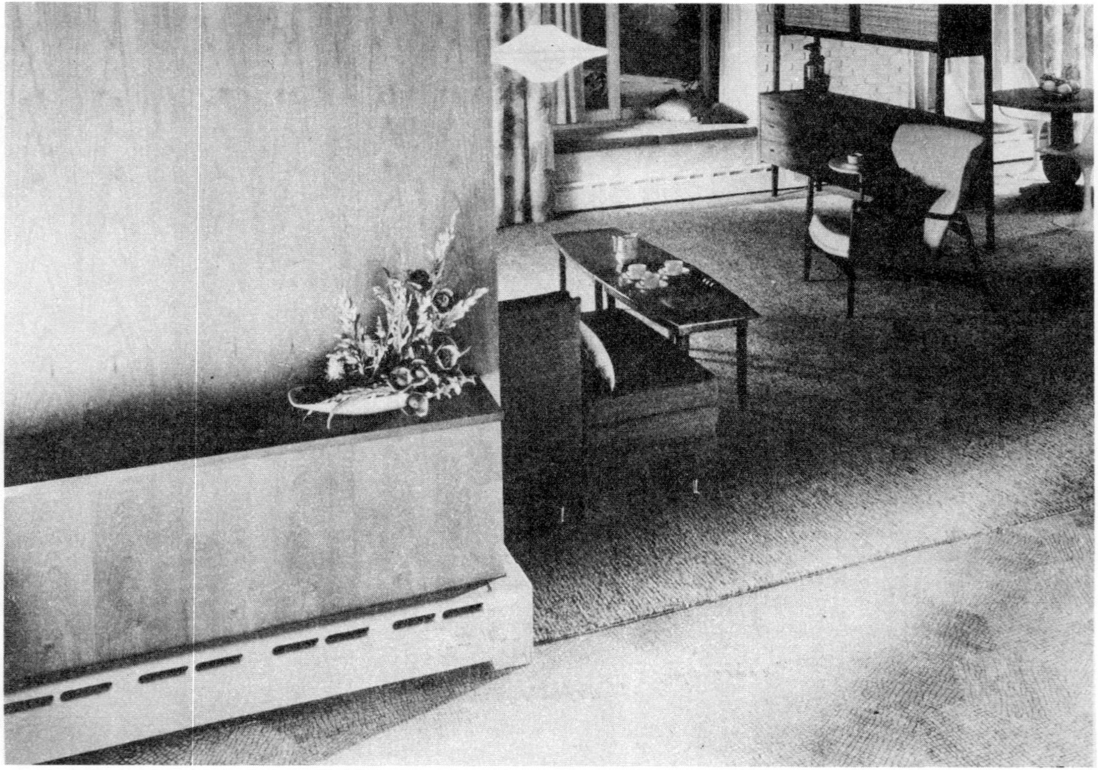

Fig. 22.

HYDRONIC (HOT WATER) BASEBOARD HEATING

Replacing old-fashioned radiators (Fig. 20) with modern hydronic (hot water) baseboard heating (Fig. 21) gives the home decorator more freedom and also provides more comfort. Since the baseboard never gets too hot to the touch, carpeting can lay wall-to-wall, and draperies can hang to floor length without fear of scorching.

Streamlined hydronic (hot water) baseboard heating units as shown in Fig. 22 replace ordinary wood baseboard to provide even, draft-free heat through the home. This type of heating is cleaner and more economical than old-fashioned methods of heating. The wall-hugging characteristics of these heating units simplify home decorating too. There is more room for furniture, draperies can be hung full length, and wall-to-wall carpeting is easily installed.

Section II

Home Heating

PERFECTION IN HOME HEATING is attained when the occupant is unaware of the heating system regardless of weather conditions. The common interpretation of heat is that we must have heat energy to keep us warm. Actually, it would be more accurate to state that heat energy is necessary to our comfort in order that the body will not lose an excessive amount of heat to surrounding objects or to the atmosphere. Thus, the body is not heated but the surroundings are heated, so that the body will lose its heat at the proper rate and thus be comfortable. For this very basic reason an even supply of heat is desirable.

Bodily loss of heat is listed as 40 percent to 60 percent by *radiation*, 15 percent to 30 percent by *convection*, and 20 percent to 30 percent by *evaporation*. Losses give us the uncomfortable feeling of being cold.

It is interesting to consider other means by which we avoid these losses. Doubtless it is a common experience to feel cold even though the thermometer indicates a high degree of temperature. One reason for this discomfort could be that the relative humidity is too low. During the winter months the air is dry and moisture must be artificially added. With a proper amount of moisture in the air, evaporation of bodily moisture is slowed down and the uncomfortable coolness caused by this evaporation is decreased. Also, if excessive air movement causes chill, we often avoid such losses by means of close fitting clothing. Body heat losses by radiation and conduction are held to a minimum by wearing clothing known for its heat insulating qualities.

Chapter 10

Systems, Fuels, and Controls

With the development of central heating, the heating plant was removed from the living quarters, taking with it the unpleasant aspects of dirt, smoke, and soot. Central heating not only made possible a cleaner, more desirable system but also offered lower operating costs as well as the advantages of space saving.

Today, heating equipment is available for hot water, steam, or warm air heating systems.

Since present day heating equipment must not only heat the space but must also offset the various heat losses due to the construction of the dwelling, it would be well to remember throughout the reading of this chapter the causes of these losses. Heat losses are caused by heat transmission through the walls, windows, doors, ceilings, floors, and by infiltration of outside air. This heat loss, as calculated by the specific method, is equal to the area (such as walls, windows, and so on) times its coefficient of heat transmission times the temperature differential. (Desired inside temperature minus outside temperature.) The infiltration losses are calculated by either the approximate method (room volume change/hour) or by the specific method which uses data based on actual air leakage per foot of crack around windows and doors.

HEAT TRANSFER

When fuel is burned in the combustion chambers the heat is transferred to some medium which carries it to the space being heated. The mediums used for heat distribution are water, steam, and air. To understand their place in space heating some fundamentals of heat transfer must be considered.

Water, steam, and air heat space in three different ways: by convection, by conduction, or by radiation.

Convection

Warm air heat is an example of heat transfer using air as its medium. Warm air heat dissipates its heat to space through the movement of the medium air. These air currents are called *convection currents.* The convection currents are caused by the fact that warm air is lighter per unit volume than colder air and, therefore, rises above the cold air. The trade winds north and south of the equator are examples of convection currents.

Convection currents are also found in water. When water is heated from the bottom the warm water moves to the top since its weight per unit volume is less than that of colder water. These currents continue movement of the water so long as any temperature differential exists. Naturally the force of the movement has a limited value. Thus, these currents extend only so far as the temperature differential and the friction-of-flow allow them. We may create convection currents in a pot of water but the same amount of heat would not extend these currents for an extreme distance. The same is true for convection currents caused by warm air. In warm air heating the friction-of-flow is very important.

Conduction

Examples of heat are easily found. Heat energy moves from the warmer end of a metal bar to the colder end. The radiator section in a warm air furnace conducts the heat from the hot gases inside the casting to the outside surface where the air for the house is warmed by its movement over the surface. Also, hot water and steam radiators conduct the boiler water heat from the inside to the outside tube surface.

Radiation

Radiant heat requires no medium of transfer. Radiant heat energy is a wave motion which moves in a straight line from its source. When radiant heat rays impinge upon a solid object it is warmed, absorbing part of the energy and reflecting the remainder. Radiant heat rays are emitted from the surface of a warm object and move toward a colder object.

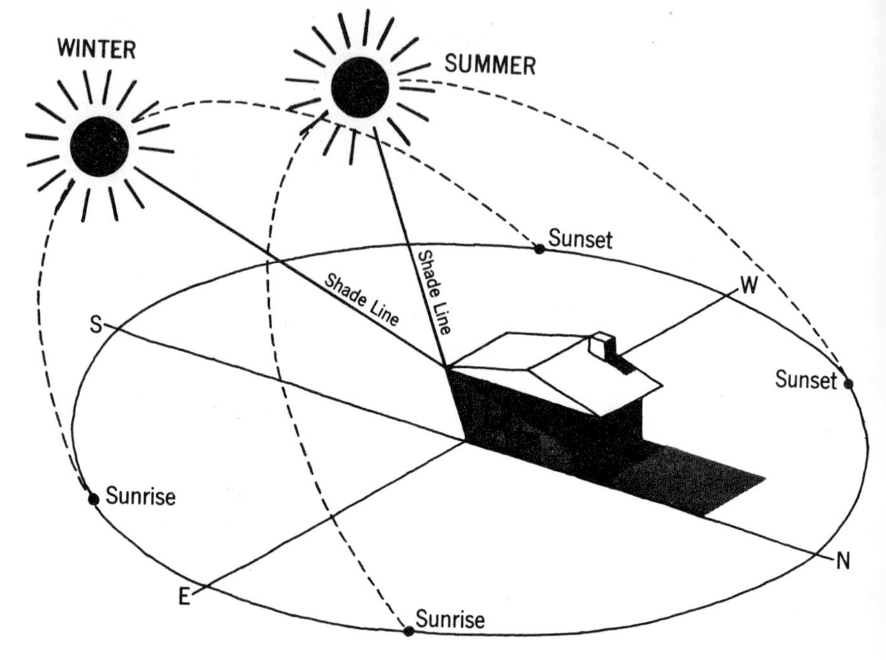

Fig. 1. Solar orientation of a house to reduce heat requirements.

HOW TO REDUCE HEAT REQUIREMENTS

Today research and the age of technology have brought dramatic changes in home heating and cooling. *Boilers* are now appliance styles and operate automatically. Trim, unobtrusive *baseboard heating panels* create ideal indoor weather in winter for the new or older home. Finally, recent developments make summer comfort a practical reality for the average homeowner with new equipment capable of providing efficient cooling for any style or size house.

Much can be done to reduce the heat requirements in your house. This in turn, can reduce heating costs and increase personal comfort.

New houses may be oriented so that the main rooms and the large windows in the rooms face south to receive maximum sunlight in the winter (Fig. 1). (In summer, the sunlight may be shaded out by trees, wide eaves, shutters, awnings, or other natural or artificial shading.)

Tight construction also reduces heat requirements. Insulate ceilings and outside walls. Calk and weatherstrip joints. Install storm sash or double- or triple-glazed windows to reduce heat loss through the windows. An old house should always be repaired and insulated before a new heating system is installed.

The chimney is a part of the heating plant; proper construction and maintenance are important. Chimneys should extend a minimum of two feet above the roof ridge. Manufacturers usually specify the size of flue required for heating equipment. Keep flues clean and free from leaks.

WARM AIR HEATING

Area heating units, which include stoves, circulator heaters, and "pipeless" furnaces, are installed in the room or area to be heated. In central systems, the heating unit is located in the basement or other out-of-the-way place and heat is distributed through ducts.

Central heating systems are the most efficient and economical method of heating.

Stoves, Circulator Heaters, and Pipeless Furnaces

Stoves are one of the simplest heating devices. Although they are cheaper than central heating systems, stoves are dirtier, require more attention, and heat less uniformly. If more than one stove is used, more than one chimney may be needed.

Wood- or coal-burning stoves without jackets heat principally by radiation. Jacketed stoves or circulator heaters heat mainly by convection and are available for burning the four common fuels—wood, coal, oil, or gas.

With proper arrangement of rooms and doors, a *circulator heater* can heat four or five small rooms, but in many instances heating will not be uniform. A small fan to aid circulation will increase efficiency. The distance from the heater to the center of each room to be heated, measured through the door opening, should be not more than about 18 feet. Doors must be left open—otherwise, grilles or louvers are needed at the top and bottom of doors or walls for air circulation.

Pipeless furnaces may be used in smaller houses. They discharge warm air through a single register placed directly over the furnace. Units that burn wood, coal, gas, or oil are available for houses with basements. Gas- and oil-burning units, which can be suspended beneath the floor, are available for houses without basements.

Small gas-fired vertical heaters are sometimes recessed in the walls of the various rooms. Such units may be either manually or thermostatically controlled. Heater vents are carried up through the partitions to discharge the burned gases through a common vent extending through the roof.

It is best to buy a heating unit designed specifically for the fuel to be used. Coal or wood burners can be converted to oil or gas but usually do not have sufficient heating surface for best efficiency.

Central Heating Systems

Forced-warm-air heating systems are more efficient and cost less to install than gravity-warm-air heating systems.

Forced-warm-air systems consist of a furnace, ducts, and registers (Figs. 2 and 3). A blower in the furnace circulates the warm air to the various rooms through supply ducts and registers. Return grilles and ducts carry the cooled room air back to the furnace where it is reheated and recirculated.

Forced-warm-air systems heat uniformly and respond rapidly to changes in outdoor temperatures. They can be used in houses with or without basements—the furnace need not be below the rooms to be heated nor centrally located. Some can be adapted for summer cooling by the addition of cooling coils. Combination heating and cooling systems may be installed (Fig. 2). The same ducts can be used for both heating and cooling. Most installations have a cold air return in each room (except the bathroom and kitchen). If the basement is heated, additional ducts should deliver hot air near the basement floor along the outside wall. In cold climates, a separate perimeter-loop heating system (Fig. 4) may be the best way to heat the basement.

The warm air is usually filtered through inexpensive replaceable or washable filters. Electronic air cleaners can sometimes be installed in existing systems and are available on specially

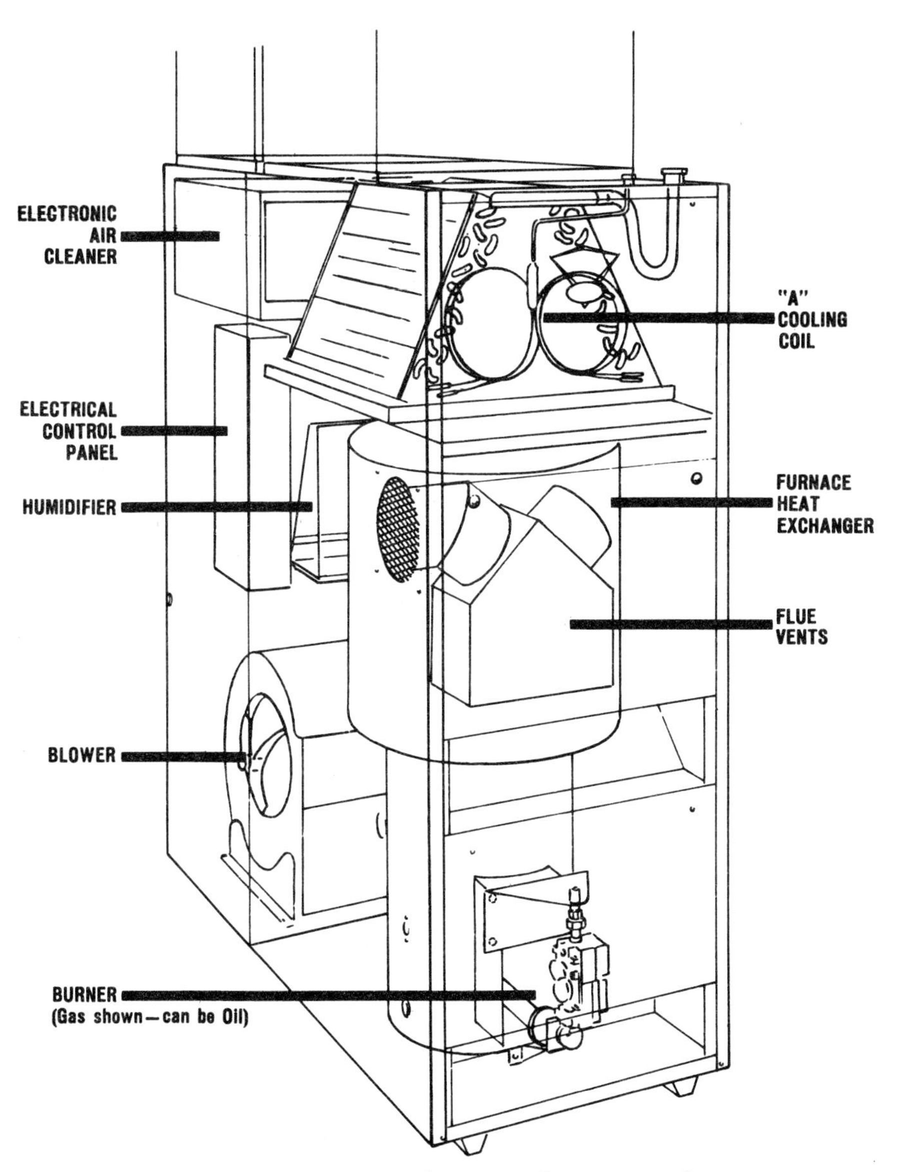

Fig. 2. Gas furnace. Modern forced-warm-air furnaces may have an electronic air cleaner for better air filtration and cooling coils for summer air conditioning.

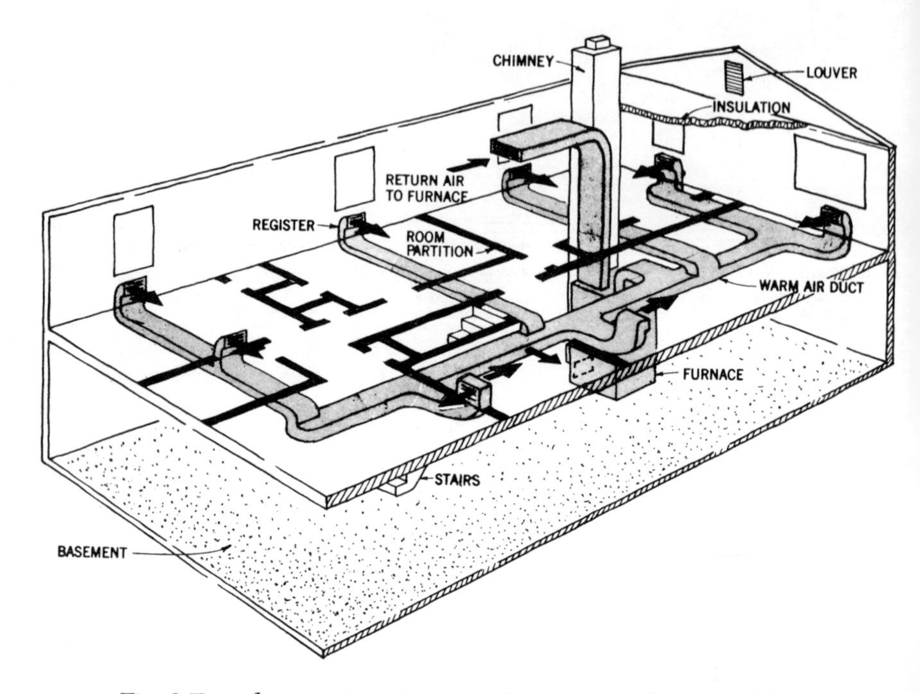

Fig. 3 Forced-warm-air systems are the most popular type of heating systems.

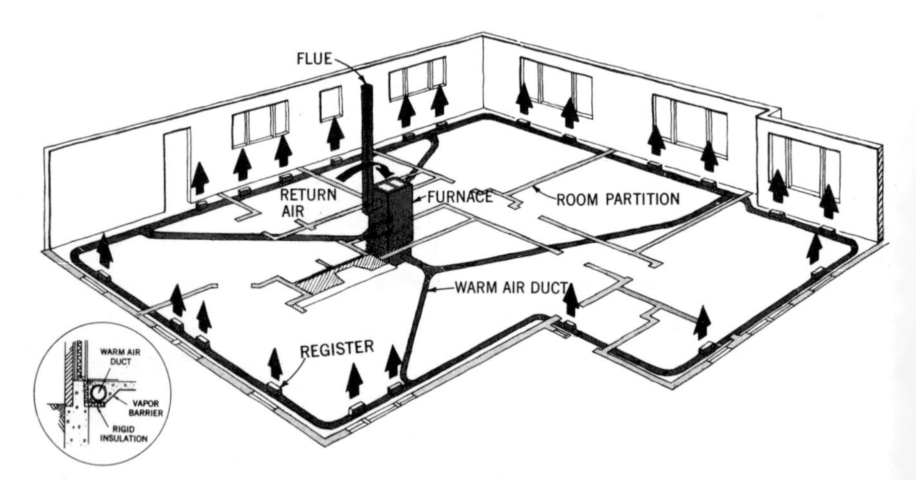

Fig. 4. Perimeter-loop heating system. The insert shows duct-slab-foundation construction details.

designed furnaces for new installations (Fig. 2). These remove pollen, fine dust, and other irritants that pass through ordinary filters and may be better for persons with respiratory ailments. The more expensive units feature automatic washing and drying of the cleaner.

A humidifier may be added to the system to add moisture to the house air and avoid the discomfort and other disadvantages of a too-dry environment.

Warm-air supply outlets are preferably located along outside walls. They should be low in the wall, in the baseboard, or in the floor where air cannot blow directly on room occupants. Floor registers tend to collect dust and trash, but may have to be used in installing a new system in an old house.

High-wall or ceiling outlets are sometimes used when the system is designed primarily for cooling. However, satisfactory cooling as well as heating can be obtained with low-wall or baseboard registers by increasing the air volume and velocity and by directing the flow properly.

Ceiling diffusers that discharge the air downward may cause drafts; those that discharge the air across the ceiling may cause smudging.

Most installations have a cold air return in each room. When supply outlets are along outside walls, return grilles should be along inside walls in the baseboard or in the floor. When supply outlets are along inside walls, return grilles should be along outside walls.

Centrally located returns work satisfactorily with perimeter-type heating systems. One return may be adequate in smaller houses as shown in Fig. 3. In larger or splitlevel houses, return grilles are generally provided for each level or group of rooms. Location of the returns within the space is not important. They may be located in hallways, near entrance doors, in exposed corners, or on inside walls.

In the crawl-space plenum system, the entire crawl space is used as an air supply plenum or chamber. Heated air is forced into the crawl space and enters the rooms through perimeter outlets, usually placed beneath windows, or through continuous slots in the floor adjacent to the outside wall. With tight, well-insulated crawl-space walls, this system can provide uniform temperatures throughout the house. (*See* Chap. 11, Plenum Heating Systems for Homes.)

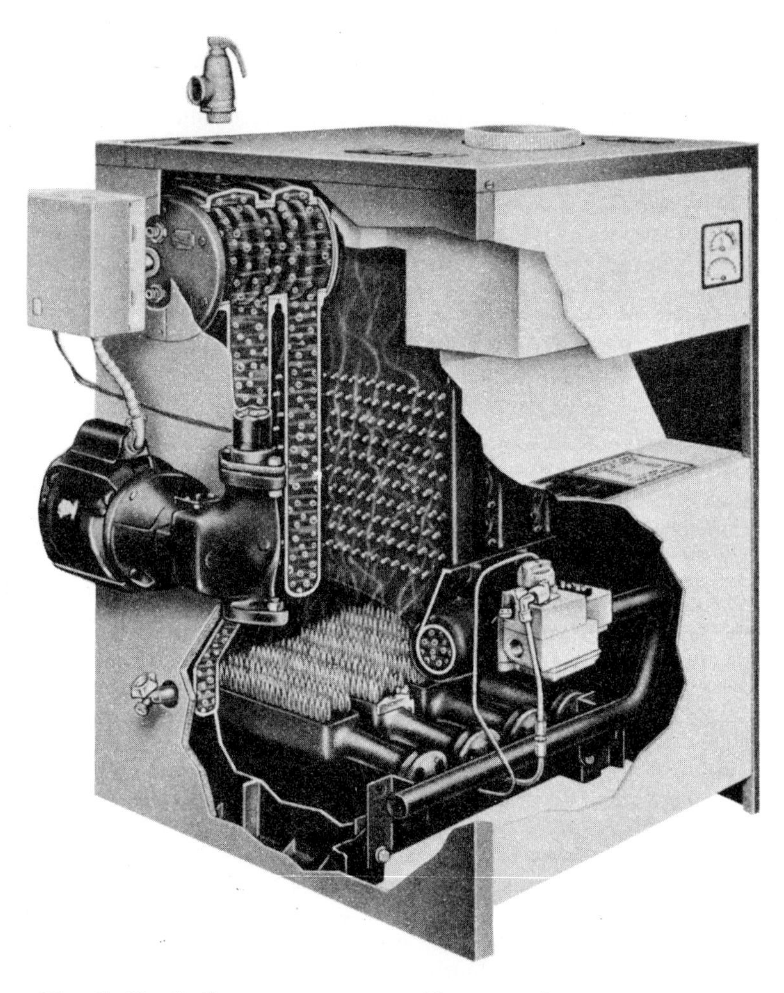

Fig. 5. Gas boilers are compact, self-contained units. Some units come equipped with a completely enclosing jacket.

In houses without basements, horizontal furnaces that burn gas or oil may be installed in the crawl space or hung from ceiling joists in the utility room or adjoining garage. The gas furnaces may also be installed in attics. Allow adequate space for servicing the furnaces. Insulate attic furnaces and ducts heavily to prevent excessive heat loss.

Vertical gas or *oil furnaces* designed for installation in a closet or a wall recess or against a wall are popular especially in small houses. The *counterflow type* discharges the hot air at the bottom to warm the floor. Some, such as the gas-fired unit provide discharge grilles into several rooms.

Upflow-type vertical furnaces may discharge the warm air through attic ducts and ceiling diffusers. Without return air ducts, these furnaces are less expensive, but also heat less uniformly.

Houses built on a concrete slab may be heated by a *perimeter-loop heating system* (Fig. 4). Warm air is circulated by a counterflow type furnace through ducts cast in the outer edge of the concrete slab. The warm ducts heat the floor, and the warm air is discharged through floor registers to heat the room.

To prevent excessive heat loss, the edge of the slab should be insulated from the foundation walls and separated from the ground by a vapor barrier.

HOT-WATER AND STEAM HEATING

Hot-water and *steam heating systems* consist of a boiler, pipes, and room heating units (radiators or convectors). Hot water or steam, heated or generated in the boiler, is circulated through the pipes to the radiators or convectors where the heat is transferred to the room air.

Boilers are made of cast iron or steel and are designed for burning coal, gas, or oil (Figs. 5 and 6). Cast-iron boilers are more resistant to corrosion than steel ones. Corrosive water can be improved with chemicals. Proper water treatment can greatly prolong the life of steel boiler tubes.

Buy only a certified boiler. Certified cast-iron boilers are stamped "I-B-R" (Institute of Boiler and Radiator Manufacturers); steel boilers are stamped "SBI" (Steel Boiler Institute). Most boilers are rated (on the nameplate) for both hot water

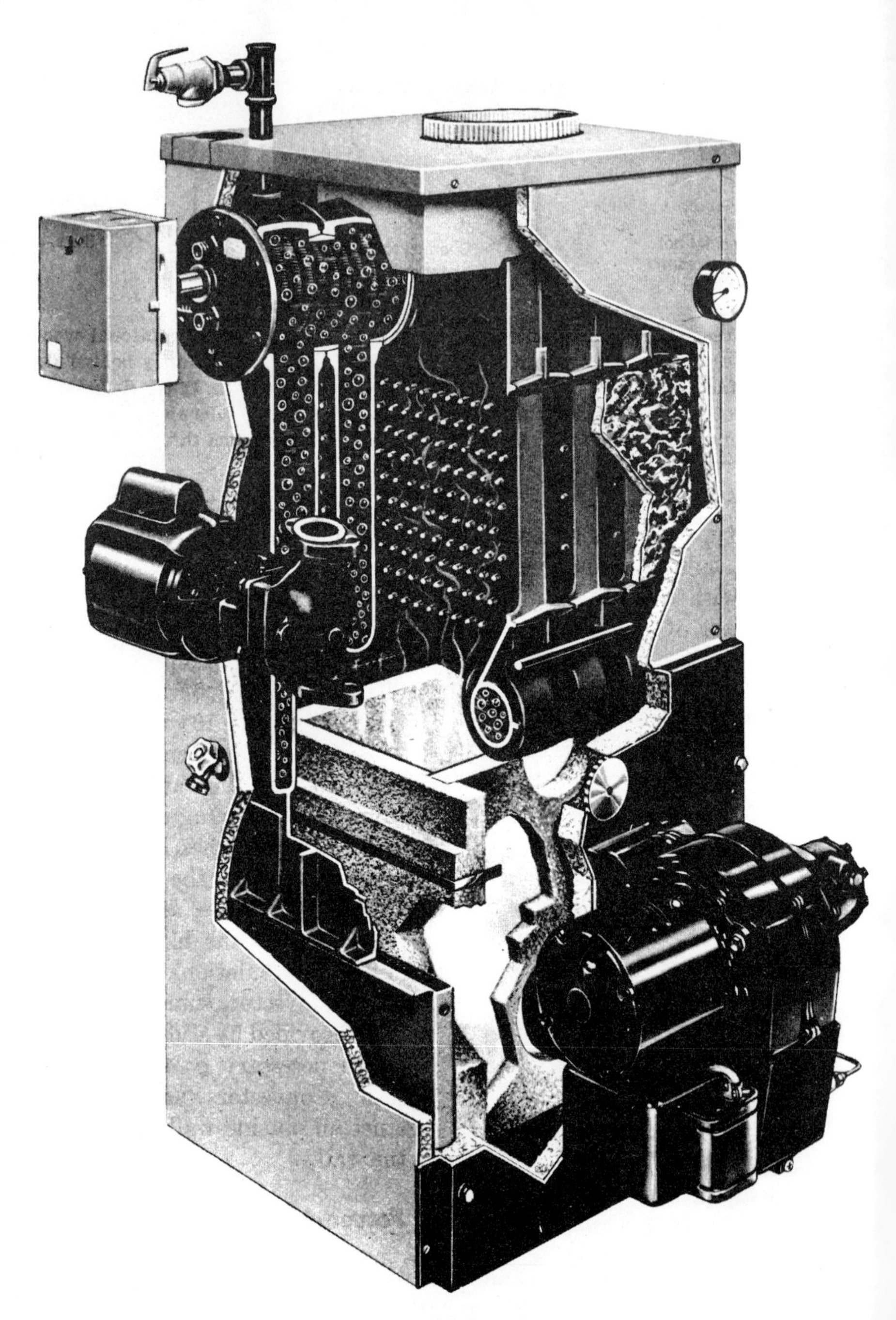

Fig. 6. Oil-fired boiler. Also available with a completely enclosing jacket.

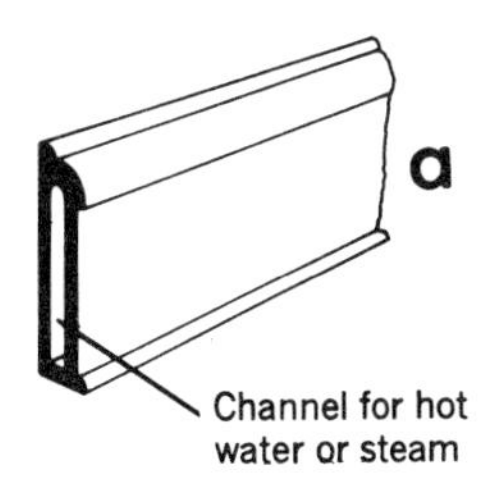

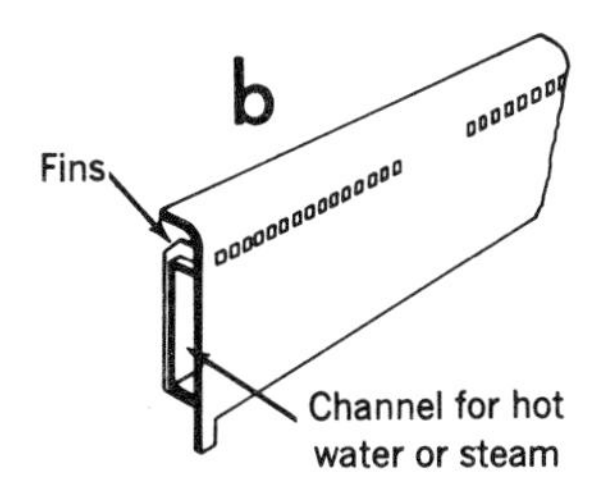

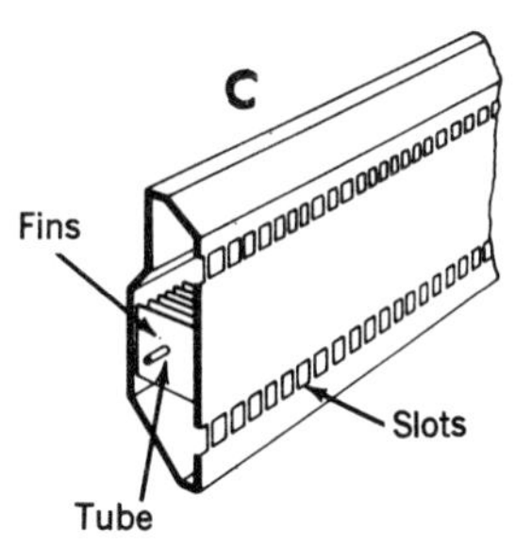

Fig. 7. Baseboard radiator units.

and steam. When selecting a boiler, consult your contractor.

Conventional radiators are set on the floor or mounted on the wall. The newer types may be recessed in the wall. Insulate behind recessed radiators with 1-inch (2.54-cm.) insulation board, a sheet of reflective insulation, or both.

Radiators may be partially or fully enclosed in a cabinet. A full cabinet must have openings at top and bottom for air circulation. Preferred location for radiators is under a window.

Baseboard radiators (Fig. 7) are designed to replace the conventional wood baseboard. In the hollow types, A and B, Fig. 7, water or steam flows directly behind the baseboard face. Heat from the surface is transmitted to the room. In the finned-tube type (C, Fig. 7) the water or steam flows through the tube and heats the tube and the fins. Air passing over the tube and fins is heated and delivered to the room through the slots. They will heat a well-insulated room uniformly, with little temperature difference between floor and ceiling.

Convectors (Fig. 8) usually consist of finned tubes enclosed

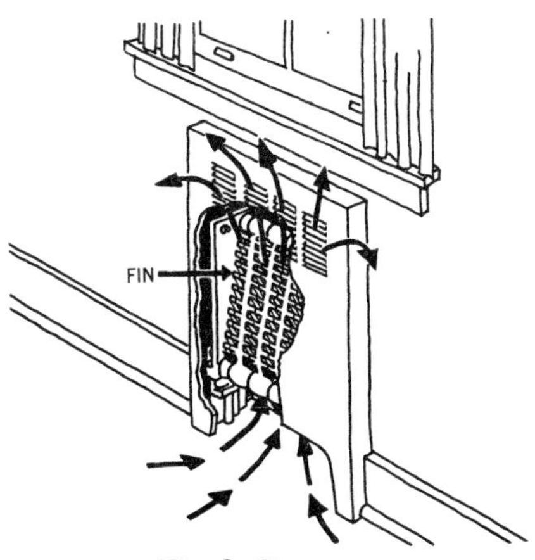

Fig. 8. Convector.

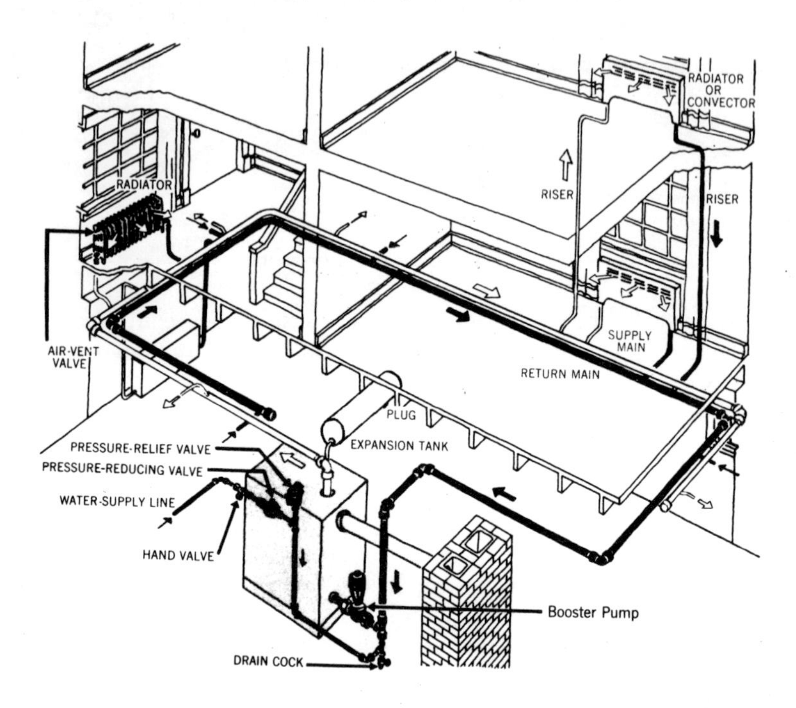

Fig. 9. Two-pipe forced-hot-water systems have two supply pipes or mains.

in a cabinet with openings at the top and bottom. Hot water or steam circulates through the tubes. Air comes in at the bottom of the cabinet, is heated by the tubes, and goes out the top. Some units have fans for forced-air circulation. With this type of convector, summer cooling may be provided by adding a chiller and the necessary controls to the system. Convectors are installed for hot-water or steam heating against the wall or recessed in the wall as shown in Fig. 8.

Forced-Hot-Water Heating Systems

Forced-hot-water heating systems are recommended over the less efficient gravity-hot-water heating systems.

In a *forced-hot-water system*, a small booster or circulating pump forces or circulates the hot water through the pipes to the room radiators or convectors (Fig. 9).

In a *one-pipe system*, one pipe or main serves for both supply and return. It makes a complete circuit from the boiler and back again. Two risers extend from the main to each room heating unit. A *two-pipe system* has two pipes or mains. One carries the heated water to the room heating units; the other returns the cooled water to the boiler.

A one-pipe system, as the name indicates, takes less pipe than a two-pipe system. However, in the one-pipe system, cooled water from each radiator mixes with the hot water flowing through the main, and each succeeding radiator receives cooler water. Allowances must be made for this in sizing the radiators—larger ones may be required further along in the system.

Because water expands when heated, an expansion tank must be provided in the system. In an "open system" the tank is located above the highest point in the system and has an overflow pipe extending through the roof. In a "closed system" the tank is placed anywhere in the system, usually near the boiler. Half of the tank is filled with air, which compresses when the heated water expands. Higher water pressure can be used in a closed system than in an open one. Higher pressure raises the boiling point of the water. Higher temperatures can therefore be maintained without steam in the radiators, and smaller radiators can be used. There is almost no difference in fuel requirements.

With heating coils installed in the boiler or in a water heater connected to the boiler, a forced-hot-water system can be used to heat domestic water year-round. If you want to use your heating plant to heat domestic water, consult an experienced heating engineer about the best arrangement.

One boiler can supply hot water for several circulation heating systems. The house can be "zoned" so that the temperatures of the individual rooms or areas can be controlled independently. Remote areas such as a garage, workshop, or small greenhouse, can be supplied with controlled heat.

Gas- and *oil-fired boilers* for hot-water heating are compact and are designed for installation in a closet, utility room, or similar space, on the first floor if desired.

Electrically heated hydronic (water) systems are especially compact, and the heat exchanger, expansion tank, and controls may be mounted on a wall (Fig. 10). Some systems have

Fig. 10. The heat exchanges, expansion tank, and controls for an electrically heated hydronic water system are compact enough to mount on a wall.

thermostatically controlled electric heating components in the hydronic baseboard units, which eliminates the central heating unit. Such a system may be a single-loop installation (Fig. 11) for circulating water by a pump, or it may be composed of individual sealed units filled with antifreeze solution. The sealed units depend on gravity flow of the solution in the unit. Each unit may have a thermostat, or several units may be controlled from a wall thermostat. An advantage of these types of systems is that heating capacity can be increased easily if the house is enlarged.

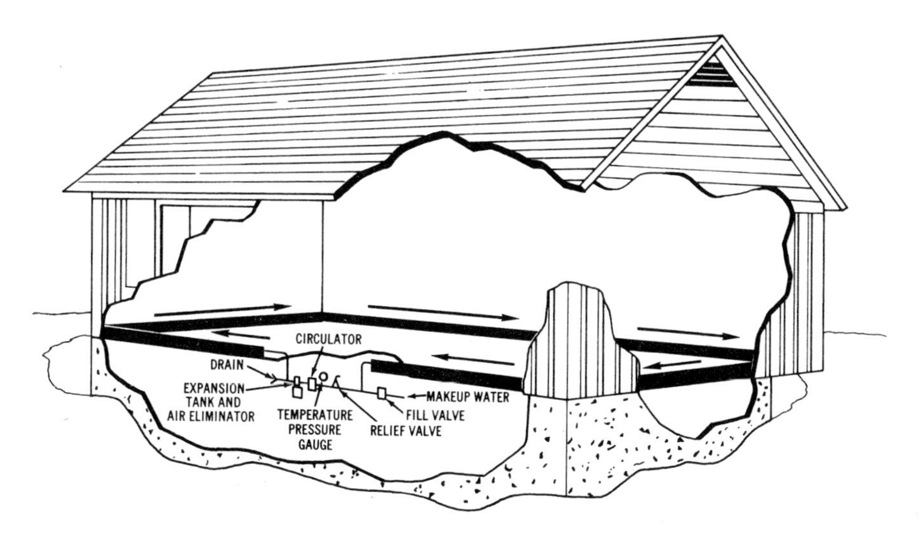

Fig. 11. Electrically heated hydronic baseboard systems are made in units.

Steam Central-Heating Systems

Steam heating systems are not used as much as forced-hot-water or warm-air systems. For one thing, they are less responsive to rapid changes in heat demands.

One-pipe steam heating systems cost about as much to install as one-pipe hot-water systems. Two-pipe systems are more expensive.

The heating plant must be below the lowest room heating unit unless a pump is used to return the condensate to the boiler.

Radiant Panel Heating

Radiant panel heating is another method of heating with forced hot water or steam. (It is also a method of heating with electricity. *See* following section on Electric Heating.)

Hot water or steam circulates through pipes concealed in the floor, wall, or ceiling. Heat is transmitted through the pipes to the surface of the floor, wall, or ceiling and then to the room by radiation and convection. No radiators are required —the floor, wall, or ceiling, in effect, act as radiators.

With radiant panel heating, rooms can be more comfortable at lower air temperatures than with other heating systems at higher air temperatures. The reason is that the radiated heat striking the occupant reduces body heat loss and increases body comfort. Temperatures are generally uniform throughout the room.

Underfloor radiant panel heating systems are difficult to design. For instance, a carpeted or bare wood floor might be very comfortable while the ceramic-tiled bathroom floor or the plastic kitchen-floor covering might be too hot for bare feet. An experienced engineer should design the system.

Panel heating in poorly insulated ceilings is not practical unless you want to heat the space above the ceiling. Exterior wall panels require insulation behind them to reduce heat loss.

ELECTRIC HEATING

Many types and designs of electric house-heating equipment are available. There is the ceiling unit, the baseboard heater, the heat pump, the central furnace, the floor furnace, and the wall unit. All but the heat pump are of the resistance type. Resistance-type heaters produce heat the same way as the familiar electric radiant heater. Heat pumps are usually supplemented with resistance heaters.

Ceiling heat may be provided with electric heating cable laid back and forth on the ceiling surface (Fig. 12) and covered with plaster or a second layer of gypsum board. Other types of ceiling heaters include infrared lamps and resistance heaters with reflectors or fans.

Fig. 12. Electric heating cable is one of the different types of electric heating used.

Baseboard heaters resemble ordinary wood baseboards and are often used under a large picture window in conjunction with ceiling heat.

The *heat pump* is a single unit that both heats and cools. In winter, it takes heat from the outdoor air to warm the house or room. In summer, it removes heat from the house or room and discharges it to the outside air. It uses less electricity to furnish the same amount of heat than does the resistance-type heater. Room air conditioners of the heat pump type are especially convenient in warmer climates where continuous heating is not needed or for supplemental heat in some areas of the house.

Either heat pumps or furnaces with resistance heaters are used in forced-air central heating systems. They require ducts similar to those discussed previously for forced warm-air heating. Hot-water systems with resistance-type heaters are also available. (*See* section on Forced-Hot-Water Heating Systems, previously described in this chapter.)

Wall units (Fig. 13), either radiant or convection, or both, are designed for recessed or surface wall mounting. They come equipped with various types of resistance heating elements. The warm air may be circulated either by gravity or by an electric fan.

Each room heated by the equipment described (with the exception of some central-heating systems) usually has its own thermostat and can be held at any desired temperature. Thermostats should be designed for long life and should be

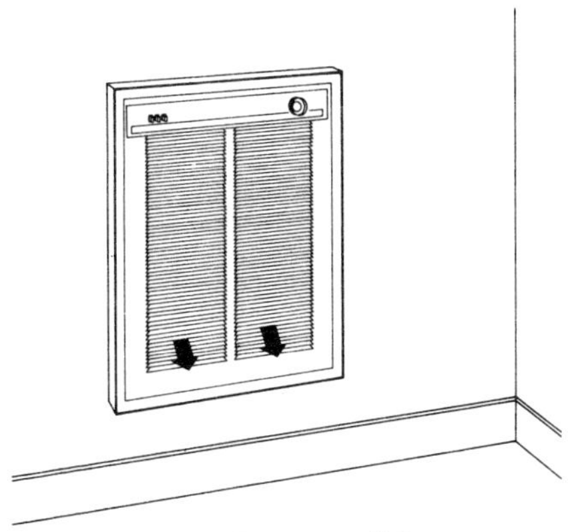

Fig. 13. Electric wall heater.

sensitive to change in temperature of one-half degree Fahrenheit, plus or minus.

FUELS AND BURNERS

The four fuels commonly used for home heating are *wood, coal, oil,* and *gas. Electricity,* though not a fuel, is being used increasingly.

Modern heating equipment is relatively efficient when used with the fuel for which it is designed. But, even with modern equipment, some fuels cost more than others to do the same job.

Fuel Costs

The therms of heat per dollar should not be the sole consideration in selecting the heating fuel. Installation cost, the efficiency with which each unit converts fuel into useful heat, and the insulation level of the house should also be considered. For example, electrically heated houses usually have twice the insulation thickness, particularly in the ceiling and floor, and, therefore, may require considerably less heat input than houses heated with fuel-burning systems. To compare costs for various fuels, efficiency of combustion and heat value of the fuel must be known.

Heating units vary in efficiency, depending upon the type, method of operation, condition, and location. (The efficiencies in utilizing fuels given in this chapter are recognized by the American Society of Heating, Refrigeration and Air Conditioning Engineers as being reasonable values where the heating equipment is properly installed and adjusted and in good condition.) Stoker-fired (coal) steam and hot-water boilers of current design, operated under favorable conditions, have 60 to 75 percent efficiency. Gas- and oil-fired boilers have 70 to 80 percent efficiency. Forced-warm-air furnaces, gas fired or oil fired with atomizing burner, generally provide about 80 percent efficiency. Oil-fired furnaces with pot-type burner usually develop not over 70 percent efficiency.

Fuel costs vary widely in different sections of the country. However, for estimates, the data given in Fig. 14 in terms of usable heat per dollar cost can be used. Here the efficiency of

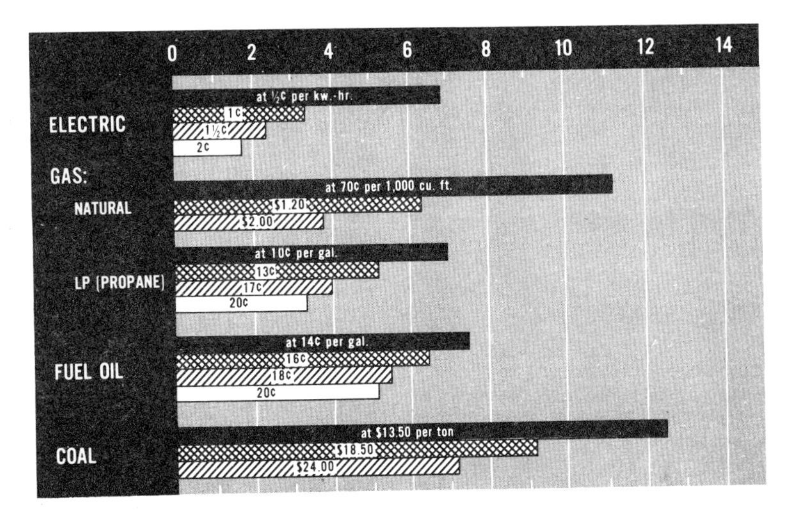

Fig. 14. Therms of heat.

electricity, gas, oil, and coal is taken as 100, 75, 75, and 65 percent, respectively. The efficiencies may be higher (except for electricity) or lower, depending upon conditions; but the values used are considered reasonable. The heat values are taken as 3413 B.T.U. per kilowatt-hour of electricity for resistance heating; 1050 B.T.U. per cubic foot of natural gas; 92,000 B.T.U. per gallon of propane (LP) gas; 139,000 B.T.U. per gallon of No. 2 fuel oil; and 13,000 B.T.U. per pound of coal. A therm is 100,000 B.T.U.

More B.T.U.s of heat per kilowatt-hour can generally be obtained with heat pump heating than with resistance heating. The difference varies depending upon the outside temperature and other factors. In warm climates heat pump heating may require about half as much electricity as resistance heating. In the extreme northern states, the consumption of electric energy may approach that required for resistance heating.

Wood

The use of wood requires more labor and more storage space than do other fuels. However, wood fires are easy to start, burn with little smoke, and leave little ash.

Most well-seasoned hardwoods have about half as much heat value per pound as does good coal. A cord of hickory, oak, beech, sugar maple, or rock elm weighs about two tons and has about the same heat value as one ton of good coal.

Coal

Two kinds of coal are used for heating homes—*anthracite* (hard) and *bituminous* (soft). Bituminous is used more often.

Anthracite coal sizes are standardized; bituminous coal sizes are not. Heat value of the different sizes of coal varies little, but certain sizes are better suited for burning in firepots of given sizes and depths.

Both anthracite and bituminous coal are used in stoker firing. Stokers may be installed at the front, side, or rear of a furnace or boiler. Leave space for servicing the stoker and for cleaning the furnace. Furnaces and boilers with horizontal heating surfaces require frequent cleaning, because fly ash (fine powdery ash) collects on these surfaces. Follow the manufacturer's instructions for operating stokers.

Oil

Oil is a popular heating fuel. It requires little space for storing and no handling, and it leaves no ash.

Two grades of fuel oil are commonly used for home heating. No. 1 is lighter and slightly more expensive than No. 2, but No. 2 fuel oil has higher heat value per gallon. The nameplate or guide book that comes with the oil burner indicates what grade oil should be used. In general, No. 1 is used in pot-type burners, and No. 2 in gun- and rotary-type burners.

For best results, a competent serviceman should install and service an oil burner.

Oil burners are of two kinds—*vaporizing* and *atomizing*. Vaporizing burners pre-mix the air and oil vapor. The pot-type burner shown in Fig. 15 is vaporizing and consists of a pot containing a pool of oil. An automatic or handset valve regulates the amount of oil in the pot. Heat from the flame vaporizes the oil. In some heaters a pilot flame or electric arc ignites the oil pot when heat is required. In others the oil is ignited manually and burns continuously at any set fuel rate between high and low fire, until shut off. There are few

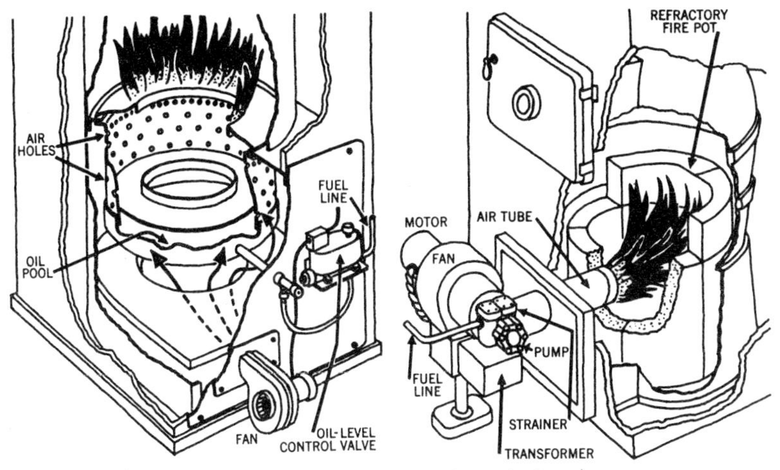

Fig. 15. Vaporizing or pot-type.

Fig. 16. Gun or pressure type oil burner.

moving parts, and operation is quiet. Some pot-type burners can be operated without electric power.

Atomizing burners are of two general types—*gun* (or pressure) and *rotary*. The gun burner (Fig. 16) is by far the more popular type for home heating. It has a pump that forces the oil through a special atomizing nozzle. A fan blows air into the oil fog; and an electric spark ignites the mixture, which burns in a refractory-line firepot.

Gas

Gas is used in many urban homes and in some rural areas. It is supplied at low pressure to a burner head (Fig. 17), where it is mixed with the right amount of air for combustion. Gas burners vary in design, but all operate on much the same principle. The controls shown in Fig. 17 are essential for safe operation.

A room thermostat controls the gas valve. A pilot light is required. It may be lighted at the beginning of the heating season and shut off when heat is no longer required. However, if it is kept burning during nonheating seasons, condensation and rapid corrosion of the system will be prevented.

The pilot light should be equipped with a safety thermostat to keep the gas valve from opening if the pilot goes out; no gas can then escape into the room. (The pilot light of all automatic gas-burning appliances should be equipped with this safety device.)

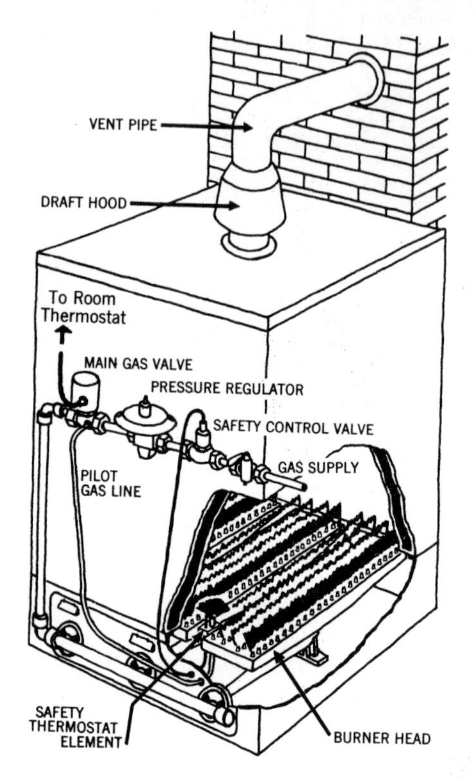

Fig. 17. Gas burner.

Three kinds of gas are used—*natural, manufactured,* and *bottled.* Bottled gas (usually propane) is sometimes called liquefied petroleum gas (LPG). It is becoming more popular as a heating fuel in recent years particularly in rural areas. Different gases have different heat values when burned. A burner adjusted for one gas must be readjusted when used with another gas.

Conversion gas burners may be used in boilers and furnaces designed for coal if they have adequate heating surfaces. Furnaces must be properly gastight. Conversion burners, as well as all other gas burners, should be installed by competent, experienced heating contractors who follow the manufacturer's instructions closely. Gas-burning equipment should bear the seal of approval of the American Gas Association.

Vent gas-burning equipment to the outdoors. Keep chimneys

and smoke pipes free from leaks. Connect all electrical controls for gas-burning equipment on a separate switch so that the circuit can be broken in case of trouble. Gas-burning equipment should be cleaned, inspected, and correctly adjusted each year.

Bottled gas is heavier than air. If it leaks into the basement, it will accumulate at the lowest point and create an explosion hazard. When bottled gas is used, make sure that the safety control valve is so placed that it shuts off the gas to the pilot as well as to the burner when the pilot goes out.

Electricity

Electric heating offers convenience, cleanliness, evenness of heat, safety, and freedom from odors and fumes. No chimney is required in building a new house, unless a fireplace is desired.

For electric heating to be more competitive economically with other types of heating, houses should be well insulated and weatherstripped, should have double- or triple-glazed windows, and should be vapor sealed. The required insulation, vapor barrier, and weatherproofing can be provided easily in new houses, but may be difficult to add to old houses.

Some power suppliers will guarantee a maximum monthly or seasonal cost when the house is insulated and the heating system installed in accordance with their specifications.

The heating equipment should be only large enough to handle the heat load. Oversized equipment costs more and requires heavier wiring than does properly sized equipment.

AUTOMATIC CONTROLS

Each type of heating plant requires special features in its control system. But even the simplest control system should include high-limit controls to prevent overheating. *Limit controls* are usually recommended by the equipment manufacturer.

The *high-limit control*, which is usually a furnace or boiler thermostat, shuts down the fire before the furnace or boiler becomes dangerously or wastefully hot. In steam systems, it responds to pressure; in other systems, it responds to temperature.

The high-limit control is often combined with the fan or pump controls. In a forced-warm-air or forced-hot-water system, these controls are usually set to start the fan or the pump circulating when the furnace or boiler warms up and to stop it when the heating plant cools down. They are ordinarily set just high enough to insure heating without overshooting the desired temperature and can be adjusted to suit weather conditions.

Other controls insure that all operations take place in the right order. *Room thermostats* control the burner or stoker on forced systems. They are sometimes equipped with timing devices that can be set to change the temperatures desired at night and in the daytime automatically.

Since the thermostat controls the house temperature, it must be in the right place—usually on an inside wall. Do not put it near a door to the outside; at the foot of an open stairway; above a heat register, television, or lamp; or where it will be affected by direct heat from the sun. Check it with a good thermometer for accuracy.

Oil-burner Controls

The *oil-burner controls* allow electricity to pass through the motor and ignition transformer and shut them off in the right order. They also stop the motor if the oil does not ignite or if the flame goes out. This is done by means of a stack thermostat built into the relay. The sensing element of the stack control is inserted into the smoke pipe near the furnace or boiler. Some heating units are equipped with electric eye (cadmium sulfide) flame detectors, which are used in place of a stack control.

Without the protection of the stack thermostat or electric eye, a gun- or rotary-type burner could flood the basement with oil if it failed to ignite. With such protection, the relay allows the motor to run only a short time if the oil fails to ignite; then it opens the motor circuit and keeps it open until it is reset by hand.

Figure 18 shows controls for an oil burner with a forced-hot-water system. The boiler thermostat acts as high-limit control if the water in the boiler gets too hot.

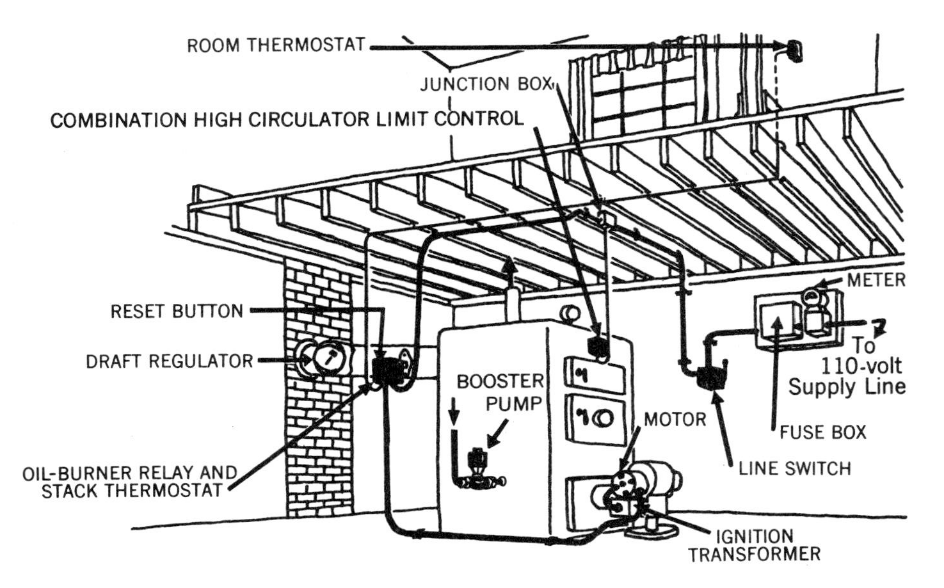

Fig. 18. Controls for an oil burner for a forced-hot-water heating system.

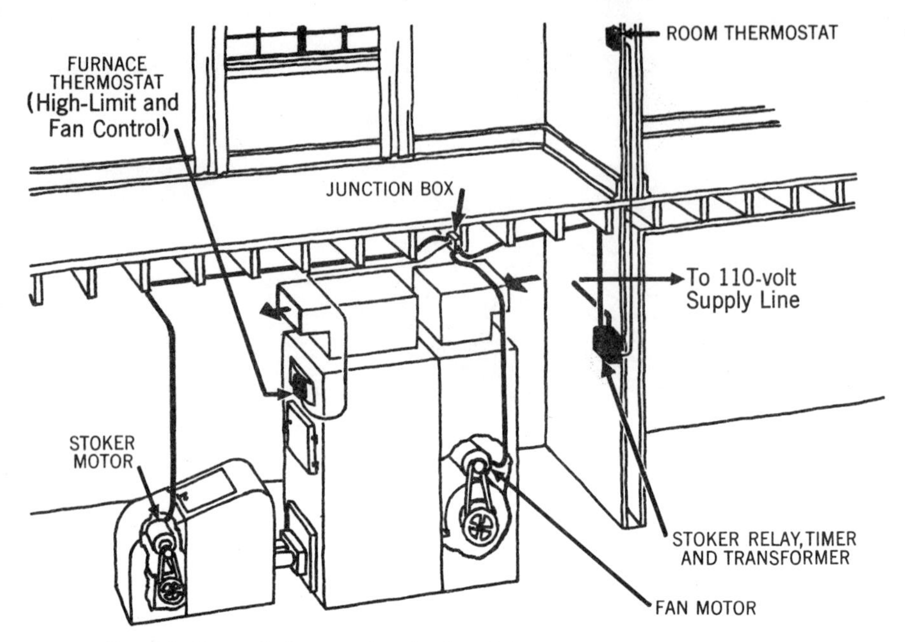

Fig. 19. Controls for a stoker-fired coal burner with a forced-warm-air heating system.

Stoker-fired Coal-burner Controls

The control system for a coal stoker is much like that for an oil burner. However, an automatic timer is usually included to operate the stoker for a few minutes every hour or half hour to keep the fire alive during cool weather when little heat is required.

A *stack thermostat* is not always used, but in communities where electric power failures may be long enough to let the fire go out, a stack thermostat or other control device is needed to keep the stoker from filling the cold fire pot with coal when the electricity comes on again. Sometimes a light-sensitive electronic device such as an electric eye is used. In the stoker-control setup for a forced warm-air system (Fig. 19), the furnace thermostat acts as high-limit and fan control.

Gas-burner Controls

Controls for the gas burner are so much a part of the burner itself that they have been described and illustrated in the section on gas, previously described in this chapter.

Other Heating System Controls

Warm-air, hot-water, or steam heat distribution systems may be controlled in other ways besides those suggested in Figs. 18 and 19. If the furnace or boiler heats domestic water, more controls are needed.

In some installations of forced hot-water systems, especially with domestic-water hookups, a mixing valve is used. The water temperature of the boiler is maintained at some high, fixed value, such as 200°F. Only a portion of this high-temperature water is circulated through the heating system. Some of the water flowing through the radiators bypasses the boiler. The amount of hot water admitted is controlled by a differential thermostat operating on the difference between outdoor and indoor temperatures. This installation is more expensive than the more commonly used control systems, but it responds almost immediately to demands; and, although it cannot anticipate temperature changes, it is in a measure regulated by outside temperatures, which change earlier than do those indoors.

The flow of hot water to each part of a building can be separately controlled. This zoning—maintaining rooms or parts of the building at different desired temperatures—can be used to maintain sleeping quarters at a lower temperature than living quarters. Electric heating is also well adapted to zoning.

Fuel savings help to offset the initial cost of the more elaborate control systems.

Chapter 11

Plenum Heating Systems for Homes

An ideal heating system must be low in cost and easy to operate. It must distribute the heat efficiently throughout the area and provide an even temperature at all times.

The objective of low cost was found in the first open fires, but they required constant attention, with low efficiency and very little comfort. To achieve efficiency, ease of operation, and better temperature distribution, more expensive systems have been developed. All improvements in heating equipment have not increased their cost, but the general trend has been better systems at higher cost. With heating becoming a larger part of the family budget, more effort has been directed toward lower cost methods of providing the desired environment in homes.

One of the systems that is directed toward this goal is the *hot-air plenum.* It utilizes the space under the structure as a duct for carrying the heated medium (air) to all the rooms in the house. Hot air is blown from the furnace into the space, creating a pressure. Registers in the floor at selected locations allow the warmed air to escape from the plenum into the room. Then this air is returned to the furnace where it is reheated and recirculated. It is a combination of a hot air and a radiant system. Warm air moving under the floor stores heat in the flooring members; this provides a large warm surface that radiates heat into the room. Warm air escaping from the plenum mixes with room air and increases the room air temperature.

In recent years several researchers have directed their efforts toward plenum floor systems.

In one system a plenum was developed over a concrete slab. Metal trusses (W-shaped) about four inches tall supported the 2 by 4 joist and subfloor above the slab. Several alternate means of supporting the floor are recommended (treated wood blocks, bricks, concrete blocks, or perforated metal studs). Warm air from the plenum is diffused into the room through a long narrow slot between the exterior walls and the floor. Effective heating and adaptability are the chief advantages of this system.

An extensive study was conducted by the Washington State University in cooperation with the National Lumber Manufacturers Association. This study was directed toward wood floors that would allow low-profile construction with an appearance similar to that of slab on grade construction. In place of a concrete slab for the bottom of this plenum, a plastic ground cover to prevent moisture buildup in the floor framing was used. Temperatures throughout the house were found to be extremely uniform. The temperature span throughout the house was usually within 2°F.

This latter study report evaluates several low-profile floor systems that have been developed. Most of these lend themselves to under-the-floor heat distribution.

1. The "Andy Place" crawl-space plenum utilizes prestressed concrete grade beams and 6-inch (15.24-cm.) steel I-beams to support a 1⅛-inch (2.8575-cm.) plywood deck.

2. The *foamed-core slab* is a sandwich panel with polystyrene core and asbestos-cement skins. The insulated core reduces heat loss through the slab and thus allows the inside surface temperature to be maintained at a higher level.

3. The one-day foundation utilizes concrete blocks and steel rods to build a grade beam to support the floor above the ground.

4. "Insta floor" is the trade name for a prefabricated floor system developed for crawl-space construction.

5. "Trofdek," "2-4-1" plywood, and "Ply-lumber" flooring are some of the systems that utilize combinations of plywood and dimensional lumber.

Canada's National Home Builders Association developed a plenum in their research house. This system uses ⅝-inch (1.5875-cm.) holes 1 inch (2.54 cm.) on center for diffusing air into the living area. Their work concentrated on the heating unit and the ventilation associated with this unit.

A report from Kansas Engineering Experimental Station shows the results of a survey on the "Effect of Heated Floor Temperature on Comfort." Since a warm floor is one of the features of a hot-air plenum, one must know what floor temperatures are applicable. Tests showed that length of exposure and body exertion affects the maximum comfortable foot temperature. Floors have been found to be comfortable at temperatures as high as 100°F, but on other occasions temperatures as low as 88°F have been determined to be uncomfortable.

The National Fire Prevention Code has set up standards for concealed-space plenums. They restrict the use of such a system to single-story structures. They limit the chamber height to 24 inches; they limit construction materials to those with a flame-spread rating of 200 (that of 1-inch lumber), and they restrict the furnace bonnet temperature to 150°F. They require blind ducts, deflection receptacles, and other items that are cumbersome and useless when a maximum bonnet temperature of 150°F is allowed.

Some of the advantages that are claimed by most researchers for their plenum systems are as follows.

1. Uniform temperature distribution.
2. Comfortably warm floors.
3. Protection of water and sewer lines from freezing.
4. Heating unit installation is usually lower in cost.
5. System is easily adaptable to several types of heating units and types of fuel.
6. With deep plenums, most of the advantages of the crawl-space type construction are accomplished.

Some of the disadvantages are as follows.

1. The National Fire Code places many restrictions on a hot-air plenum.
2. Entire operation dependent on the blower operation.
3. House must be underpinned.
4. Access doors to plenum or crawl space must be kept closed.
5. Direct opening from living area to plenum.
6. A cold house does not warm up fast.
7. More fuel may be necessary.

HOT-AIR PLENUM

Low-profile Plenum

A new bedroom built onto the experimental house D

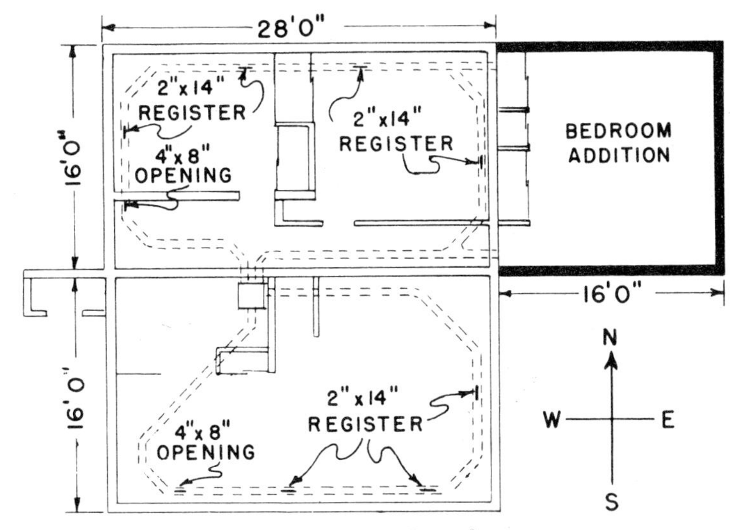

Fig. 1. House floor plan.

provided an opportunity to compare a plenum heating system installed in the addition with the perimeter-duct heating system existing in the original part of the house (Fig. 1).

Installation of Plenum. The plenum was developed on a 4-inch concrete slab that was insulated and moistureproofed according to common practice (Fig. 2). Regular brick placed on their flat face in a bed of mortar acted as spacers to support the floor and its frame two inches above the slab. The frame was constructed of 2- by 4-inch timbers laid flat. The subfloor was 1⅛-inch tongue-and-groove plywood in 4- by 8-foot sheets. Vinyl tile was glued to the plywood subfloor as a decorative and wearing surface.

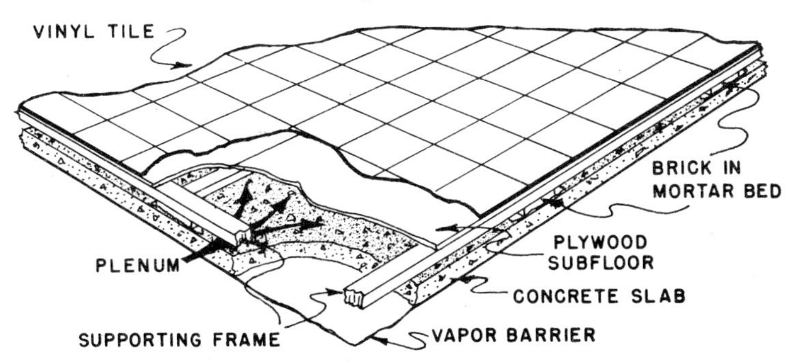

Fig. 2. Hot-air plenum construction in addition to house.

The perimeter-duct system already in use was tapped to obtain a heat supply for the plenum (Fig. 3). Registers were omitted because they concentrate the heat flow into the room and create air currents. To overcome this undesirable feature (registers), which is common to most heating systems, a perimeter slot was designed to distribute air flowing into the room (Fig. 4). A long, narrow opening was developed between the floor and outside wall by a loose-fitting subfloor. This opening ranged from 0 to ½ inch in width, but its effective width was further reduced by the baseboard that shielded it. Air entering the room through this opening was evenly distributed and annoying air currents were avoided.

Testing Periods and Conditions. Tests were conducted over a 3-year period in the house D.

Each room is different in orientation, construction of walls, and exposure (Fig. 5).

The master bedroom is on the northwest corner of the house. It has two exposed walls, one to the north and one to the west. Each exposed wall has a large window extending from the floor to the plate.

The center bedroom has one wall exposed on the north side. It has one window with the same dimensions as those in the master bedroom.

The new bedroom has three exposed walls, one to the north, one to the south, and one to the east. One medium-size

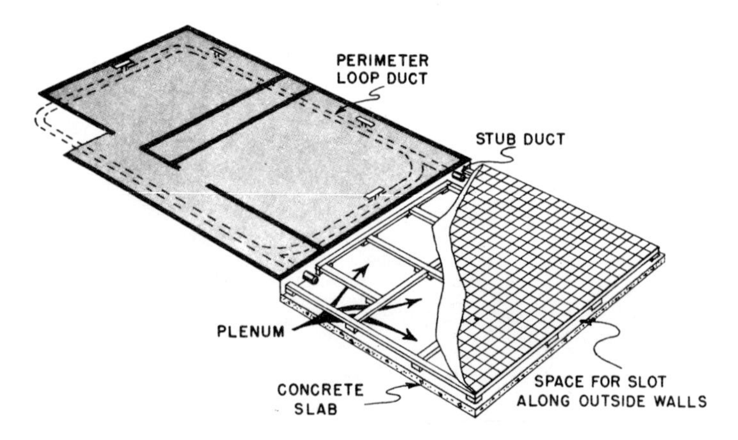

Fig. 3. Perimeter-loop ducting and underfloor plenum in house D.

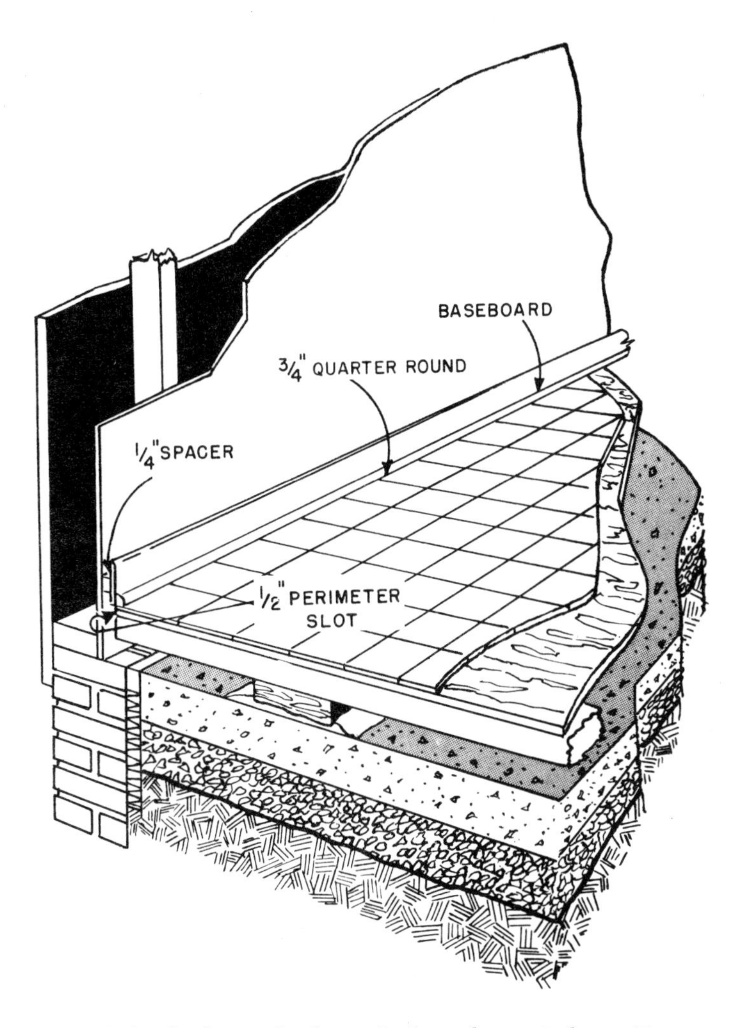

Fig. 4. Air discharge slot for underfloor plenum in house D.

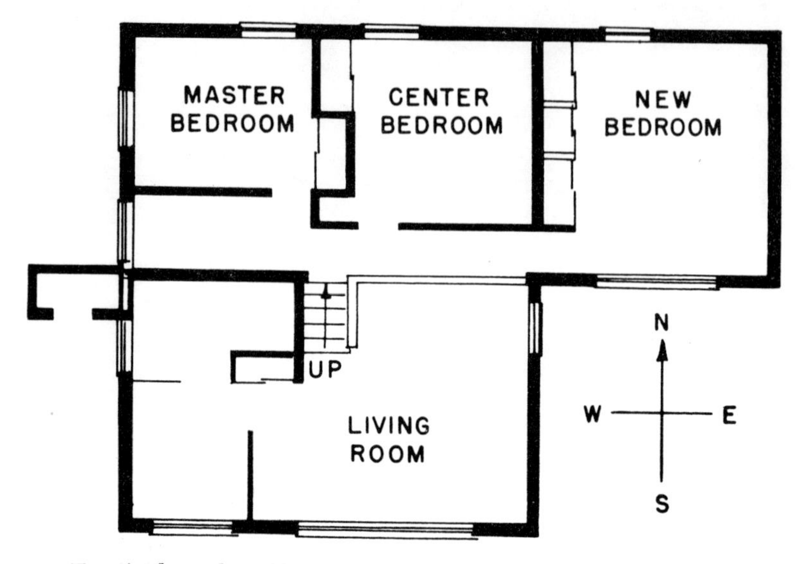

Fig. 5. Floor plan of house D showing the orientation of each of the rooms being tested.

window is located near the top and center of the north wall. One long flat window is located in the upper center of the south wall.

The north wall of the new bedroom is SCR brick with two inches of furring, two inches of fiberglass insulation, and a ⅜-inch gypsum board interior surface. (SCR brick have a nominal size, which is 12 inches long, 6 inches wide, and 2⅔ inches thick.) The east and south walls are standard 2- by 4-inch framing 16 inches on center with two inches of fiberglass insulation. The exterior walls are textured plywood and the interior is ⅜-inch gypsum board.

Originally the walls of the master and the center bedrooms consisted of one layer of SCR brick. These walls were not altered during the first season.

During the second year, the walls of the center and the master bedrooms were insulated and finished in successive steps (*see* section on Second-season tests later in this chapter) in an effort to have more uniform testing conditions in three bedrooms. Each step provided a new series of tests.

The test for the third season was designed to confirm data collected during the two previous years.

The heating system in the master and the center bedroom

is a perimeter loop with air traveling in the loop in the direction of least resistance. The heater is an oil furnace and it is operated by on-off wall thermostat.

The duct was tapped in two places to obtain a supply of heated air for the plenum of the new bedroom (Fig. 3). Other outlets were located under each window in the master room (two each), one in the bathroom, and under the window and along the east wall of the center bedroom (two each). Many variables (air temperatures and velocity, inside temperature, and ground temperature) were presented throughout the testing season.

Objectives of Tests. The objectives of these tests were to study the performance of heated plenum as a system for house heating, to determine its faults and its merits, to determine if and how it can be improved, and to test ideas that suggest better performance.

First-season Tests. To determine uniformity of heat distribution in the room or rooms from time to time and point to point, temperatures were measured at three elevations in each room (4, 48, and 90 inches) and about one foot from the outside wall. These temperatures were sensed by copper-constant thermocouples, which were supported at the proper elevation by telescoping aluminum poles.

Thermocouples were also located in the plenum chamber, in the soil under the chamber, and at various other points in the structure.

Temperatures were recorded by a 16-point potentiometer that was scheduled to record every two hours. Since there were more than 16 thermocouples in the structure and all thermocouples could not be recorded at the same time, it was necessary to group the thermocouples for alternate tests.

The first series of tests were to determine the temperature distribution in the plenum and in the slab. The temperatures recorded by the four thermocouples located in the corners of the plenum are shown in A, Fig. 6. There was considerable variation in plenum temperature from point to point at the same time and from time to time at the same point.

The slab temperatures four inches below the plenum (at the base of the slab) in each corner of the room addition are shown in B, Fig. 6. The variation from point to point was fairly large, but each point showed little fluctuation from time to time. The temperatures were found to be higher where

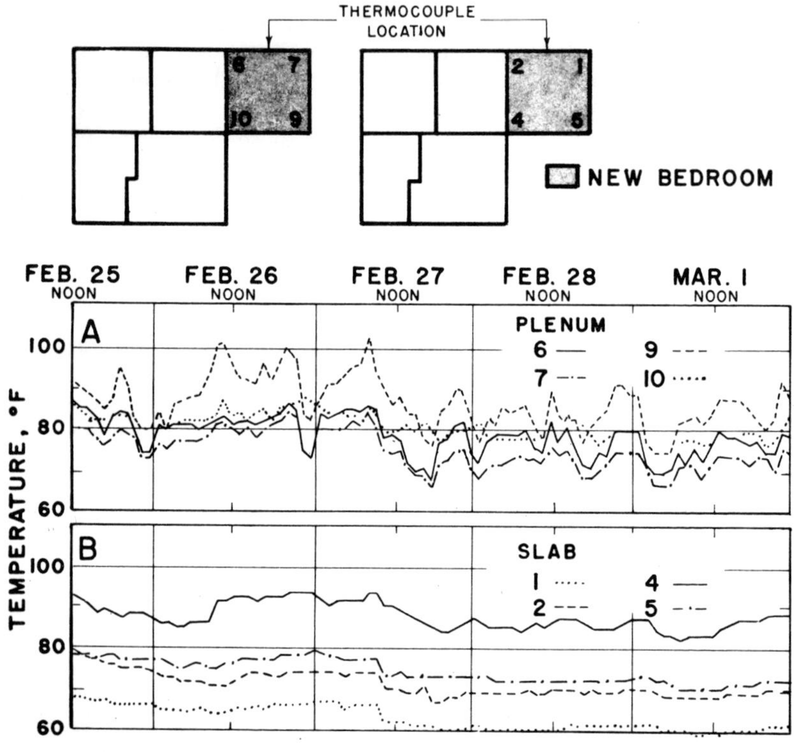

Fig. 6. Temperature in bedroom addition to house D. A, In corners of plenum chamber; B, in corners at base of concrete slab.

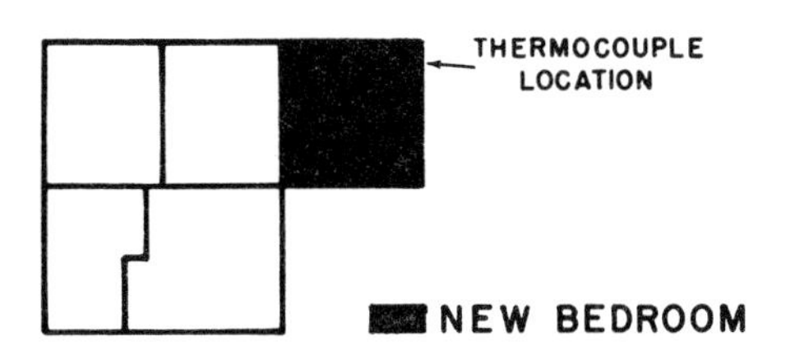

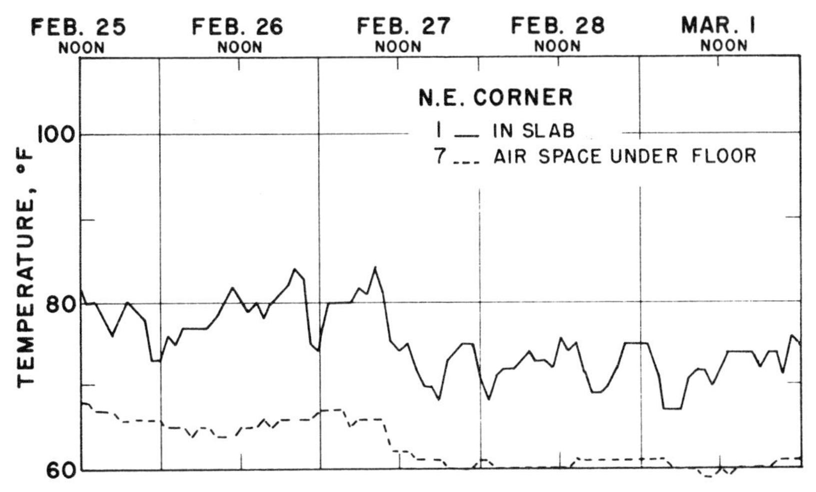

Fig. 7. Temperature in northeast corner of plenum chamber and in concrete slab immediately below it.

the air entered the plenum, and these temperatures dropped off in the corners away from the heat supply. The temperatures along the south wall were higher than those along the north wall. This was probably due to the shorter run of piping from the furnace to the plenum on the south side.

Figure 7 is a comparison of the plenum and the slab temperatures for one corner of the plenum. The air temperatures fluctuated considerably, but the slab, with a greater mass, showed little fluctuation and remained about 10°F. colder than the air temperature. In case of system failure the slab would act as a heat source to help maintain the house temperature.

In the second series of tests, the air temperatures at 4, 48, and 90 inches above the floor in four rooms of the house were recorded. These tests were to compare the temperature distribution of the room with the hot-air plenum with that of the rooms with the perimeter loop.

The temperatures in the new bedroom and the living room are plotted in A, Fig 8. The temperatures in both rooms varied in a daily cycle. The differential between the temperatures at the 4-inch and the 90-inch level was greater in the living room, and the hourly fluctuation was sharper in the living room. The new bedroom showed more uniform temperatures, but the room averages were about the same for both rooms.

When the same comparison was made between the center bedroom and new bedroom, the differential between the temperatures at the 4-inch level and the 90-inch level was found to be about the same for both rooms (B, Fig. 8). The temperatures maintained a stable daily cycle in both rooms; but the average temperature in the center bedroom was several degrees colder than that of the new bedroom.

When the temperatures of the new bedroom at the 4- and 90-inch levels were compared with those of the master bedroom, the differential was found to be about twice as large in the master bedroom (C, Fig. 8). The temperatures in both rooms maintained a stable daily cycle with very little hourly fluctuation. The average temperature of the master bedroom was several degrees colder than that of the new bedroom.

Second-season Tests. For the second season the walls of the center and the master bedrooms were insulated and finished inside. The wall construction was changed six times before it was completed, with each change constituting the basis for a new test. Throughout these tests, the walls of the bedroom addition were unchanged. The following list indicates the wall treatments of the master and center bedrooms for each of the seven tests.

1. Bare brick wall (no treatment).
2. One-inch furring with ½-inch Homosote finish.
3. One-inch furring with ¼-inch Upson board finish.
4. One-inch furring with ⅜-inch gypsum board finish.
5. Master bedroom—2-inch furring with ¼-inch Upson board finish.

   Center bedroom—1-inch furring with ⅜-inch gypsum board finish.
6. Master bedroom—2-inch furring with 2-inch wool insulation and ¼-inch Upson board finish.

   Center bedroom—2-inch furring with 2-inch wool insulation and ⅜-inch gypsum board finish.

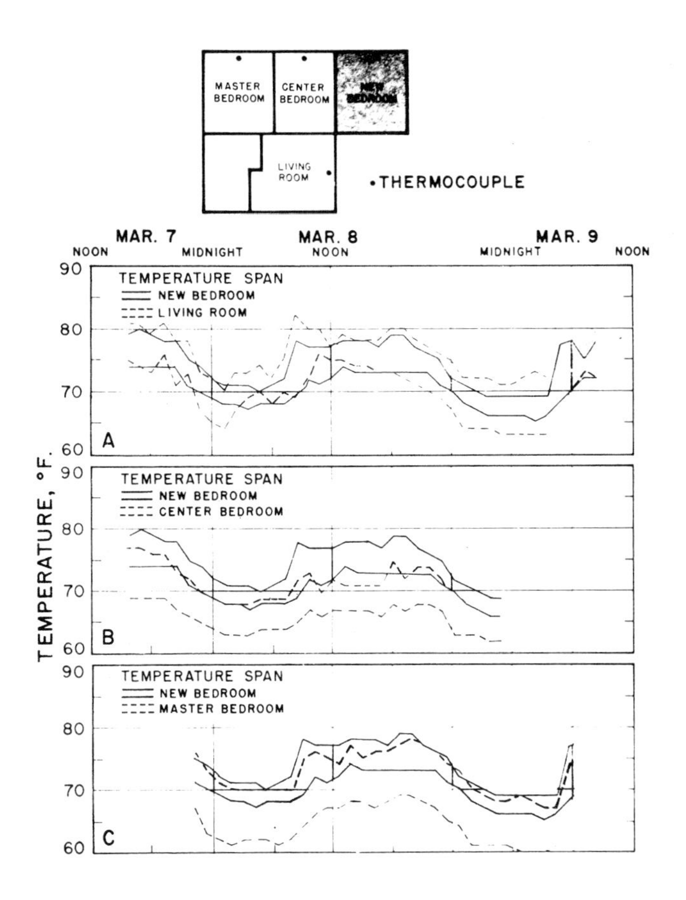

Fig. 8. Comparison of temperatures at the 4- and 90-inch levels in new bedroom with those in other rooms of house D. A, In living room; B, in center bedroom; and C, in master bedroom.

7. Master bedroom—same as No. 6 except 4-mil. vapor barrier added.

Center bedroom—same as No. 6.

Data were taken in the same manner and with the same equipment used for the first season.

Results of Tests. The average temperature in each of the three bedrooms at the 4-inch (floor), 48-inch, and 90-inch (ceiling) levels are given in Table 13. At the floor level there was a radical difference in temperature from room to room. At the 48-inch level the temperature difference from room to room was only a few degrees, and at the ceiling level it was about the same in all rooms with little variation.

Data (Table 13) on the average temperature 4 inches above the floor in the master bedroom and the average outside temperature indicate a direct relation between these inside and outside temperatures for the first five comparisons. For test No. 6, two insulations were added to the master bedroom wall. The continuity of the relation was broken by the change in the thermal resistance of the wall, but it was regained and continued at a new relative differential.

Table 14 shows the comparison of the average temperature difference between the 4-inch (floor) and 90-inch (ceiling) level for the three bedrooms, with the outside temperature.

Figure 9 shows the average temperature spans between the 4-inch (floor) and 90-inch (ceiling) level in the center bedroom under treatments 1 and 2 and comparison of those in the center bedroom under treatments 3 to 7 with those in the new bedroom.

In the master bedroom there was no relation between the temperature differential and the outside temperature when the outside temperature was above 40°F. (*See* Table 14.) However, when the average outside temperature dropped below the level, the differential seemed to increase rapidly. After the wall was insulated with two inches of batt insulation, there seemed to be no relation at temperatures as low as 32°F.

In the center bedroom there was a constant trend toward a lower differential as insulation was added. (*See* Table 14 and Fig. 9.) The effect of insulation is illustrated in tests 6 and 7 when the outside temperature was quite cold. (*See* Table 14.) The differential dropped from about 4.5 in test 5 to about 2.0 in tests 6 and 7.

TABLE 13

COMPARISON OF AVERAGE TEMPERATURE IN EACH OF THE THREE BEDROOMS AT THE 4-INCH (FLOOR), 48-INCH, AND 90-INCH (CEILING) LEVELS IN EACH OF THE TESTS, AND OUTSIDE TEMPERATURES.

Wall treatment[1]	Outside temperature	Average temperature at floor			Average temperature at 48 inches			Average temperature at ceiling		
		Master bedroom	Center bedroom	New bedroom[2]	Master bedroom	Center bedroom	New bedroom[2]	Master bedroom	Center bedroom	New bedroom[2]
	°F.	°F.	°F.	°F.	°F.	°F.	°F.	°F.	°F.	°F.
1	54.1	69.3	68.4	---	73.2	72.5	---	73.9	73.9	---
2	50.8	67.2	69.5	---	73.1	73.2	---	74.2	74.4	---
3	46.5	66.2	70.1	69.7	72.9	72.3	71.4	73.8	73.2	71.9
4	40.9	66.3	69.1	69.2	72.4	71.2	71.0	73.7	72.1	71.5
5	32.8	61.9	67.5	68.1	71.7	71.0	70.0	73.5	72.0	70.7
6	32.3	65.6	71.4	69.2	73.4	72.5	71.1	73.2	73.3	71.9
7	46.0	66.7	71.6	69.4	73.4	73.0	71.5	74.0	73.7	72.2

[1] See wall treatment of master and center bedrooms previously described.
[2] See wall construction of new bedroom previously described.

TABLE 14

COMPARISON OF THE AVERAGE TEMPERATURE OF DIFFERENCE BETWEEN THE 4-INCH (FLOOR) AND THE 90-INCH (CEILING) LEVEL FOR THE THREE BEDROOMS, WITH THE OUTSIDE TEMPERATURE.

Wall treatment[1]	Average temperature difference from floor to ceiling			Outside temperature
	Master bedroom	Center bedroom	New bedroom[2]	
	°F.	°F.	°F.	°F.
1	4.6	5.5	--	54.2
2	7.0	4.9	--	50.6
3	7.6	3.1	2.2	46.5
4	7.4	3.1	2.3	40.9
5	11.6	4.6	2.6	32.8
6	7.6	1.9	2.7	32.3
7	7.0	2.1	2.8	46.0

[1] See wall treatment of master and center bedrooms previously described.
[2] See wall construction of new bedroom previously described.

In the new bedroom no patterns were noticed. (*See* Table 14 and Fig. 9.) At lower outside temperatures, the differential was smaller in the new bedroom than in the center bedroom until the center bedroom had an equal amount of insulation. Then the differential was about the same in both rooms.

Figure 10 shows the average temperature spans between the 4-inch (floor) and 90-inch (ceiling) level in the master bedroom under treatments 1 and 2 previously described and comparison of those in the master bedroom under treatments 3 to 7 with those of the new bedroom. The differences between these temperatures are given in Table 14. During the tests the temperature differences between the floor and ceiling in the new master bedroom was greater than the temperature differences between the floor and the ceiling. (*See* Table 14 and Fig. 10.) As insulation was added to the walls in the master bedroom, the differential became smaller but remained larger than that of the new bedroom. The floor-level temperature fluctuated severely in the master bedroom during the cold days, but the ceiling-level temperature maintained a stable daily cycle.

Table 15 compares the average temperature at 4 feet above the floor with the temperature at the ceiling level in each of

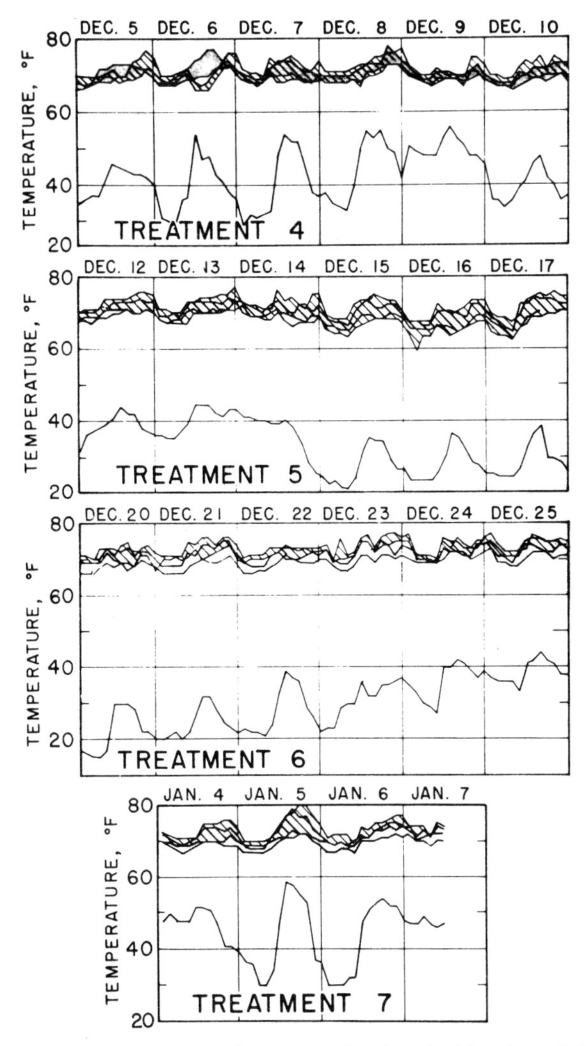

Fig. 9. Temperature spans between the 4-inch (floor) and 90-inch (ceiling) level in center bedroom under wall treatments 3 to 7 with those in new bedroom. Outside temperatures are also shown.

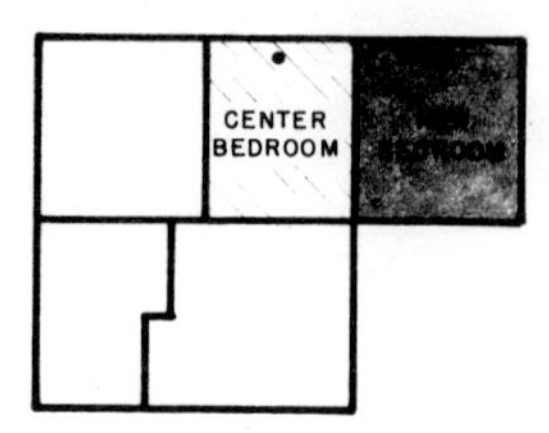

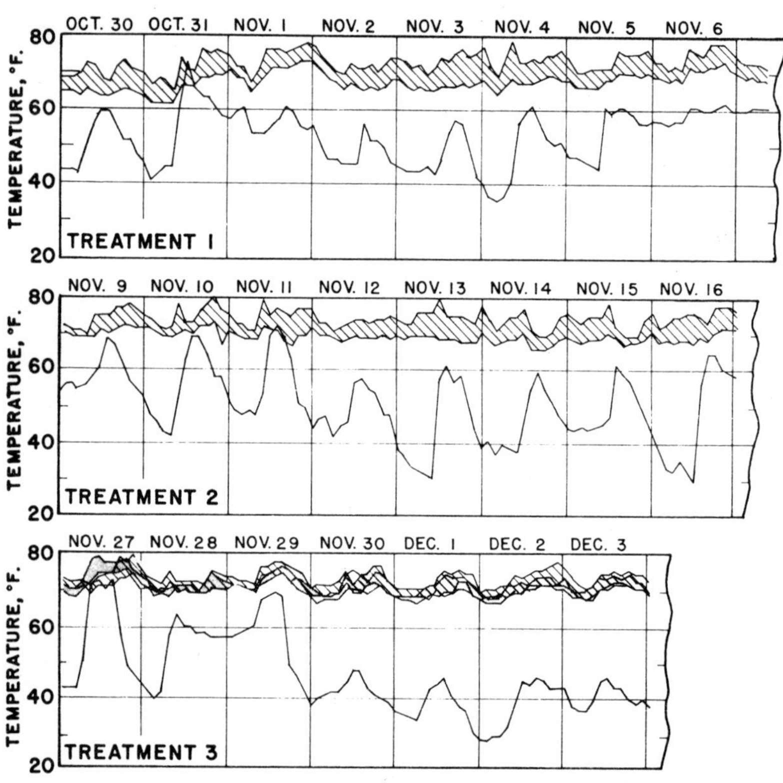

Fig. 9 (continued).

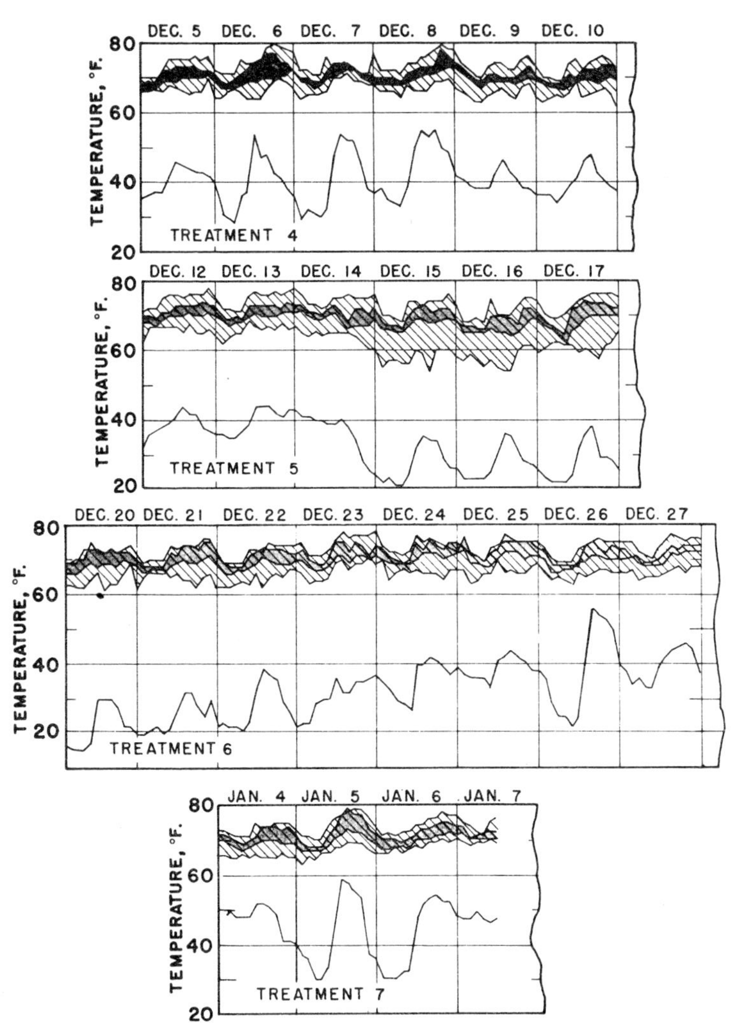

Fig. 10. Temperature spans between the 4-inch (floor) and 90-inch (ceiling) level in master bedroom under wall treatments 1 and 2 (Table 14) and a comparison of temperature spans in master bedroom under treatments 3 to 7 with those in new bedroom. Outside temperatures are also shown.

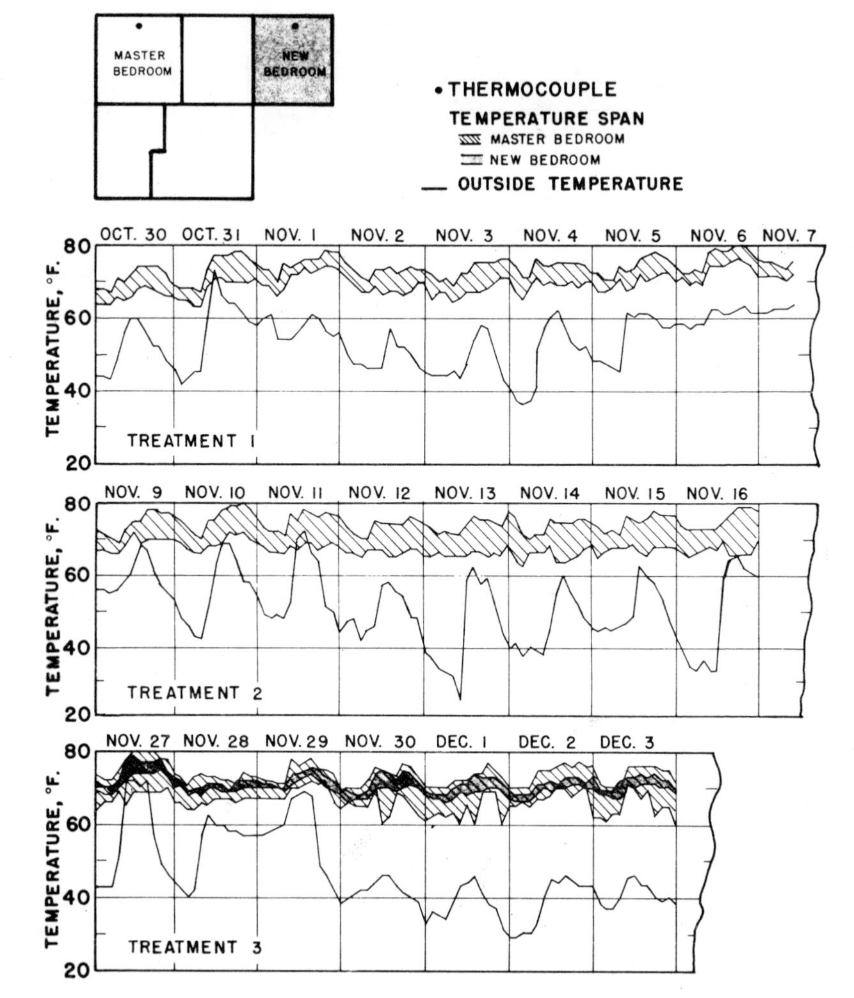

Fig. 10. (continued).

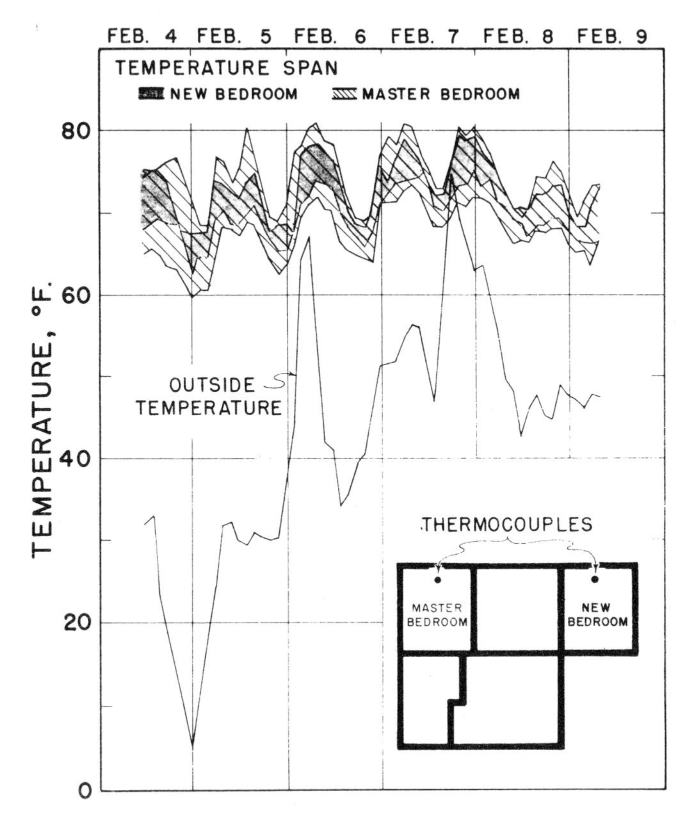

Fig. 11. Temperature span between the 4-inch (floor) and 90-inch (ceiling) level in master bedroom, with the perimeter-loop heating system, compared with that in the new bedroom, with the hot-air plenum system. Outside temperatures are also shown.

the three rooms. This shows that the temperature at the 4-foot level and that at the ceiling level was usually within 2°F. of each other.

Third-season Tests. Tests conducted during the third season were designed to confirm the data from the two previous tests. The data were collected in the same manner as for the first two seasons.

Results of Tests. Figures 11 and 12 compare the temperature of the master bedroom and the new bedroom. In Fig. 11 the

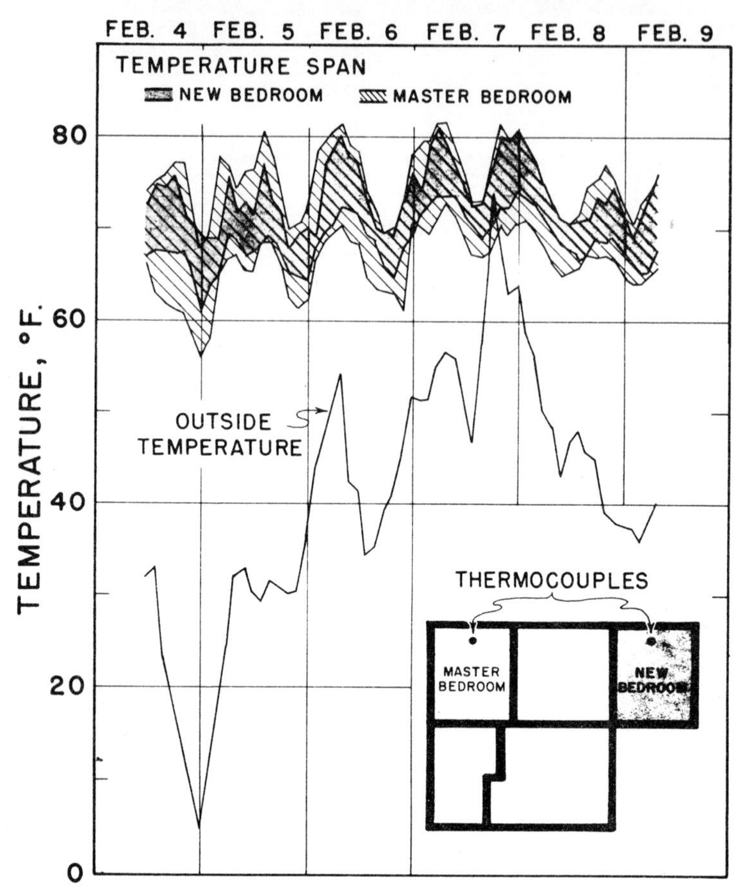

Fig. 12. Maximum temperature span in master bedroom, with the perimeter-loop heating system, compared with that in new bedroom, with the hot-air plenum system. Outside temperatures are also shown.

average temperature at 90 inches above the floor and the average temperature at four inches above the floor were plotted for the two rooms every two hours. The curve formed is similar to those of previous tests, which showed greater variation and fluctuation of temperature in the master bedroom with the perimeter-duct heating than in the new bedroom with the hot-air plenum.

To show that there were no extreme temperatures recorded in the room, Fig. 12 was plotted to show not the average but the maximum 90-inch temperature and the minimum 4-inch temperature recorded in each room at the same 2-hour intervals. This curve fluctuation is more severe than that of the average temperatures, but it is not much different from the average curve in Fig. 11. This indicated that the average temperature curve is representative of the whole-room temperature.

Evaluation of Tests. The different wall treatments did not greatly affect the temperature relations in the different rooms. However, some evidence of the effect of wall covering is found when the two inches of wool insulation were added to the master bedroom wall. The temperature at the 4-inch level rose several degrees with the outside temperature remaining near the same level. Another indication became evident when the temperature differential between the floor and the ceiling gradually became smaller as the insulation of the wall was increased in successive treatments.

Outside temperature did not severely affect the inside temperature relations between the 4-inch and the ceiling level when the walls were well insulated (2 inches of batt insulation with inside and outside finishes). However temperatures at floor level tended to cycle with the outside temperature.

The difference in room performance is due to orientation, exposure, wall construction, and the heating system. But since the new bedroom has more wall exposure and the orientation is no better than the other rooms, most of the differences can be attributed to the heating systems.

CRAWL-SPACE PLENUM

Since tests conducted in house D proved successful, plans were made to install a plenum to heat an entire house. The plenum was designed for a pole-frame house, 24 feet 8 inches by 33 feet in size (Fig. 13).

This plenum differed from the one in house D in the following ways.

1. It served an entire house.
2. It was a crawl-space plenum (18 inches deep instead of 4 inches).
3. It used a plastic ground cover without the concrete slab.
4. The perimeter slot was wider.

Installation of Plenum. Figure 14 shows the plenum construction. It is constructed by placing nine 4-inch (10.16-cm.)

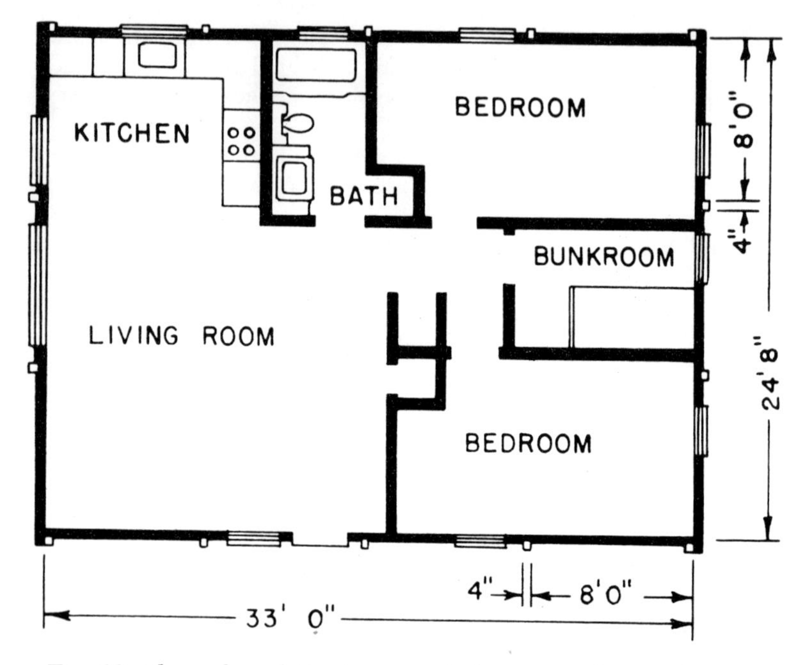

Fig. 13. Floor plan of the house used for testing the crawl-space plenum.

by 6-inch (15.24-cm.) by 2-foot (60.96-cm.) posts (built up from 2 by 4's) on concrete pads. These nine posts support three main beams that were built up from 2-inch (5.08-cm.) by 8-inch (20.32-cm.) timbers. These three beams support 2- by 6-inch (5.08- by 15.24-cm.) floor joists placed 16 inches (40.64-cm.) on center. The floor deck is ½-inch (1.27-cm.) plywood with a decorative and wearing surface of ⅛-inch (.3175-cm.) asphalt tile.

Heat was supplied to the plenum by a counterflow furnace, which was countersunk through the floor near the center of the house. Air taken into the furnace near the ceiling was blown over the furnace bonnet and discharged into the plenum. The air flowed unobstructed through the plenum to the perimeter of the structure where it was allowed to escape back into the living area through a slot in the baseboard.

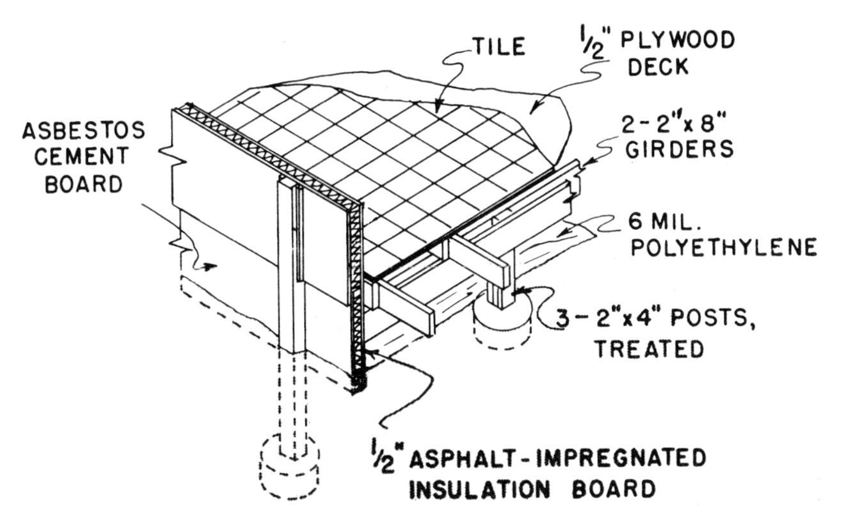

Fig. 14. Crawl-space plenum construction.

Testing Procedure. The basis of comparison (temperature uniformity) for evaluating this heating system was the same as that used in house D. The temperatures were again measured at three elevations in each room (4, 48, and 90 inches above the floor). Identical telescoping aluminum poles were used to support the copper-constantan thermocouple sensing elements and the 16-point potentiometer was used to record the temperatures.

Objective of Tests. A series of tests was conducted shown in the floor plan of the house (Fig. 13) to determine the effect of various construction features on temperature uniformity from point to point at any particular time and from time to time at any particular point. Also a test was set up to determine the effect of reduced fan speed on temperature distribution.

Results of Tests. Examination of the data indicated that graphs showing the maximum and the minimum temperatures recorded in the house would most appropriately show the temperature uniformity with the various construction features.

The first test was conducted with baseboard installed in the living room only. A plot of the extreme temperatures in the house (Fig. 15) shows that the temperature span was from 2°

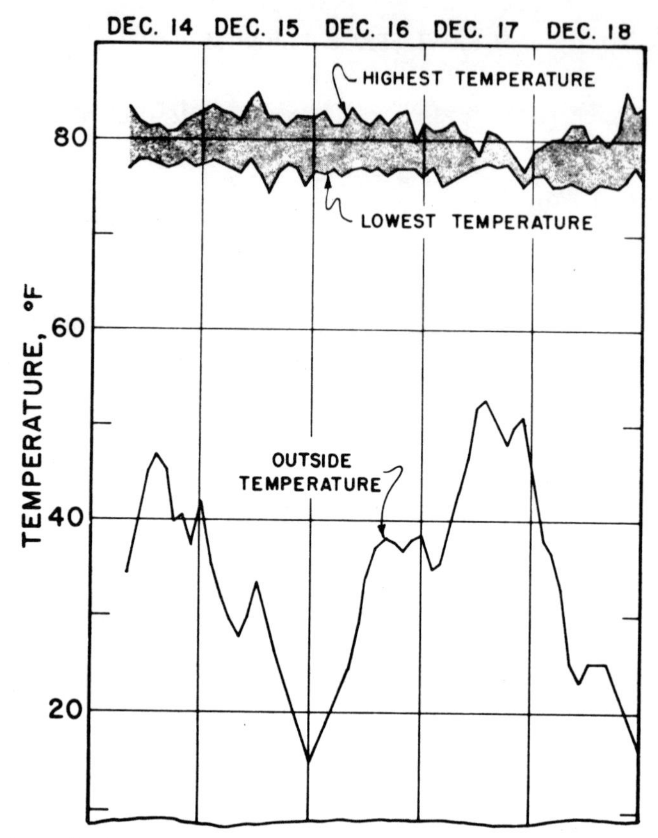

Fig. 15. Temperature span in the house when baseboard was installed in living room only. Outside temperatures are also shown.

to 10°F. Much of this span is due to the irregular width of the perimeter slot from room to room. The temperature span within any one room was only about 2°, and the rooms with wider slot had higher temperature (not shown in Fig. 15).

The second test was conducted with baseboard installed in all rooms except the bunkroom. A plot of the extreme temperatures in the house (Fig. 16) shows a temperature difference of about 5°F. The reduction in the temperature span (compared with test No. 1) can be attributed to the more uniform perimeter slot.

A third test was conducted after the prime doors were hung to determine the effect of the additional insulation. Storm doors had been initially installed and the prime doors were

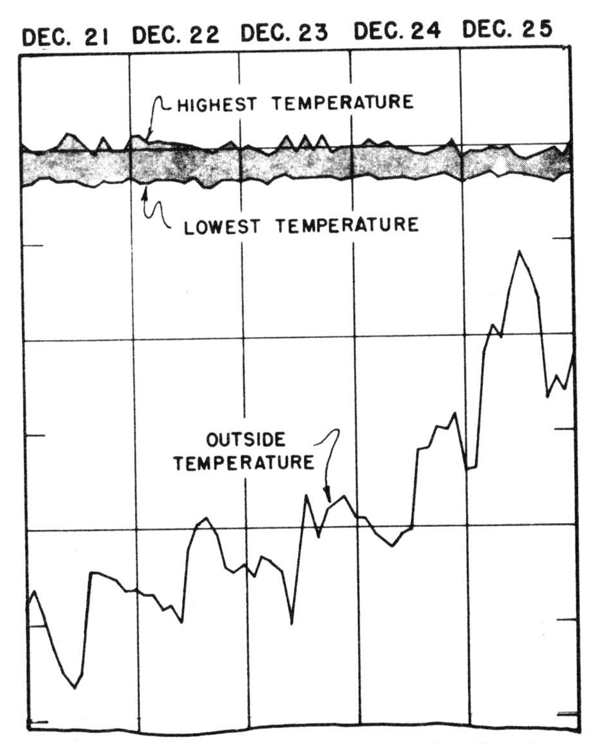

Fig. 16. Temperature span in the house after baseboard was added in all rooms except the bunkroom. Outside temperatures are also shown.

hung at this time. The maximum temperature span was again about 5°F. (Fig. 17) a year later.

Tests 4, 5, and 6 were conducted to evaluate the effect of window insulation on temperature uniformity. The temperature control throughout these tests was extremely uniform and additional layers of plastic over the windows did not improve temperature uniformity (Figs. 18, 19, and 20).

A seventh test was set up to determine the effect of reduced fan speed on temperature distribution. Although outside temperatures were extremely low, good temperature distribution was maintained with fan speed reduced to about two-thirds of its previous speed (Fig. 21). The new airflow rate is about 600 cubic feet per minute.

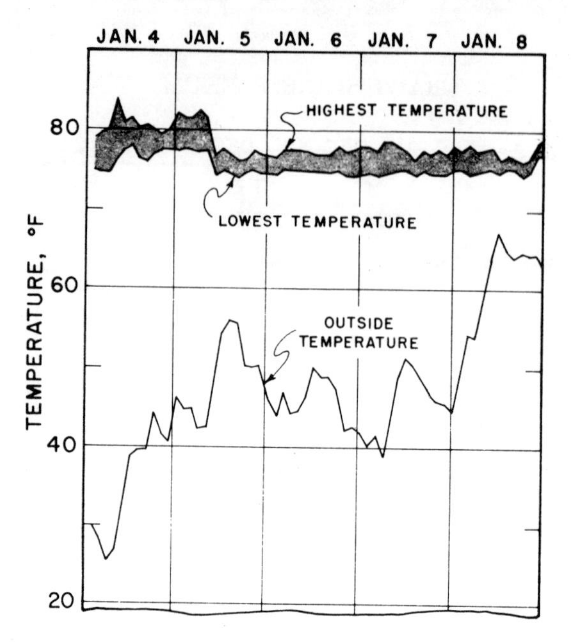

Fig. 17. Temperature span in the house with both prime doors and storm doors in place. Outside temperatures are also shown.

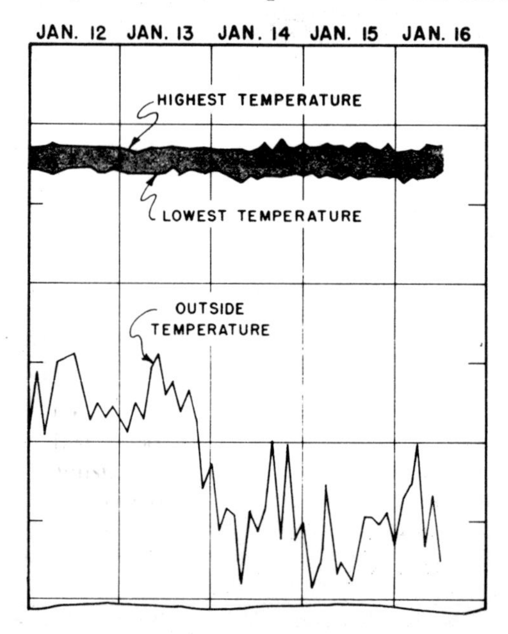

Fig. 18. Temperature span in the house after one layer of polyethylene was stretched over the windows. Outside temperatures are also shown.

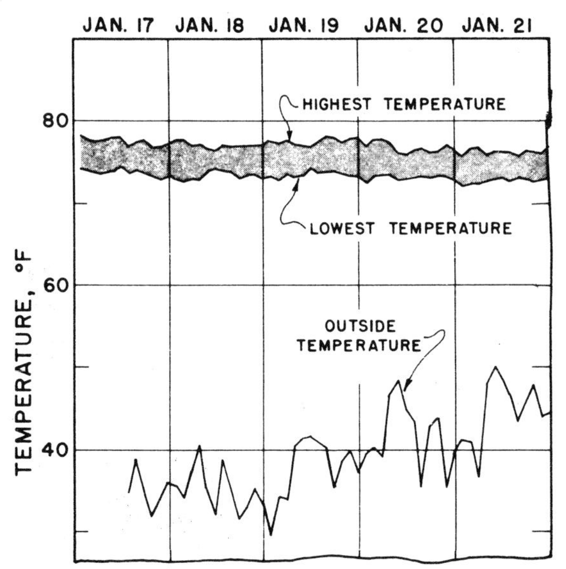

Fig. 19. Temperature span in the house after two layers of polyethylene was stretched over the windows. Outside temperatures are also shown.

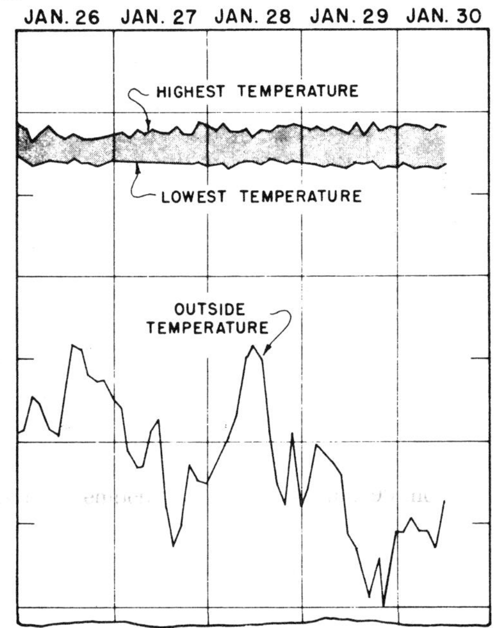

Fig. 20. Temperature in the house after three layers of polyethylene was stretched over the windows. Outside temperatures are also shown.

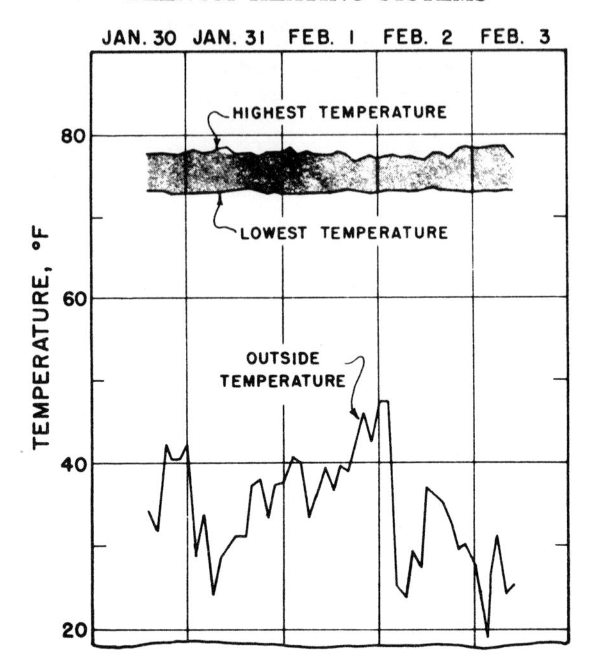

Fig. 21. Temperature span after fan speed had been lowered to 600 c.f.m.

EVALUATION OF HOT-AIR PLENUM

The hot-air plenum with a perimeter slot affords the occupants maximum comfort (uniform temperature distribution). The floor is a heat duct, thus cold floors are eliminated. Warm air from the duct (plenum) flows evenly up across the wall and maintains the wall near room temperature. At the same time the air is cooled so it does not appear as a blast of hot air. The perimeter slot diffuses the air into the room and distributes it evenly. Occupants are comfortable at lower temperatures with this system because less heat is radiated from their bodies to the walls and the floors. A hot-air plenum keeps the substructure warm and dry and eliminates freezing of water and sewer systems.

The National Fire Code places so many restrictions on a hot-air plenum that it is impossible to utilize the full potential of the system.

With the hot-air plenum the entire operation of the system is dependent on the blower operation; therefore, a failure of the electric system would cause failure of the heating system.

The house must be underpinned if it is to serve as a duct, or plenum. This will not be costly in the well-constructed house, which should be fairly tight except for ventilation vents. There will be a greater tendency to heat the soil with this system than with some other systems; thus, more fuel may be necessary. Another increase in fuel may be necessary, since more heat is lost through the sidewalls as the hot air flows up across them.

Another feature of this system that proved to be objectional was the absence of a warm spot in the house where persons coming in from outside could warm their hands or where persons could receive immediate heat while the house was heating up after a period of non-use.

The cost of the heating system is not expensive compared with most central heating systems. However, it may be beyond the means of the person building a low-cost house. In the test house it constituted one-seventh of the total cost.

CIRCULATION PLENUM

In an effort to overcome some of the objections to the hot-air plenum a new concept of plenum use was initiated. The plenum was to be used as a circulation, or turbulation, center instead of a hot-air distribution center. Since the hot-air plenum already utilizes the plenum as a circulation center, the only change to be made in the system is the location of the heat source (Fig. 22).

The use of the plenum as a circulation, or turbulation, center separates it from the heating system as such. It employs no temperatures above those normally found in the living area of the house and its association with the heating system is only incidental. For these reasons the circulation plenum brings the comforts of central heating to a house with a wood or coal radiant heater without the accompanying hazards of the hot-air plenum. Since this plenum only circulates room-temperature air throughout the plenum and the house and there is no flame or heat source as a part of this pattern, it does not create the need for the restrictions placed on hot-air plenums by the National Fire Code.

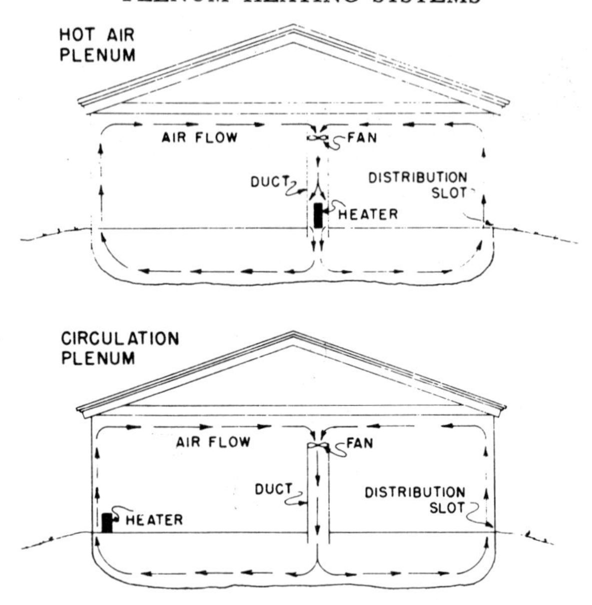

Fig. 22. Schematic hot-air plenum and of the circulation plenum.

In the hot-air plenum the heat source is located so that most of the heat is injected directly into the plenum. If the heat source is removed from the air duct leading into the plenum and placed in another area of the house, then the plenum ceases to be a hot-air plenum or even a return duct. Its only function would be to circulate the air in the house. The purpose of this circulation would be to pull air from all sections of the house, mix it, and force it into the plenum for redistribution. This process should tend to cool warm areas and to warm cold areas in the structure. Therefore, a heat source at any location in the house will provide warm air from recirculation into the other areas of the house.

Contrary to what one might imagine, the temperature distribution should be as uniform in the rooms with this heat source as it would be with a hot-air plenum. And there would be one hot spot or area in the house, which as previously indicated, may be desirable. If the blower operation should be interrupted by power failure, the system could still provide heat to part of the house. The furnace-type system would be of no value during a power failure.

The cost of the heating system could vary over a wide range. A simple wood or coal stove could easily serve as the heat source. Circulation could be accomplished with a blower mounted on a duct near the ceiling of the house that would collect air from the living area and force it into the plenum. A series of tests was set up to evaluate the circulation plenum.

An oil radiant heater was installed in the prototype house. The oil burner was shut off on the furnace, but the blower was allowed to operate continuously. This was the only change necessary to convert the house heating system from a hot-air plenum to a circulation-type plenum. The same instrumentation and the same method of evaluation as those used for testing the hot-air plenum were used for testing the circulation plenum.

Results of Tests. Figure 23 shows the extreme temperatures recorded in the house. The highest temperature was recorded at ceiling height in the room with the heater. Since this temperature was several degrees higher than the temperature at any other area of the house the second highest temperature was also plotted for comparison purposes. The irregular temperature at the ceiling of the living room is an indication of the high and low flame operation of the oil radiant heater. Examination of the original data shows the living room temperature at four inches and 48 inches above the floor was several degrees lower than the temperature at the ceiling. Evidently the air flowing through the perimeter slot picks up heat from the heater and rises to the ceiling. This warm air moves along the ceiling to the blower at the center of the house where it is mixed with air from the rest of the house and blown into the plenum. One can sense this heat stream by placing the hand near the ceiling. There is a sharp change in air temperature 6 to 8 inches below the ceiling. Most of the heat is being channeled to the plenum without overheating the living area.

NATURAL-CONVECTION CIRCULATION

In an effort to demonstrate the simplicity of the plenum circulation system and the independence of the heating source from the circulation system, a test was conducted in the test house without blower operation. This operation would be similar to that of a circulation plenum during a period of power

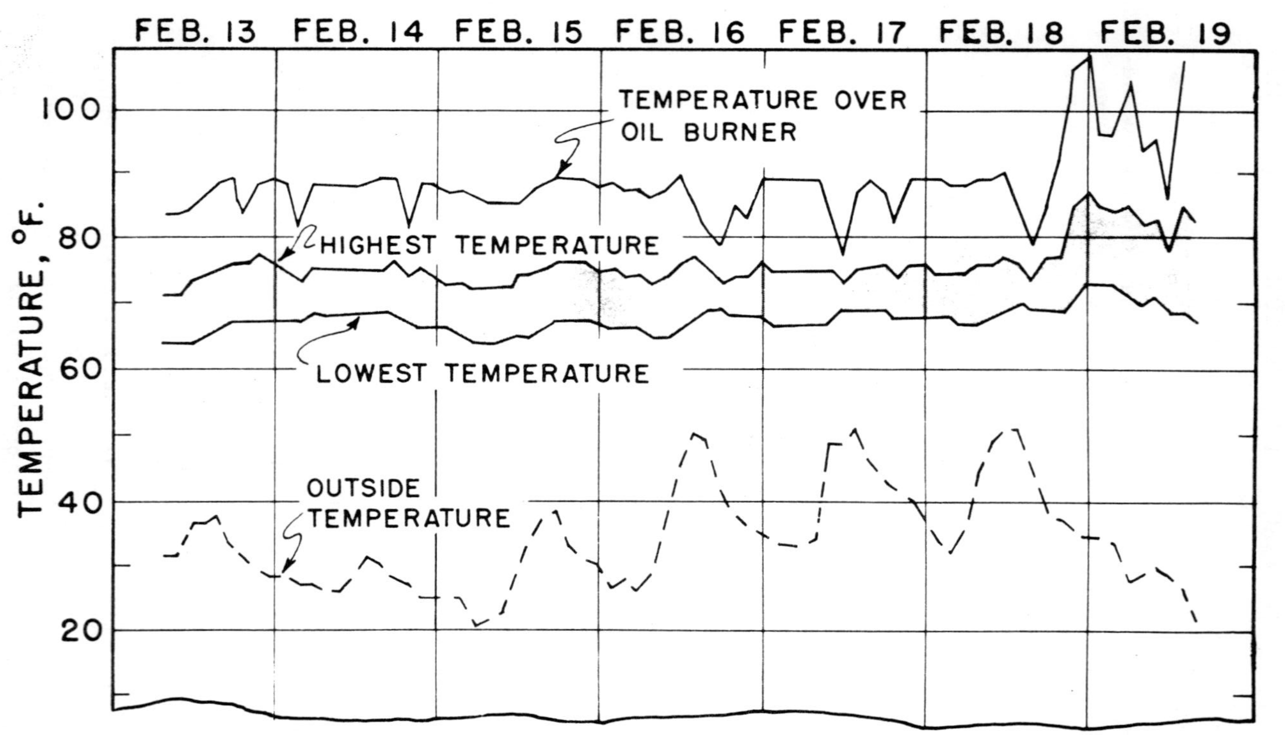

Fig. 23. Comparison of the highest temperatures in the house, except those over oil burner, and the lowest with circulation plenum; temperature over oil burner are also given. Outside temperatures are also shown.

failure. This would also be similar to that which might be used when only a small part of the house is to be heated from time to time. Although the heat source was controlled by an electric thermostat, an on-off manual control would produce the same relative situation.

In this test the high temperature was near the ceiling of the living room in the vicinity of the oil heater (Fig. 24). Since this space was receiving heat directly from the heater, the second highest temperature was plotted for comparison purposes. Since the blower did not operate, the heat was moved slowly through the house. The heater operated in a more abrupt on-off cycle, which caused considerable variation in the living room temperature.

COMPARISON OF THE THREE HEATING SYSTEMS

Figure 25 shows the heat distribution for the three circulation systems: (1) hot-air plenum; (2) circulation plenum; and (3) natural-convection circulation. These graphs show the temperature spread at 4, 48, and 90 inches through the house for each heating system.

With the hot-air plenum (January 30 and 31) there is evidently little difference in temperatures from floor to ceiling; most of the variation is from room to room. One room with a proportionally large perimeter slot showed temperature 3° to 5°F. higher than the other rooms.

With the plenum circulation system (radiant oil heater) there is a greater temperature span, especially at the 4- and 48-inch elevations. The extremely high temperature over the heater is plotted here, but it must be disregarded for comparison purposes, since it is not representative of the 90-inch temperature in most areas of the house and this air has not been circulated through the system. Examination of the graph (B, Fig. 25) shows that there is some indication of stratification but that this is very limited.

The natural-convection system shows definite stratification of the house temperature. Here the temperature span at 4 and 48 inches is much larger than it was in the other tests. This indicates that heat is reaching some areas of the house fast enough to maintain the temperature while supplying the

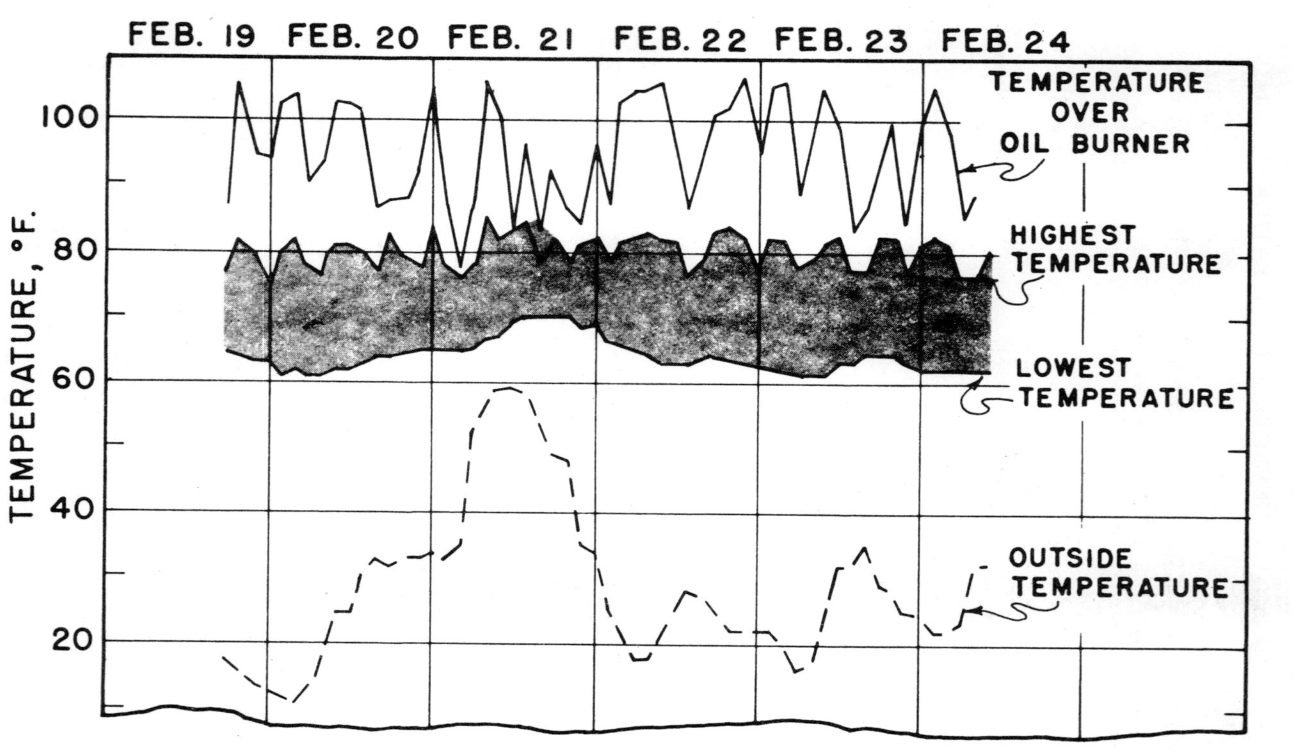

Fig. 24. Comparison of the highest temperature in the house, except those over oil burner, and the lowest with natural convection circulation. Temperatures over oil burner are given. Outside temperatures are also shown.

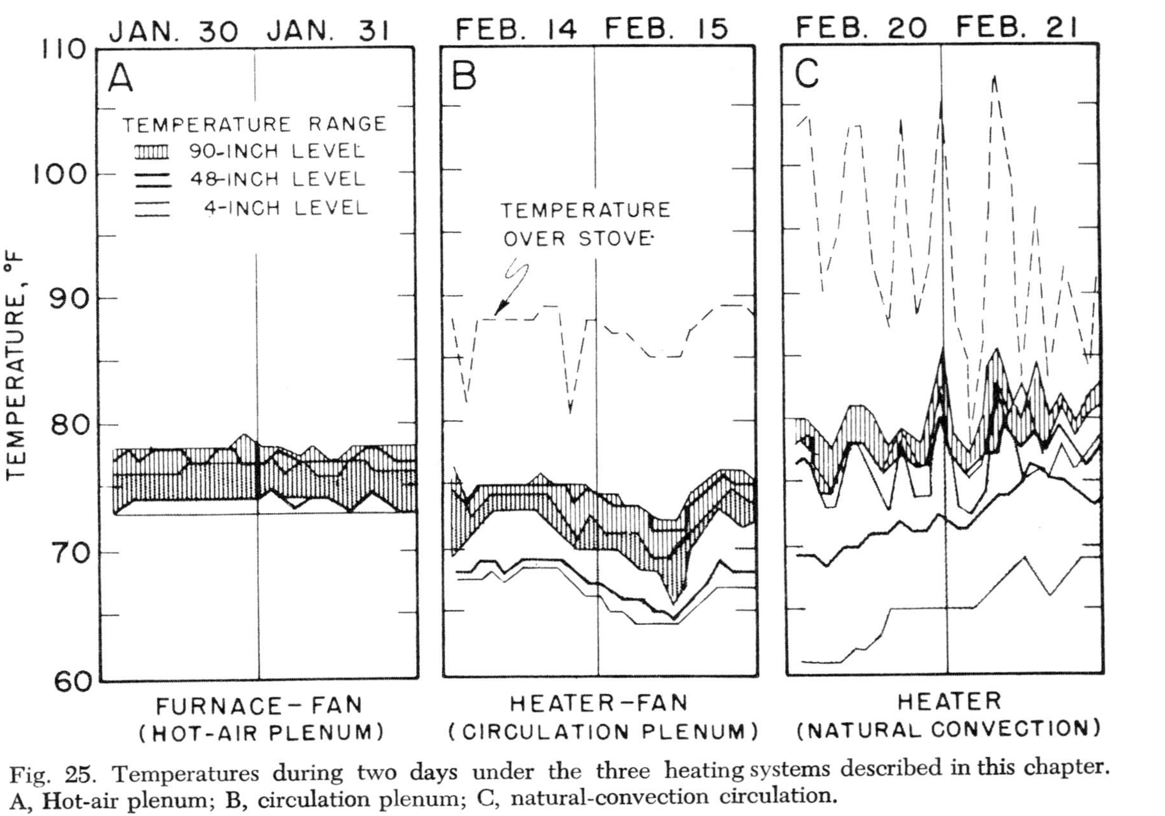

Fig. 25. Temperatures during two days under the three heating systems described in this chapter. A, Hot-air plenum; B, circulation plenum; C, natural-convection circulation.

losses, but in other areas the temperature level is not being maintained. Heat distribution is best at the 90-inch (ceiling) level. Here there is little variation from room to room at any one time, but because of the slow movement of heat away from the heat source the on-off operation of the heater causes considerable fluctuation in the ceiling temperature from hour to hour. The 90-inch (ceiling) temperature over the heater was exceedingly high when compared with other 90-inch temperatures; therefore, although it was plotted, it should not be used for comparison purposes because it is not representative of 90-inch temperatures throughout the house. The second highest 90-inch temperature was plotted for this comparison.

SUMMARY

Tests conducted in house D and in prototype house 1 have shown that a hot-air plenum can give excellent temperature distribution if the inlets into the rooms are properly sized. The slot size is not a critical factor. The only rooms that became much warmer than the others were those with slot openings two or three times wider than those in other rooms.

The constant circulation of air has given excellent distribution in the individual rooms. Usually the temperature range from floor to ceiling in a room is not over 2°F.

The circulation plenum has been found almost equally as good as the hot-air plenum. Its hot spot for quick warm up, its lower cost, its independence from public utilities, and its aesthetic appeals are advantages that outweigh the slightly more uniform distribution with hot-air plenum.

If there were no need for heat in rooms other than that with the heat source, it would not be necessary to operate the blower with the circulation plenum.

TABLE 15

COMPARISON OF THE AVERAGE TEMPERATURES AT 4 FEET ABOVE THE FLOOR AND AT THE 90-INCH (CEILING) LEVEL IN EACH OF THE THREE BEDROOMS.

Wall treatment[1]	Average temperatures at 4-foot and ceiling levels and difference								
	Master bedroom			Center bedroom			New bedroom[2]		
	4 feet	90 inches	Δt	4 feet	90 inches	Δt	4 feet	90 inches	Δt
	°F.	°F.	°F.	°F.	°F.	°F.	°F.	°F.	°F.
1	73.9	73.2	0.7	73.9	72.5	1.4	---	---	---
2	74.2	73.1	1.1	74.4	73.2	1.2	---	---	---
3	73.8	72.9	.9	73.2	72.3	.9	71.9	71.4	0.5
4	73.7	72.4	1.3	72.1	71.2	.9	71.5	71.0	.5
5	73.5	71.7	1.8	72.0	71.0	1.0	70.6	69.9	.7
6	73.2	73.4	−.2	73.3	72.4	.9	71.9	71.1	.8
7	74.0	73.4	.6	73.7	73.0	.7	72.2	71.5	.7

[1] See wall treatment of master and center bedrooms previously described.
[2] See wall construction of new bedroom previously described.

Chapter 12

Peripheral Circulation System in Old Houses

Satisfactory central heating with low-cost, even distribution may be provided to new and old houses by the peripheral circulation system. The system is simple to install and operate. It utilizes all sources of heat in the house, keeps all the air in the house in constant motion, eliminates stagnant or dead air pockets and stratification, and maintains even temperature throughout the house. This chapter describes the installation and use of this system in an existing house with a shallow crawl space.

A *peripheral circulation system* forces all the air in the house to move in a definite pattern (Fig. 1). A blower pulls air from all parts of the house into a centrally located duct near the ceiling. This air is mixed in the duct to a uniform temperature and forced into a plenum or crawl space that serves as a duct, carrying the air to any desired point under the floor. A continuous slot or a series of holes along the entire outside wall allows the air to escape from the plenum through the floor and back into the living area. A thin film of air flows up along the outside wall and gradually moves toward the central duct for remixing and redistribution.

Heat produced in any part of the house raises the temperature of the surrounding air. The hot and cold air from all parts of the house are mixed together at the central duct to an average temperature, then returned to all parts of the house. The flexibility of peripheral circulation allows many different combinations of the aforementioned assets. Among the various kinds of heat-producing equipment are a potbellied stove, a furnace, baseboard heaters, or radiant panels. Heat from cooking, light bulbs, and electric motors is also uniformly distributed.

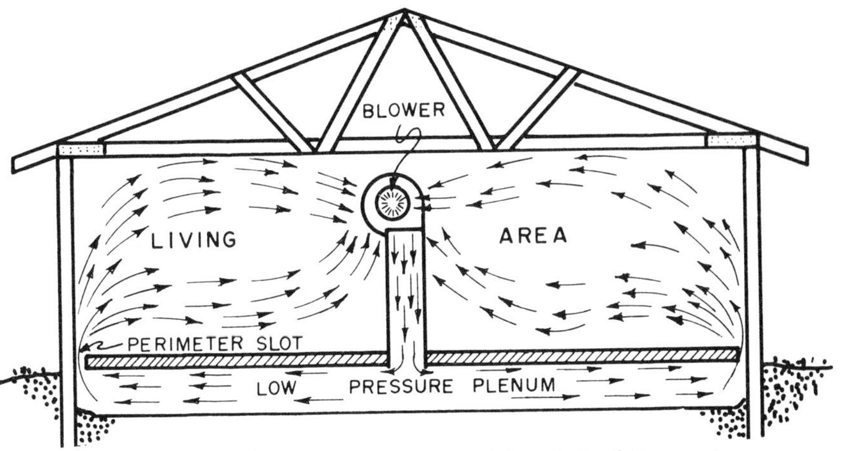

Fig. 1. Pattern of air movement in a peripheral circulation system.

The criteria for home heating vary. What one person likes may be very objectionable to another. Even the most antiquated heating systems are preferred by some. Features of good heating proclaimed by many and strived for here are as follows.

1. Uniform temperature distribution.
2. A hotspot for quick warmup.
3. Economy of operation.
4. Dependability of operation.
5. Low initial cost.
6. Cleanliness.
7. Independent operation.

In most heating systems, some of these features are sacrificed for others. In an effort to achieve as many of these features as possible, several installations utilizing the peripheral circulation distribution system have been studied in the past few years.

Most installations of peripheral circulation have been in new houses; however, most heating problems occur in older houses. Because of the low cost of installing peripheral circulation and because of the predominance of old homes among low-income families, the possibility of adding peripheral circulation to existing heating systems is under investigation.

Because tests in new houses have shown that peripheral circulation can maintain uniform temperatures with simple heating equipment, it was believed that this system could be effective in older houses.

WOOD FRAMED OLD HOUSE

The *wood framed old house* being described has lap siding and drywall interior surfaces. The L-shaped floor plan of this three-bedroom house is shown in Fig. 2. The total floor area is 856 square feet. The floors are hardwood over plywood on 2- by 8-inch (5.08- by 20.32-cm.) floor joists, and the roof is constructed of standard rafters with plywood sheathing and 235-pound shingles.

The foundation is concrete blocks to which sills are anchored with ½-inch (1.27-cm.) bolts. There is a crawl space 12 to 24 inches (30.48 to 60.96-cm.) deep under the entire house. The house is insulated with fiberglas—2 inches (5.08-cm.) thick in the sidewalls and 4 inches (10.16-cm.) thick in the ceiling. Before the installation, there was no vapor barrier to control the flow of moisture from the house into the walls

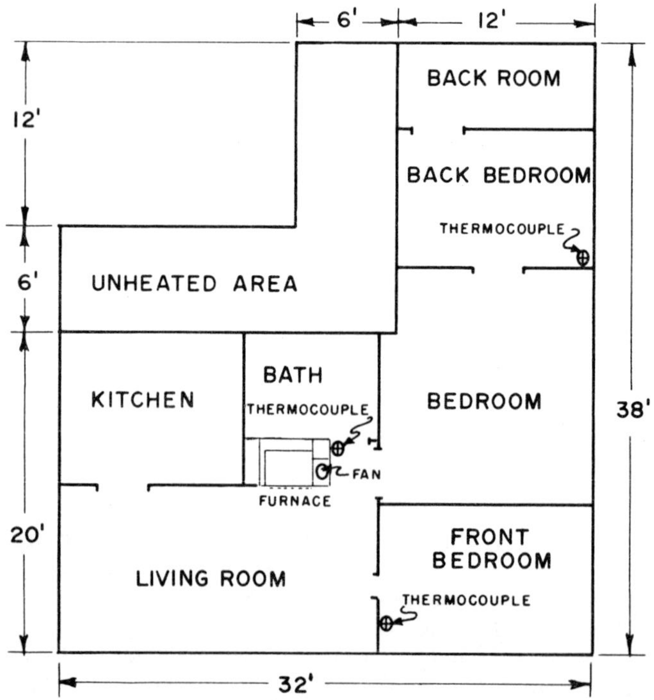

Fig. 2. Floor plan of house showing location of temperature–sensing thermocouples.

nor from the ground to the crawl space. The windows are woodframed and double-hung with a single thickness of glass. Some also have a storm sash.

HOUSE CONDITION WHEN PERIPHERAL CIRCULATION HEAT WAS INSTALLED

In general, the paint on this 4-year-old house was in good condition except for the lower siding boards that extended below the floor level. The paint on these boards had blistered and peeled considerably as shown in Fig. 3. This damage indicated that moisture was migrating from the crawl space through the lap siding. Because the moisture could not pass through the paint film, it condensed under the paint and caused the paint to loosen. Gradually more and more moisture collected under the paint film, pushing it away from the board.

The heating system was converted during the first week of January. The house had not been occupied for several days, but the living room space heater was operating. The outside temperatures ranged from 4°F. to 32° over a 5-day period during the conversion.

The house had a musty odor, which indicated the growth of fungi and mold. There was some evidence of the foundation settling, and some of the doors did not swing level.

Fig. 3. Moisture migrating from crawl space had condensed under paint and caused blisters.

Under the house the ground was muddy, and under the back bedroom frost had formed on the joists and headers adjacent to the outside wall (2 feet around perimeter). Water had condensed and was dripping from the headers under the front bedroom. Mold or fungi was evident on all floor joists under the back bedroom, especially on those timbers within 2 feet of the outside wall. Some of the timbers were rotten. Floor joists under the back bedroom contained as much as 49 percent moisture.

The crawl space did have vents in the foundation wall, but most of them were not open. In some places the grade along the house was sloped toward the structure causing water to flow along and seep into the crawl space.

EXISTING HEATING SYSTEM

The house was heated with a 65,000 B.T.U. hour oilstove that was recessed in the center wall of the living room in a metal-lined compartment (Fig. 4). The burner operated from a high flame to a low flame by the signal from the wall-mounted thermostat. A small fan was mounted in the stove compartment and thermostatically controlled to prevent overheating of the stove; it created a turbulence in the compartment but was not large enough to distribute air throughout the house.

A grille at the top of the stove allowed hot air to escape from the metal-lined compartment, and a similar grille below allowed cool air to enter the compartment. There was no positive air movement in the house, and heat was distributed by the natural forces of convection.

PHYSICAL CONSIDERATIONS

The settling floor, the rotting floor joists, and the peeling paint in the house indicated that the structure was damaged. However, it was believed that this damage could be corrected satisfactorily at reasonable cost. Since the grade around the house did not adequately divert the water from the foundation wall, additional grading was desirable.

Cracks along the top of the foundation wall and at other points in the crawl space would allow considerable air leakage should the chamber be used as a pressurized plenum for air

Fig. 4. Wall stove, which was the original heating system.

Fig. 5. Central duct and fan installation.

distribution. This air leakage could be reduced by calking these cracks or by installing some type of seal on the crawl space wall. Calking was used as the most expedient method.

The uninsulated foundation wall would allow considerable heat loss by conduction. The wall could be insulated by covering it with pressed fiberboard or other insulating material.

The next problem was to find a suitable spot for a central duct. A small built-in bookcase containing several shelves was located beside the wall stove near the center of the house (Fig. 5). The shelves could be removed to allow the central duct to be installed. The blower could be mounted at the floor level in the duct, at the top of the duct, or in the crawl space. The most suitable place, and that selected in this house, was at the floor level where the duct could be enlarged to accommodate the blower. The shallow crawl space was not selected because it would be difficult to service the fan. Locating the fan at the top of the duct would cause excessive vibration and noise.

In new houses a slot is formed around the outside wall during construction, but a suitable air return had to be developed for the old house. Care was required to avoid unsightly or expensive construction.

After studying all these conditions, it was decided to install the system as follows.

Shelves were removed from the bookcase in the living room, and an opening was prepared in the floor at its base. Then rubber mounts were installed to support the fan and to reduce vibration. A 1,200-cubic-feet-per-minute blower was installed over the opening in the floor and provided with electric power. Except for the 144-square-inch intake at the top, the front of the bookcase was enclosed to develop a chamber around the fan and to form the central duct. A 77-square-inch opening was made between the top of the central duct and the stove cabinet to form a short circuit duct to allow for the hot air to enter directly into the distribution chamber. The crawl space wall was sealed by calking all cracks, and a plastic vapor barrier was spread over the soil. In the living area the shoe molding was removed along all outside walls, and ¾-inch (1.905-cm.) holes were drilled 4 inches (10.16-cm.) on center under the shoe molding. Then the shoe molding was blocked up ⅛ inch (.3175-cm.) and nailed in place (Fig. 6).

TEST SCHEDULE FOR RECORDING ENVIRONMENTAL AND PHYSICAL CONDITIONS IN THE HOUSE

1. Before circulation is started:
   A. Record moisture content of—
      1. Floor joist
      2. Floor
      3. Center beam
      4. Sill

Fig. 6. Three-quarter-inch holes drilled under baseboard through the floor.

5. Sill header

B. Record humidity in plenum and living area.

C. Record temperature patterns throughout the house at one point at least in each room and at three elevations at each point.

2. After circulation is started proceed as follows.

A. Until conditions stabilize, record moisture content three to seven times per week of—

1. Floor joist
2. Floor
3. Center beam
4. Sill
5. Sill header

B. Keep continuous record of relative humidity in crawl space and in certain rooms until conditions stabilize.

C. Keep continuous record of temperatures at four locations throughout the house and at three elevations at each location.

D. Record occupant reaction.

E. Keep continuous record of temperature under house.

INSTRUMENTATION

To measure the effect of peripheral circulation on the house the following procedure was used.

1. A hydrothermograph was installed in the crawl space several days before the circulating blower was started. The hydrothermograph was a circular chart, continuous recording type that makes one rotation each day. This instrument was operated after the blower operation started until the relative humidity stabilized under the house.

2. An instrument that records relative humidity on a strip chart was installed in the living area. Recordings were made until conditions stabilized.

3. Temperatures throughout the house were recorded by a 12-point continuous recording potentiometer, which utilized copper constantan thermocouple sensing elements. Sensing elements were placed at three locations in the house (Fig. 2) and at three different elevations (4, 48, and 90 inches above the floor).

4. An electrical resistance moisture meter was used to mea-

sure the moisture content of the framing timbers of the floor. Several points were selected under the house to measure the moisture content at regular intervals.

RESULTS

Before the blower was started, the back bedroom temperature was quite low. Near the floor, temperatures of 40°F. were not uncommon when outside temperatures were below 20°. Temperatures near the ceiling of the living room above the heater were as much as 110°, or a 70° difference.

On January 13 the fan was started at 11:30 A.M. There was an immediate drop in the ceiling temperature of the living room and a fairly rapid increase in the temperature of the back bedroom. Twelve hours after the blower was started, the low temperature was 60°F. and the high was 84°. Temperatures in this range continued for the next several days, when it was decided that the volume of airflow was not great enough to maintain uniform temperatures (within about 5°).

On January 16 additional holes were drilled under the shoe molding in the back bedroom. The short circuit duct was also enlarged to increase the direct distribution of heat. Again there was an immediate response to the increased volume of air. The living room ceiling temperature dropped to about 72°F., and the back bedroom temperature rose to between 68° and 70°. For the next several days temperatures were uniform, ranging between 74° and 80°.

On January 22 there was evidence that the house was occupied, because the temperatures began to fluctuate, but the variation continued to be about 6°F. On January 24 there was evidence that the blower was not operating for about 36 hours and the temperature range increased. Since this was a warm period, the maximum temperature variation in the house was only 26°. From this date until January 28, temperatures remained fairly constant, again with about a 6° variation. Tests started on February 29 show a noticeable increase in temperature variation. An inquiry was made, and it was found that the occupants had reset the shoe molding tight against the floor. Thus the circulation in all rooms except the back bedroom was shut off. Since the door to this room was frequently closed, heat distribution was not very effective from this date until the outside temperature began to rise in late March and early April.

RELATIVE HUMIDITY IN LIVING AREA

A record was made of the temperature and relative humidity in several parts of the house for several days before and after the installation of the peripheral circulation system. Examination of the plot showed that the relative humidity fluctuated, with an average reading of about 40 percent. A sudden rise in temperature would cause a sudden drop in the relative humidity.

MOISTURE CONTENT OF THE TIMBERS IN CRAWL SPACE

The initial investigation of the house showed that timbers in the crawl space were dripping wet in places, and an abundance of fungi was growing. The initial moisture content of the timbers ranged from 18 to 49 percent. Table 15 shows a series of moisture readings of several different locations in the crawl space. On January 22 the moisture content of all the timbers had dropped below 15 percent, which is an acceptable level. Readings taken after January 22 showed that the moisture content continued to drop until February 20, when it was less than 8 percent in all locations. There was a drop in moisture content of timbers between January 3 and January 12 because the system was operated off and on during installation.

SUMMARY

It has been proven that peripheral circulation can be easily and economically installed in houses with crawl spaces. Temperature records showed that the variations in temperatures were effectively reduced from an intolerable difference of 70°F. to a comfortable 5° to 7° range.

Wet, unsanitary crawl spaces can be effectively dried out and maintained in good condition by the proper circulation of warm air. Heat can be effectively distributed without undesirable drafts.

Section III

Plumbing, Heating, and Other Projects

THE HOMEOWNER should have a working knowledge of details for plumbing, heating, and other utilities so that he or she can assist in the designing, follow the work closely, and properly inspect and maintain the completed job. In many instances the homeowner may have the know-how to do some of the work personally.

Chapter 13

Framing Details for Plumbing, Heating, and Other Utilities

When framing a house, it is desirable to limit cutting of framing members for installation of plumbing lines and other utilities. A little planning before framing is started will reduce the need for cutting joists and other members. This is more easily accomplished in one-story houses, however, than in two-story houses. In a *single-story house*, many of the connections are made in the basement area. In *two-story houses* they must be made between the first-floor ceiling joists. Therefore, it is sometimes necessary to cut or notch joists, but this should be done in a manner least detrimental to their strength.

PLUMBING STACK VENTS

One wall of the bath, kitchen, or utility room is normally used to carry the water, vent, and drainage lines. This is usually the wall behind the water closet where connections can be easily made to the tub or shower and to the lavatory. When 4-inch (10.16-cm.) cast iron bell pipe is used in the soil and vent stack, it is necessary to use 2- by 6- or 2- by 8-inch (5.08- by 15.24- or 5.08- by 20.32-cm.) plates to provide space for the pipe and the connections. Some contractors use a double row of studs placed flatwise so that no drilling is required for the horizontal runs (A, Fig. 1).

Building regulations in some areas allow the use of 3-inch (7.62-cm.) pipe for venting purposes in one-story houses. When this size is used, 2- by 4-inch (5.08- by 10.16-cm.) plates and studs may be employed. However, it is then necessary to reinforce the top plates, which have been cut, by using a *double scab* (B, Fig. 1). Scabs are well-nailed on each side of the

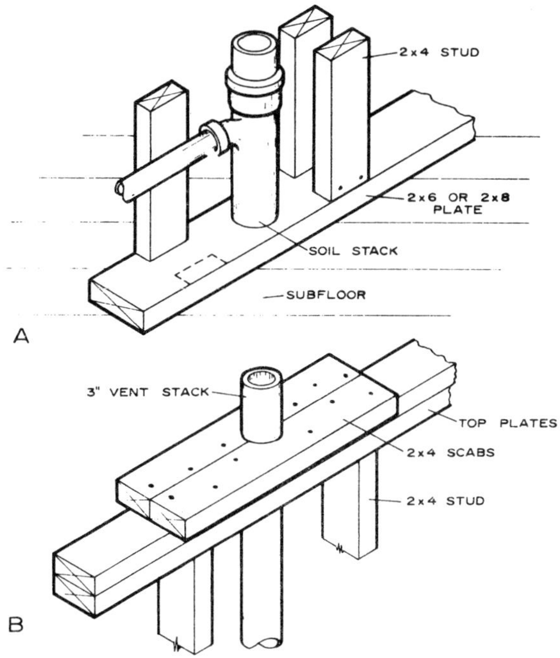

Fig. 1. Plumbing stacks. A, 4-inch cast iron stack; B, 3-inch pipe for vent.

stack and should extend over two studs. Small angle irons can also be used.

BATHROOM FRAMING

A bathtub full of water is heavy; therefore floor joists must be arranged to carry the load without excessive deflection. Too great a deflection will sometimes cause an opening above the edge of the tub. Joists should be doubled at the outer edge (Fig. 2). The intermediate joist should be spaced to clear the drain. Metal hangers or wood blocking support the inner edge of the tub at the wall line.

CUTTING FLOOR JOISTS

Floor joists should be cut, notched, or drilled only where

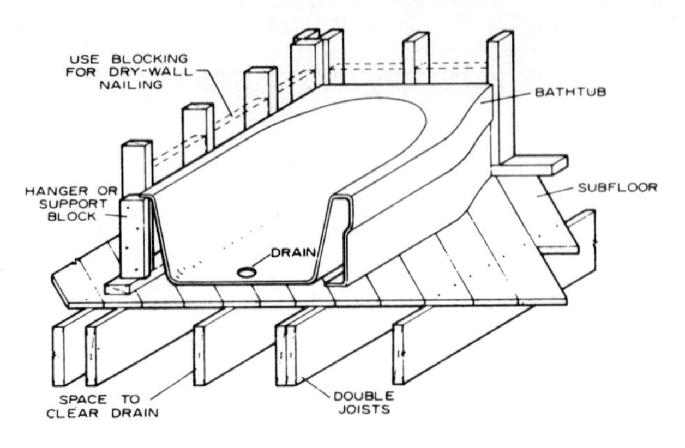

Fig. 2. Framing for bathtub.

the effect on strength is minor. While it is always desirable to prevent cutting joists whenever possible, sometimes such alterations are required. Joists or other structural members should then be reinforced by nailing a reinforcing scab to each side or by adding an additional member. Well-nailed plywood scabs on one or both sides of altered joists also provide a good method of reinforcing these members.

Notching the top or bottom of the joist should only be done in the end quarter of the span and not more than one-sixth of the depth. When greater alterations are required, headers and tail beams should be added around the altered area. This may occur where the closet bend must cross the normal joist locations. In other words, it should be framed out similar to a stair opening.

When necessary, holes may be bored in joists if the diameter is no greater than 2 inches (5.08-cm.) and the edge of the hole is not less than 2½ to 3 inches (6.35 to 7.62-cm.) from the top or bottom edge of the joists (Fig. 3). This usually limits the joist size to a 2 by 8 or larger member. Such a method of installation is suitable where joist direction changes and the pipe can be inserted from the long direction, such as from the plumbing wall to a tub on the second floor. Connections for first-floor plumbing can normally be made without cutting or drilling of joists.

ALTERATIONS FOR HEATING DUCTS

A number of systems are used to heat a house—from a multi-controlled hot-water system to a simple floor or wall furnace.

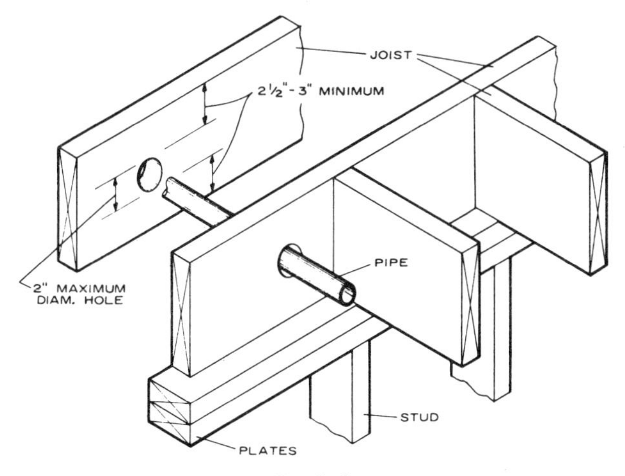

Fig. 3. Drilled holes in joists.

Central air conditioning combined with the heating system is becoming a normal part of house construction. Ducts and piping should be laid out so that framing or other structural parts can be adjusted to accommodate them. However, the system which requires heat or cooling ducts and return lines is perhaps the most important from the standpoint of framing changes required.

Supply and Cold Air Return Ducts

The installation of ducts for a forced-warm-air or air-conditioning system usually requires the removal of the sole-plate and the subfloor at the duct location. *Supply ducts* are made to dimensions that permit them to be placed between studs. When the same duct system is used for heating and cooling, the duct sizes are generally larger than when they are designed for heating alone. Such systems often have two sets of registers—one near the floor for heat and one near the ceiling for more efficient cooling. Both are furnished with dampers for control.

Walls and joists are normally located so that they do not have to be cut when heating ducts are installed. This is especially true when partitions are at right angles to the floor joists.

When a load-bearing partition requires a doubled parallel floor joist as well as a warm-air duct, the joists can be spaced

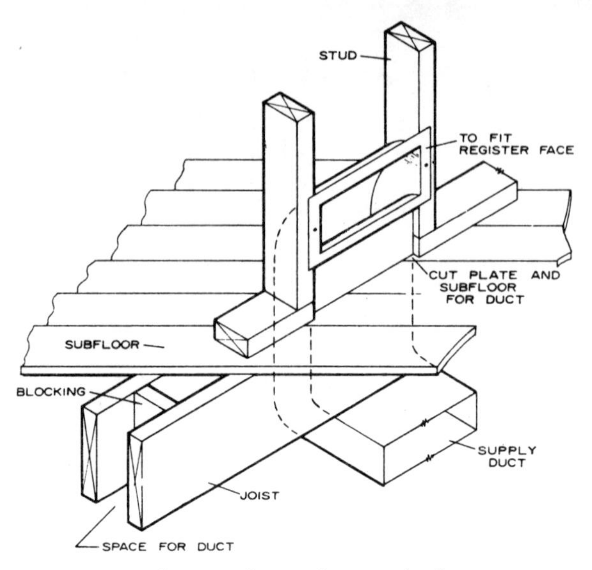

Fig. 4. Spaced joists for supply ducts.

apart to allow room for the duct (Fig. 4). This will eliminate the need for excessive cutting of framing members or the use of intricate pipe angles.

Cold-air returns are generally located in the floor between joists or in the walls at floor level (Fig. 5). They are sometimes located in outside walls, in which case they should be lined with metal. *Unlined ducts* in exterior walls have been known to be responsible for exterior-wall paint failures, especially those from a second-floor room.

The elbow from the return duct below the floor is usually placed between floor joists. The space between floor joists, when enclosed with sheet metal, serves as a cold-air return. Other cold-air returns may connect with the same joist-space return duct.

FRAMING FOR CONVECTORS

Convectors and hot-water or steam radiators are sometimes recessed partly into the wall to provide more usable space in the room and improve appearance by the installation of a decorative grille. Such framing usually requires the addition of a doubled header to carry the wall load from the studs above (Fig. 6). Size of the headers depends on the span and should be designed the same as those for window or door

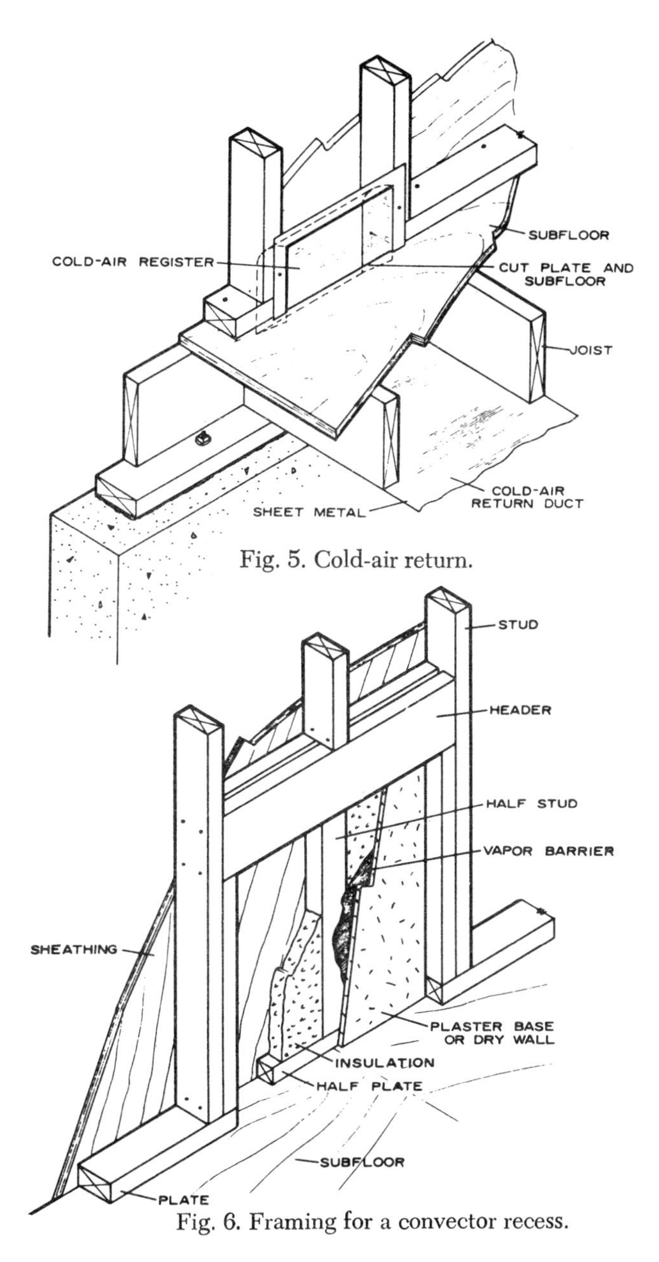

Fig. 5. Cold-air return.

Fig. 6. Framing for a convector recess.

openings. (*See* Table 17 for size of headers.) Because only 1⅝ inches (4.1275-cm.) of space in the wall is available for insulation, a highly efficient insulation (one with a low "k" value) is sometimes used.

TABLE 16
MOISTURE CONTENT OF CRAWL SPACE TIMBERS.

Date in 1968	Moisture reading		
	Under bathroom	Under living room	Under back bedroom
	Percent	*Percent*	*Percent*
January:			
3	18	22	49
12	11	19	26
14	7½	9	21
19 [1]	7	7	25
22	....	7½	14
25	([2])	7	9
27	([2])	7	10
29	([2])	7	10
31	([2])	7	9.5
February:			
3	([2])	7	9.2
7	([2])	([2])	8.8
10	([2])	([2])	7.5
14	([2])	([2])	7.5

[1] Snow melting.
[2] No indication of moisture.

WIRING

House wiring for electrical services is usually started some time after the house has been closed in. The initial phase of it, termed "roughing in," includes the installation of conduit or cable and the location of switch, light, and outlet boxes with wires ready to connect. This roughing-in work is done before the plaster base or dry-wall finish is applied, and before the insulation is placed in the walls or ceilings. The placement of the fixtures, the switches, and switch plates is done after plastering.

Framing changes for wiring are usually of a minor nature and, for the most part, consist of holes drilled in the studs for the flexible conduit. Although these holes are small in diameter, they should comply with locations shown in Fig. 6. Perhaps the only area which requires some planning to prevent excessive cutting or drilling is the location of wall switches at entrance door frames. By spacing the doubled framing studs to allow for location of multiple switch boxes, little cutting will be required.

Switches or convenience outlet boxes on exterior walls must be sealed to prevent water vapor movement. Sealing of the vapor barrier around the box is important.

TABLE 17
SIZE OF HEADERS.

Maximum span (*Ft.*)	*Header size* (*In.*)
3½	2 by 6
5	2 by 8
6½	2 by 10
8	2 by 12

Chapter 14

Chimneys and Fireplaces

Chimneys are generally constructed of masonry units supported on a suitable foundation. A chimney must be structurally safe, capable of producing sufficient draft for the fireplace, and capable of carrying away harmful gases from the fuel-burning equipment and other utilities. Lightweight, prefabricated chimneys that do not require masonry protection or concrete foundations are now accepted for certain uses by fire underwriters. Make certain, however, they are approved and listed by Underwriters' Laboratories, Inc.

Fireplaces should not only be safe and durable but should be so constructed that they provide sufficient draft and are suitable for their intended use. From the standpoint of heat-production efficiency, which is estimated to be only ten percent, they might be considered a luxury. However, they add a decorative note to a room and a cheerful atmosphere. Improved heating efficiency and the assurance of a correctly proportioned fireplace can usually be obtained by the installation of a factory-made circulating fireplace. This metal unit, enclosed by the masonry, allows air to be heated and circulated throughout the room in a system separate from the direct heat of the fire.

CHIMNEYS

The *chimney* should be built on a concrete footing of sufficient area, depth, and strength for the imposed load. The footing should be below the frostline. For houses with a basement, the footings for the walls and fireplace are usually poured together and at the same elevation.

The size of the chimney depends on the number of flues,

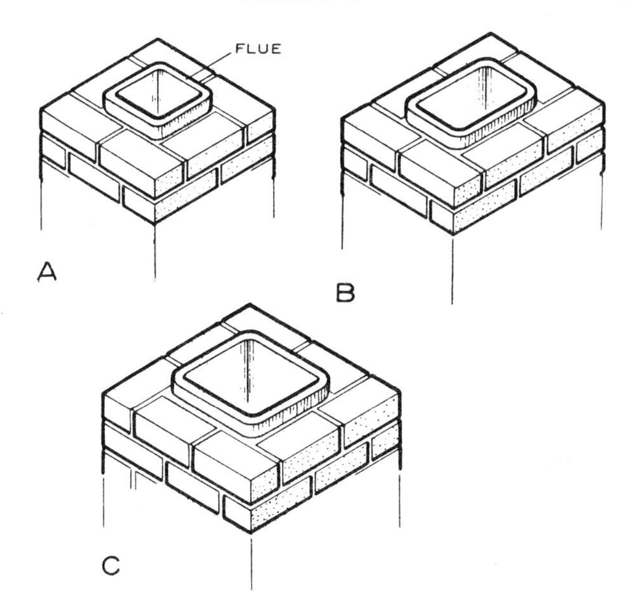

Fig. 1. Brick and flue combinations.

the presence of a fireplace, and the design of the house. The house design may include a room-wide brick or stone fireplace wall which extends through the roof. While only two or three flues may be required for heating units and fireplaces, several "false" flues may be added at the top for appearance. The flue sizes are made to conform to the width and length of a brick so that full-length bricks can be used to enclose the flue lining. Therefore an 8- by 8-inch (20.32- by 20.32-cm.) flue lining (about 8½ by 8½ inches [21.59 by 21.59-cm.] in outside dimensions) with the minimum 4-inch (10.16-cm.) thickness of surrounding masonry will use six standard bricks for each course (A, Fig. 1). An 8- by 12-inch (20.32- by 30.48-cm.) flue lining (8½ by 13 inches [21.59 by 33.02 cm.] in outside dimensions) will be enclosed by seven bricks at each course (B, Fig. 1), and a 12- by 12-inch (30.48- by 30.48-cm) flue (13 by 13 inches [33.02 by 33.02-cm.] in outside dimensions) by eight bricks (C, Fig. 1), and so on. Each fireplace should have a separate flue and, for best performance, flues should be separated by a 4-inch-wide (10.16-cm.) brick spacer (width) between them (A, Fig. 2).

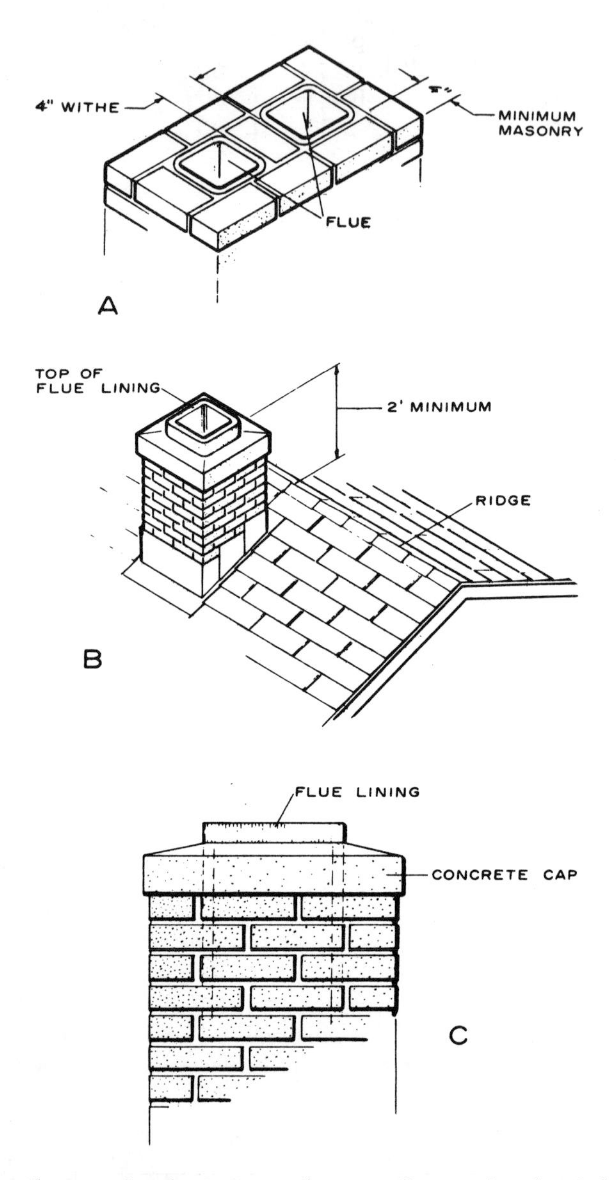

Fig. 2. Chimney details. A, Spacer between flues; B, height of chimneys; C, chimney cap.

The greater the difference in temperature between chimney gases and outside atmosphere, the better the draft. Thus, an interior chimney will have better draft because the masonry will retain heat longer. The height of the chimney as well as the size of the flue are important factors in providing sufficient draft.

The height of a chimney above the roofline usually depends upon its location in relation to the ridge. The top of the extending flue liners should not be less than two feet above the ridge or a wall that is within 10 feet (B, Fig. 2). For flat or low-pitched roofs, the chimney should extend at least three feet above the highest point of the roof. To prevent moisture from entering between the brick and the flue lining, a concrete cap is usually poured over the top course of brick (C, Fig. 2). Precast or stone caps with a cement wash are also used.

Flashing for chimneys is shown in Fig. 1, Chapter 15. Masonry chimneys should be separated from wood framing, subfloor, and other combustible materials. Framing members should have at least a 2-inch (5.08-cm.) clearance and should be firestopped at each floor with asbestos or other types of noncombustible material (Fig. 3). Subfloor, roof sheathing,

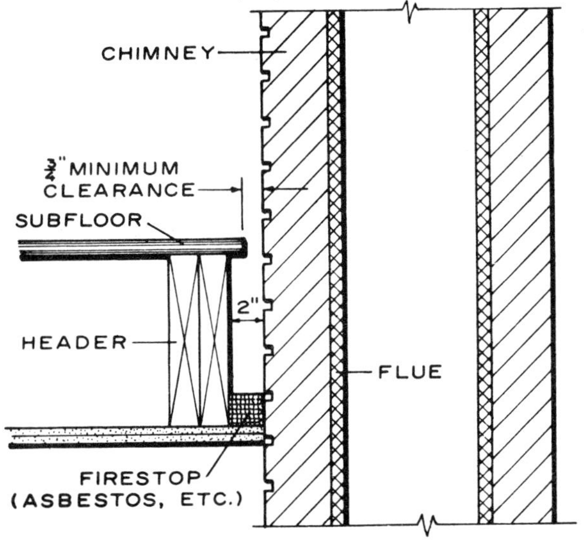

Fig. 3. Clearances for wood construction.

and wall sheathing should have a ¾-inch (1.905-cm.) clearance. A cleanout door is included in the bottom of the chimney where there are fireplaces or other solid fuel-burning equipment as well as the bottom of other flues. The cleanout door for the furnace flue is usually located just below the smokepipe thimble with enough room for a soot pocket.

FLUE LININGS

Rectangular fire-clay flue linings (previously described) or *round vitrified tile* are normally used in all chimneys. *Vitrified* (glazed) *tile* or a *stainless steel* lining is usually required for gas-burning equipment. Local codes outline these specific requirements. A fireplace chimney with at least an 8-inch-thick (20.32-cm.) masonry wall ordinarily does not require a flue lining. However, the cost of the extra brick or masonry and the labor involved are most likely greater than the cost of flue lining. Furthermore, a well-installed flue lining will result in a safer chimney.

Flue liners should be installed enough ahead of the brick or masonry work, as it is carried up, so that careful bedding of the mortar will result in a tight and smooth joint. When diagonal offsets are necessary, the flue liners should be beveled at the direction change in order to have a tight joint. It is also good practice to stagger the joints in adjacent tile.

Flue lining is supported by masonry and begins at least eight inches (20.32-cm.) below the thimble for a connecting smoke or vent pipe from the furnace. In fireplaces, the flue liner should start at the top of the throat and extend to the top of the chimney.

Rectangular flue lining is made in two-foot lengths and in sizes of 8 by 8, 8 by 12, 12 by 12, 12 by 16, and up to 20 by 20 inches. Wall thicknesses of the flue lining vary with the size of the flue. The smaller sizes have a ⅝-inch-thick (1.5875-cm.) wall, and the larger sizes vary from ¾ to 1⅜ inches (1.905 to 3.4925-cm.) in thickness. Vitrified tiles, 8 inches (20.32-cm.) in diameter, are most commonly used for the flues of the heating unit, although larger sizes are also available. This tile has a bell joint.

FIREPLACES

A *fireplace* adds to the attractiveness of the house interior, but one that does not "draw" properly is a detriment, not an asset. By following several rules on the relation of the fireplace opening size to flue area, depth of the opening, and other measurements, satisfactory performance can be assured. Metal circulating fireplaces, which form the main outline of the opening and are enclosed with brick, are designed for proper functioning when flues are the correct size.

One rule which is often recommended is that the depth of the fireplace should be about two-thirds the height of the opening. Thus, a 30 inch high fireplace would be 20 inches deep from the face to the rear of the opening.

The flue area should be at least one-tenth of the open area of the fireplace (width times height) when the chimney is 15 feet or more in height. When less than 15 feet, the flue area in square inches should be one-eighth of the opening of the fireplace. This height is measured from the throat to the top of the chimney. Thus, a fireplace with a 30-inch width and 24-inch height (720 square inches) would require an 8- by 12-inch flue, which has an inside area of about 80 square inches. A 12- by 12-inch flue liner has an area of about 125 square inches, and this would be large enough for a 36- by 30-inch opening when the chimney height is 15 feet or over.

The back width of the fireplace is usually 6 to 8 inches narrower than the front. This helps to guide the smoke and fumes toward the rear. A vertical back-wall of about a 14-inch height then tapers toward the upper section or "throat" of the fireplace (Fig. 4). The area of the throat should be about 1¼ to 1⅓ times the area of the flue to promote better draft. An adjustable damper is used at this area for easy control of the opening.

The *smoke shelf* (top of the throat) is necessary to prevent back drafts. The height of the smoke shelf should be 8 inches above the top of the fireplace opening (Fig. 4). The smoke shelf is concave to retain any slight amount of rain that may enter.

Steel angle iron is used to support the brick or masonry over the fireplace opening. The bottom of the inner hearth,

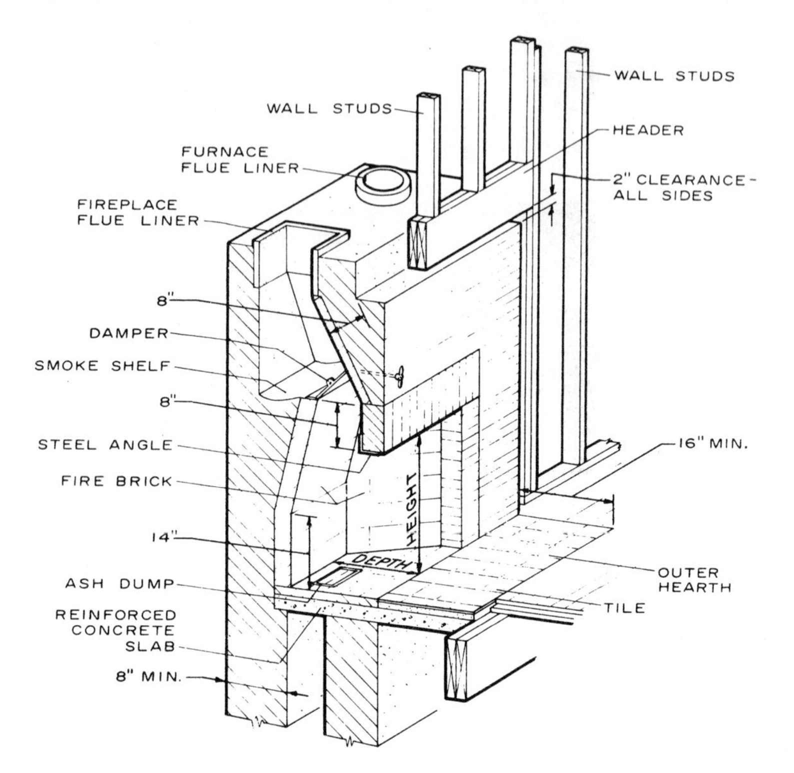

Fig. 4. Masonry fireplace.

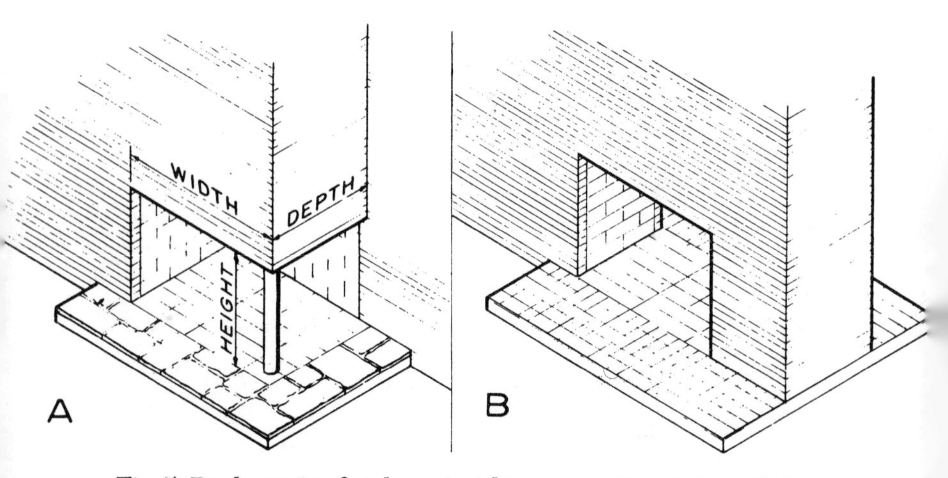

Fig. 5. Dual-opening fireplace. A, Adjacent opening; B, through fireplace.

the sides, and the back, are built of a heat-resistant material such as firebrick. The outer hearth should extend at least 16 inches out from the face of the fireplace and be supported by a reinforced concrete slab (Fig. 4). This outer hearth is a precaution against flying sparks and is made of noncombustible materials such as glazed tile. Other fireplace details of clearance, framing of the wall, and cleanout opening and ash dump are also shown. Hangers and brackets for fireplace screens are often built into the face of the fireplace.

Fireplaces with two or more openings (Fig. 5) require much larger flues than the conventional fireplace. For example, a fireplace with two open adjacent faces (A, Fig. 2) should require a 12- by 16-inch flue for a 34- by 20- by 30-inch (width, depth, and height, respectively) opening. Local building regulations usually cover the proper sizes for these types of fireplaces.

PREFABRICATED FIREPLACES AND CHIMNEYS

Prefabricated fireplace and chimney units—all parts needed for a complete fireplace-to-chimney installation—can be bought at most local dealers (Fig. 6).

Such units offer the following features.

1. Wide selection of styles, shapes, and colors.
2. Pretested design that is highly efficient in operation.
3. Easy and versatile installation—can be installed free-standing of flush against a wall in practically any part of a house.
4. Light in weight.
5. Lower cost than comparable masonry units.

The basic part of the prefabricated fireplace is a specially insulated metal firebox shell. Since it is light in weight, it can be set directly on the floor without the heavy footing required for masonry fireplaces.

Prefabricated chimneys can be used for furnaces, heaters, and incinerators as well as for prefabricated fireplaces. The chimneys are tested and approved by the Underwriters' Laboratories, Inc., and other nationally recognized testing laboratories, and are rapidly being accepted for use by building codes in many communities.

Fig. 6. Prefabricated fireplace.

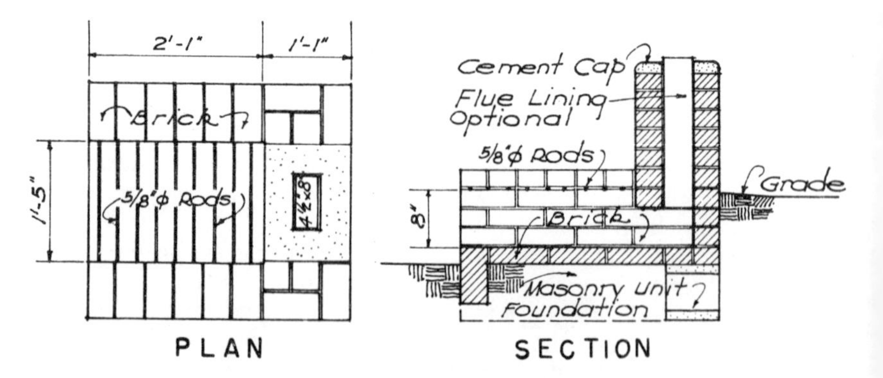

Fig. 7. An outdoor fireplace of modest design.

OUTDOOR FIREPLACES

Outdoor fireplaces range from simple makeshift units to elaborate structures designed to harmonize with and enhance the appearance of the house and the landscape.

If the fireplace is to be built with local labor and materials, a relatively simple design, such as that shown in Fig. 7 is advisable.

Built-in features, such as ovens, cranes, grilles, storage compartments, sinks, and benches, add to the appearance and convenience of fireplaces. Dealers in outdoor-fireplace equipment usually have catalogs listing types and sizes of accessory equipment.

Elaborately designed fireplaces that include many built-in features or that are an integral part of a building should be built by skilled labor.

Chapter 15

Flashing, Gutters, and Downspouts

In house construction, the sheet-metal work normally consists of flashing, gutters, and downspouts, and sometimes attic ventilators. *Flashing* is often provided to prevent wicking action by joints between moisture-absorbent materials. It might also be used to provide protection from wind-driven rain or from action of melting snows. For instance, damage from ice dams is often the result of inadequate flashing. Thus, proper installation of these materials is important, as well as their selection and location.

Gutters are installed at the cornice line of a pitched-roof house to carry the rain or melted snow to the downspouts and away from the foundation area. They are especially needed for houses with narrow roof overhangs. Where positive rain disposal cannot be assured, downspouts should be connected with storm sewers or other drains. Poor drainage away from the wall is often the cause of wet basements and other moisture problems.

MATERIALS

Materials most commonly used for sheet-metal work are galvanized metal, terneplate, aluminum, copper, and stainless steel. Near the seacoast, where the salt in the air may corrode galvanized sheet metal, copper or stainless steel is preferred for gutters, downspouts, and flashings. Molded wood gutters, cut from solid pieces of Douglas fir or redwood, are also used in coastal areas because they are not affected by the corrosive atmosphere. Wood gutters can be attractive in appearance and are preferred by some builders.

Galvanized (zinc-coated) *sheet metal* is used in two weights of zinc coatings—1.25 and 1.50 ounces per square foot (total weight of coating on both sides). When the lightly coated 1.25-ounce sheet is used for exposed flashing and for gutters and downspouts, 26-gage metal is required. With the heavier 1.50-ounce coating, a 28-gage metal is satisfactory for most metal work, except that gutters should be 26-gage.

Aluminum flashing should have a minimum thickness of 0.019 inch, the same as for roof valleys. Gutters should be made from 0.027-inch-thick metal and downspouts from 0.020-inch thickness. Copper for flashing and similar uses should have a minimum thickness of 0.020 inch (16 ounces). Aluminum is not normally used when it comes in contact with concrete or stucco unless it is protected with a coat of asphaltum or other protection against reaction with the alkali in the cement.

The types of metal fastenings, such as nails and screws, and the hangers and clips used with the various metals, are important to prevent corrosion or deterioration when unlike metals are used together. For aluminum, only aluminum or stainless steel fasteners should be used. For copper flashing, use copper nails and fittings. Galvanized sheet metal or terneplate should be fastened with galvanized or stainless-steel fasteners.

FLASHING

Flashing should be used at the junction of a roof and a wood or masonry wall, at chimneys, over exposed doors and windows, at siding material changes, in roof valleys, and other areas where rain or melted snow may penetrate into the house.

Material Changes

One wall area which requires flashing is at the intersection of two types of siding materials. For example, a stucco-finish gable end and a wood-siding lower wall should be flashed (A, Fig. 1). A wood molding such as a drip cap separates the two materials and is covered by the flashing which extends behind the stucco. The flashing should extend at least 4 inches (10.16-cm.) above the intersection. When sheathing paper is used, it should lap the flashing (A, Fig. 1).

When a wood-siding pattern change occurs on the same wall, the intersection should also be flashed. A vertical board-

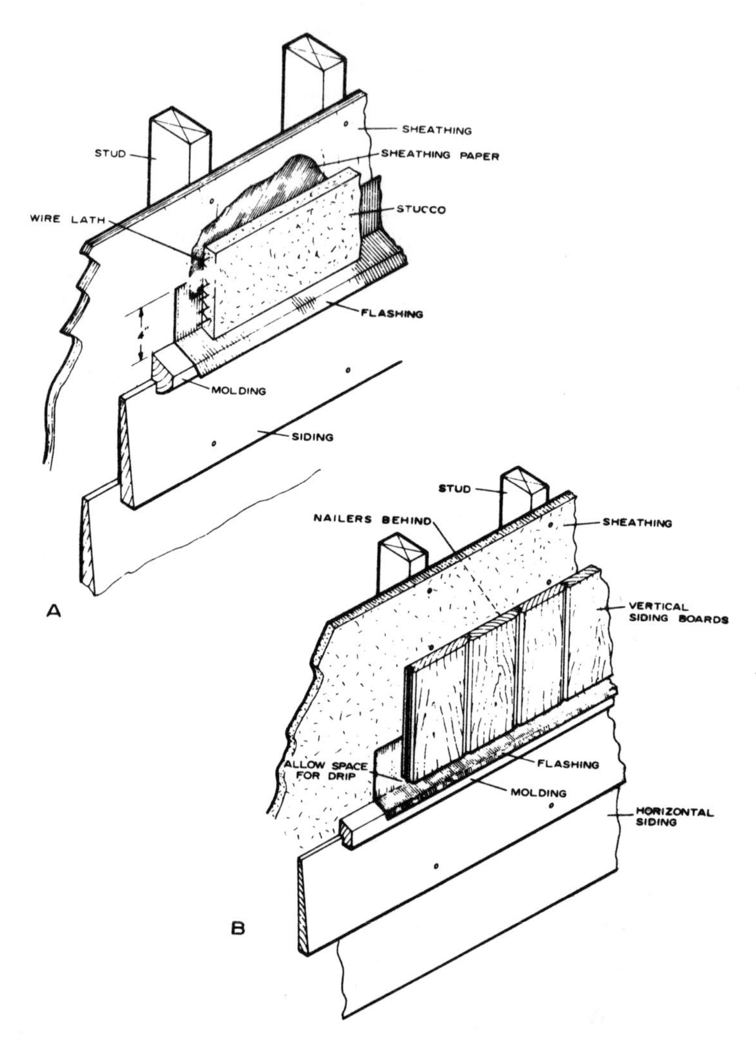

Fig. 1. Flashing at material changes.

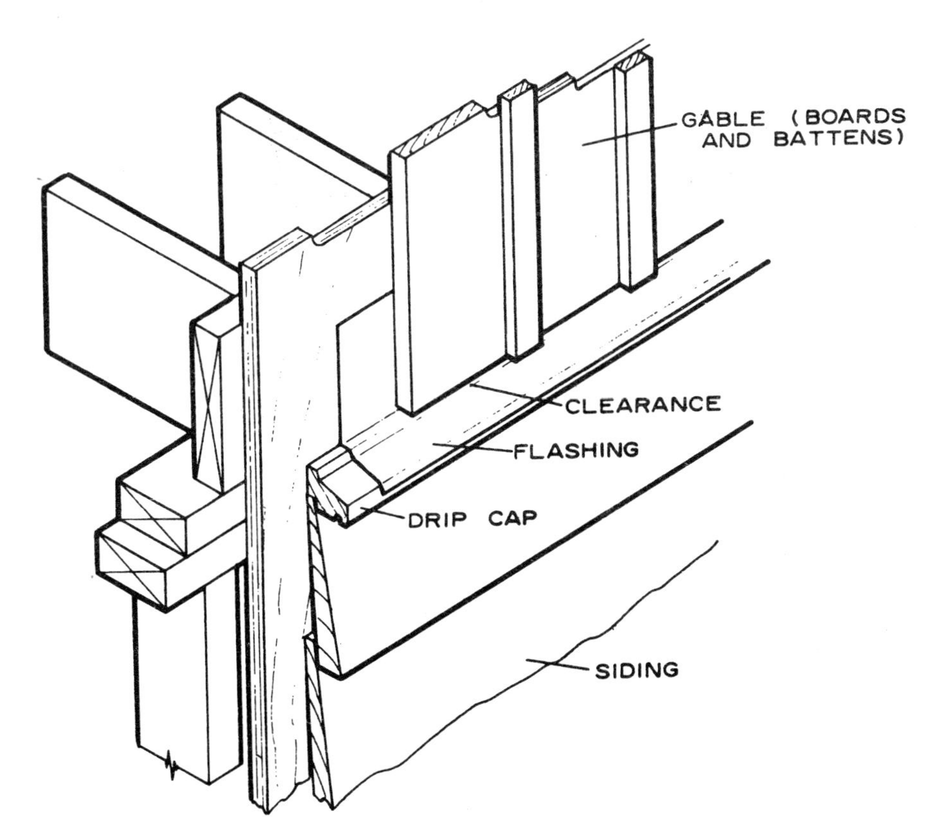

Fig. 2. Gable-end finish (material transition).

sided upper wall with horizontal siding below usually requires some type of flashing (B, Fig. 1). A small space above the molding provides a drip for rain. This will prevent paint peeling which could occur if the boards were in tight contact with the molding. A drip cap is sometimes used as a terminating molding (Fig. 2). When the upper wall, such as a gable end, projects slightly beyond the lower wall, flashing is usually not required.

Doors and Windows

The same type of flashing shown in A, Fig. 1 should be used over door and window openings exposed to driving rain. However, window and door heads protected by wide overhangs in a single-story house with a hip roof do not ordinarily require such flashing. When building paper is used on the sidewalls, it should lap the top edge of the flashing. To protect the walls behind the window sill in a brick veneer exterior, flashing should extend under the masonry sill up to the underside of the wood sill.

Flat Roof

Flashing is also required at the junctions of an exterior wall and a flat or low-pitched built-up roof (Fig. 3). When a metal roof is used, the metal is turned up on the wall and covered by the siding. A clearance of 2 inches (5.08-cm.) should be allowed at the bottom of the siding for protection from melted snow and water.

Ridge and Roof

Ridge flashing should be used under a Boston ridge in wood

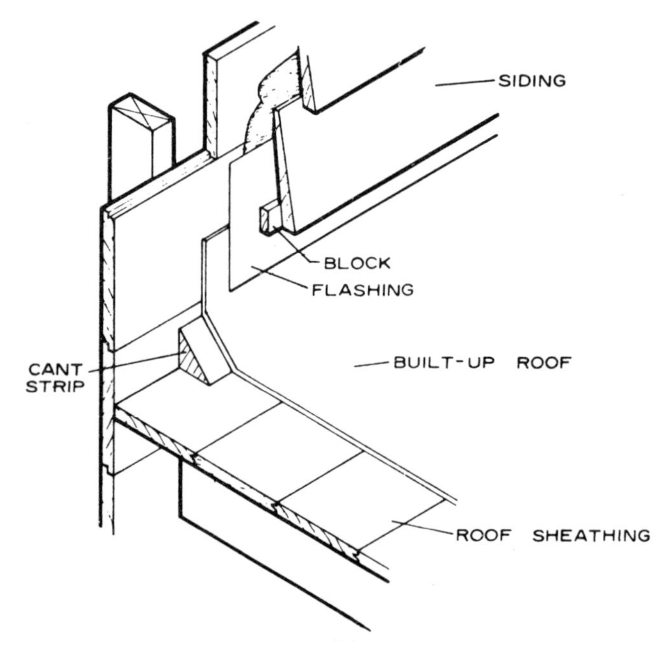

Fig. 3. Built-up roof.

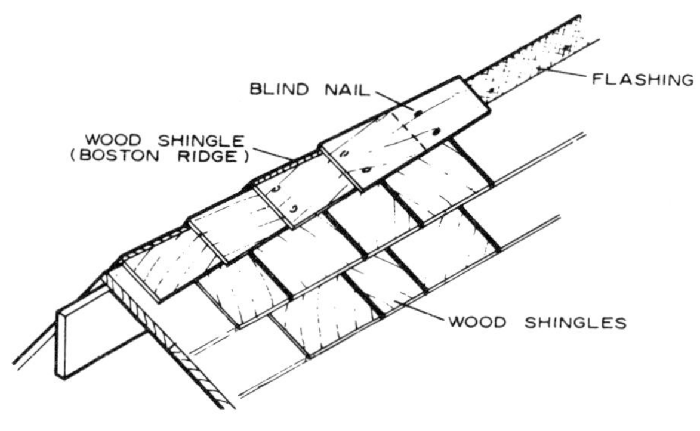

Fig. 4.

shingle or shake roofs to prevent water entry (Fig. 4). The flashing should extend about 3 inches (7.62-cm.) on each side of the ridge and be nailed in place only at the outer edges. The ridge shingles or shakes, which are 6 to 8 inches (15.24 to 20.32-cm.) wide, cover the flashing.

Stock vents and roof ventilators are provided with flashing collars which are lapped by the shingles on the upper side. The lower edge of the collar laps the shingles. Sides are nailed to the shingles and calked with a roofing mastic.

Valley

The *valley* formed by two intersecting rooflines is usually covered with metal flashing. Some building regulators allow the use of two thicknesses of mineral-surfaced roll roofing in place of the metal flashing. As an alternate, one 36-in-wide (91.44-cm.) strip of roll roofing with closed or woven asphalt shingles is also allowed. This type of valley is normally used only on roofs with a slope of 10 in 12 or steeper.

Widths of sheet-metal flashing for valleys should not be less than:

(a) 12 inches (30.48-cm.) wide for roof slopes of 7 in 12 and over.

(b) 18 inches (45.72-cm.) wide for 4 in 12 to 7 in 12 roof slopes.

(c) 24 inches (60.96-cm.) wide for slopes less than 4 in 12.

The width of the valley between shingles should increase from the top to the bottom (A, Fig. 5). The minimum open width at the top is 4 inches (10.16-cm.) and should be in-

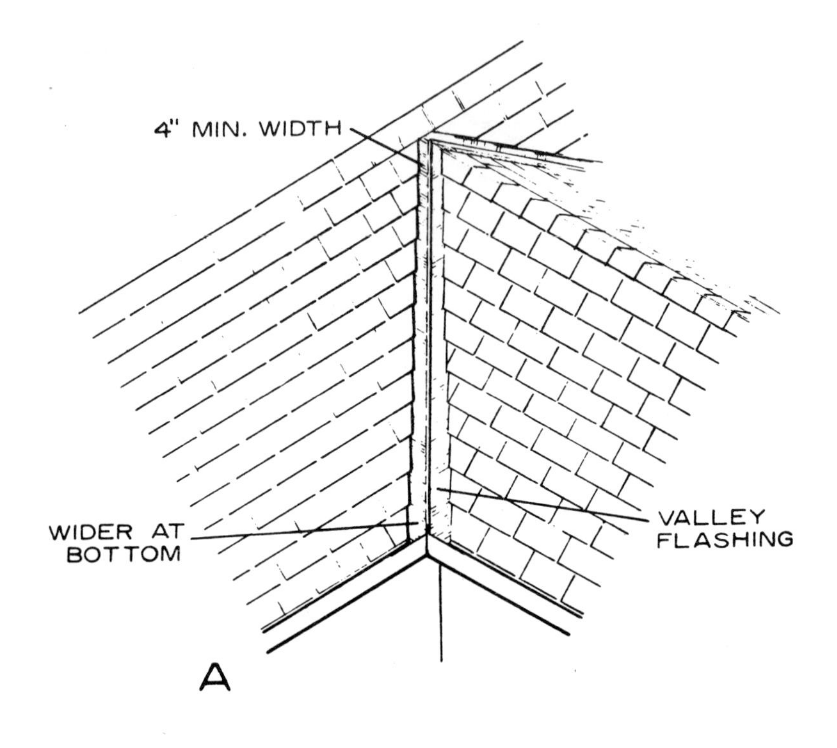

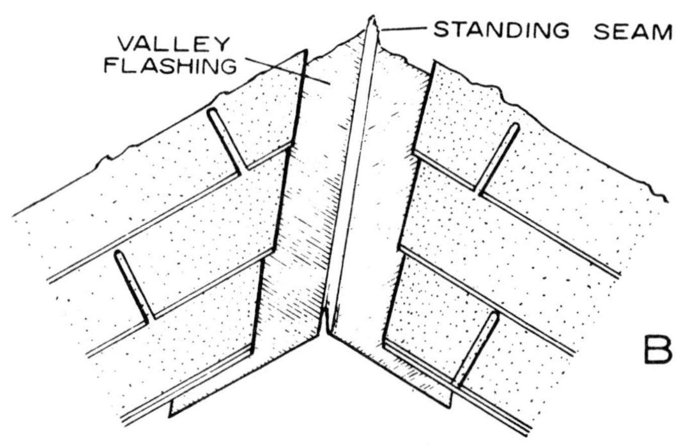

Fig. 5. Valley flashing, A, Valley; B, standing seam.

creased at the rate of about ⅛ inch per foot (.3175-cm. per 30.48-cm.). These widths can be chalklined on the flashing before shingles are applied.

When adjacent roof slopes vary, such as a low-slope porch roof intersecting a steeper main roof, a 1-inch (2.54-cm.) crimped standing seam should be used (B, Fig. 5). This will keep heavy rains on the steeper slopes from overrunning the valley and being forced under the shingles on the adjoining slope. Nails for the shingles should be kept back as far as possible to eliminate holes in the flashing. A ribbon of asphalt-roofing mastic is often used under the edge of the shingles. It is wise to use the wider valley flashings supplemented by a width of 15- or 30-pound asphalt felt where snow and ice dams may cause melting snow water to back under shingles.

Roof-Wall Intersections

When shingles on a roof intersect a vertical wall, shingle flashing is used at the junction. These tin or galvanized-metal shingles are bent at a 90-degree angle and extend up the side of the wall over the sheathing a minimum of 4 inches (10.16-cm.) (A, Fig. 6). When roofing felt is used under the shingle, it is turned up on the wall and covered by the flashing. One piece of flashing is used at each shingle course. The siding is then applied over the flashing, allowing about a two-inch (5.08-cm.) space between the bevel edge of the siding and the roof.

If the roof intersects a brick wall or chimney, the same type of metal shingle flashing is used at the end of each shingle course as described for the wood-sided wall. In addition, counterflashing or brick flashing is used to cover the shingle flashing (B, Fig. 6). This counterflashing is often performed in sections and is inserted in open mortar joints. Unless soldered together, each section should overlap the next a minimum of 3 inches (7.62-cm.) with the joint calked. In laying up the chimney or the brick, the mortar is usually raked out for a depth of about 1 inch (2.54-cm.) at flashing locations. Lead wedges driven into the joint above the flashing hold it in place. The joint is then calked to provide a watertight connection. In chimneys, this counterflashing is often preformed to cover one entire side.

Around small chimneys, chimney flashing often consists of

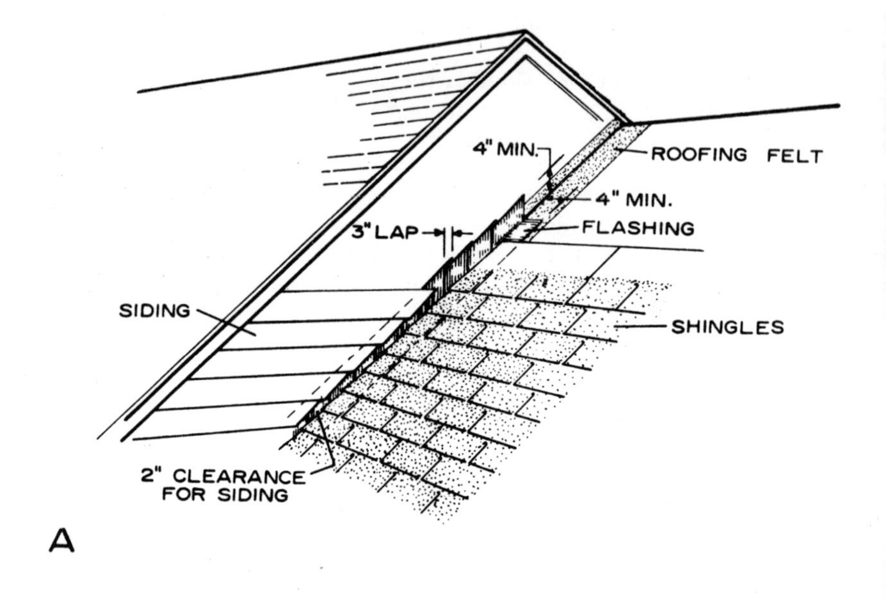

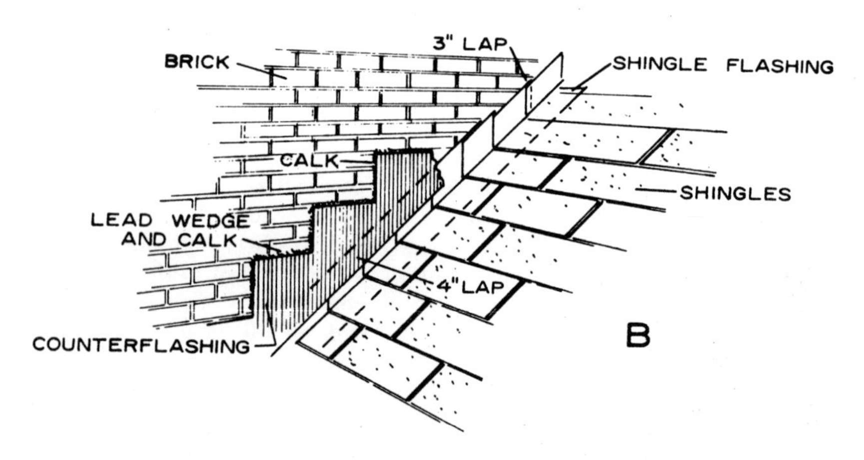

Fig. 6. Roof and wall intersection. A, Wood siding; B, brick wall.

simple counterflashing on each side. For single-flue chimneys, the shingle flashing on the high side should be carried up under the shingles. The vertical distance at top of the flashing and the upturned edge should be about 4 inches (10.16-cm.) above the roof boards (A, Fig. 7).

A *wood saddle* usually constructed on the high side of wide chimneys for better drainage, is made of a ridgeboard and post and sheathed with plywood or boards (B, Fig. 7). It is then covered with metal, which extends up on the brick and under the shingles. Counterflashing at the chimney is then used (as previously described) by lead plugging and calking. A very wide chimney may contain a partial gable on the high side and be shingled in the same manner as the main roof.

Roof Edge

The cornice and the rake section of the roof are sometimes protected by a metal edging. This edging forms a desirable drip edge at the rake and prevents rain from entering behind the shingles.

At the eave line, a similar metal edging may be used to advantage (A, Fig. 8). This edging, with the addition of a roll roof flashing (B, Fig. 8) will aid in resisting water entry from ice dams. Variations of it are shown at B and C, Fig. 8. They form a good drip edge and prevent or minimize the chance of rain being blown back under the shingles. This type of drip edge is desirable whether or not a gutter is used.

GUTTERS AND DOWNSPOUTS

Types of Gutters and Downspouts

Several types of gutters are available to guide the rainwater to the downspouts and away from the foundation. Some houses have built-in gutters in the cornice. These are lined with sheet metal and connected to the downspouts. On flat roofs, water is often drained from one or more locations and carried through an inside wall to an underground drain. All downspouts connected to an underground drain should contain basket strainers at the junction of the gutter.

Perhaps the most *commonly used gutter* is the type hung from the edge of the roof or fastened to the edge of the cornice

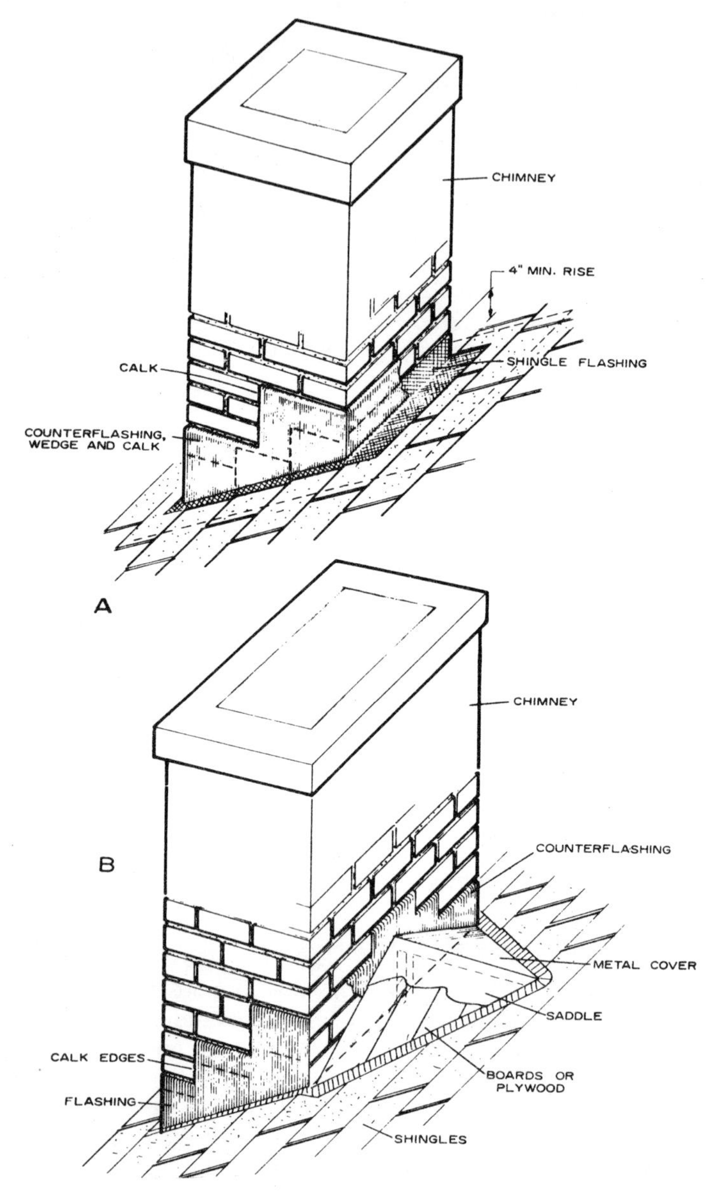

Fig. 7. Chimney flashing. A, Flashing without saddle; B, chimney saddle.

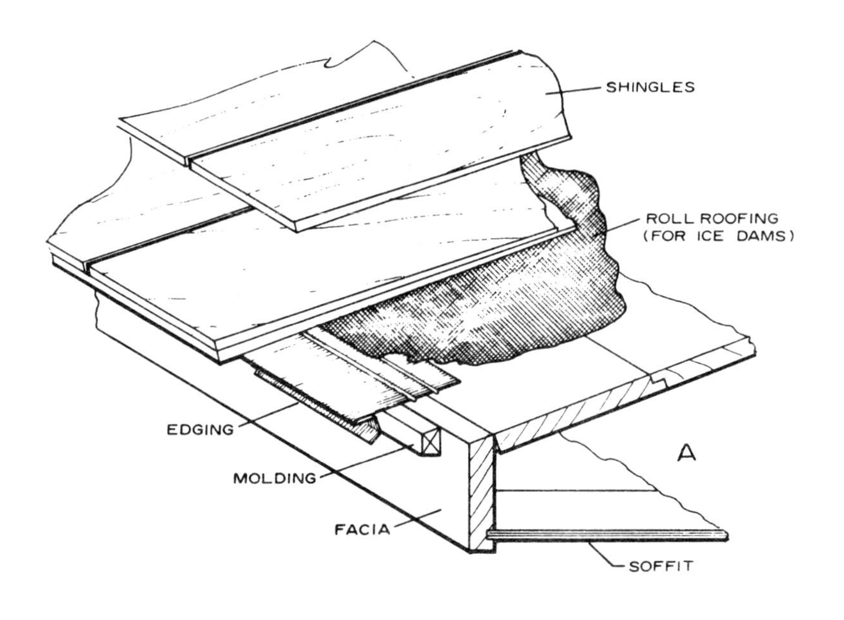

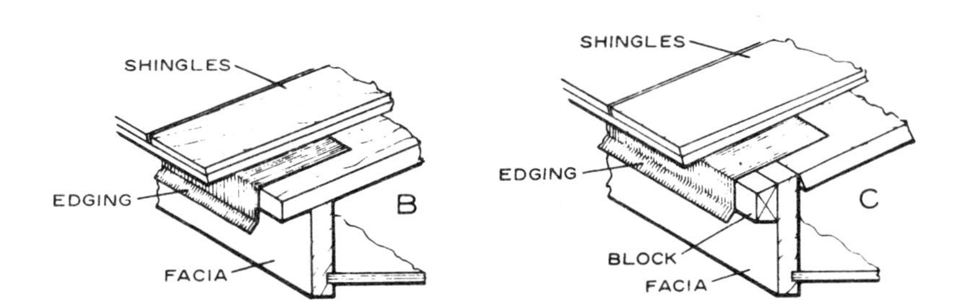

Fig. 8. Cornice flashing. A, Formed flashing; B, flashing without wood blocking; C, flashing with wood blocking.

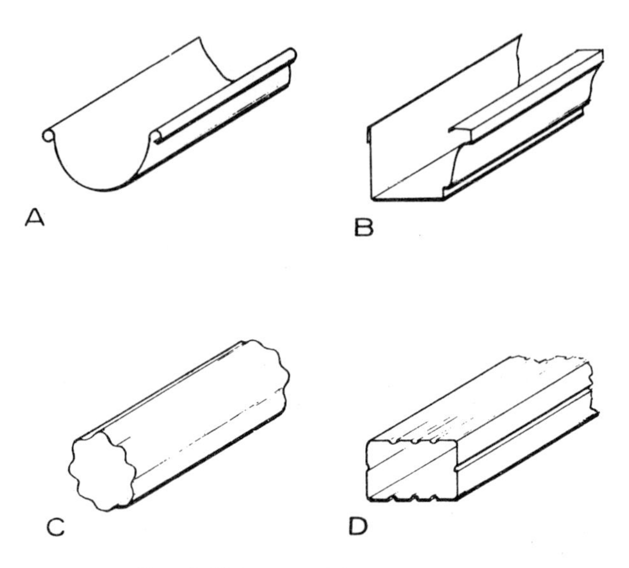

Fig. 9. Gutters and downspouts.

facia. Metal gutters may be the half-round (A, Fig. 9) or the formed type (B, Fig. 9) and may be galvanized metal, copper, or aluminum. Some have a factory-applied enamel finish.

Downspouts are round or rectangular (C and D, Fig. 9), the round type being used for the half-round gutters. They are usually corrugated to provide extra stiffness and strength. Corrugated patterns are less likely to burst when plugged with ice.

Wood gutters have a pleasing appearance and are fastened to the facia board rather than being carried by hangers as are most metal gutters. The wood should be clear and free of knots and preferably treated, unless made of all heartwood from such species as redwood, western red cedar, and cypress. Continuous sections should be used wherever possible. When splices are necessary, they should be square-cut butt joints fastened with dowels or a spline. Joints should be set in white lead or similar material. When untreated wood gutters are used, it is good practice to brush several generous coats of water-repellent preservative on the interior.

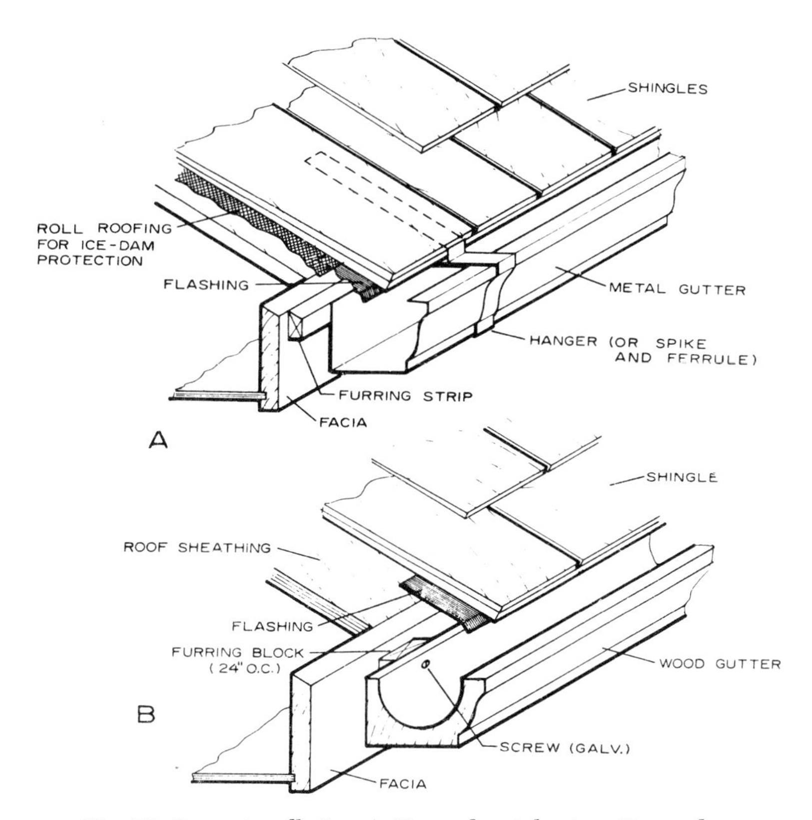

Fig. 10. Gutter installation. A, Formed metal gutter; B, wood gutter.

Size of Gutters

The size of gutters should be determined by the size and spacing of the downspouts used. One square inch of downspout is required for each 100 square feet of roof. When downspouts are spaced up to 40 feet apart, the gutter should have the same area as the downspout. For greater spacing, the width of the gutter should be increased.

Installation

On *long runs of gutters*, such as required around a hip-roof house, at least four downspouts are desirable. Gutters should be installed with a slight pitch toward the downspouts. *Metal gutters* are often suspended from the edge of the roof with hangers (A, Fig. 10). Hangers should be spaced 48 inches

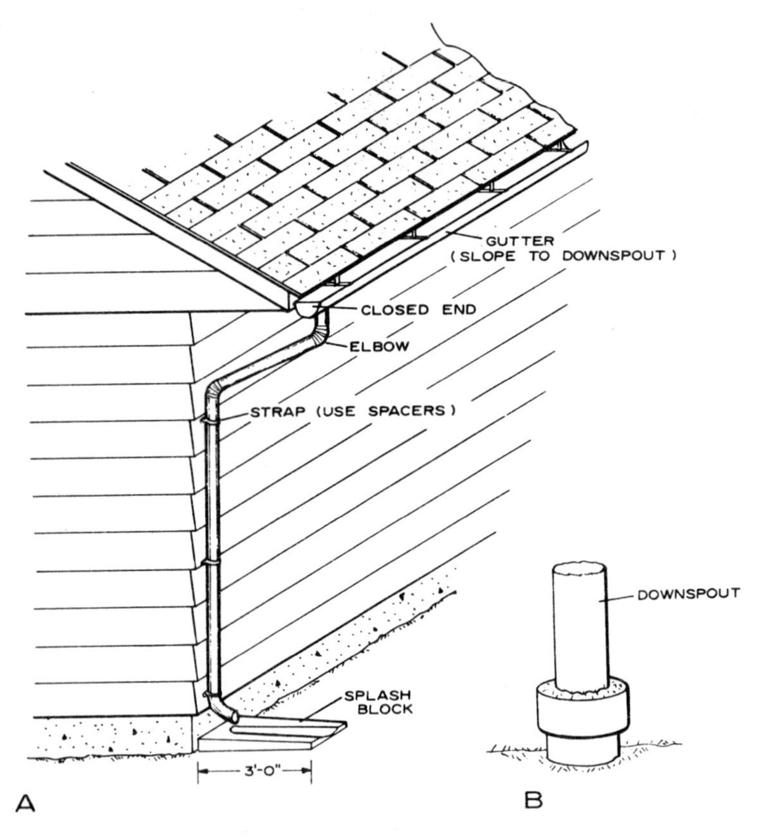

Fig. 11. Downspout installation. A, Downspout with splash block; B, drain to storm sewer.

apart when made of galvanized steel and 30 inches apart when made of copper or aluminum. *Formed gutters* may be mounted on furring strips, but the gutter should be reinforced with wrap around hangers at 48-inch intervals. Gutter splices, downspout connections, and corner joints should be soldered or provided with watertight joints.

Wood gutters are mounted on the facia using furring blocks spaced 24 inches apart (B, Fig. 10). Rust-proofed screws are commonly used to fasten the gutters to the blocks and facia backing. The edge shingle should be located so that the drip is near the center of the gutter.

Downspouts are fastened to the wall by straps or hooks (A, Fig. 11). Several patterns of these fasteners allow a space between the wall and downspout. One common type consists of a galvanized metal strap with a spike and spacer collar. After the spike is driven through the collar and into the siding and backing stud, the strip is fastened around the pipe. Downspouts should be fastened at the top and bottom. In addition, for long downspouts a strap or hook should be used for every six feet of length.

An elbow should be used at the bottom of the downspout, as well as a splash block, to carry the water away from the wall. However, a vitrified tile line is sometimes used to carry the water to a storm sewer. In such installations, the splash block is not required (B, Fig. 11).

Chapter 16

Protection Against Fire

Fire hazards exist to some extent in nearly all houses. Even though the dwelling is of the best fire-resistant construction, hazards can result from occupancy and the presence of combustible furnishings and other contents.

The following tabulation showing the main causes of fires in one- and two-family dwellings is based on an analysis made of 500 fires by the National Fire Protection Association.

Cause of Fire	*Percent of Total*
Heating equipment	23.8
Smoking materials	17.7
Electrical	13.8
Children and matches	9.7
Mishandling of flammable liquids	9.2
Cooking equipment	4.9
Natural gas leaks	4.4
Clothing ignition	4.2
Combustibles near heater	3.6
Other miscellaneous	8.7
	100.0

Fire-protection engineers generally recognize that a majority of fires begin in the contents, rather than in the building structure itself. Proper housekeeping and care with smoking, matches, and heating devices can reduce the possibility of fires.

FIRE STOPS

Fire stops are intended to prevent drafts that foster movement of hot combustible gases from one area of the building

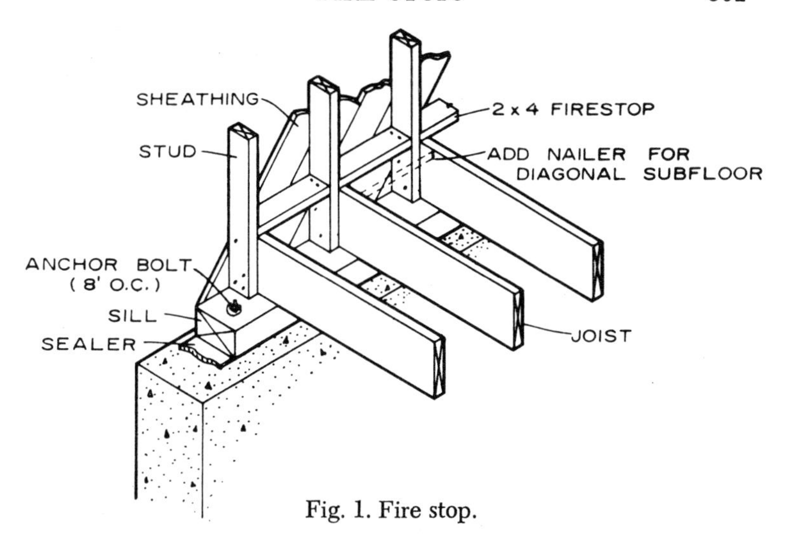

Fig. 1. Fire stop.

to another during a fire. Exterior walls of wood frame construction should be fire stopped at each floor level, at the top-story ceiling level, and at the foot of the rafters. *See* Figs. 1, 2, and 3.

Fire stops should be of noncombustible materials or wood not less than two inches (5.08-cm.) in nominal thickness. The fire stops should be placed horizontally and be well fitted to completely fill the width and depth of the spacing. This applies primarily to balloon-type frame construction. Platform walls are constructed with top and bottom plates for each story (Fig. 4). Similar fire stops should be used at the floor and ceiling of interior stud partitions, and headers should be used at the top and bottom of stair carriages (Fig. 5).

Noncombustible fillings should also be placed in any spacings around vertical ducts and pipes passing through floors and ceilings, and self-closing doors should be used on shafts, such as clothes chutes.

When cold-air return ducts are installed between studs or floor joists, the portions used for this purpose should be cut off from all unused portions by tight-fitting stops of sheet metal or wood not less than two inches (5.08-cm.) in nominal thickness. These ducts should be constructed of sheet metal or other materials no more flammable than 1-inch (2.54-cm.) (nominal) boards.

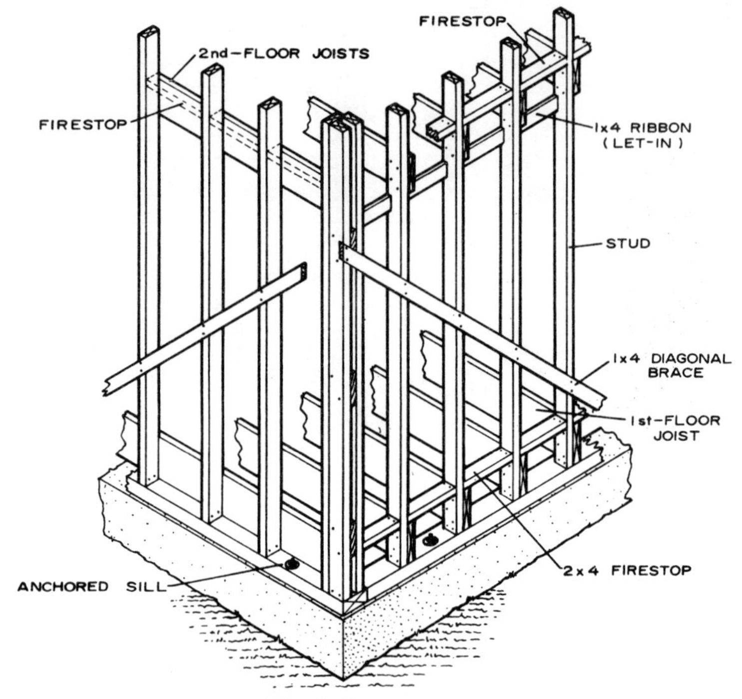

Fig. 2. Fire stops.

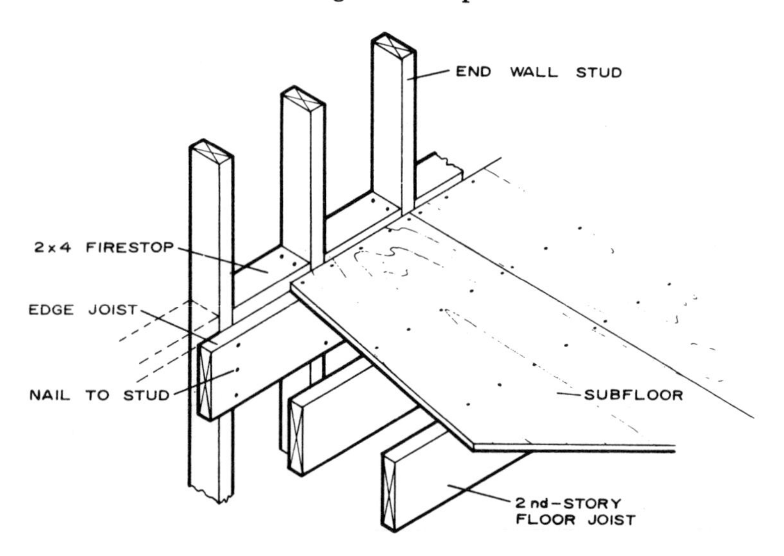

Fig. 3. Fire stop in balloon construction.

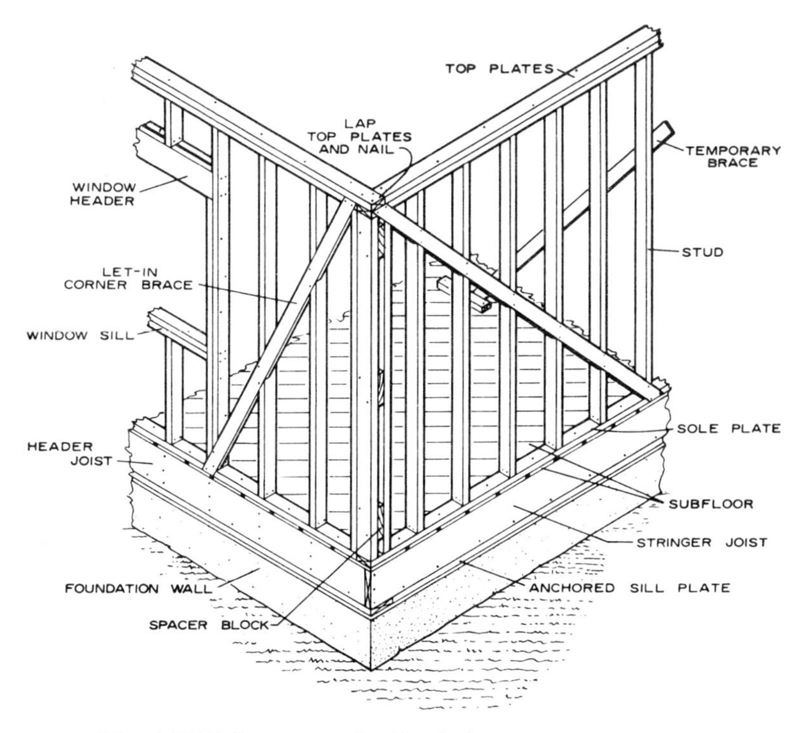

Fig. 4. Wall framing used with platform construction.

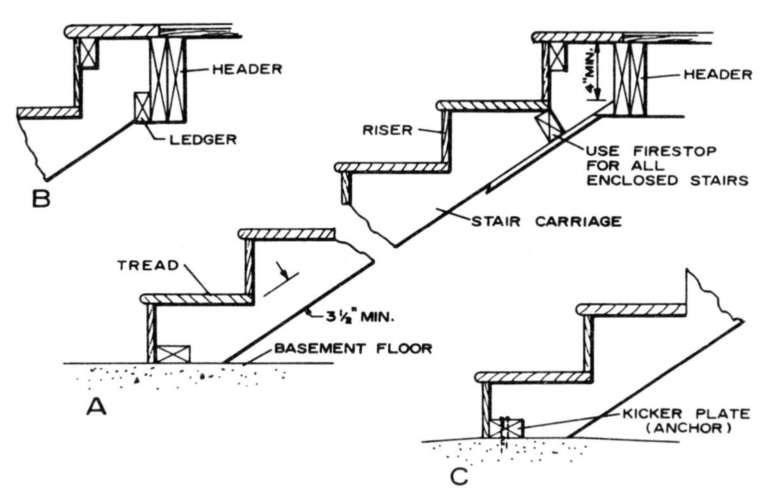

Fig. 5. Fire stops and headers.

Fire stops should also be placed vertically and horizontally behind any wainscoting or paneling applied over furring, to limit the formed areas to less than ten feet in either direction.

With suspended ceilings, vertical panels of noncombustible materials from lumber of 2-inch (5.08-cm.) nominal thickness or the equivalent, should be used to subdivide the enclosed space into areas of less than 1,000 square feet. Attic spaces should be similarly divided into areas of less than 3,000 square feet.

CHIMNEY AND FIREPLACE CONSTRUCTION

The fire hazards within home construction can be reduced by insuring that chimney and fireplace constructions are placed in proper foundations and properly framed and enclosed. (*See* Chapter 14, Chimneys and Fireplaces.) In addition, care should be taken that combustibles are not placed too close to the areas of high temperature. Combustible framing should be no closer than two inches (5.08-cm.) to chimney construction. However, when required, this distance can be reduced to ½ inch (1.27-cm.), provided the wood is faced with a ¼-inch-thick (.635-cm.) asbestos sheeting.

For fireplace construction, wood should not be placed closer than four inches (10.16-cm.) from the backwall nor within eight inches (20.32-cm.) of either side or top of the fireplace opening. When used, wood mantels should be located at least 12 inches (30.48-cm.) from the fireplace opening.

HEATING SYSTEMS

Almost 25 percent of fires are attributed to faulty construction or to improper use of heating equipment, and the greater proportion of these fires originate in the basement. Combustible products should not generally be located nearer than 24 inches to a hot-air, hot-water, or steam-heating furnace. However, this distance can be reduced in the case of properly insulated furnaces or when the combustible materials are protected by gypsum board, plaster, or other materials with low flame spread. Most fire-protection agencies limit to 170°F. the temperature to which combustible wood products should be exposed for long periods of time, although experimentally, ignition does not occur until much higher temperatures have been reached.

In confining a fire to the basement of a home, added protection can be obtained with gypsum board, asbestos board, or plaster construction on the basement ceiling, either as the exterior surface or as backings for decorative materials. These ceiling surfaces are frequently omitted to reduce costs, but particular attention should be given to protection of the wood members directly above and near the furnace.

FLAME SPREAD AND INTERIOR FINISH

In some areas of a building, flame-spread ratings are assigned to limit spread of fire on wall and ceiling surfaces. Usually, these requirements do not apply to private dwellings because of their highly combustible content, particularly in furnishings and drapes usually found in this type of structure.

To determine the effect of the flammability of wall linings on the fire hazards within small rooms, burn-out tests have been made. For this purpose, an 8- by 12- by 8-foot high room was furnished with an average amount of furniture and combustible contents. This room was lined with various wall panel products, plywood, fiber insulation board, plaster or fiberboard lath, and gypsum wallboard. When a fire was started in the center of the room, the time to reach the critical temperature (when temperature rise became very rapid) or the flashover temperature (when everything combustible burst into flame) was not significantly influenced by either combustible or noncombustible wall linings. In the time necessary to reach these critical temperatures (usually less than ten minutes) the room would already be unsafe for human occupancy.

Similar recent tests in a long corridor, partially ventilated, showed that the "flashover" condition would develop for 60 to 70 feet along a corridor ceiling within five to seven minutes from the burning of a small amount of combustile contents. This flashover condition developed in approximately the same time, whether combustible or non-combustible wall linings were used, and before any appreciable flame spread along wall surfaces.

Wood paneling, treated with fire-retardant chemicals or fire-resistant coatings as listed by the Underwriters' Laboratory, Inc. or other recognized testing laboratories, can also be used in areas where flame-spread resistance is especially critical. Such treatments are not considered necessary in dwellings, nor can the extra cost of treatment be justified.

FIRE-RESISTANT WALLS

Whenever it is desirable to construct fire-resistant walls and partitions in attached garages and heating rooms, information on fire resistance ratings using wood and other materials is readily available through local code authorities. Wood construction assemblies can provide ½ hour to 2 hour fire resistance under recognized testing methods, depending on the covering material.

Chapter 17

Soldering with Soft Solders

Man has been doing soldering for many centuries. The process has remained fundamentally the same, but a few improvements in methods and tools have lightened the work considerably. The greatest improvement has been in the heating methods. The modern heating methods such as the gasoline blowtorch, automatic alcohol torch, blowpipe, gas oven, and the electric soldering copper and tools used are discussed in this chapter.

SOLDERING

Soldering is the simplest of the joining processes. It is a process of joining cast or fabricated metals at low temperature. This means the fastening together of two like or unlike metals by using another metal entirely different from one or both of the base metals without melting either one.

SOLDERS

Solders are used for joining iron, nickel, lead, tin, copper, zinc, and many of their alloys; also to make metal joints and seams leakproof. They are classified as soft or hard solders. *Soft* solders differ from *hard* solders (brazing solders and silver solders) in that the former are alloys that are fusible (capable of being melted) at temperatures below 700°F.

One of the differences between the soldering process and the welding process is that in soldering the metals to be joined are not heated to the melting point, whereas in welding they are heated to this point or above it. Therefore the melting point of solder must be lower than that of the metals to be joined.

Soft solders are of low strength and are easily applied because of their low melting point.

Solder is an alloy of two metals—lead and tin. *Lead* melts at 621°F. *Tin* melts at 465°F. But, tin-lead (50/50 solder) melts at 450°F., a lower temperature than the melting temperature of either tin or lead. This half-and half solder is the most effective for use in sheet metal work. Solder containing 60 percent tin and 40 percent lead has a still lower melting point, 390°F. For this reason it is especially well suited for work on light gage metal. The lower the temperature at which a solder can be worked, the less likely one is to heat the metal to a point that would cause rapid expansion, known as heat buckling.

Solders are further classified according to their tin content as *common, medium,* and *fine*. *Common solder* has the least percent of tin, and *fine solder* the greatest percent of tin. Common solder has the highest melting point and is cheaper. For that reason it is used most frequently in plumbing work and for splicing the covering of lead cables. *Fine* and *medium solders* have lower melting points and are used for electrical work.

Commercial bar solder is identified by numbers giving the percentages of tin and lead. The first number is the percentage of tin contained in the alloy; for example, 30/70 indicates that the bar of solder is made up of 30 percent tin and 70 percent lead. Alloy designated 50/50 is called half-and-half and is sometimes labeled that way. It is preferred for most sheet metal jobs. A 30/70 alloy is in common use, however, because it is cheaper.

Scientists have discovered that *molten solder* sticks to the surface of the base metal by means of molecular attraction. A very strong bond is formed because of the friendliness of the molecules of the solder for the molecules of the base metal. This molecular attraction is known as *adherence*.

For such jobs as securing electrical connections, joining sheet metal where great strength is not required, or making watertight riveted or lap joint sheet metal containers soldering is the answer.

PREPARATION OF SURFACES

The first step in the preparation of surfaces to be soldered is *cleaning*. The strength of the joint depends on the adherence of the solder to the metal to be joined. To secure good adherence, the surface of the metal and of the solder must be free of oxide, grease, dirt, and other foreign substance. All metals are normally covered with oxides which increase as the metal is heated to the soldering temperature. Cleaning may be done mechanically or chemically.

In *mechanical cleaning* the surfaces are scraped, filed, or rubbed with sandpaper or emery paper. If a power buffer or emery wheel is available, it may be used effectively. Cleaning of tough jobs on repair work is made easier by warming the area to be repaired almost to the temperature required to melt the solder. After the metal is heated with a torch, it is a good idea to rub the surface with a wire brush, sandpaper, or steel wool to remove oxide that may have formed during the heating process.

Surfaces are usually cleaned with chemicals. Cleaning by chemical action with pickling solution of acid is known as *fluxing*. Ordinarily, pastes or solutions that contain zinc chloride are used for soft soldering. The material holding the flux is evaporated by the heat of the soldering operation, leaving a film of the flux on the work. At the soldering temperature, as the molten solder is applied, the flux melts and dissolves the oxides from the solder and the work. As the process is completed, a thin film is formed, protecting the work from further oxidation. A complete discussion of fluxes is presented later in this chapter.

Frequently, metal parts to be soldered have a coat of tin. If this is not the case, you may find it easier to solder by applying a coat of pure tin or solder metal to the surfaces of the metal parts to be joined. This process is called *tinning*. Use the same method described later in this chapter for tinning a soldering copper.

METHODS OF APPLICATION

Most common metals and their alloys can be readily soldered. Several factors are involved in determining the method of soldering to be used. These include character of metals to be soldered, their position, size of the parts to be joined, speed with which job must be completed, and the shape, tensile strength, and appearance required of the finished job. Surfaces to be soldered must always be heated to the melting point of the solder to be used.

Different methods for heating these surfaces are—

1. By the wiping method.
2. By use of a soldering copper.
3. By use of a blowtorch.
4. By solder bath.
5. By sweating.

WIPING METHOD

The *wiping method* of soldering is used for joining sections of lead pipe; joining lead pipe with brass, bronze, or copper fittings; and joining lead covered electrical cables using a lead sleeve.

The steps in soldering by the wiping method are as follows.

1. Heat the solder until it is in a semiliquid state.
2. Hold cloth or wiping pad under the joint to be soldered.
3. Pour solder slowly over the joint, catching overflow on cloth.
4. Press the hot solder caught on cloth against the joint and work until smooth.
5. Continue to pour and work until sufficient heat is applied to form the joint properly.
6. Allow to cool slowly, wiping off excess solder and shaping joint.

This particular method requires considerable skill but you will find it convenient for soldering in places too tight for a soldering copper.

If joints are large, repeated pouring may be necessary to attain proper heat.

Stearin (stearic acid) in stick form is sometimes used for fluxing when wiping joints. Wiping solders usually contain about 30 percent tin and 70 percent lead, and you have a range of about 140° in which to work them.

SOLDERING COPPERS

A *soldering copper*, sometimes incorrectly referred to as a *soldering iron*, consists of a forged piece of copper connected

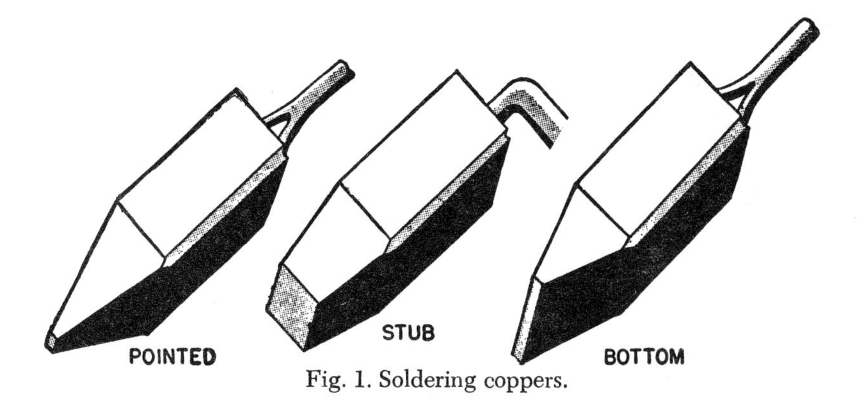

Fig. 1. Soldering coppers.

by an iron rod to a handle. The copper end is called the head or bit. There are several types of soldering coppers, but the most frequently used are the pointed copper, the stub copper, and the bottom copper (Fig. 1). The *pointed copper* is used for general soldering work. The *stub copper* is used for flat seams requiring considerable heat. The *bottom copper* is used for soldering seams of pails, pans, trays, and other tough spot jobs. *Electric soldering coppers* (Fig. 2) are built

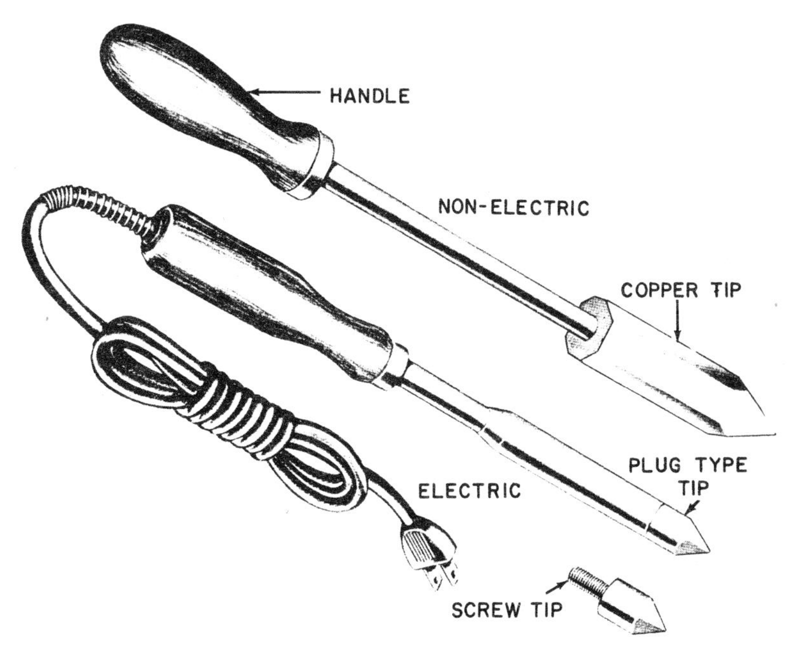

Fig. 2. Soldering irons.

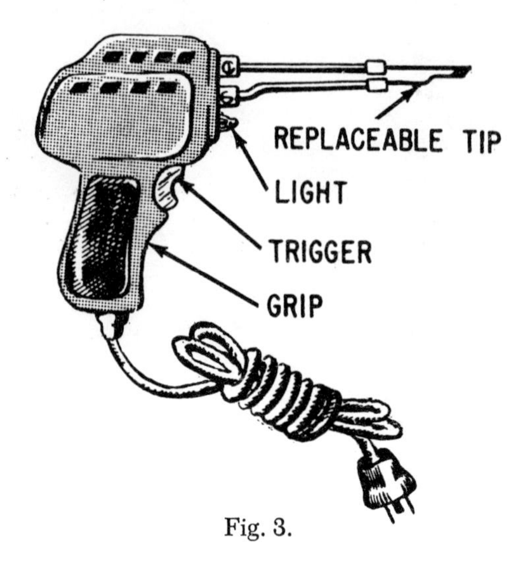

Fig. 3.

with internal heating coils; also shown in Fig. 2 is a non-electric soldering iron. The heads are removable and interchangeable, thus allowing the use of various shaped tips. Electric coppers may be used for light work, and are especially good for work on electrical connections. An electric soldering gun is shown in Fig. 3.

HANDLES FOR SOLDERING COPPERS

Both wood and fiber handles are available for soldering coppers. The standard wood handle is forced on the rod. Some fiber handles are screwed on. Two coppers should be used while soldering. One is heated while the other is being used. Their sizes are determined by the weight in pounds for two copper ends. For example, a pair of 2-pound (2-lb.) coppers means that each weighs 1 pound. The common sizes are 1-lb., 1½-lb., 2-lb., 3-lb., 4-lb., and 6-lb. per pair.

It is not desirable to use a light copper for heavy gage metal since it will not hold enough heat to heat the metal properly or allow the solder to flow. Neither should heavy coppers be used for light gage metal as they are awkward to use and result in poor work. Also, there is the danger of excess heat causing the metal to buckle.

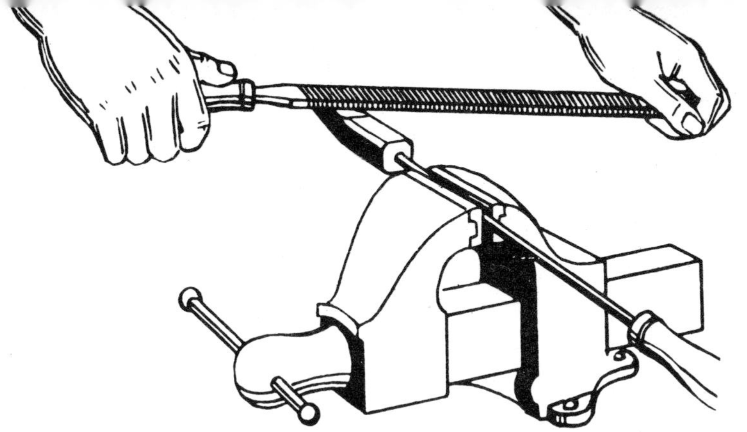

Fig. 4. Filing a copper.

FILING AND TINNING A SOLDERING COPPER

Before starting a soldering job see that your copper is properly filed and tinned (coated with solder). For filing and tinning a soldering copper proceed as follows.

1. Heat the copper to be tinned until it is cherry red.
2. Clamp the copper in a vise as shown in Fig. 4.
3. Using a 12-inch single cut bastard file, bear down on the forward stroke and release pressure on the return stroke. Do not rock the file.
4. File the tapered sides of the copper until they are bright and smooth. Be sure you keep your hands off the hot copper.
5. Smooth off sharp edges and point.
6. Reheat sufficiently to melt solder.
7. Rub each filed side back and forth across a cake of sal ammoniac, adding a little solder to the copper until it is coated (tinned) (Fig. 5).

Powdered rosin placed on a brick (Fig. 6) may be used in-

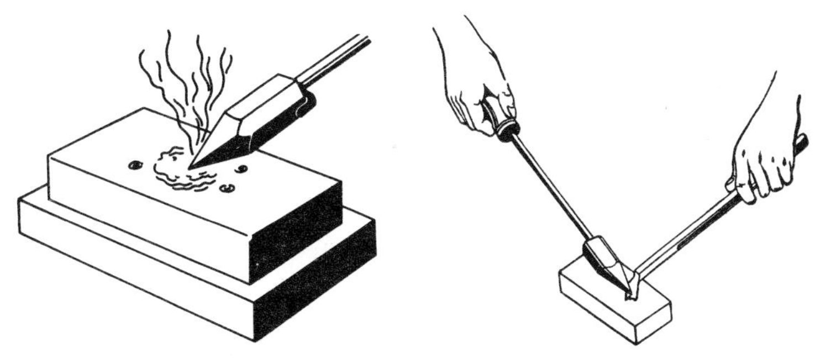

Fig. 5. Tinning with sal ammoniac. Fig. 6. Tinning with rosin.

stead of the cake of sal ammoniac. The same method is used. This method for tinning is used when sal ammoniac is not available or when rosin is to be used as a flux on tin plate.

Every time a copper is overheated it must be retinned. The soldering copper is filed to remove burnt substances from it and to keep it smooth. This must be done often. If the soldering copper becomes too blunt or if it is not the correct shape for the job to be done, it must be forged and filed. For forging a soldering copper proceed as follows.

1. Heat the copper to a bright red.
2. File until burnt tinning and pits are removed.
3. Reheat copper to bright red.
4. Holding the copper on an anvil, forge it to the required shape by striking it with a ball peen hammer (Fig. 7). The point should be blunt and the head should not be tapered too much or the end of the copper will cool too rapidly. Hammer the point back as the forging progresses. Turn the copper often to produce a square surface. Reheat as often as necessary.
5. Reheat to bright red and take as many of the hollows out of it as possible with a flat faced hammer.
6. File and tin copper.

Dipping Solution. A pair of soldering coppers will last a long time if they are not overheated and if cleaned just before soldering. To clean, wipe the point on a damp rag, clean the point on sal ammoniac, or dip the point in a *dip cup.* Any container holding a dipping solution is called a dip cup.

A *dipping solution* for cleaning a soldering copper may be made by dissolving a small amount of powdered sal ammoniac in water. *Soldering salts* dissolved in water may also be used as a dipping solution. Never use as a dipping solution the acid or flux that is used for the soldering process. Your copper will color the flux and stain the job.

Fig. 7. Forging a soldering copper.

SOLDERING RULES

Now that you have the principle of soldering well in mind, follow these rules for a topnotch job.

1. Select the proper pair of soldering coppers, observing size, and shape. File and tin according to previous instructions.
2. Heat the copper. Do not allow it to become red hot as this burns the tinning and oxidizes the copper, requiring retinning. Keep one copper heating while the other is in use.
3. Select the proper flux. (*See* the discussion of fluxes later in this chapter.)
4. Place the job to be soldered in the proper position on a suitable support.
5. Apply the flux with swab or brush, one or two strokes is sufficient.
6. Clean the soldering copper by dipping the tinned surface in dipping solution or by wiping off the tip with a damp cloth.
7. Pick up solder with the copper.
8. Hold the copper at one end of the seam until heat from the copper penetrates through the metal. The metal must absorb enough heat from the copper to melt the solder or the solder will not stick.
9. If the job is a seam, tack it together at enough places to hold the pieces in position.
10. Start at one end of the seam and hold the copper with a tapered side of the head flat along the seam until the solder starts to flow freely into the seam.
11. Draw the copper with a slow, steady motion toward you along the seam, and add as much solder as necessary without raising the soldering copper from the work.
12. Heat the copper as often as necessary, but remember that the best job can be done if the seam can be started and finished without lifting the copper from the surface to be soldered, and without retracting completed work.
13. If raw or killed acid is used as a flux, wipe the excess acid off the seam with a clean damp cloth.
14. Allow the seam to cool undisturbed and the solder to set before moving the job.
15. When soldering a grooved seam, tacking is not necessary, as the seam is held together by the lock. Nor is it necessary to tack a riveted seam.

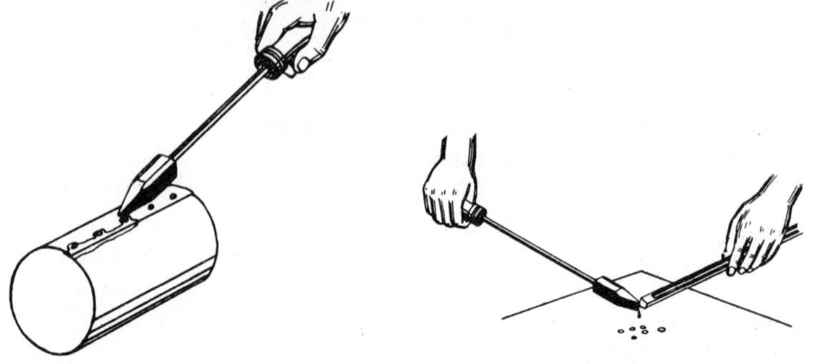

Fig. 8. Soldering around rivets.

Fig. 9. Making solder beads.

16. For a watertight job the rivets must be soldered (Fig. 8).

17. To solder a square, rectangular, or cylindrical bottom, first make solder beads (sometimes called shots) by holding solder against a hot copper and allowing the beads to drop on the bench (Fig. 9). Flux the seam and drop one of the cold beads of solder in the bottom of the container. Heat, clean, and dip the soldering copper and place it in the bottom, as shown in Fig. 10. Hold the soldering copper in a stationary position until the solder starts to flow freely into the seam. Draw the copper slowly along the seam, rolling the job on the edge of the bottom at the same time. Add more beads as needed and reheat the copper when necessary.

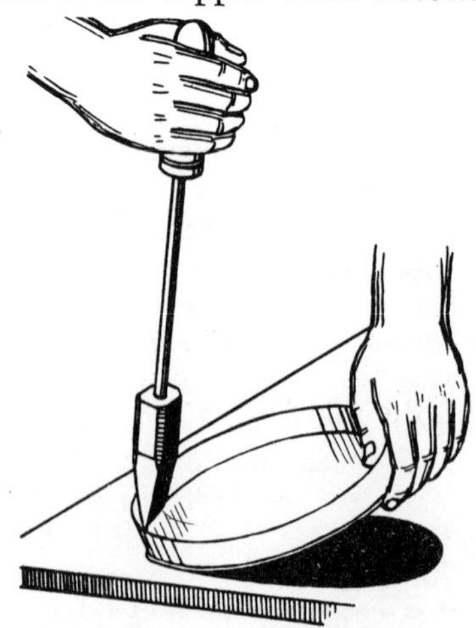

Fig. 10. Soldering a bottom.

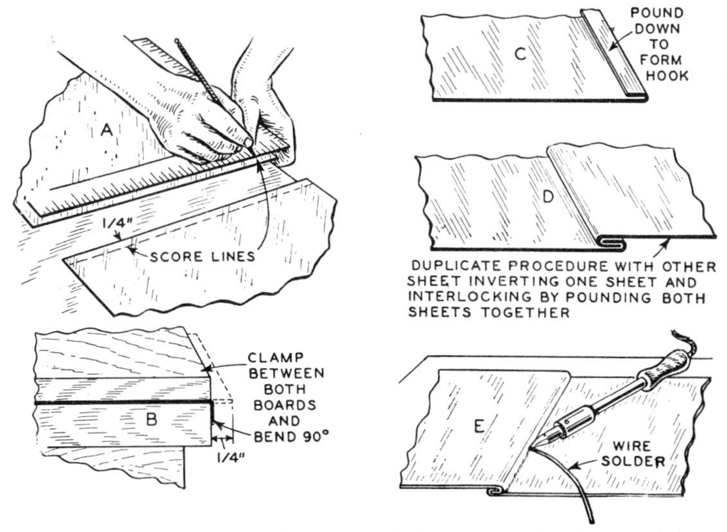

Fig. 11. Making and soldering interlocking seams in sheet metal.

SOLDERING SEAMS IN SHEET METAL

To solder seams in sheet metal, mark or score a line approximately ¼ inch (.635-cm.) to ½ inch (1.27-cm.) from the edges of both pieces to be joined (A, Fig. 11). These guide lines indicate where to make the necessary bends before interlocking. Clamp the sheets of metal between two boards, with the parts to be bent projecting (B, Fig. 11). Bend the edges at right angles with a wooden block or mallet. Then remove the sheets from the boards and lay them on the workbench. Bend each edge over completely by pounding it with a mallet (C, Fig. 11), then interlock the folds and pound them together with the mallet (D, Fig. 11).

Place the interlocked pieces of metal on sheets of newspaper or on a piece of asbestos paper to help keep an even temperature when the soldering iron is used, and secure them so that they will be stationary.

Apply the proper flux to the seam, the type of metal that is being soldered.

Heat the soldering iron and insert the edge of one of the tinned faces in one end of the seam fold. Hold a piece of wire solder against both the seam and the iron (E, Fig. 11).

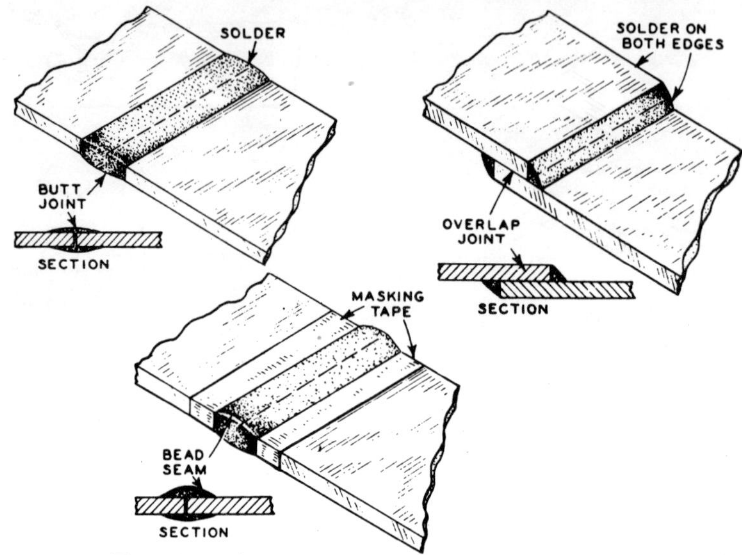

Fig. 12. Methods of reinforcing sheet metal joints.

When some of the solder melts and the seam attains a sufficient temperature so that the solder runs freely into it, draw the iron very slowly toward the other end of the seam. As the iron moves along, keep feeding additional wire solder so that it penetrates and runs or sweats in the seam without piling up. Move the iron very slowly so that every part of the seam attains the proper temperature.

SOLDERING REINFORCED JOINTS IN SHEET METAL

Where it is impractical or impossible to make interlocking bends in sheets or rods that are to be joined, there are several methods of reinforcing a butt joint, as shown in Fig. 12.

A simple method of reinforcing a butt joint on sheet metal is shown in Fig. 12. This type of seam or joint, called a bead seam, is formed by first soldering the under portion of the metal in the same manner as that used for an interlocked bend, then forming a reinforced strip of solder in the form of a bead seam on the upper portion of the two pieces of metal.

To make a bead seam, paste a piece of masking tape or Cellotape on each of the sheets parallel to the two edges and

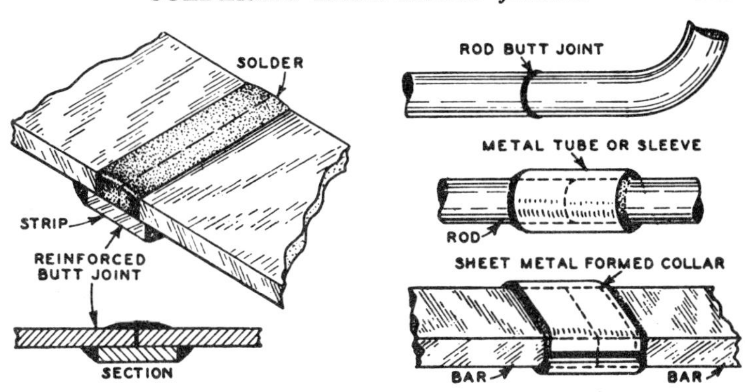

Fig. 13. Methods of reinforcing joints in sheet metal, rods, and bars.

approximately half the width of the bead from each edge. Apply flux to the space between the pasted-down strips. With a hot soldering iron, apply enough solder to form a bead. Place the edge, not the flat side, of the soldering iron in the center of the seam, and move it along as fast as the solder melts. The adhesive strips serve to keep the molten solder from spreading. To make a bead requires a little practice. At first the solder will be quite rough. Run the iron over the seam several times until a smooth bead is formed.

Another method of reinforcing seam joints is with a lap seam shown in Fig. 12. The two pieces of metal to be joined are overlapped and soldered in the position shown. The solder flows between the two pieces and along each of the edges. This type of joint is considered stronger than the butt or the bead seam. When making a lap seam, use the flat side of the iron.

Another method of reinforcing a butt joint is to solder an additional strip of metal between the two sheets of metal lengthwise with the joint (Fig. 13.).

SOLDERING REINFORCED JOINTS IN RODS AND BARS

Rods and bars, whether circular, square, or rectangular, must be reinforced with a collar of metal when butt-jointed. A short section of brass or copper tubing just large enough to slip over the rod can be used as a collar for a circular rod.

For a square or rectangular rod, bend a piece of sheet metal to the proper shape to form the reinforcing collar. When fitting the collar, allow enough space to permit the solder to flow freely between the collar and the rod.

Thoroughly clean and polish both the rod and the collar with emery paper and apply the flux to both before putting the collar on the rod. Then solder the work as shown in Fig. 13.

SOLDERING ALUMINUM

To solder aluminum, a special flux and solder must be used. The aluminum flux, which is available in powder form, must be immediately sprinkled over the cleaned surface to prevent the formation of oxide. If an electric soldering iron is used, it should be at least a 200-watt iron.

Hold the flat surface of the iron on the fluxed area. As the iron attains the proper temperature, the flux will melt, bubble and then smoke. When it smokes, do not remove the iron from the work, but apply the special aluminum solder to the iron and melt it over the fluxed surface of the aluminum. If additional solder is needed to reinforce or strengthen the joint, apply it with the tip of the iron.

When soldering other metals to aluminum, it is necessary to use aluminum flux for the aluminum and whichever flux is required for the other metal. Wherever possible, apply the heat to the underside of the aluminum.

SOLDERING CAST IRON

Soldering cast iron is not especially recommended, but sometimes it is necessary to make temporary repairs in cracked cast-iron parts. Widen the top of the crack with a cold chisel or a file, or by grinding into a V-shaped groove. The groove should be made wide and deep enough to permit a sufficient amount of solder to enter and close the crack. Clean all grease, dirt, rust, or paint from the cast iron. Most of the aluminum fluxes also can be used for cast iron, but better results are obtained by using a special cast-iron flux. Place the flux in the groove and on the surface surrounding the groove. Apply heat until the flux smokes. The peculiar structure of iron castings makes it necessary sometimes to repeat the fluxing operation several times before applying the solder. Melt the solder and run it into the groove with a hot iron until the entire surface has been soldered.

SOLDERING BRASS AND BRONZE

Brass is an alloy of copper with zinc or another metal. A flux of zinc chloride or rosin is used for brass. Apply it in the usual manner to the cleaned surface, and be careful not to use too much heat. Then solder the parts in the manner described previously for the type of work. If zinc chloride has been used as a flux, the joined parts must be washed clean of all traces of it with water containing soap and washing soda.

Bronze is the general term used for various alloys of copper and tin. Some types of bronzes also contain zinc, silicon, lead, and nickel. Bronze is fluxed and soldered as described for brass.

SOLDERING IRON AND STEEL

Zinc chloride flux and tin-leaded solder are used for soldering iron and steel, but the special commercial aluminum solders and fluxes give better results. The soldering procedure is the same as that for other metals, according to the type of work.

SOLDERING STAINLESS STEEL

Stainless steel can be soldered if certain important factors are considered. Stainless steel is a poor heat conductor, so an iron of high wattage must be used, and it must be held in one place on the stainless steel longer than is necessary when soldering other metals. Too much heat will change the color of stainless steel.

Special commercial stainless-steel fluxes are available, but muriatic acid used undiluted is considered the best flux for the purpose. After cleaning the metal, brush the acid on the surfaces to be soldered and allow to remain there for several minutes before beginning to solder. The ordinary half-and-half type of solder can be used on stainless steel.

After the acid has been allowed to bite into the material,

apply the solder with the heated iron, moving the iron along the joint or seam very slowly. In fact, the iron should be held on each spot before moving it along. After the soldering has been completed and the metal has cooled, wash off the excess flux with a solution of washing soda and soapy water.

SOLDERING LEAD, TIN, PEWTER, AND ZINC

Lead has a very low melting point (621°F.) so the soldering iron must not be too hot. The flux and solder also must have a low melting point.

In *lead soldering*, the joints must be scraped bright and fluxed before the solder is applied. An ideal flux for lead is ordinary tallow. Scrape a few shavings of tallow on the joint and melt it into the joint. A special commercial tin-lead-bismuth solder, which has an extremely low melting point, has to be used. Keep the soldering iron moving to avoid melting the surounding metal.

Tin has a low melting point, and the procedure for lead should be followed in soldering this metal.

Pewter is an alloy of tin and lead. It also has a low melting point, very near that of lead, and the procedure for soldering pewter is the same as for lead.

Zinc also has a low melting point, approximately 775°F. Use zinc chloride as a flux, and the same solder and procedure as for lead. After soldering zinc, wash off the excess flux with a solution of soapy water and washing soda.

SOLDERING WHITE METALS

White metals are usually combinations of lead, tin, antimony, and other metals that have a very low melting point. They vary to such an extent that specific directions for soldering them cannot be given without knowing which metals have been combined. As a rule, no flux is used on white metals. Clean the metal by sanding or scraping. Apply commercial tin-lead-bismuth solder with a wire brush before using the soldering iron. The general procedure for soldering white metals is the same as for lead.

USING A BLOWTORCH

On some jobs, surfaces are not flat or in position for use of a soldering copper. On others, the metals to be joined are not large enough in diameter or rigid enough to permit the wiping method. If the wiping method or the use of a soldering copper is not practicable, soldering may be done by playing the flames of a blowtorch directly on the surfaces and then applying the solder cold in bar or wire form. The heated surfaces melt the solder and excess solder is removed by wiping before the solder has had time to harden completely. This method is most frequently used for soldering wire joints and sweating on lugs.

Gasoline Blowtorch. The gasoline blowtorch (Fig. 14) is most commonly used in soldering, and its operation is simple. Fill the tank about two-thirds full of clean, unleaded gasoline.

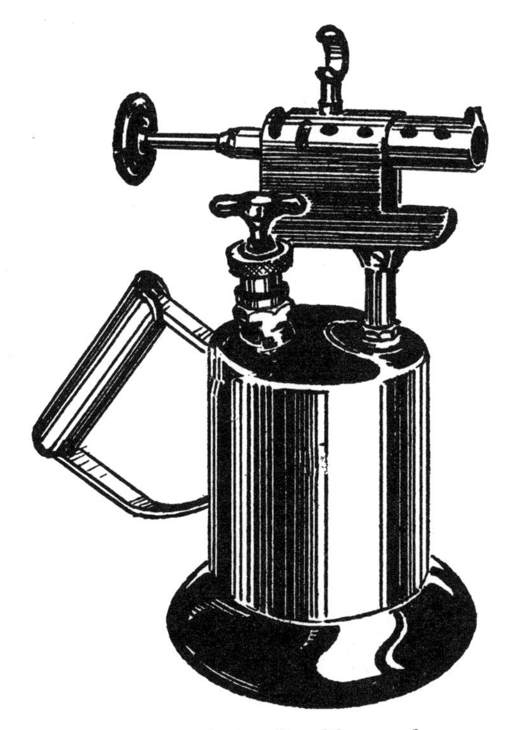

Fig. 14. Gasoline blowtorch.

Unleaded gasoline contains no tetraethyl lead. Pump until sufficient pressure is built up in the tank to cause the gasoline to flow when the valve is opened. Do not pump too tight. Excessive pressure may rupture the torch bottom.

With the valve open, liquid gasoline will flow from the jet of the torch and drip into the priming pan. When the pan is about three-fourths full, close the valve and light off the gasoline with a match. The flame from this burning gasoline heats the perforated nozzle (heating tube). Most of the trouble you will have in lighting and in using torches will be caused by insufficient heating before opening the valve to start the torch. Give it time, and keep it out of a draft. When the nozzle is hot, open the valve slowly, allowing the gasoline vapor which has been formed to flow from the nozzle. It will burn with an almost colorless flame and with considerable force. By working the valve, you can adjust this flame to any desired intensity. Adjust to produce a clean, pale blue flame for best results. When heating a soldering copper, the flame will show green when the copper has reached proper temperature for soldering work. If the pressure decreases, it is permissible to pump the torch while it is operating. To secure, close the valve sufficiently tight to prevent gas from escaping. Do not close the valve too tightly. Remember the metal is hot and will contract when it cools, causing the valve to stick.

There is very little maintenance to the gasoline torch providing you use only clean, clear, unleaded gasoline. If you use leaded gasoline, a compound will form that will stop up the gasoline passages. The torch will be a source of trouble from then on, as it is almost impossible to clean these passages completely.

Gas Blowpipe or Alcohol Torch. For direct flame soldering on small jobs use a *gas blowpipe* (Fig. 14) or an *alcohol torch* (Fig. 15). They are designed for the job. The *gas blowpipe* is

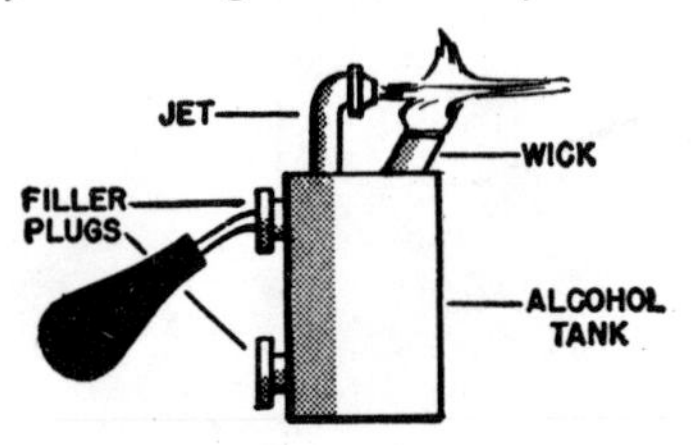

Fig. 15. Alcohol torch.

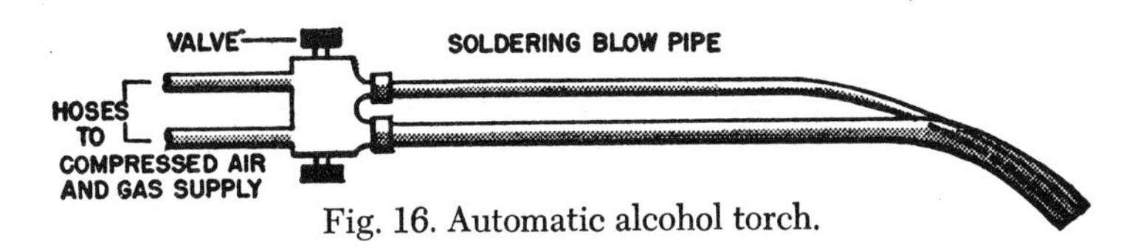

Fig. 16. Automatic alcohol torch.

an attachment similar to the one used for welding but smaller. It operates on the same principle. They have two tubes, one furnishing either natural or manufactured gas, and the other supplying compressed air or oxygen that run together to form the blowpipe.

Automatic Alcohol Torch. The *automatic alcohol torch* works on the same principle as the blowpipe (Fig. 16). It is a self-contained unit and creates its own pressure. The tank is filled about two-thirds full with alcohol and sealed. To operate this torch, remove the cap and light the wick. The flame from the wick heats the jet tube, causing the liquid alcohol in the container to vaporize and expand. The expansion forces the alcohol vapor from the jet openings where it is ignited to form a hot, light blue flame.

USING A GAS OVEN

Another heating device commonly used is a *gas oven* (Fig. 17). If one is available, it is used to heat soldering coppers. These ovens have one, two, or three burners, on which heat is hand controlled by means of valves.

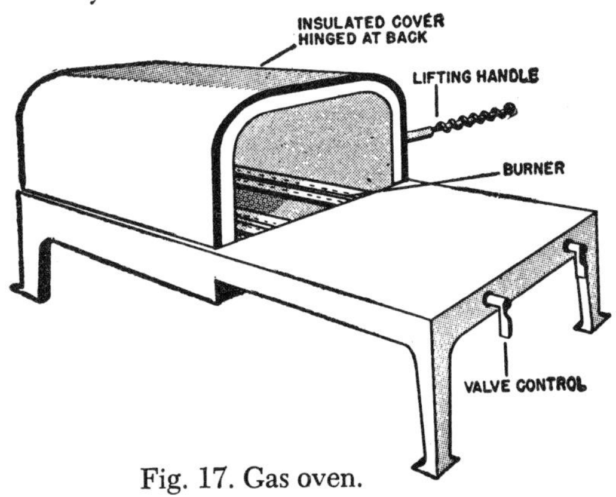

Fig. 17. Gas oven.

The gas oven, sometimes called a gas furnace, is the best device for heating soldering coppers. A refractory lining that resembles cement protects the metal part of the furnace from being burned, and reflects the heat over the soldering coppers making it possible to produce a temperature of 1800°F. which is more than is needed to heat 10-pound soldering coppers. Some ovens are provided with a pilot light which is allowed to continue to burn during the working period. To secure, turn off pilot light valves and main valves. Check the valves before and after using, to avoid accumulations of gas and resulting danger of explosion.

SOLDER BATH

In some instances, where parts are too small or too numerous to be soldered by one of the methods previously mentioned, the parts are joined by means of dipping in molten solder. For making the *solder bath*, solder may be obtained in slabs weighing from 15 to 35 pounds. These slabs are usually 50/50 or 40/60 tin-lead (40 percent tin and 60 percent lead).

SOLDERING BY SWEATING

In electrical work sweating is the best method for securing a terminal lug to a cable. To sweat a lug on a cable, clean and tin the end of the cable (Fig. 18). Flux the terminal lug and fill it with molten solder. Insert the cable in the lug. Hold fast in place, giving ample time for the solder to cool and set.

WORKING HEAT

Molten solder should never be raised much beyond the necessary working temperature. If the soldering job is big, a lot of heat will be needed to raise the temperature of the metal sufficiently to receive the solder. It is better to use a large amount of solder at about the usual soldering temperature than to raise the temperature of the solder.

As the temperature of molten solder is increased, the rate of oxidation is increased. When molten solder is overheated in the air, more tin than lead is lost. If it is necessary to heat

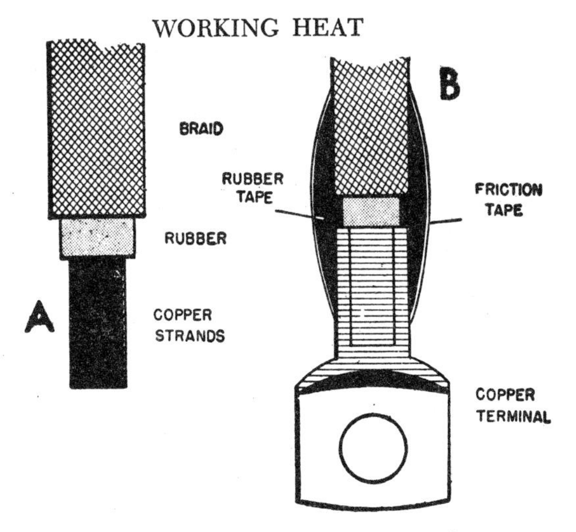

Fig. 18. Sweating terminal lug on cable.

a pot of solder excessively, for example when the solder has to be carried a long distance from the place where it is heated to the place where it is used, be sure to protect the exposed surface with a hard soldering flux such as that made of powdered borax, charcoal, and soda. Never stir or skim overheated solder, as this causes excessive loss by oxidation.

SOLDERING FLUXES

A *soldering flux* is a material which will remove the oxide film already present and protect the surface of both metal and solder while they are heated to the soldering temperature.

The strength of a soldered joint depends on the adherence of the solder to the metal being joined. To secure good adherence it is necessary that the surface of the metal and of the solder be free from dirt, grease, and oxides. Familiar examples of oxide on metals are rust on iron, corrosion on copper, and the white film visible on aluminum that has been exposed to the atmosphere for some time.

There are two classes of fluxes—*corrosive* and *noncorrosive*. The *corrosive fluxes* eat away the soldered metals unless they are thoroughly washed off after soldering. The fluxes ordinarily used for soft soldering are solutions or pastes that contain zinc chloride. *Zinc chloride* and *sal ammoniac* are corrosive fluxes. The solvent or other medium holding the flux material is evaporated by the heat of the soldering operation, leaving an unbroken layer of the solid flux on the work. At the soldering temperature, the solid flux is melted and partially decomposed with the liberation of hydrochloric acid. This acid then dissolves the oxides from the surfaces of the solder and the work. The melted flux also forms a protective film on the work that prevents further oxidation from taking place.

Corrosive fluxes should not be used in electrical work, as there is a possibility of some flux remaining on the joint and eventually breaking the circuit by eating through the metal conductor. These fluxes also lower the insulation resistance of the circuits and increase electrical leakage, since all types of corrosive fluxes are excellent conductors.

Because zinc chloride and sal ammoniac exhibit corrosive action, it is sometimes necessary to use a noncorrosive flux. *Rosin* is the most commonly used *noncorrosive flux*. Rosin is the amber-colored chemical compound and remains after turpentine has been removed from the sap of certain pine trees. Rosin is used as a flux when soldering tin plate and lead. It does not clean the surface of the work, but it does prevent oxidation during soldering by covering the surface with a protective film. Rosin may be obtained in powdered, liquid, or paste form. Some core solders contain rosin. *Core solder* is a small tube made of solder which contains flux in the center, and is also available on spools.

In choosing fluxes, attention should be given to the kinds of metal to be joined and to their massiveness. It takes more heat to solder iron or steel than to solder lead. When a soldering copper is used, the metal is heated in one spot and the heat may be conducted away rapidly. The iron therefore must be exceptionally hot unless the job has been preheated. In such a case the use of rosin is unsatisfactory since this material carbonizes or chars at high temperature, preventing rather than aiding soldering. When you are soldering large pieces of metal together, pre-tin the parts to be soldered before assembling them. Use the same method previously described to tin a soldering copper, as this makes soldering easier and makes joints stronger. Sometimes it is best to preheat the job. Some experts recommend that cast iron be treated with a cold pickling bath of five percent hydrofluoric acid before it is soldered. *Use hydrofluoric acid with care* as it is very corrosive, attacking the flesh and forming painful sores which heal slowly. The vapor is extremely dangerous if inhaled. This acid must be kept in a wax or lead container as it will eat away glass or most metals.

Muriatic acid (also called raw acid) is the commercial form of hydrochloric acid and is yellow in color. This raw acid may be used as a flux when soldering galvanized iron, but it is good practice to "kill" or "cut" the acid by adding scraps of zinc. After the zinc has dissolved, the acid may be weakened to the correct strength by pouring it into an equal amount of water. This forms zinc chloride. Here is what happens: The acid dissolves the zinc, forming zinc chloride and releasing hydrogen gas. When the reaction stops, the solution should be strained. *Remember that the fumes are injurious when inhaled.* They are inflammable, and will cause metal to corrode. Zinc chloride should be prepared in the open or near openings to the outside. Keep it in closed glass containers away from machines and tools.

Sal ammoniac, another of the corrosive fluxes, is muriate of ammonia, or ammonium chloride. It can be obtained in a high refined state in crystal form practically free from metallic impurities, and also in one-pound brick form under such trade names as "Salamac" and "Speco." When a hot copper is placed in contact with sal ammoniac, the copper is cleaned by the chemical, thus enabling solder to adhere to the copper. (*See* Table 18 for fluxes used for common metals.)

TABLE 18
FLUXES FOR COMMON METALS.

Metal	*Flux*
Brass, copper, tin	Rosin
Lead	Tallow, Rosin
Iron, steel	Borax, Sal ammoniac
Galvanized iron	Zinc chloride
Zinc	Zinc chloride
Aluminum	Stearine, special flux

Stearic acid or *stearine* may be used in soldering aluminum. But a special aluminum solder is now available which makes the use of flux unnecessary.

Paste fluxes, commercially manufactured, are usually available in one-pound cans. They contain grease for counteracting corrosion. This grease sticks to the metal and collects dirt. Past fluxes are safe to use and are particularly useful for odd-job soldering, or at times when it is inconvenient to mix and use zinc chloride.

SAFETY PRECAUTIONS

Do not overheat your soldering copper as this will shorten its life.

Do not dip your soldering copper into soldering flux, as it will stain your work.

Do not forget to use a separate dipping solution. It is essential to good work.

Do not leave soldering tools adrift. You might start a fire.

Do not drive handles on soldering coppers by bumping them against an anvil, as handles split easily.

Do not inhale fumes from zinc chloride, as they are injurious.

Do not leave acid in open containers. The fumes are corrosive.

Do not spill raw acid on your clothes or body.

Do not treat gasoline carelessly, as this is dangerous.

Section IV

Home Plumbing and Water System, Heating and Ventilating, Piping Installation Care and Repair

Both old and new homes, even though carefully built, need occasional repair and maintenance to keep them in good condition and to retard depreciation. Like other kinds of property, houses will deteriorate and decline in value if they are not properly cared for. Therefore, alertness to signs of wear in the plumbing, water system, in heating and ventilating, and piping installation; and in other utilities can often eliminate the need for major repairs that stem from neglected minor problems. Thus, repairs of faulty plumbing and heating should not be neglected. Difficulties can be avoided by proper inspection and prompt correction.

Extensive repairs or alterations in the plumbing and heating systems, and in some utilities require authorization from local authorities, and possibly inspection of the completed work. Repairs that require special knowledge and skill should be performed only by qualified persons. However, the homeowner can save money and avoid delays by making some repairs and alterations if he or she is handy with tools.

Chapter 18

Plumbing and Water System Care and Repair

The plumbing of a residence includes pipes for distributing the water supply, fixtures for using water, and drainage pipes for removing waste water and sewage, together with fittings and accessories of various kinds.

Each part of a plumbing system is designed for a specific purpose and should be used only for that purpose.

Grease or refuse should not be thrown into closet bowls, sinks, or lavatories; faucets should be tightly closed when not in use; and waste pipes should be flushed frequently with hot water to keep them in good working order.

Some local or state plumbing regulations permit only licensed plumbers to make major installation or repairs. However, although the average homemaker cannot be expected to perform all of the work of an experienced plumber, it is possible for such a person to make many small repairs. Trouble and unnecessary expense will be avoided by promptly repairing a leaky faucet, or by cleaning out a fixture drain at the first signs of clogging. If difficulty involving the pipes arises, a plumber should be called in.

SHUTTING OFF WATER

The flow of water in a house is controlled by means of stopcocks or shut-off valves in the pipes. To shut off the main supply of water as it enters the house, it is necessary to close the wall cock or valve which is usually in the main pipe in the basement. This cock may have a handle, or a wrench may be needed to turn it. It may be of the ground-key type with a small hole bored in its side for draining the pipes after the water is shut off, or it may be a compression-stop type with a cap nut covering the drain opening. In either case, this opening should be closed when the water is turned off, for, if not closed, a stream of water will shoot from the hole with considerable force.

Where means have not been provided for shutting off a drain opening, a small wooden peg driven into it will temporarily stop the flow of water until the pressure is relieved by draining the pipes through the faucets.

If a house is left vacant during the winter months, it is safest to have the water shut off at the curb cock to prevent freezing between the cellar wall and the main shut-off or wall cock.

Separate shut-off cocks are sometimes provided below the sink, lavatory, water closet, or other fixtures for convenience when repairs are to be made, so that the flow of water may be cut off from any one fixture without disturbing the flow to the other parts of the system (Fig. 1). It is important that all members of the family know where these various shut-offs are located, especially for use in cases of emergency.

Fig. 1. Typical water shut-off cock.

DRAINING PIPES AND SYSTEMS

For minor repairs, such as the replacing of a washer in a faucet, the temporary shutting off of the water from the branch shut-off previously described is sufficient. Where extensive repairs such as pipe changes are proposed, the main water supply should be cut off and the pipes drained. If a house is to be vacated, with no heat provided during cold weather, it it is advisable to drain the entire water system. A reliable plumber should be engaged to perform this service and to do whatever else is necessary to protect the piping and fixtures against freezing and possible damage.

To drain the pipes, first shut off the water, as previously described. Then, starting at the top floor, open all faucets on the way down. When water ceases to run from the faucets, the small cock or cap in the main pipe valve may be opened or the plug removed to allow what little water remains to drain into a bucket or tub.

In addition to draining the pipes, the water that remains in the traps under sinks, water closets, tubs, lavatories, and showers, should be removed by opening the traps and draining them, by forcing the water out with a force pump, or by drawing it out with a suction pump or siphon hose.

The water-closet tanks should also be emptied by flushing after the water has been turned off, and any surplus water taken out with a sponge or cloth. The water-closet traps can be cleared of water by means of a sponge tied to a stick or wire.

The traps may then be filled with kerosene, crude glycerine, or some similar nonfreezable liquid to form a seal against bad odors from waste pipes. Alcohol and kerosene mixed is a good solution to use, since the kerosene will rise to the top and prevent evaporation of the alcohol.

If a house is to be left unoccupied and there is danger of freezing, the hot-water supply tank should be emptied by opening the faucet at the bottom of the tank. To facilitate the flow of water all hot-water faucets should be open while tank is being drained.

If the house is heated by hot water or steam and is to be left unoccupied during winter months, the system should be drained unless an automatic oil or gas burner is used to maintain above-freezing temperatures in the house. To drain the system the fire should be out, the main water supply shut off at the wall or curb, and the water from the boiler drawn off by opening the draw-off cock at the lowest point in the heating system. The water-supply valve to the boiler should be opened so that no water will remain trapped above it. After that, in any hot-water system, beginning with highest radiators, the air valves on all radiators should be opened as fast as the water lowers. In a one-pipe steam system every radiator valve should be opened to release the water or condensation. After a heating system has been drained, never start a fire under the boiler until it has been properly refilled. (*See* Chapter 20, Heating and Ventilating Care and Repair, section on Gravity or Pressure Hot-Water System.)

If a house is to be left vacant in the summer, it is not necessary to drain the water supply pipes or system although the water should be shut off at the basement wall as a precaution against waste of water from a dripping faucet or possible leak in the piping.

FAUCETS

Faucets generally used in dwelling houses are of three types: compression, Fuller ball, and ground key.

The *compression type* of *faucet* is usually fitted with a lever, T, or four-ball handle which offers firm resistance to efforts to turn down the spindle much beyond the point where the flow of water stops. The stem of the spindle may be seen to move in or out of the body of the faucet when the handle is turned. A self-closing faucet is usually of the compression type.

Fuller ball faucets are generally fitted with a lever handle and the stem does not move in or out of the body of the faucet when the handle is turned. When a Fuller ball is in good condition, the handle should require a one-quarter turn only to open or close the flow opening.

The *ground-key faucet* is easily distinguished by the lever handle and plunger, which is made in one piece, and by the exposed nut or screw at the lower end which holds the plunger in place.

It is sometimes impossible to determine the type of faucet from the outside appearance. If so, the only way to find out will be to dismantle it. At the same time, the kind and size of washer or Fuller ball may be determined and the condition of the brass screw examined to see whether or not it needs replacing.

Compression-type Faucet

In the ordinary *compression-type faucet* shown in Fig. 2, the flow of water is regulated by turning a lever, T, or four-ball handle which is attached to a threaded spindle. When the spindle is turned down, the washer or disk attached to its lower end is pressed tightly against the smoothly finished ring or "ground seat" which surrounds the "flow opening," thus, shutting off the flow of water. If the washer and the seat do not make a firm contact at all points, water will leak through and drip from the faucet. A leak usually results from a worn-out washer. If washers wear out rapidly, it may be because a poor grade of washer is being used, because the ground seat has become sharp and rough as a result of corrosion, or because the seat has become scratched or worn by grit.

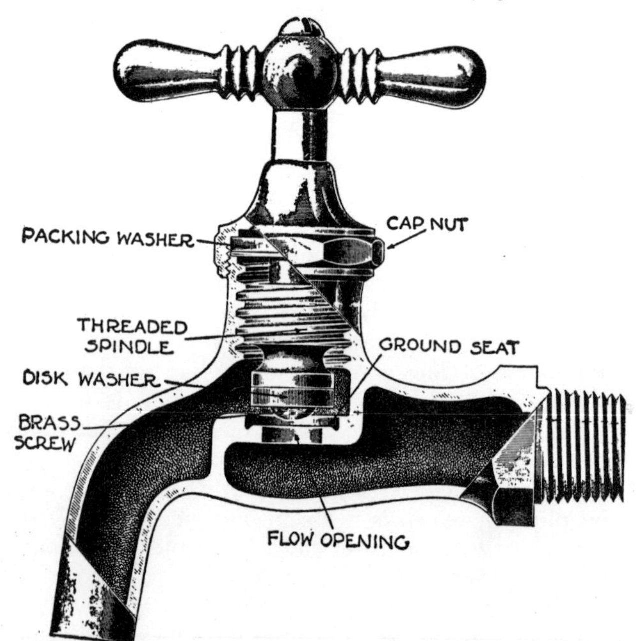

Fig. 2. Cross section of compression-type faucet.

Moderate force on the handle of a compression-type faucet in good repair should stop all flow and drip. If a leak develops, it may be caused by faulty washers which are not difficult to replace. It is important that faucets be tightly closed after they are used because dripping faucets tend to produce or aggravate leaks, waste water, and result in rust stains on porcelain surfaces. Soon after a hot-water faucet has been shut off and the water cools, contraction takes place which may cause a drip to develop. Should this occur, the spigot handle should be tightened without opening while the faucet is still cool.

The following tools and materials are needed: Monkey wrench, screwdriver, and pliers; fiber or special composition washers for compression-type faucets.

To avoid frequent renewals, a good grade of washer should be selected. The sizes most frequently used are ⅜- (.9525-), ½- (1.27-), and ⅝-inch (1.5875-cm.) and it is well to have a supply of each size on hand. *Composition washers* have one side flat and the other side slightly rounded. A good contact is made with this type of washer because, by fitting partly down into the seat of the faucet, it is subject to both horizontal and vertical pressure. Some faucets require specially shaped washers the size and type of which should be determined for replacement.

To renew a washer, shut off the water directly below the fixture or in the main water supply pipe. If the water is shut off by the valve in the main pipe and there are fixtures located higher than the one in which the washers are to be replaced, the riser pipes to the higher fixtures should be drained before disassembling the faucet. If this is not done, it may be impossible to control the flow of water issuing from the faucet when taken apart. If shut-offs located directly below the fixture are used, this precaution will not be necessary. Then, with a wrench (using a cloth to protect the fixture from being marred), unscrew the cap nut of the faucet to allow the spindle to be unscrewed and removed. Carefully remove the brass screw that holds the washer to the bottom of the spindle, and replace the worn washer with a new one. If the head of the brass screw is bady worn, it will be difficult to remove and may be twisted off, unless handled carefully. A drop or two of kerosene and gently tapping the screw may help to loosen it in the stem. The screwdriver should have a good square edge and should be turned with a strong steady pressure. If the head of the screw chips off or breaks so that it does not hold the screwdriver, the slot will have to be deepened by cutting into the head with a hacksaw. A badly worn screw should always be replaced with a new one.

A worn or roughened washer seat can often be ground true and smooth with a faucet seat-dressing tool. Such a tool is inexpensive and will probably more than pay for itself within a reasonable time. One type consists of a stem with a cutter at the lower end and a wheel handle at the top to rotate the tool. It is fitted with a spiraled cone to be inserted into the body of the faucet and screwed down firmly for the purpose of centering and holding the cutter on the washer seat. When the tool is properly placed, it should be carefully rotated back and forth several times with the wheel handle until the seat is ground free of irregularities. When the grinding is finished, all metal cutting should be wiped out with a cloth before the faucet is reassembled. If the seat is in such bad condition that it does not respond to this treatment and continues to cut the washers, it will be necessary to substitute a new faucet.

If water leaks around the stem when the faucet is open, it may frequently be stopped by tightening the cap nut, but the nut should not be made so tight as to cause the faucet to bind. If tightening does not stop the leak, it is probable that the packing washers under the cap nut are worn out and need renewing. To put in new washers, remove the handle and cap nut and substitute new washers for the old. To stop the leakage temporarily, wrap a small piece of oil-soaked candle-wicking or soft string around the stem under the cap nut where the stem enters the body of the faucet.

Fuller Ball

In the Fuller ball faucet, a hard rubber or composition ball-like stopper, known as the *Fuller ball,* is fastened by a small nut or screw to a shaft with an eccentric end (Fig. 3). When the faucet handle is closed, this ball is drawn firmly against the opening, shutting off the flow of water. When the faucet handle is opened, the ball is pushed away from the opening, allowing the water to pass through. The best grade of Fuller ball should be used. They range in sizes from ¾ to 1 inch (1.905 to 2.54-cm.).

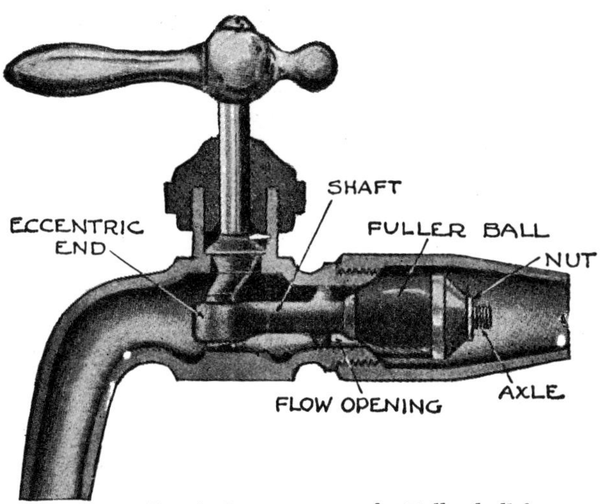

Fig. 3. Cross section of a Fuller-ball faucet.

Replacing a Fuller Ball. To replace a Fuller ball the water should be shut off and the faucet unscrewed and separated from the supply pipe. The nut or screw should be taken off with pliers or a screwdriver and the ball removed and replaced with a new one.

Sometimes the metal axle which holds the Fuller ball or the eccentric part becomes worn, making it impossible to pull the ball tight against the seat, and allowing leakage between the ball and the seat. If this happens, it will be necessary to purchase new metal parts to replace the worn ones. If water leaks out around the stem when the faucet is open, repairs can be made in the same manner as prescribed for similar leakage in compression-type faucets.

Ground-key

The *ground-key faucet* has a tapered cylindrical brass plunger or plug which should fit snugly into a sleeve, bored vertically through the body of the faucet. The plunger, which is rotated by a handle, has a hole or slot bored horizontally through it, to coincide with a similarly shaped horizontal opening in the body of the faucet. When the handle that rotates the plunger is parallel to the body of the faucet, the two openings are in line with each other and allow the water to pass through. A short turn of the handle to the right or left throws the opening out of line and cuts off the flow.

The plunger or its sleeve may become grooved or worn by sand particles rubbing against the metal and may allow the water to leak through. This requires repolishing of the rubbing surfaces. Also, the nut or screw at the bottom may become loose, permitting the plunger to move out of its proper position, allowing leakage. On the other hand, if the nut or screw is too tight, the plunger will bind and will be difficult to turn.

Noise in Faucets

Sometimes when a faucet is partly turned on or suddenly closed, a water hammer, tapping, or pounding noise is heard. *In a compression-type faucet*, this may be caused by a loose cap nut, a worn spindle, or a defective washer. *In a Fuller ball faucet*, the ball may become loose, or the metal eccentric connecting the handle to the Fuller ball may become worn.

The only tools needed are the following: Monkey wrench, screwdriver, and pliers.

To eliminate noise in a compression-type faucet, shut off the water and remove the spindle and washer so that they may be examined. If the washer is found to be loose, the brass screw should be tightened; if the washer is worn, the brass screw should be removed and a new washer attached. If the threads on the spindle or in the body of the faucet are badly worn, letting the spindle rattle, it will be necessary to purchase a new faucet.

If the faucet is of the Fuller ball type, shut off the water and tighten the small nut or screw which holds the Fuller ball; if the ball is badly worn it should be replaced. If parts of the eccentric are worn and tend to rattle, the faucet should be taken to a plumber. If the eccentric is beyond repair and new parts cannot be obtained, it will be necessary to install a new faucet.

PROTECTION OF PIPES

Water pipes which are exposed to freezing temperatures should be covered, especially if located out of doors or in unheated spaces. Smaller water pipes are more likely to freeze than larger waste or sewer pipes, since the latter carry water which has usually been warmed to some extent, and which flows off quickly leaving the pipe empty.

Insulating Pipes

For pipes located in an unheated basement, attic, garage, or out of doors where they are likely to freeze, it is well to apply insulation similar to that used on hot-water and steam-heating pipes. (*See* Chapter 19, section on Insulating Heating Systems.) In severe climates, it is advisable to apply two thicknesses of this insulation and to break or stagger the joints and seams in order to make the insulation tight. If exposed to the weather, insulation should be protected by wrapping it spirally with cotton fabric tape or strips of saturated felt, followed by two coats of asphalt-varnish. Insulation may also be protected from the weather by constructing a water-tight box around the pipe.

Underground Pipes

When pipes are laid underground they should be buried deep enough to be protected from damage by heavy vehicles passing over them and to eliminate the danger of freezing. The depth depends on climate and local soil conditions.

Frost protection in the central and northern latitudes of the United States usually requires that pipe be placed from 2½ to 3 feet in the ground. In the extreme Northern states, it is well to go as deep as 4 to 6 feet, and in the Southern states a depth of from 1½ to 2 feet is usually enough covering to protect the pipe from damage.

Thawing Pipes and Drains

If water-supply pipes become frozen, they should be promptly thawed out to avoid possible bursting. In *lead* and *soft copper pipes*, a bulge in the pipe will disclose the location of the frozen area, whereas, in *other metals*, no such bulge will appear.

Some form of heat will be required to melt the ice in the pipes. The heat may be applied to the exterior of a frozen pipe by *electrical resistance, direct flame*, or *hot applications* of water or steam. In thawing out water pipes, it is best to work toward the supply end keeping a faucet open to indicate when the flow starts. When thawing out a waste or sewer pipe, it is best to start at the lower end and work upward, to allow the water to flow off as the ice is melted.

In heating frozen water-supply pipes by *electrical resistance*, a source of low voltage, such as a welding generator, should be connected directly to the water pipe with the two electrical conductors clamped to the pipe to span the frozen section. As soon as a section has been thawed out, the conductors should be moved along the pipe to thaw another section. A welding shop, or plumber who has welding equipment, and the necessary plumbing experience, should be called to perform this work.

Direct flame may be applied to frozen pipes with a gasoline blowtorch, provided there is no danger of burning the adjoining woodwork. The flame should be played gradually along the pipe to spread the heat evenly.

Hot applications on a frozen pipe do not produce as quick results as direct flame, but are much safer because they lessen the fire hazard and the possibility of bursting the pipe.

Other methods of hot application may be used if the water pipes are accessible. They can either be wrapped with cloths and saturated with boiling water or the boiling water can be poured directly over the pipe. In both cases, a receptacle should be placed below the pipe to catch the water.

Steam, used by plumbers to thaw out pipes, is provided by a steamer resembling a 5-gallon oil can, having a hose and nozzle attached. The steamer is heated by a plumber's furnace.

Frozen traps, waste pipes, drains, and *sewer pipes* may be opened by pouring boiling water into them through the drain opening or trap. If this is not successful, a can of lye or drainpipe cleaner dissolved in two gallons of cold water in a porcelain container should be poured carefully into the drain opening or trap. Do not use hot water with lye or cleaner and avoid splashing the solution on face, hands, or clothing.

Condensation on Pipes

It may be desirable in some cases to insulate cold-water pipes to prevent condensation of moisture on them in hot humid weather. One type of insulation for this purpose is a *cylindrical-shaped split pipe covering* of wool insulating felt with a canvas jacket. This pipe covering comes in 3-foot lengths, is of various thicknesses, and is made for standard pipe sizes. It may be applied in the same manner as the insulation for hot-water and steam heating pipes. (*See* Chapter 19, section on Insulating Heating System.)

To prevent water vapor from reaching the outside surface of the pipe, a *vapor-resistant covering* should be applied to the surface of the canvas jacket.

If appearance is important, two coats of spar-varnish aluminum paint can be applied to the canvas jacket followed by one or two coats of paint in any desired color. Instead of aluminum paint, the jacket may be wrapped with aluminum foil, such as is sold for kitchen use, and painted. If the basement is damp, the paint should contain a fungicide to prevent mildew.

If appearance is not important, the canvas jacket may be wrapped and sealed with aluminum foil or with asphalt-impregnated paper, or it may be painted with an unbroken coating of asphalt.

In addition to the above, *tape-form insulating coverings* are also available which may be wrapped spirally around the cold-water pipes to a thickness of about ¼ inch (.635-cm.), as well as thick paints mixed with insulating materials to be applied to the pipes in a coating of about ¼ inch (.635-cm.) thickness.

DRAINAGE SYSTEM

Plumbing fixtures from which water is discharged are equipped with *traps* designed to retain enough water to form a seal and prevent gases and bad odors from entering the house through the drain. In general they cause little trouble, but occasionally they may become clogged by accumulations of hardened grease and dirt that need to be removed.

Traps

Before traps were manufactured as separate fittings, a single or double U-bend was made in the lead drain pipe below the fixture to provide for a water seal. Plumbing fixtures are now equipped with either a P or S *form* of separate trap fitted with a screw clean-out plug on the lower side of the bend (Fig. 4).

In the manufacture of water closets, suitable bends are cast in the lower portion of the bowl.

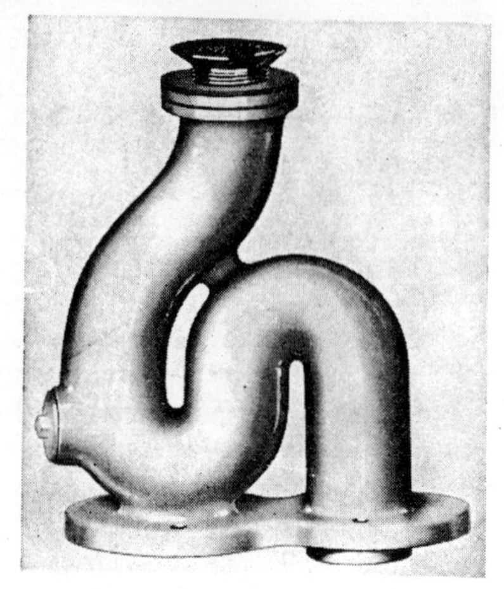

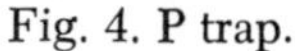

Fig. 4. P trap.

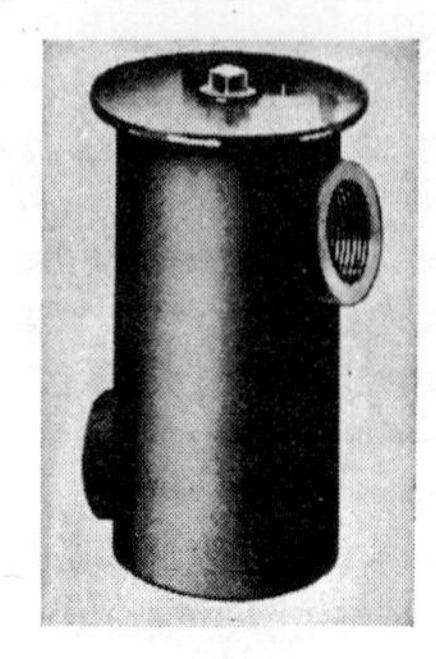

Fig. 5. Drum trap.

Drum and *bottle-type traps* (Fig. 5) for bathtubs or kitchen sinks consist of a cylindrical-shaped metal box or settling basin attached to the water pipe. They are generally provided with a screw-cap cover, which can be removed when cleaning is necessary.

Causes of Stoppage

When waste water gurgles and seeps away slowly from a sink, washbowl, or bathtub, or backs up in the water-closet bowl, there is evidently foreign matter in the waste line that is retarding the flow of water. For example, the trap on the kitchen sink may be filled with hardened grease and settlings, or the waste pipe beyond the trap may be clogged by more solid material. Water-closet drains are often stopped up by toilet articles or other objects, which have been accidentally dropped into the bowl, or by cloth, or heavy paper.

For repair work, the following tools and materials are needed: Monkey wrench, screwdriver, stiff wire with hook bend at one end, force cup or "plumber's friend," coil spring steel auger, small funnel, and galvanized water bucket; also a commercial drainpipe cleaner.

Removal of Stoppage

Stoppages in traps and waste lines may be removed by means of a force cup or plumber's friend, chemical drainpipe cleaners, by opening the trap, or by using a flexible coil spring steel auger.

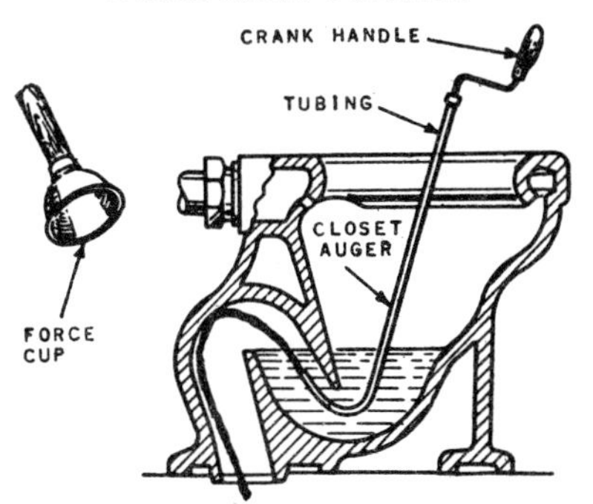

Fig. 6. "Plumber's friend" or rubber force cup, and closet auger.

Plumber's Friend

To open a clogged waste line, it is usually best to try a force cup or plumber's friend first. The plumber's friend is a stiff rubber bell-shaped cup about five inches in diameter fastened by its top to a stick about the size of a broom handle (Fig. 6). In using this tool, the sink or bowl should be partly filled with water, and the rubber cup placed over the mouth of the drain opening. After this the stick of the plunger should be worked forcibly up and down several times. The alternate compression and suction thus created will usually loosen minor obstructions.

The operation should be repeated until the pipe is cleared, after which boiling water should be poured into the drain to thoroughly clear the waste line. The force cup or plumber's friend is equally effective in removing obstructions from water-closet waste lines. Place the cup over the outlet in the bowl, and force the handle up and down until the passage is clear.

Chemical Cleaners

If flow through a trap, waste line, or drain becomes sluggish, and the use of a plumber's friend fails to remove the obstruction, a chemical drainpipe cleaner may be used.

Drainpipe cleaners may be lye (caustic soda or sodium hydroxide), either as is or mixed with aluminum or zinc-coated aluminum turnings or chips. Mixtures of lye with sodium nitrate and aluminum turnings are also used for this purpose. Lye acts on the accumulated grease and insoluble soap curds that have formed in and are clogging the pipe. When water is added to mixtures containing aluminum and lye, they react vigorously, forming a gas, the stirring or agitating effect of which facilitates removal of the waste matter.

Caution: Lye (caustic soda or sodium hydroxide) is a caustic poison. While handling this material or drainpipe cleaners containing it, the eyes, skin, and clothing should be protected from splashings. Be sure to keep water out of the can and do not let the material get on wood, painted surfaces, floor coverings, or aluminum. Flush with cold water only, and do not use a plunger.

If drainpipe cleaner gets into the eyes or into the mouth, call a physician at once and apply the following emergency treatments. For the eyes—flush copiously and quickly with water and then wash with a five percent solution of boric acid in water. For internal treatment—drink one-fourth cup of vinegar diluted with two cups of water or two glasses of lemon, orange, or grapefruit juice, following this with two or three tablespoons of olive oil, butter, or other cooking oil. For external treatment—flood with water, wash with vinegar, and apply vegetable oils or butter.

If drainpipe cleaner has been spilled on floors, floor coverings, or clothing, the spot should be quickly flooded with water, then treated with vinegar, and finally rinsed with water.

The Federal Caustic Poison Act requires products containing caustic soda to be labeled "Poison," with directions for treatment of external or internal personal injury to be printed on the container.

Trap Clean-out Plug

If use of the plumber's friend or chemical drainpipe cleaner does not clear the waste line, it is possible that the trap is clogged with refuse which must be removed. For cleaning purposes, a *removable clean-out plug* is sometimes provided in the lower side of the bend in the trap (Fig. 4). Before opening the trap, a bucket should be placed under it to catch surplus water. The plug should then be unscrewed with a wrench and when the water has drained out, a bent wire should be inserted into the trap opening to pull out the accumulated grease and dirt. If possible, the trap should be brushed out well with a bottle brush and flushed by pouring boiling water into the fixture. If the washer or gasket around the plug is broken, it should be renewed before the plug is replaced.

Where the trap has *no clean-out plug*, it should be removed at the slip joint provided for that purpose. The trap and adjoining pipes should then be cleaned out with the wire hook and bottle brush, as previously described.

Coil Spring-steel Auger

In most cases, cleaning out the trap will allow water to flow freely, but there may be an obstruction in the pipe beyond the trap. In this case, a coil spring-steel auger should be inserted through the trap opening, and the obstruction either pulled out or bored through and forced out. The *coil spring-steel auger* is especially effective in opening clogged water-closet traps, drains, and long sections of waste pipe lines (Fig. 6).

Cleaning Drum or Bottle-type Trap

In a *drum or bottle-type trap*, the screw cap is fastened in either the top or bottom, and may be removed when cleaning is necessary. In some cases, this cap is just as accessible as the plug on a U-bend trap but, when used in connection with bathtubs, the trap may be placed below the level of the floor and covered by a metal plate, which must first be removed before the cap is exposed. When the cap has been taken off, the grease and dirt can be removed and the trap and its openings washed out with boiling water.

The various methods recommended for cleaning out U-bend traps can be used with equal success on the drum or bottle trap.

Toilets

If water continues to run into the closet bowl after a toilet is flushed, it is evident that some part of the mechanism of the flush tank or flush valve is out of order and needs adjustment or renewal.

Flush Tank. The supply pipe that fills the flush tank is fitted with a lever ball cock, which operates somewhat like a compression-type faucet having a plunger with a rubber

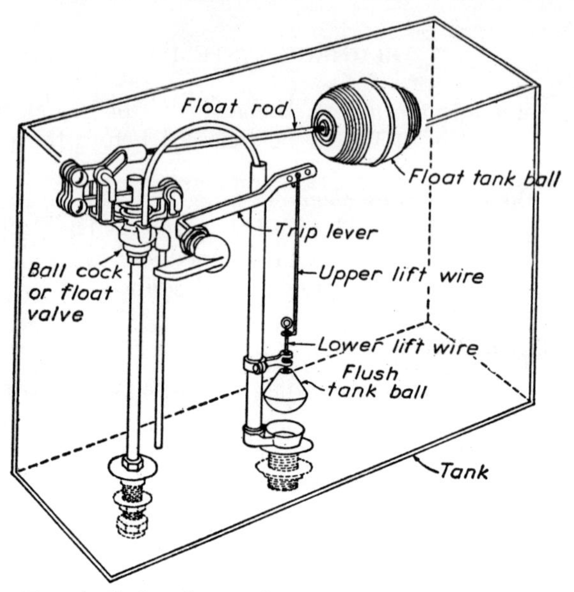

Fig. 7. Flush valve mechanism for filling flush tank.

washer to close the opening (Fig. 7). The water supply to the tank is turned on and off by the raising and lowering of a hollow ball float connected to the plunger by a lever mechanism. When the flush valve is opened, and the water level goes down, the ball float goes down with it and the lever raises the plunger, allowing a fresh supply of water to enter the tank. When the flush valve is again closed, the incoming water gradually refills the tank and in rising carries the float up with it, slowly lowering the plunger into its seat and stopping the filling process. An overflow tube or pipe is provided in the flush tank to carry off the water should it rise above its proper level. This pipe empties into the closet bowl.

The *opening* and *closing* of the *outlet* from the tank to the closet bowl is accomplished by the raising and lowering of a rubber ball stopper suspended by lift wires, which are attached by another lever to the push button or handle on the outside of the flush box. When the button is pressed or the handle turned, the rubber ball stopper is lifted from its seat on the outlet pipe allowing the water in the tank to rush into and flush the toilet bowl. In the meantime, the stopper remains suspended until the tank is empty, when it again sinks to its seat where it is held in place by water pressure.

Leakage may occur either from the supply cock or the outlet valve and is usually caused by improper seating of the plunger in the first case or of the rubber ball stopper in the other.

If the water in the tank rises high enough to flow off through the overflow tube, the supply cock is out of order; if water leaks past the rubber ball stopper and out through the outlet valve, the fault lies with the valve. In either case, water will continue to run into the closet bowl after it has been flushed.

The tools needed in repair work are a small monkey wrench, pliers, and a screwdriver.

A *rubber ball stopper* that does not fit tightly over the top of the outlet pipe, a defective ball, an irregular stopper seat, or bent lift wires may cause an outlet valve to leak. Sometimes a ball may be covered with a slimy coating which can be wiped off. If the ball is worn, is out of shape, or has lost its elasticity and fails to drop tightly into the hollowed seat, it will have to be replaced. To *replace* the *ball*, empty the tank and, if there is no supply shut-off, place a prop under the lever arm of the copper-ball float to hold it up, thereby shutting off the intake cock and preventing the tank from refilling. Then unscrew the ball from the lower lift wire and attach a new ball of the same diameter as the old one.

It may be found that the top of the outlet pipe is corroded or covered with grit in such a way as to make an irregular seat for the stopper ball. If this is the case, the valve seat should be made smooth if possible, by means of emery cloth or other abrasive, or replaced if this is not successful.

Sometimes the handle and lever fail to work smoothly, or the lift wires get out of plumb, causing the ball stopper to remain suspended or to incompletely cover the outlet pipe opening. The lift wires should be straightened and made plumb so that the ball will drop squarely into the hollowed seat of the outlet. The lower lift wire can be adjusted by means of the adjustable guide arm, which is usually fastened to the overflow pipe. The thumbscrews should be loosened to raise, lower, or rotate the guide arm until it is centered directly over the outlet seat. There are various ways in which the intake cock can get out of order. The seat washer on the bottom of the plunger may become worn; the seat itself may be irregular; or the intake cock may be in good condition, but may not work properly because of faults in the copper-ball float or its attachments.

A leaky water-logged float will hold the plunger up and fail to shut off the water completely. If there is a leak in the ball, it is advisable to buy a new one. If the rod that connects the float to the plunger lever has become bent, it may allow leakage by not lowering the plunger sufficiently to completely shut off the flow of water. In this case, the rod should be removed and straightened.

Sometimes, because of faulty installation, the tank will overflow or may not fill sufficiently. This can be corrected without disturbing the supply cock by bending up or down the rod which is attached to the copper float. If the rod is bent up, the water will rise higher in the tank, but if bent down, the water will not rise so high.

To replace a washer on the plunger of the intake cock, the water should be shut off by means of the shut-off valve, which is usually located below the flush tank, after which the tank can be drained and bailed out. The two thumbscrews should then be unscrewed to release the plunger. The old washer, held in place by a nut and brass ring cap, can be removed and the washer replaced. The brass ring cap, into which the washer fits, may be corroded and break while being taken off. If this occurs, a new ring should be substituted. Be sure that the seat upon which the plunger rests is free from nicks and grit; if not, it should be smoothed off with emery cloth or other abrasive.

Flush Valve. In some houses, a flush valve instead of a tank is provided to flush the water-closet bowl. There are several kinds of flush valves, but the adjustments required for each are much the same. The type shown in Fig. 8 is one commonly used and will therefore be taken as an example.

In this type, the rubber segment diaphragm A-56 separates the valve into an upper and lower chamber, with the pressure the same on both sides equalized by the bypass A-24. The slightest touch of handle B-3 in any direction pushes in plunger B-8, which tilts relief valve A-19, releasing the pressure in the upper chamber. The pressure below then raises the entire working parts including relief valve (A-19), disk A-15, segment diaphragm (A-56), and guide A-13, allowing the water which flushes the bowl to go down through the barrel of the valve. While this is occurring, a small amount of water passes up through bypass (A-24) and gradually fills the upper chamber and closes the valve.

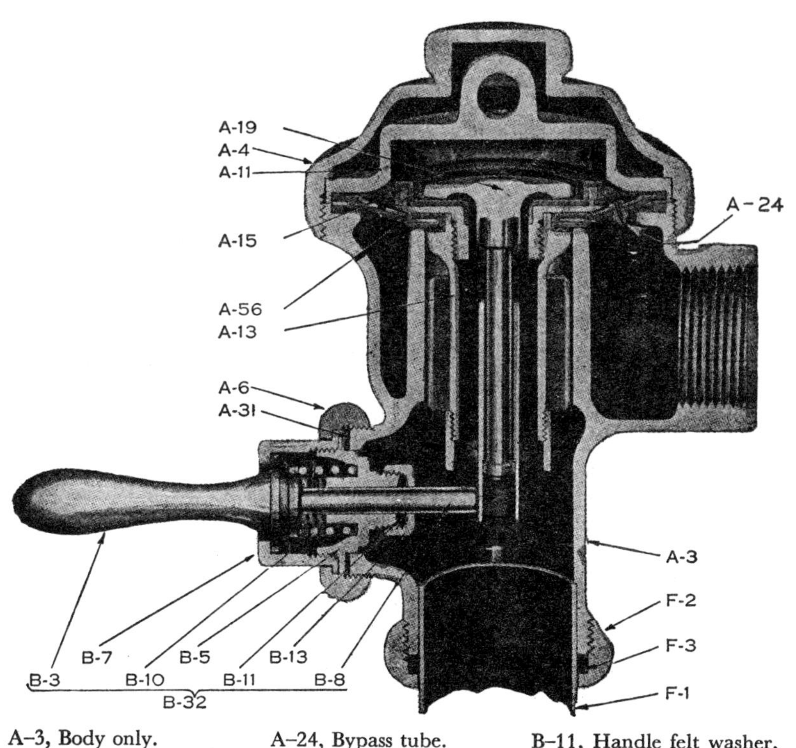

A–3, Body only.
A–4, Outside cover.
A–6, Handle coupling.
A–11, Inside cover.
A–13, Guide.
A–15, One-piece molded disk.
A–19, Relief valve.
A–24, Bypass tube.
A–56, Segment diaphragm.
A–31, Handle gasket.
B–3, Metal grip.
B–5, Bushing.
B–8, Handle plunger.
B–10, Handle spring.
B–11, Handle felt washer.
B–13, Handle packing nut.
B–32, Metal handle assembly.
F–1, Outlet.
F–2, Outlet coupling assembly.
F–3, Friction ring.

Fig. 8. Cross section of a water-closet flush valve.

If the *valve is out of order*, water may continue to run into the bowl after the handle has been pressed. In valves similar to the type shown in Fig. 8, this flow may result from stoppage of the bypass (A-24) or from a deposit of grit on the relief valve seat. If the *bypass* is clogged, water cannot pass into the upper chamber to close the valve. If there is sediment on the relief valve seat or if the seat is badly worn, the valve may not close tightly, allowing the water to escape. The segment diaphragm (A-56) may also deteriorate in time and need to be replaced. The relief valve seat or washer and the segment diaphragm are made of rubber and are usually sold together, as it is generally advisable to replace both at the same time.

To repair a flush valve, it is not necessary to cut off the entire water supply if the water-closet supply alone can be cut off by means of a shut-off valve in the supply pipe near the fixture.

To reach the parts mentioned, unscrew the outside cover A-4 being careful not to mar the nickel finish. The inside cover A-11 can then be lifted out together with the relief valve (A-19) complete.

The bypass (A-24) and the corresponding hole in cover (A-11) can be cleaned by running a fine wire through the openings.

To *replace* the *rubber washer* (relief valve seat) insert a screwdriver under the washer at the hole in the center and pull the washer out. Then, with a spanner wrench, unscrew the disk ring which holds the washer in place and clean the surface of the seat on which the washer rests. Insert a new

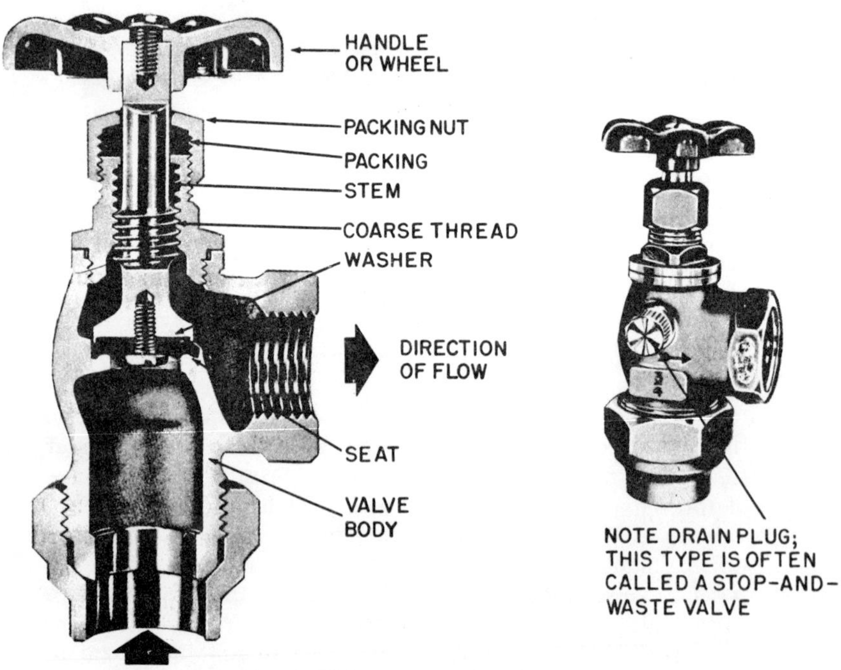

Fig. 9. Globe type angle valve.

washer and replace the disk ring, screwing it down until firm, but not too tight.

When the disk (A-15) is unscrewed, the segment diaphragm (A-56) can be lifted out. In replacing the diaphragm, it should be laid in with the cup down and the copper gasket on the under side. It will be noted that the dowels and tube holding the diaphragm in position are unequally spaced, to prevent it from being placed in the valve upside down. It may require a few trials to find the correct position, but when that position has been determined, the segment diaphragm should be fastened by screwing the disk (A-15) into guide (A-13).

If the above suggestions are not applicable in a particular case, directions for repairing the valve in question may be obtained from the local dealer or from the manufacturer.

Where a vacuum breaker is located between a flush valve and water closet, a leak may develop due to improper functioning of internal parts. If this occurs, do not tape up the air ports, since that will render the device ineffective and may possibly permit back siphonage to take place if a partial vacuum develops in the supply line. In this case, a plumber should be called for maintenance or replacement.

REPAIRING VALVES

Shutoff valves (Fig. 9) commonly used in home water systems are very similar in construction to water faucets. Your valves may differ somewhat in general design from the one shown in Fig. 9, because valves come in a wide variety of styles.

If a shower head drips, the supply valve has not been fully closed or the valve needs repair.

After extended use and several repairs, some valves will no longer give tight shutoff and must be replaced. When this becomes necessary, it may be advisable to upgrade the quality with equipment having better flow characteristics and longer life design and materials. In some cases, ball valves will deliver more water than globe valves. Some globe valves deliver more flow than others for identical pipe sizes. Y-pattern globe valves, in straight runs of pipe, have better flow characteristics than straight stop valves. Figure 10 shows the

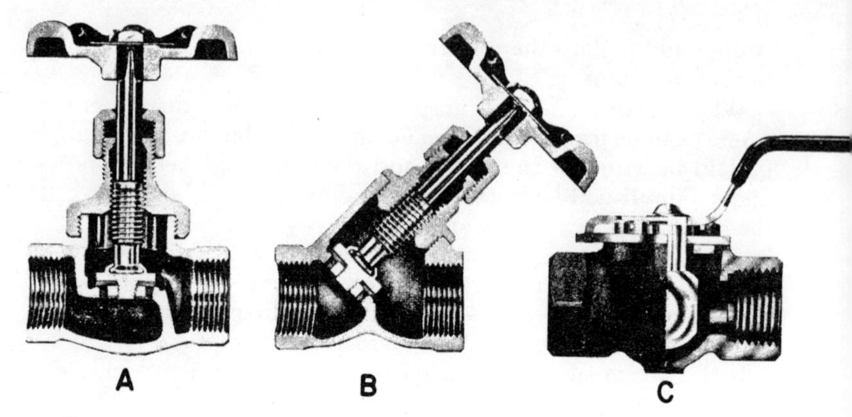

Fig. 10. Various types of valves. A, Glove valve (note large passages of water); B, Y-pattern globe valve (the flow is almost straight); C, ball valve (straight flow).

features of different types of valves. Some valves are available with the port in the ball the same diameter as the pipe.

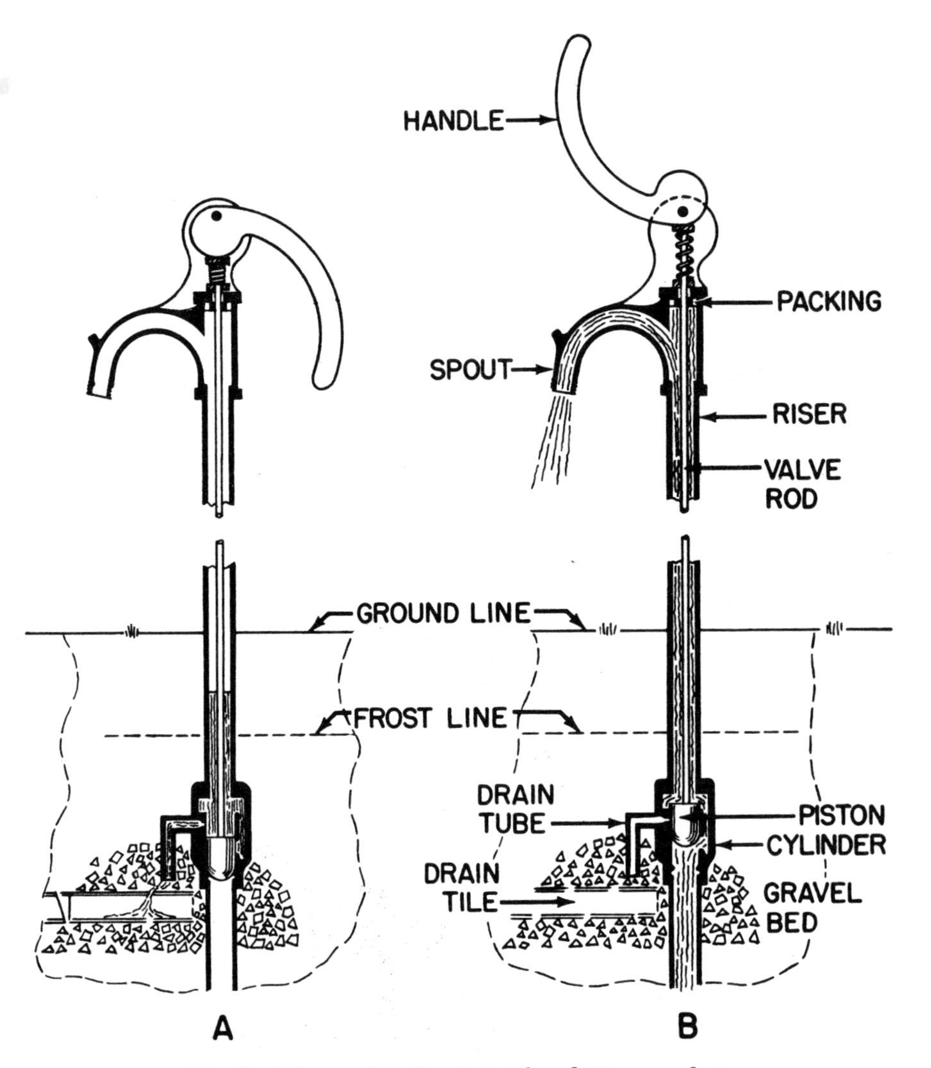

Fig. 11. Frostproof hydrant. A, Closed; B, opened.

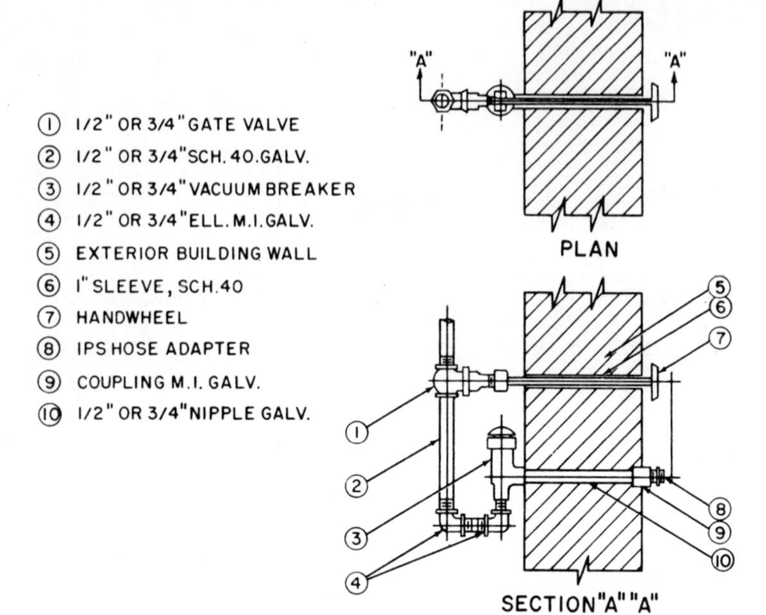

Fig. 12. Vacuum breaker arrangement for outside hose hydrant.

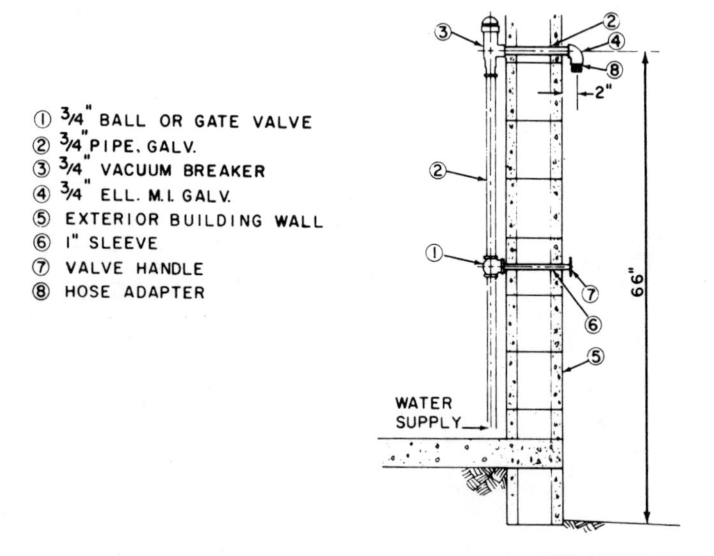

Fig. 13. Protected wall hydrant suitable for filling agricultural sprayers.

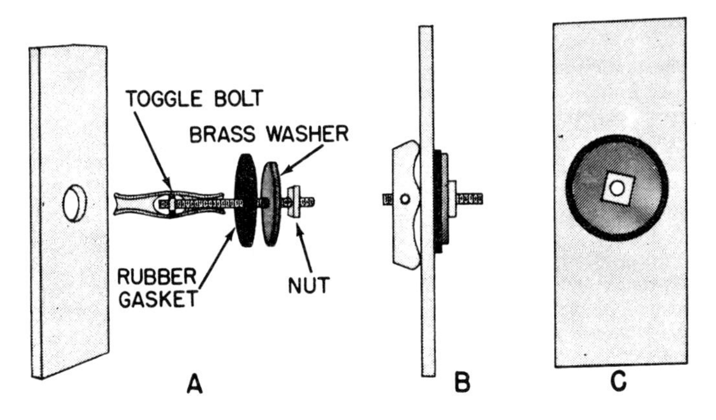

Fig. 14. Closing a hole in a tank.

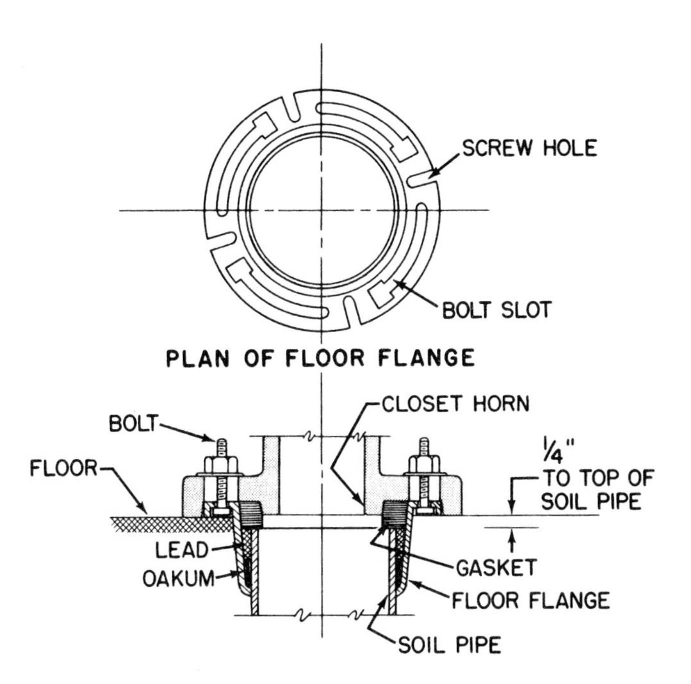

Fig. 15. Connection of water closet to floor and soil pipe.

Chapter 19

Heating and Ventilating Care and Repair

Many kinds of heating and ventilating equipment, including both hand-fired and automatic fuel-burning devices, are in domestic use. In general, these systems give satisfactory service if their capacity is adequate to serve the heat demands of the house, if they have been properly installed, and if they are efficiently maintained and operated.

To be sure that a heating system will work successfully, the homeowner or occupant should procure fuel suitable for his particular equipment and be familiar with its mechanical operation. He or she should not tamper unnecessarily with the controls or mechanism of his installation but, if competent, can make minor adjustments and, sometimes, small repairs. Major repairs, replacements, or seasonal overhauling, should be done by reliable and qualified heating mechanics.

GRAVITY OR PRESSURE HOT-WATER SYSTEM

Ordinarily, it is desirable to leave the water in a steam or hot-water heating system from one year to the next to minimize corrosion, to prevent the introduction of air, and to minimize the accumulation of salts or sediment in the system. Air may accumulate in some radiators of a hot-water system during normal operation. If the accumulation becomes too great, circulation is retarded and inadequate heating of the room or house may result.

Filling a hot-water heating system is easier if two persons are present, one to control the flow of water into the boiler and the other to operate the radiator relief valves.

All radiator shut-off valves should be opened, being sure that the air valve on each radiator is closed. The draw-off cock at the lowest point in the system should be closed, and the valve in the supply pipe which feeds the boiler opened. As soon as the water in the pipes begins to rise, the air valve on each radiator should be opened, beginning at the one nearest the boiler, in order to release the air so that the radiators may fill with water. When water begins to spurt from an air valve, it should be shut and the operation repeated until all radiators become free from air and full of water.

When adding water to a boiler with fire under it, the fire should be low and the water should be allowed to flow in gradually, since a large volume of cold water suddenly injected into a hot boiler might cause it to crack.

Expansion Tanks

To fill a hot-water heating system equipped with an *expansion tank*, the water supply valve should be shut off when the water has risen to such a height that the expansion tank at the top of the system is about one-third full. The expansion tank is usually located on an upstairs closet shelf or in the attic, preferably near the chimney, to protect it from freezing. The height of water in the tank is usually indicated on a water-gage glass attached to it. The water should be kept at the level mentioned to insure complete circulation throughout the system. An overflow pipe attached to the expansion tank and leading to the outside of the house or to a drain should be provided to carry off excess water.

Reducing Valves

In water systems operating under pressure, two automatic valves are usually provided to control the pressure of water in the system. A *reducing valve* is used to admit city water when the pressure in the heating system falls below the normal level, and a relief valve is used to discharge a small amount of water from the system automatically when the pressure becomes higher than normal due to the expansion of the water upon being heated. Occasionally, these valves become slightly corroded and fail to work properly. Under such conditions, they should be dismantled and the moving parts polished with fine emery cloth, then cleaned and reassembled.

Because of the danger that someone may build a fire in a dry boiler, it should be kept filled with water when not in use. Air can be expelled from a steam boiler by raising the water level to the steam outlet. A hot-water system is usually left filled to the expansion tank. Rusting or corrosion is most severe at the water surface so that keeping the boiler completely filled during the summer will prolong its life.

Altitude or Pressure Gage

The purpose of the altitude or dial gage on a hot-water boiler is to indicate the level of the water in the system. On the first filling, the water level is raised to the proper height in the expansion tank and this is checked by inspection. The red hand on the dial gage is then set in the same position as the black hand. Thereafter, proximity of the black hand to the red hand will indicate proper filling of the system.

Steam-heating and Hot-water Boilers

During operation, *steam boilers* should be kept filled with water at least to the center of the water-gage glass or to the level indicated by the manufacturer. The instruction card furnished by the manufacturer should be kept nearby and directions followed carefully.

All accessories of the boiler should be in good working order and regulator parts oiled. A coat of paint applied to the external parts after the boiler is thoroughly cleaned will enhance the appearance and promote durability of the metal. Silicone-aluminum and other suitable paints are available for this purpose.

Boiler-water Treatment

In localities where the water supply is unusually hard or where large amounts of fresh water are introduced into the system, *boiler water* may require *treatment*. Commercial compounds are available for this purpose but should be used with discretion, depending upon the hardness of the water supply.

Fresh water is frequently treated with lime and soda ash (sodium carbonate) to precipitate scale-forming salts, and disodium phosphate or trisodium phosphate may be added to boiler water to produce nonscale-forming precipitates. Commercial water-treating compounds often contain some of these chemicals.

Blowing Down a Boiler

Sometimes an excessive amount of dissolved salts or the presence of oil or organic matter in the boiler water causes foaming or priming of the water—that is, the carrying of small drops of water out of the boiler with the steam. Trouble thus caused can usually be relieved by replacing part of the boiler water with fresh water. Foaming may also be eliminated by blowing the surface water along with the foam from the boiler, while it is steaming, through a special pipe or hose that has been connected to a threaded opening in the boiler at or near the level of the water line. This process should be continued until no visible foam is discharged from the boiler. This work should probably be done by an experienced heating mechanic.

Repairing Boiler Sections

Occasionally, sections of a cast-iron boiler crack from sudden heating or other causes. If this occurs, it is sometimes possible to mend a crack by *brazing* or *welding*, particularly if it does not pass through a machined surface. Brazing is more often used since the welding of cast iron is comparatively difficult. It is usually desirable to employ experienced workmen to repair a cracked boiler.

For *sealing leaks* in boilers, there are many effective compounds on the market similar to those used in automobile radiators, but they should be regarded as temporary expedients. Flaxseed meal has also been used for this purpose.

Repacking a Leaky Radiator Valve

If a *radiator valve leaks around the stem*, it should be promptly attended to, in order to avoid damaging the floor and possibly the ceiling below the radiator. Worn or insufficient packing inside the nut or a loose packing nut at the base of the stem may cause a leak (Fig. 1). To remedy this, the nut should be tightened. If the leak does not stop, it will be necessary to repack the valve.

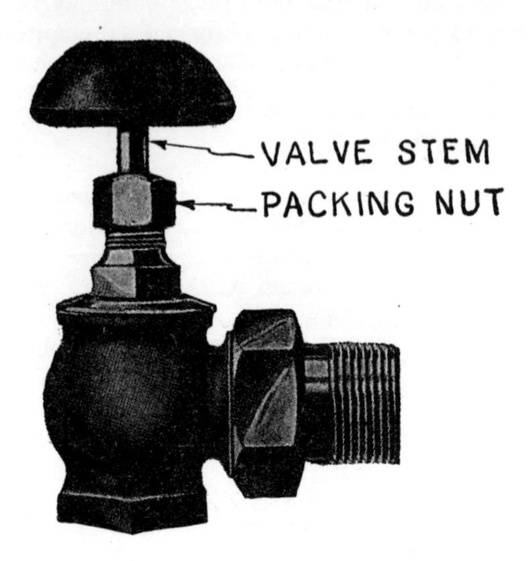

Fig. 1. Typical radiator valve.

Many radiator valves are so constructed that the *packing nut* can be raised without lowering the pressure in the radiator. In a hot-water system where the type of valve used permits water or steam to escape after the valve is closed and the packing nut loosened, the level of the water must be lowered below the height of the valve by opening the drain cock of the system. With a steam system, the pressure should be reduced by allowing the boiler to cool.

Two forms of packing may be used to pack a valve stem—*washers* of different sizes or *packing cord.* If washers are used, the valve handle should be removed by loosening the screw that holds it. After the handle is removed, the packing nut can be withdrawn from the stem. The old packing should be taken out and the new packing washers slipped over the stem. The new washers must be of the right number and size to fill the packing space in the nut. If cord is used, a sufficient amount should be wrapped around the valve stem to fill the packing space in the nut. The nut should be tight enough to prevent water and steam from escaping, but not so tight as to produce excessive friction on the stem when the valve is turned.

For a *steam-heating system,* it is advisable to let the fire go out, or at least to have a low head of steam, before starting work on the valves. After the system is cool enough to work on, the valve should be closed tightly and the packing nut unscrewed at the base of the stem to permit packing the space between the inside of the nut and the stem with metallic packing compound, using a small screwdriver for the purpose. This compound may be purchased in small quantities from a dealer in heating supplies. To do this, remove the handle and lift the nut from the stem. After the nut has been well packed, screw it down tightly and refill the system, as described previously in this chapter.

CARE OF HEATING SYSTEM

A *heating system* should be maintained in good condition not only for operating economy but also to prolong the life of the equipment. Even during the heating season, the interior of the furnace should be inspected occasionally and cleaned if enough soot and dust have accumulated to materially lower the efficiency of the furnace. If the amount of soot collected seems excessive, it may be that the wrong kind of fuel is being used, that the furnace is not adjusted properly, or that the dampers are not set to produce complete combustion.

Furnace and Chimney

At the end of the heating season, all heating surfaces of a furnace should be cleaned thoroughly of soot, ash, and other residue. The smoke pipe should also be cleaned and the chimney, if necessary. In most localities, heating and plumbing concerns clean furnaces with vacuum systems which do not spread dust to other parts of the house. Professional chimney sweeps may be employed to inspect, clean, and repair chimneys.

The smoke pipe of a furnace should be inspected for holes or perforations caused by the corrosive action of flue gases. It should be assembled with sheet metal screws and held in place, if necessary, by wire or other incombustible supports. After cleaning, the smoke pipe should be placed in a dry place until the furnace is to be used again. Unless properly cared for, it may be necessary to replace a smoke pipe every two or three years.

The inside of the furnace should be cleaned through the clean-out door with a wire brush and scrapper. After cleaning inside and out, the heating surfaces of steel boilers should be given a coat of lubricating oil on the fire side to prevent rust, provided the furnace is not to be used during the summer. If desired, the oil may be applied by spraying. All machined parts should be coated with oil or grease. Broken or defective parts should be replaced and all loose joints made tight. Furnace door hinges should be inspected, and warped or broken grates replaced. The grates and ashpit should be cleaned.

Clogged Grate

If a furnace burns coal, the fire should be shaken down at least twice a day to remove the ashes and to afford a better draft through the grates. The ashpit should be cleaned out daily, because if it is permitted to fill to a level where ashes touch the grate bars, there is danger of warping or burning out the grates.

In most furnaces and stoves, there are clean-out doors just above the grate surface through which a poker may be inserted to break up and pull out clinkers. If unusually large or hard clinkers lodge between the grate bars so that it is impossible to shake the grate or turn it over, the fire should be allowed to die out and the fire pot cleaned to give free access to the grate and permit removal of the obstruction. When the fire pot has been emptied, the poker can be used to better advantage and the obstruction dislodged or broken up. The grate should not be turned forcibly, as this might break it.

Soot Removers

Soot removers are intended to free heating surfaces of the furnace or boiler, the smoke pipe, and the chimney from soot. Certain metallic chlorides, which can be vaporized by a hot fire and deposited on the surface of soot to lower its ignition temperature, may be used for this purpose. These chlorides are copper chloride, lead chloride, tin chloride, zinc chloride,

chloride of lime, and sodium chloride (table salt), of which copper chloride is the most effective and sodium chloride the least.

Finely divided metals, such as lead, copper, or tin may also be used to reduce the amount of soot, but they are not as effective as the chlorides of the metals.

Some commercial soot removers contain a certain percentage of the metallic chlorides mentioned previously and will assist in the reduction of soot in the heating plant, but will seldom have much effect in the smoke pipe or chimney.

The burning of soot in the furnace, smoke pipe, and chimney should preferably be done on a rainy day by someone familiar with such work. The removal of soot requires such a high temperature that unless the soot fire is carefully handled, it can easily get out of control. Combustible materials in the attic and shingles on the roof should be carefully watched for an hour or two after application of the soot-burning treatment to observe any indications of fire and to be in position to take whatever action may be necessary, in case fire should occur.

Furnace Water Pipe

The *water pipe* frequently found in the combustion chamber of coal furnaces for heating domestic hot water may need replacement at intervals of a few years. These pipes may become burned on the outside from overheating or clogged on the inside with scale precipitated from the water as it is heated. They may be renewed by draining the domestic hot-water system and replacing the original coil with lengths of standard pipe, bent or coiled to fit.

Repairs to Chimneys

Most cities require that masonry chimneys with four-inch (10.16-cm.) wall thickness be lined with a fire-clay lining, but unlined chimneys may be used if the walls are eight inches (20.32 cm.) thick or more. The chimney should be examined to determine the soundness of mortar joints. Sometimes mortar crumbles, leaving openings in the joints and, unless these openings are repaired, they become a dangerous fire hazard, especially in cold weather when drafts are strong and the furnace is being fired to capacity. Joints, in which mortar has crumbled, should be raked out to solid mortar before repointing. If the mortar has deteriorated too much, the chimney should be taken down and rebuilt.

Flue Openings

Only *flue openings* that are in use should be left open—no more than one opening per floor should be used. It is common practice, however, to connect gas and oil hot-water heaters to the same flue that serves the house-heating equipment even when they are on the same floor. Hot-water heaters should be connected to the flue from 18 to 24 inches (45.72 to 60.96 cm.) above the connection for the heating plant. If the smoke pipe extends into the flue space, it should be cut back flush with the flue wall. If the smoke pipe enters the chimney on a descending slope, its position should be changed to enter at an ascending slope.

INSULATING HEATING SYSTEMS

Insulation on the *boiler* and *pipes* of a heating system will increase the efficiency of the system and reduce the cost of operation. Heat loss from a furnace or piping that is not properly insulated is not entirely lost as it warms the basement and first floor and, in some cases, may contribute a large share of the heat required in a house. However, in summer, if a boiler is operated to provide domestic hot water, such heat losses serve no useful purpose and are, in fact, undesirable and uneconomical, making insulation advantageous.

The small amount of heat that escapes through the insulating material on hot-water pipe, combined with radiation from doors or other exposed parts of the boiler, will warm the basement slightly, if it is well constructed and properly weatherproofed, but ordinarily not enough to provide a comfortable temperature in cold weather.

The extent of insulation is usually governed by the type of heating system and the conditions that exist. The boiler and pipes through which water or steam is distributed to radiators should be covered, but the advisability of covering the return pipes depends upon the type of system and the amount of heat desired in the basement.

In hot-water heating systems, covering the return pipes is recommended in order that water may be returned to the boiler with the minimum of heat loss. In vacuum and vapor steam-heating systems, it is considered better to leave pipes bare, to aid in the condensation of any steam which might escape into the returns through defective thermostatic traps.

Properties of Insulating Materials

Insulating materials for covering a furnace or hot-water distribution piping should be fire-resistant and poor conductors of heat. Materials used for the purpose are corrugated or air-cell asbestos, 85 percent magnesite asbestos, and mineral wool and similar coverings. The insulation may be obtained to preformed shapes, as flexible rolls, or as a dry powder to be mixed with water to form a plastic cement coating which dries in place.

A description of how to apply some of these materials follows. An *air-cell pipe covering* is made of layers of corrugated asbestos sheet wrapped in canvas. It is manufactured in sections three feet long in the shape of hollow cylinders split lengthwise on one side so that each section may readily be placed around the pipe. The covering is made in several thicknesses and for various sizes of pipe. Each section has a canvas lap to be pasted over the longitudinal joint and a canvas flap at one end to be pasted over the joint between sections. To further bind the covering and make it neat-looking, metal bands are placed about 18 inches apart, over the joints between sections and around the middle of each section.

For insulating boilers and pipe fittings, such as valves, Ls and Ts, where the use of fabricated coverings is not practicable, *asbestos cement* or *other refractory insulating cements* may be used. They serve the same purpose as the pipe coverings because of their insulating and fire-resistant qualities.

The following tools and materials are needed: Steel tape measure, plasterer's trowel, handsaw, sharp pocket knife, pliers, metal tub or similar container for mixing cement, pan or dish for paste, and small, flat paint brush; asbestos air-cell covering for insulating pipes, asbestos cement, wire mesh, canvas, and paste.

The asbestos air-cell covering should be four-ply or one inch thick. To estimate the quantity needed, the pipe between fittings to be covered should be measured and the measurements combined for each size of pipe to obtain the total linear feet for each size of covering. Metal band fasteners are furnished with the covering.

To estimate the quantity of asbestos cement needed to cover the boiler and the pipe fittings, the entire surface of both the boiler and fittings should be computed. A 100-pound bag of cement will cover from 20 to 25 square feet of surface to a thickness of one inch. Some brands of cement are also available in 10-pound bags.

Sufficient one-inch wire mesh or "chicken wire," to cover the surface of the boiler is necessary; the canvas should be of the same weight as that on the air-cell covering and large enough to enclose the cement covering around pipe fittings; the paste is the type sold by manufacturers of the covering material for pasting the canvas laps on air-cell coverings and for fastening canvas jackets over pipe fittings.

The paste may be purchased ready-to-mix to which cold water is added. If preferred, a paste may be made by mixing one part of powdered alum and 50 parts of sifted white flour in a small amount of cold water until smooth and then adding hot water gradually to the mixture; then it should be boiled until it reaches a paste-like consistency.

Asbestos Cement

Asbestos cement should be mixed thoroughly in a tub or large container, using only enough water to make the mixture workable. At least two coats should be applied to the boiler and pipe fittings and this should be done at a time when the pipes are warm to insure best results. The first coat on the boiler should be one-inch thick and the second coat one-half inch thick; each coat on the pipe fittings should be one-half inch thick. In all cases, the first coat should be applied roughly with the hands or with a plasterer's trowel, scratching the surface to insure a good bond with the second coat.

Boiler Insulation

When the first coat of asbestos cement on the boiler is fairly dry and before applying the second coat, the wire netting should be stretched and fastened over the surface to hold the first coat. This will probably cause some cracking, but the wire will serve as reinforcement for both coats.

Insulating Heating Pipes

Before hot-water and steam-heating pipes are covered, they should be clean and in good condition. The canvas lap on the pipe covering should be located and brushed along the edge with paste to refasten it. The pipe should then be encased with a section of covering placed with the open side up and with the end which has no canvas-joint overlap placed tight against the fitting and pressed closely together, pasting the lap securely over the longitudinal. The second section should be applied in the same manner, and pushed tightly against the first. The joint between the two sections should be sealed by pasting the overlap attached to the first section over the joint. The pipe covering should be continued in this way until the next fitting is reached. When a short section is needed, the covering can be cut with a sharp knife and handsaw.

For the fittings, the first coat of asbestos cement should be applied with the hands to a thickness of about one-half inch. The second coat on the boiler should be troweled smooth as it dries.

The next step is to finish covering the fittings by the application of a half-inch second coat or one of the same thickness as the pipe covering. This coat should be troweled smooth and beveled down to meet the surface of the pipe covering.

Asbestos cement on the fittings will be protected and will present a neat appearance if covered with a canvas jacket. The canvas should be of the same weight as that used on the pipe covering, and should be pasted down smoothly. To look well and preserve the canvas, it should be sized and painted with two coats of lead and oil in a suitable color.

Metal bands to hold the pipe covering should be applied after all other work is completed, so they will be clean and present a neat finished appearance. The bands should be placed about 18 inches apart, over the joints and midway between. They should be pulled up tightly and fastened with the pliers.

Warm-air Furnaces and Ducts

A warm-air furnace and its ducts may be insulated to prevent the basement or utility room from getting too warm. Should a warm basement be desired, the supply ducts may be left uninsulated. In any case, it is not necessary to cover the return ducts.

The following tools and materials are needed: Steel tape measure, pencil, heavy shears or sharp knife, wire-cutting pliers, small trowel, and container for mixing asbestos cement; corrugated asbestos pipe covering, black or galvanized wire, and asbestos cement.

The quantity of corrugated asbestos paper should be sufficient to provide a three-layer covering for all pipes and the furnace, if not already insulated. The material is sold in rolls about 37 inches wide, containing about 250 square feet of material. For the average seven- or eight-room house two rolls are sufficient.

The wire should be No. 16 or No. 18 gage, black or galvanized. One small roll will be needed to hold the covering in place. A 100-pound bag of asbestos cement should be sufficient to cover the sloping shoulder of the furnace—about 20 to 25 square feet of surface to a thickness of one inch. Some brands of asbestos cement are also sold in 10-pound bags.

Insulating Materials for Warm-air Systems

Insulating materials suitable for use on warm-air heating systems may be corrugated-asbestos paper or similar insulators designed to form a tight-fitting jacket and asbestos insulating cement to cover the irregular surfaces of the furnace.

The corrugated-asbestos paper should be wrapped around the ducts and the cylindrical surface of the furnace and laid on top of it. Asbestos insulating cement may be used to cover the sloping shoulder of the furnace.

Application to Straight Ducts

Before ducts are covered, they should be clean and in good condition. To get the measurement of the covering, add an extra 1½ inches to the distance around the duct to allow for thickness of the material and cut a strip of that length from the roll, wrapping it around the duct and tying it in place with a cord or wire in the middle and at each end. After this, measure around the outside of the covered duct and add 1½ inches to get the length of the second piece. Tie this piece around the duct over the first piece, staggering the end and longitudinal joints, and proceed in the same way to apply the third or top layer. Finally, the three layers should be bound together by fastening bands of wire around the covering at intervals of about 18 inches.

Bends in Ducts

It is not absolutely necessary to cover the bends in ducts because they form but a small part of the system, but it is not a difficult task and will add to the efficiency and appearance of the covering. A 90-degree bend will require two or three pieces cut especially to fit, depending upon the number of separate sections in the bend. These pieces should be diamond-shaped to conform to the surface to be covered. The dimensions may be obtained by measuring the widest and the narrowest parts of the bend (Fig. 2).

Furnace Insulation

To cover the vertical surface of the furnace, wrap from one to three layers of corrugated-asbestos paper covering around it, and fasten with bands of tie or stove wire. The ends of the wires may be attached to bolts or other projections at or

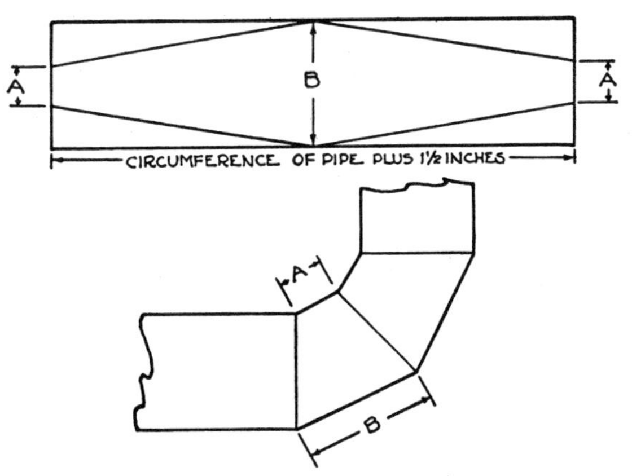

Fig. 2. Pattern for cutting pipe-bend covering.

near the furnace doors. Obviously, holes for doors and other attachments should be cut in the covering.

The top of the furnace may have a hollow shape and be filled with sand. The furnace top can also be insulated by laying three or four layers of corrugated asbestos paper over it. These pieces may be held in place by extending the asbestos-cement shoulder coating up over the edges of the covering.

The sloping shoulder of the furnace from which the ducts lead can be insulated with a covering of asbestos cement to a thickness of one-half to one inch. The cement, which comes in powdered form, should be mixed with just enough water to make it workable, applied to the furnace, and troweled smooth.

INSULATING HOT-WATER TANKS

Hot-water tanks heated by pipe coils in a coal furnace or by a waterback in a coal cooking range should not be insulated because such tanks are more likely to become overheated than uninsulated tanks.

Domestic hot-water heaters are usually insulated, thicker insulation being used on electric rather than on gas or oil heaters because electricity is ordinarily more expensive than gas or oil. If a domestic hot-water tank is not well insulated, it may cause discomfort in summer because of heat radiation.

Tank covers may be bought ready-made to fit tanks of standard sizes. They are made of incombustible heat-insulating materials similar to those used for pipe and boiler coverings. Some have the appearance of a large section of pipe covering, split lengthwise on one side, so that they may be readily wrapped around the tank. If connection pipes are in the way, openings for them can easily be cut in the covering at the joint edge. Metal bands, laces, or other forms of fasteners are furnished to hold the covering together. The top of the tank may be covered with asbestos insulating cement.

EQUIPMENT FOR WARM-AIR SYSTEMS

Filters

Filters are often used in forced warm-air heating systems to insure cleaner air in the living space. They require occasional cleaning, however, or replacement if they are the throw-away type. Should the filters become overloaded with dust and dirt, the amount of air circulation will be diminished, the air overheated and, in severe cases, the efficiency of the heating plant will be reduced.

The frequency of cleaning or replacement of filters will depend upon the amount of dust which accumulates. Permanent-type filters may be cleaned by tapping the filter frame to shake the dust out or by washing the filter medium with soap and water or cleaning fluid. Throw-away type filters are not intended to be cleaned and are made of inexpensive materials so that they may be replaced at nominal cost.

In *gravity warm-air heating systems* where air filters cannot be used or in forced warm-air systems that have no filters, dust may accumulate on the heating surfaces of the furnace, and in the warm-air supply ducts. In such cases, the accumulated dust on the heating surfaces may give off an unpleasant odor when the furnace is being fired to capacity. If this odor becomes objectionable, the furnace and piping should be cleaned by commercial vacuum cleaning methods. If such services are not available, it may be necessary to dismantle the furnace and piping in order to properly clean them. Where possible, an air filter should be installed to avoid recurrence of this trouble.

Humidifier Pans

Manually and automatically supplied *humidifier pans* are used in warm-air heating systems. In the automatic type, the mechanism for regulating the water flow to the pan sometimes fails to work properly because of the continuous evaporation of water. Precipitates, scale, or solids may form on parts of the mechanism, eventually preventing free movement of the levers and pins. Thorough cleaning of all parts should restore normal operation. The valve that controls the flow of water into the pan may also become worn after long usage and require replacement. If excessive dust, dirt, or scale have accumulated in a humidifier pan it should be cleaned to avoid unpleasant odors.

Grilles

Floor grilles or *registers* may require frequent cleaning, since dust and small objects can easily fall into a register. This cleaning can usually be done from above the floor with a vacuum cleaner. If large objects have become lodged in the warm-air supply pipe, however, it may be necessary to dismantle the pipe in order to remove the objects.

Wall grilles or *registers* are not directly subjected to dust or other accumulation as are floor grilles, but may require dismantling for cleaning if they become clogged.

HEATING EQUIPMENT

Hand-fired heating equipment seldom needs adjustment other than to keep the system in good working order. *Automatic heating equipment*, such as hot-water heaters, coal stokers, oil burners, and gas-fired furnaces have more complicated mechanisms and sometimes need adjustment or repair that the householder is unable to make. If trouble occurs, the adjustments and repairs should preferably be made by the contractor or dealer who installed the equipment or by the utility company that services it.

Electric Motors for Automatic Systems

Electric motors used for driving blowers, pumps, or oil burners should be inspected and oiled at least once every heating season. Some single-phase motors, which are the type generally provided, use a pair of brushes in the starting mechanism—they may wear out and require replacement after several years' operation. Motors that are not totally enclosed often accumulate dust inside the casing which, in time, may interfere with the operation of the starting mechanism. Sometimes they can be cleaned sufficiently by forcing a jet of air through the casing, but often the motor has to be disassembled and washed with a solvent, such as carbon tetrachloride, to remove the accumulated dust and grease on the interior.

Blowers also require seasonal lubrication and may require cleaning where filters are not used in the system. The belts connecting blowers to motors may wear out in a few years' time and need replacement.

Coal Furnaces or Boilers

Too much fly ash in the flue gas passages and smoke pipe may cause smoking and improper heating. After several years' operation, it is not unusual to burn holes in the heating surface of a warm-air furnace allowing smoke or flue gas to mix with the warmed air which is circulated throughout the house.

If the water level is allowed to become low in a steam or hot-water boiler, the heating surface of the boiler directly over the fire may be burned or warped, causing a serious leakage.

Ashes should not be permitted to accumulate in the ashpit to a depth where they can touch the bottom of the grate, as by so doing the passage of air will be blocked. This may cause the grate to overheat and become warped or burned.

With certain kinds of coal, clinkers may form in the firebox and, if not removed, can appreciably reduce the amount of combustion air passing through the grate so that the heating plant will not properly heat the house.

Formation of clinkers is caused by excessive stirring of the fire in such a way that the ashes become mixed with the active fuel. Care should be exercised in removing clinkers so as not to damage the grates. Some furnaces have clinker doors just above the grate level, but in others, clinkers have to be removed through the firing door.

Automatic Coal Stoker

Stokers are intended to feed coal automatically into a furnace or boiler. The most common residential stoker is the underfeed type where a coal-feed screw, driven by an electric motor, supplies fresh coal from either a hopper or a storage bin into the firepot of the furnace. Air necessary for combustion is forced by a motor-driven fan through openings in the firepot.

A shear pin is usually provided in the shaft of the coal-fired screw to protect the other parts of the mechanism in case the feed screw becomes jammed with large pieces of coal or other solid material which may be in the coal. If this occurs, the obstruction should be removed and the shear pin replaced.

Sometimes the "hold-fire" control, the purpose of which is to maintain the fire in the firepot, whether or not heat is required, feeds too little coal allowing the fire to go out, or feeds too much coal causing the house to be overheated in mild weather. Adjustments to correct this condition should be made by someone experienced in such matters.

While an anthracite stoker is in operation, the ashes are being pushed to the outside edge of the fuel bed to fall into the ashpit for removal. If bituminous coal is used in the stoker, however, clinkers which form may be removed through the furnace door.

When the heating season is over, coal, ash, and clinkers should be removed from the system and the stoker cleaned. The coal-feed screw and inside surfaces of the hopper should be coated with oil to prevent rust. Before the system is again put in operation, the stoker should be inspected, repaired if necessary, and adjusted by a competent serviceman.

Automatic Oil-burning Equipment

For mechanical service during the heating season, firms that supply oil for domestic burners have a yearly basis plan for repair and replacement of worn burner parts by which they agree to respond to emergency service calls, inspect boiler and burner, vacuum clean boiler, adjust controls, clean strainers, lubricate burner, analyze flue gases, adjust burner flame, and inspect burner for leaks.

Gun- and Rotary-type Oil Burners

If the mechanical service is not available, the homeowner will find that the usual sources of trouble with gun- and rotary-type burners are clogged strainers, nozzles, and fuel lines, and improperly located ignition electrodes. Most oil burners have a strainer or filter attachment in the oil line at the inlet to the burner, the filter element or strainer of which is readily removed by taking off the housing or cover plate after the oil supply line has been closed. Use of a wire brush or jet of compressed air often proves the most effective means of removing the foreign matter from the strainer. Filters are not often cleanable and usually require replacement of the filter element or strainer. The nozzle and electrodes in most gun-type burners can be removed as one assembly through the rear of the burner. The nozzle can be removed by the use of suitable wrenches and should be carefully cleaned and reassembled. The electrode spacing is often about one-fourth inch, but is not the same for all burners and the manufacturer or his representative should be consulted for exact information. The electrodes should be near the oil spray, but far enough forward and above or below the nozzle so that the spray will not strike them.

Pot-type Oil Burners

To avoid a smoky, sooty flame in a vaporizing *pot-type oil burner*, the proportion of fuel and air should be properly adjusted. In the case of improper draft or wrong adjustment of the combustion air, the burner may need cleaning more frequently than once a season, and sometimes as often as every few weeks. A chimney of at least 15 or 20 feet in height is usually required to produce enough natural draft to operate a pot-type burner. An automatic draft regulator, placed between the oil burner and the chimney, is often used to maintain a steady draft at the proper level. Draft regulators will not function, however, if the chimney is not of sufficient height, in which case a small forced-draft fan can be used. The pot type burner is usually cleaned by hand through the inspection door of the heater.

Soot in Oil Burners

Commercial soot removers, as previously described, may assist in removing the soot from the burner and combustion chamber, but ordinarily they will not remove the hard carbon that forms on the bottom of the firepot. The oil-feed pipe between the float valve and the burner sometimes becomes stopped with carbon and must be cleaned by forcing a rod through it. In any case where a soot remover is used to burn the soot from a heater that has a smoke pipe, the chimney should be carefully watched for an hour or so because a soot fire may develop in the chimney which might ignite combustible materials adjacent to the chimney.

Gas Burners and Automatic Gas Furnaces

The most frequent difficulties with *gas burners* are the sticking of the plunger of the main gas valve, the accumulation of gum or other foreign matter in the pressure regulator, and the extinguishing of the pilot light. Repairs to the gas valve and regulator should be made only by the utility company's representative. A pilot light may be relighted by the homeowner after ample opportunity has been given for the combustion chamber to be aired out and after making certain that the main gas valve of the appliance has been closed. In case the pilot light becomes extinguished for an unknown reason, the furnace operation should be watched carefully for a time after relighting it to determine whether further maintenance or repair may be necessary.

Another source of trouble is improper adjustment of the primary air nozzle. If too much air is supplied, the flame will burn above the burner ports and not be in contact with the burner ports as it should be. If, on the other hand, too little air is supplied, the flame tips may become yellowish. Adjustments of the primary air shutters should be made by a representative of the utility company.

Pilot lights on gas-fired furnaces are sometimes left lighted during the summer to prevent condensation and rusting inside the furnace.

Gas Heaters

While *unvented gas heaters* are permitted for some installations, and those approved by the American Gas Association are not expected to produce enough carbon monoxide to be hazardous, it is recommended that unvented gas heaters not be kept lighted in bedrooms during the night or when the occupants are sleeping. Asphyxiation can occur without arousing the sleeper.

For repairs to the piping, radiators, and duct work, a plumber or heating mechanic should be called.

Electrical Heating Equipment

In most localities, heating the entire house by electricity is too expensive to be generally used. *Portable electric heaters*, however, are a convenient source of heat for bathrooms or other areas where occasional heat is needed.

Electric heaters require little maintenance except for the replacement of burned-out heating elements or defective switches. In some cases, switches may be repaired by polishing the contacts that have become pitted or corroded by repeated arcing. Portable electric heaters should be supplied with extension cords of ample capacity, having adequate protection for the wire. *Portable electric steam radiators* should be checked occasionally to determine whether they contain sufficient water to cover the electric heating element. *Radiant electric heaters* should not be located too near furniture or drapes as the radiation may overheat and ignite combustible materials. Laundry should not be dried over radiant heaters because of the fire hazard involved.

STOVES

Cracks in a Stove

A crack in the iron casing of a stove can be repaired by filling the crack with stove putty or commercial iron-repair cement made of iron filings and water glass (silicate of soda). Enough of the filings should be used to form a thick paste. The paste should be forced well into the crack with the aid of a small trowel or putty knife, and the surface of the crack plastered over with the same material. Heat from the stove will harden the cement and make a tight joint.

Another iron mender can be made of iron filings, flowers of sulphur, and water, mixed to a stiff paste and applied to the crack in the same manner as previously described. The mixture burns when heated and turns into iron sulphide, which fuses and welds into one mass with the iron of the stove.

Summer Storage

Stoves for heating are usually stored during the summer months. Before being placed in storage, however, they should be cleaned and polished and, if possible, wrapped with newspaper, burlap, or old carpet to protect them from dust and rust. They should then be stored in a dry place.

It is well to examine the grates and lining as soon as the stove is taken down and to have any needed repairs made at that time. If parts are found to be defective, an order should be placed promptly for replacements. The make and number of the stove is usually marked on the part which needs replacing and the identification should be given to the hardware or heating-equipment dealer when the order for the new parts is placed.

Stovepipe

Stovepipes need frequent cleaning, especially if the draft is poor. Soot collects in the pipe, particularly if soft coal is burned. Before taking down the pipe, it is well to cover the floor beneath and around the stove with newspapers or a drop cloth to protect the floor covering. The pipe should then be taken out of doors and away from the house before cleaning it of soot. Care should be taken, when handling the pipe, not to pound, dent, or bend the ends so as to make it difficult to fit them together again.

Stovepipe is usually made of sheet iron and should be kept polished to prevent rusting. When being put away for the summer, each length should be wrapped in paper and stored in a dry place.

Chapter 20

Piping Installation, Repair, and Maintenance

Installation, repair and maintenance work, especially if urgent, frequently requires shutting down equipment. Therefore the job must be planned and performed with greatest efficiency. There are many things you can do—to help yourself and the job. Following are a few suggestions—a sort of checklist you can use to advantage on every job.

1. Find out exactly what is wrong or what is to be done.
2. Arrange for the best time to do the job and to shut down any equipment if necessary.
3. Make a list of all the materials and tools required.
4. Determine how much of the work is to be done in your workshop, and what must be done on the job. Carefully planning here of all fabrication and assembly work that can be done at your workbench will save time and effort.
5. See if you can do any work such as installing pipe supports before shutting down the equipment.
6. Know exactly how you will proceed once you start the job.
7. When the job is completed, check up on all details to be sure they have been properly carried out.
8. Test the work if possible. Re-open any lines shut off to do the job.
9. Be a good housekeeper. Return all materials and tools to their respective places. You will need them again.
10. Making a good job is largely a matter of learning good practices and making them a habit. No book or no course of instruction can give you all the piping pointers there are. But, making the best use of material in this book and studying it and keeping it handy for reference—plus your own resourcefulness and ingenuity, will surely lead to success in your work.

SAFETY RULES

There are certain general rules designed to meet specific working directions if you are to be a safe worker, and should always be observed. A few are given here.

1. Before breaking into a line, shut it off and drain. Failure to do this may result in a shower of water, but it might have been steam.

2. Always work from above a line when possible.

3. When loosening a flanged joint, start with the bolts farthest from you. In case of drip or spray, it will not be likely to leak on you.

4. Look out for hot lines. Just touching hot temperature piping may mean a bad burn, and the shock may cause dropping of tools, leading to other accidents.

5. Always use chain blocks, cable slings, rope falls, or rope slings when raising or lowering a piece of heavy piping. Make sure hoists are strong enough and that hitches will not slip.

6. Never leave a job without blanking off an open line. Someone may unknowingly turn it on and do damage. Best practice is to chain and lock shut-off valves and to hang signs on lines that are being repaired.

7. Tightening flange bolts with the line under pressure is not good practice. It may cause excessive strain resulting in rupture.

8. Always see that overhead piping is well supported with approved hangers—not makeshifts. Safest policy is to use hangers every 10 feet. Heavy valves may need additional support.

9. Always put valves where they can be operated easily without reaching or using ladders. Where that cannot be done, install chain wheels or other means of remote control.

10. Keep your work bench clean of "odds and ends." Wipe up oil spots on the floor to avoid slipping. Use safety goggles on chisel or grinding work.

PIPING INGENUITY

A few examples of *piping ingenuity* follow. But, they are just a start toward what can be done when the need arises. Keep them in mind—learn others and work out more. They will come in handy when you least expect it.

When You Are Short a Globe Valve

When a *globe valve* is not available for the place it is wanted, perhaps an angle valve will serve as well, and at the same time eliminate two joints and an elbow. Such installations are generally satisfactory because usually there is a turn in the line close to where the valve is to be located. (*See* Figs. 1 and 2.)

Bushings

When you cannot get the exact fitting you want, use the nearest size that is available, straight or reducing, and bring it down to size with a *bushing* (Fig. 3).

Emergency Welded Joints with Screwed Fittings

Before socket-weld fittings were made, it was common practice to use steel screwed fittings and seal-weld them around the threads. While this is not considered best practice now, in an emergency when socket-weld fittings are not on hand, *seal-welding with steel screwed fittings* will enable completion of a job (Fig. 4).

Fig. 1. Globe valve.

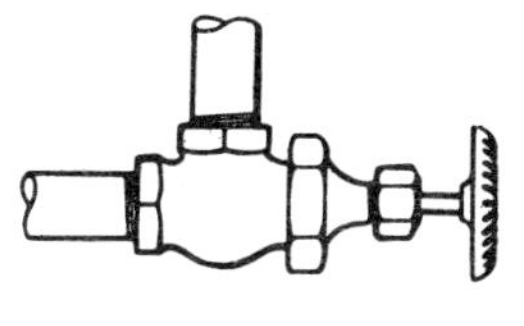
Fig. 2. Angle valve.

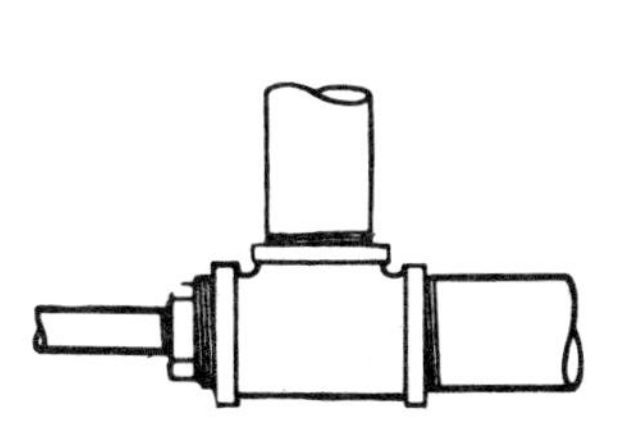
Fig. 3. Bushings.

Fig. 4. Emergency welded joints with screwed fittings.

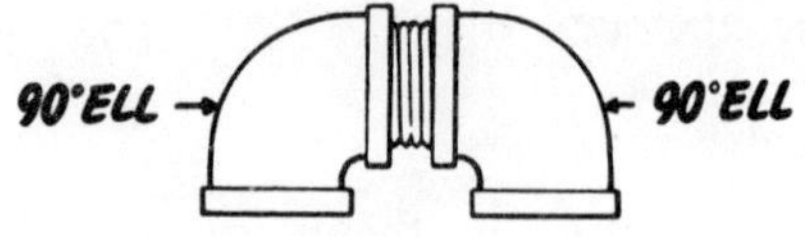

Fig. 5. Use of two ells and a nipple.

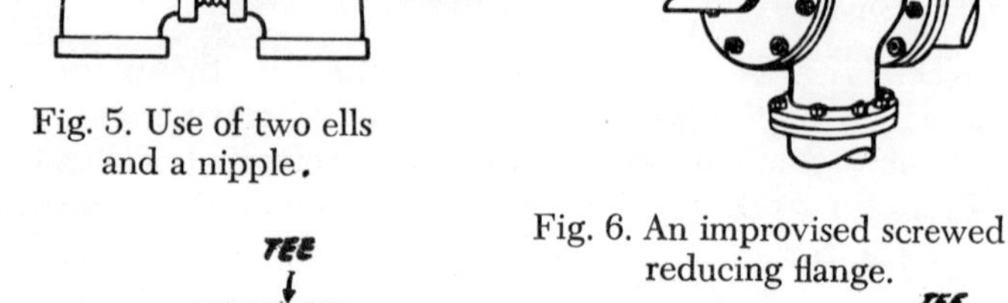

Fig. 6. An improvised screwed reducing flange.

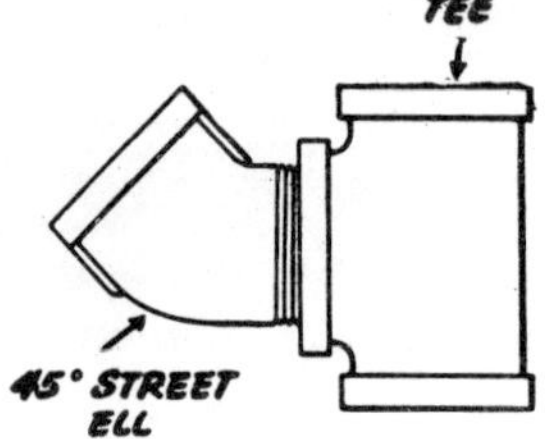

Fig. 7. An emergency Y branch.

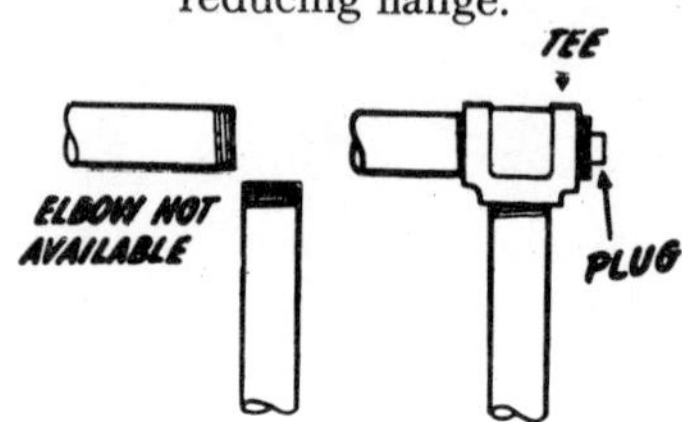

Fig. 8. Use of tees.

Return Bend

When you discover that you do not have the *return bend* that is needed for an emergency in installation, join two ells and a nipple to make the good substitute shown in Fig. 5.

An Improved Screwed Reducing Flange

A *blind flange* can be tapped—either concentric or eccentric as needed—to make a *reducing flange* (Fig. 6). This is practical as long as the tapped opening is not too large in pipe size to prevent sufficient thread length for a safe joint. Another way is to use a bushing in a straight flange.

An Emergency Y-branch

When you find yourself short of a Y-*branch fitting*, think of the following substitute. A tee and a 45-degree street ell, joined as shown in Fig. 7, will serve the same purpose.

A Tee for an Elbow

Use *tees or crosses* with plugs to make 90-degree turns in a line when you do not have elbows to complete a rush job (Fig. 8). Either will work satisfactorily. In lines requiring frequent cleaning, tees or crosses at turns make rodding out easier by simply removing the plugs.

Two Tees Make a "Cross"

Suppose, in an emergency, you needed a *cross of a given type and size* which was not available—do not give up. You may have two tees with which to make an offset cross (Fig. 9). True, it may not look as well as a cross fitting and makes extra joints, but it serves the purpose and may save time.

RANGE OF MATERIALS

Knowing the range of materials from which valves, fittings, and pipe are usually made is a must for piping. Each material has its limitations of pressure and temperature, and to use it for services beyond the recommended maximum may be highly unsafe.

The piping equipment commonly used in plumbing and heating systems falls into three basic material groups: (1) brass, (2) iron (cast, malleable, ferrosteel), and (3) steel (cast, forged, alloy). In each of these groups are one or more variants, each having individual service characteristics.

When and where to use each type of material is best determined by following the markets on the equipment, or by referring to the manufacturer's service recommendations. As a general guide, however, the following rules should be applied without exception.

(1) Do not use *brass* for temperatures exceeding 550°F.

(2) Do not use *iron* for temperatures exceeding 500°F.

(3) Use *steel* for all services above 550°F. and for all lower temperatures where working conditions, either internal, such as extremely high pressure, or external, such as shock or vibration, may be too severe for brass or iron.

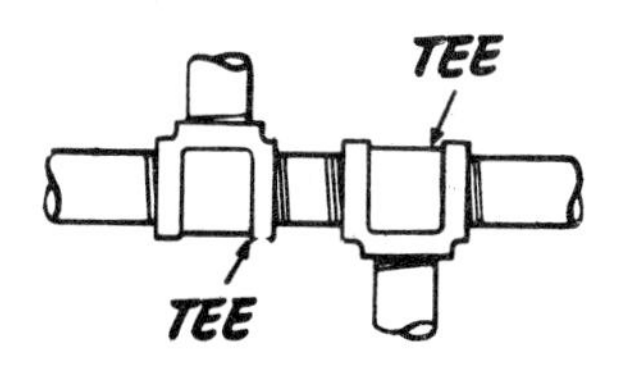

Fig. 9. Making an offset cross.

Typical Brass Globe Valve

Fig. 10.

Standard Iron Gate Valve

Fig. 11.

150-pound Steel Gate Valve

Fig. 12.

SERVICE RATING MARKS

In addition to the maker's brand and size mark, there appears on most valves and many fittings a *basic service rating.* Pressure and temperature ratings are always expressed in steam ratings unless otherwise indicated. Steam ratings are used as a basis since temperature is the factor which determines suitability of a material for a given application. Where no temperature is present, the safety factor of a material in terms of pressure is much greater than indicated by the steam rating.

For example, the brass valve shown in Fig. 10 is rated at 125 pounds saturated steam pressure at its equivalent temperature of 344°F. But for services with no temperature or very little temperature, actually this valve is suitable for pressures up to 200 pounds.

The Meaning of "Wog"

Some general purpose valves may show two service ratings. For example, the iron body gate valve shown in Fig. 11 is marked "125 S." This is the universal language for 125 pounds steam working pressure. Cold service rating is designated by the mark "200 WOG," which means 200 pounds cold water, oil, or gas, non-shock pressure. (*See* Fig. 12.)

Steel Ratings Are Different

Cast and forged steel valves and fittings bear a mark such as 150, 300, 600, and so on. These figures denote the maximum pressure at a given maximum temperature for which an item is suited. A certain 600-pound steel valve, for example, may be suited for 600-pound pressure at temperatures up to 900°F. But, if the temperature exceeds that point, up to 1000°F., let us say, the valve is unsafe at pressures over 380 pounds. This important effect of temperature makes it imperative to know both pressure and temperature conditions of a service, and to consult the manufacturer's service recommendation tables before selecting steel materials for it.

Temperature and pressure, however, are not always the only factors influencing the use of steel materials. Frequently, they are used for their structural ability to meet severe external conditions of shock and vibration in lines, which brass or iron materials cannot handle.

Formerly, steel valves and fittings were known by series numbers, as Series 15 for 150 pound valves, Series 30 for 300 pound valves, and the like. Under present standards, they are marked with *pressure classifications* in which they are regularly available, namely: 150, 300, 600, 900, and 1500 pounds.

Catalogs

When in doubt regarding any marking on valves and fittings, it is best to refer to the catalog. If the exact maker's catalog is not handy, almost any one will do, since pressure and temperature ratings for piping materials are standardized by all manufacturers.

TYPES OF END CONNECTIONS USED IN PIPING

You cannot begin to know piping until you know the various end connections it employs to join valves and fittings with other elements of piping. Knowing the common types is important because materials are identified by their type of end connection.

Valves and fittings are always selected with the type of *end connection* desired. Pipe may be ordered with threaded, flanged, or welding ends, or with plain ends and be prepared

Fig. 13. Screwed ends.

Fig. 14. Welding ends.

Fig. 15. Brazing ends.

Fig. 16. Solder end.

Fig. 17. Flared end.

Fig. 18. Hub end.

on the job for the type of connection required.

Figures 13, 14, 15, 16, 17, and 18 show the types of end connections most commonly used on valves and fittings, although not all materials are available in all types of end connections. Each type has one or more variants, with steel flange joints perhaps having the greatest variety.

Screwed Ends

Screwed end materials are by far most widely used (Fig. 13). They are suited for all pressures, but are not so frequently used for lines larger than six inches, because the larger the pipe, the more difficult it is to make up the joint.

Welding Ends

For lines not requiring frequent dismantling, the trend today is to *welding end connections.* Available in steel materials only, they are used mainly for higher pressure-temperature services. Welding end materials are made in two types —*butt-welding* and *socket-welding.* Valves and fittings for butt-welding come in all sizes; socket-welding ends are usually limited to sizes up to 2 inches (5.08 cm.). (*See* Fig. 14.)

Brazing Ends

Brazing end connections (Fig. 15) are available on brass materials. The joint comes equipped with an alloy brazing ring insert. When heated with a welding torch, the brazing material forms a tight seal between the pipe and valve or fitting.

Solder Ends

Solder-joint valves and *fittings* are used with copper tubing for plumbing and heating lines and for lubricating oil lines around machines (Fig. 16). The joint is made on the socket principle and soldered. Its use under pressure and temperature is limited.

Flared Ends

Flared end connections are commonly used on valves and fittings for copper and plastic tubing up to two-inch (5.08-cm.) diameter. The end of tubing is skirted or flared and a ring nut is used to make a union-type joint (Fig. 17).

Hub Ends

Hub end connections (Fig. 18) are generally limited to valves for water supply and sewage piping. The joint is assembled on the socket principle, with pipe inserted in hub end of valve or fitting, then calked with oakum and sealed with molten lead.

FLANGED END MATERIALS

Flanged end materials, although made in sizes as small as

½ inch (1.27 cm.), are generally used for lines above four inches (10.16 cm.) because of their easier assembly and take-down. (*See* Fig. 19.) Flange joints are made up with much smaller tools than required for screwed connections of comparable size. Flanges which are separate pieces can be attached to pipe in several ways. A few are shown in Fig. 20.

VALVE DESIGNS AND HOW THEY OPERATE

Two types of valves are most commonly used: (1) *gate valves*, which get their name from their gate-like disc that moves across the path of flow; (2) *globe valves*, which are so named for the globular shape of their body.

Flow through a gate valve travels in a straight line. When the valve is wide open, there is very little resistance to flow since seat openings are approximately the same size as the inside diameter of pipe. In a *globe valve* the direction of flow

Fig. 19. Flanged ends.

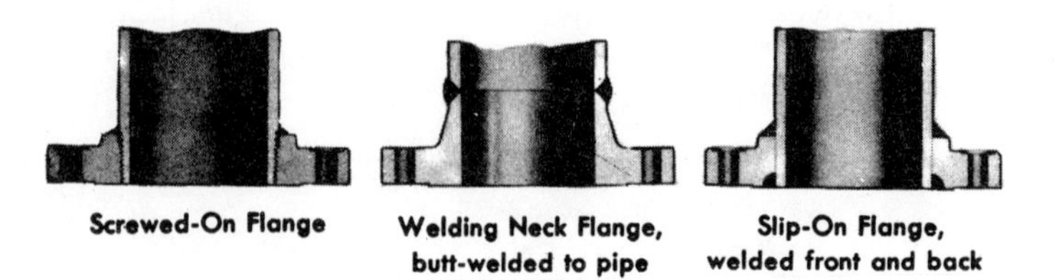

Fig. 20. Types of flanges.

passing through the body is changed, and this results in greater resistance to the movement of fluids. Each of these designs has certain advantages in specific applications as described in this chapter.

Both gate and globe valves are made in varying patterns. While the variants adhere to their respective design principle, their difference is in construction details mainly in the disc, stem, and bonnet.

Angle valves in basic design are the same as globe valves, except that their pipe openings are at right angles to each other. As their name implies, they are used at 90-degree turns in piping where they save a nipple and an elbow. By reducing the number of joints, they cut installation time as well as lessen restriction to flow.

Check valves perform the single function of checking or preventing reversal of flow in pipe lines. They are made in a wide variety of patterns, but in operating principle they conform to two basic designs, (1) swing and (2) lift.

GATE VALVES

Figure 21 shows a cross-section of standard *iron-body wedge disc gate valve* with outside screw and yoke. The disc is in fully opened position. Whether wedge-shaped or double-disc, the disc in a gate valve moves up and down at right angles to the path of flow, between two perpendicular seat rings against which it is seated to shut off flow. Gate valves have an all-metal disc; its travel is actuated by the stem screw. (*See also* Fig. 22.)

Note the straight-through course of flow. This construction reduces pressure drop to the absolute minimum. That is why gate valves are best for liquid lines, pump lines, main supply lines, and the like—in which unrestricted flow is usually required.

Gate valve design is not practical for throttling (flow regulating) services. The velocity of flow against a partly opened disc may cause vibration and chattering and result in damage to the seating surfaces. Also, when throttled, the disc is subjected to severe wire-drawing erosive effects at the lower edge.

In the main, gate valves are best for services in which valves are infrequently operated, and where the disc is kept in fully opened or closed position. They come in all sizes.

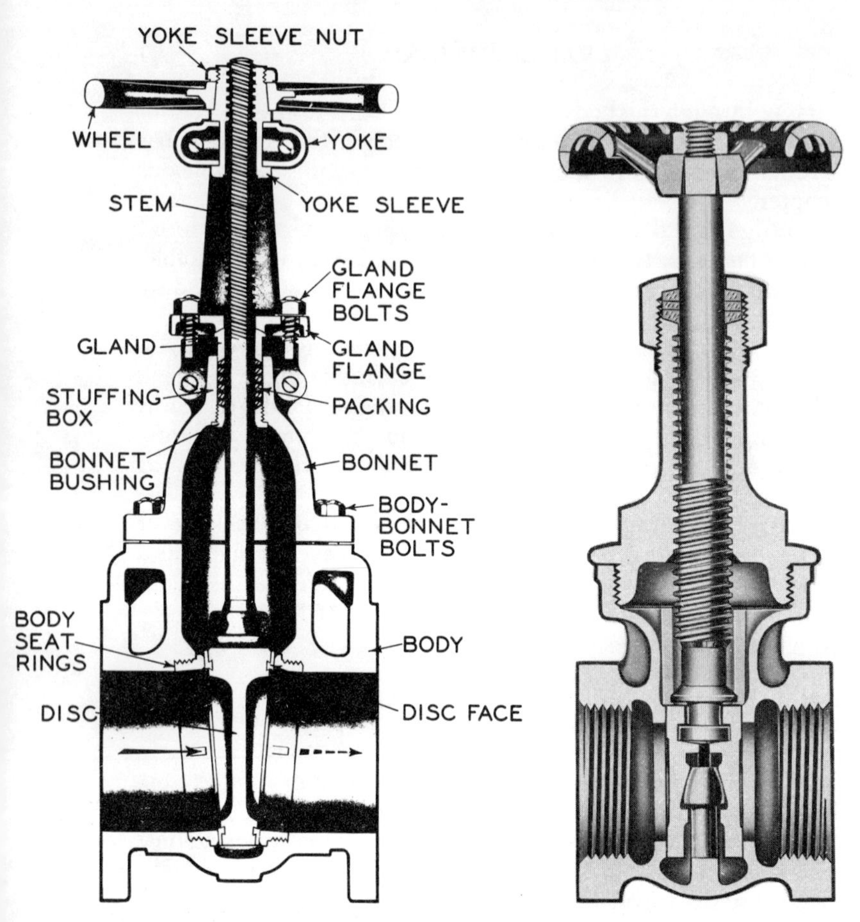

Fig. 21. Cutaway view of an iron body wedge gate valve.

Fig. 22. Double-disc gate valve.

TYPES OF DISCS IN GATE VALVES

There are two types of discs in gate valves—the wedge disc and the double disc.

Wedge Disc

Of the two disc designs used in gate valves, the *wedge-shaped* or *tapered disc* is by far the most common. Simpler design is its primary advantage. It has no loose internal moving parts.

Where gate valves cannot be installed with stem pointing upward as is usually recommended, the wedge disc eliminates the danger of disc jamming which has been known to occur in double disc valves.

Wedge disc seating is available in brass, iron, and steel gate valves and is highly recommended for all gate valve services.

Double Disc

Like the wedge disc, the *all-metal double-disc* or *parallel seat disc* makes closure by descending between two perpendicular seats in the valve body. But instead of wedging between the seats, its parallel faces, after being lowered into position, are seated by being spread against the body seats. A disc-spreader is used between the two halves of the disc. In closing, the spreader makes contact with a stop in the bottom of the body and forces the discs apart.

Double disc-gate valves are widely used on non-condensing gas and liquid services at normal temperatures. They are best known in the water works and sewerage fields.

GLOBE VALVES

Figure 23 shows a cross section view of a *brass plug type disc globe valve with disc* in fully opened position. It typifies the operating design of the three types of discs commonly employed in globe valves.

Course of flow through the globe valve body is indicated by the heavy black arrows. Note how the direction of flow changes, accounting for the greater resistance to flow than in a gate valve. The disc is actuated by a stem screw and handwheel.

Disc and seat in globe valves can be quickly and conveniently regrouped or replaced. This feature makes them ideal for services necessitating frequent valve maintenance. Shorter disc travel in globe valves saves operators' time when valves must be operated frequently.

Globe valves are ideally suited for throttling services (flow regulation). They are available in brass, iron and steel. They are seldom used in sizes above 12 inches owing to the difficulty of opening and closing larger sizes, and making them tight.

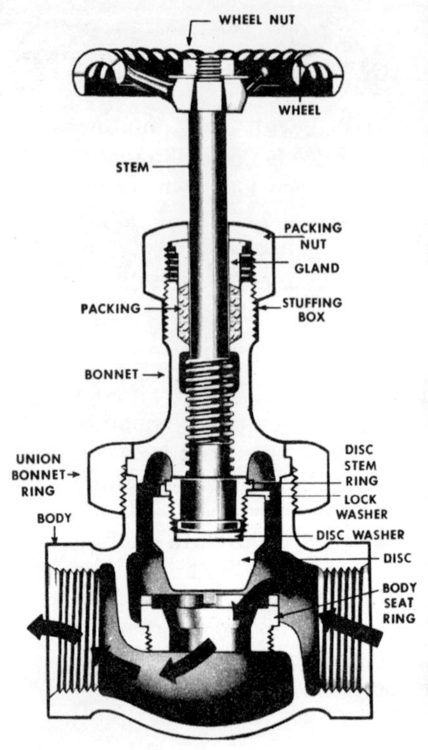

Fig. 23. Cutaway view of brass plug type disc globe valve.

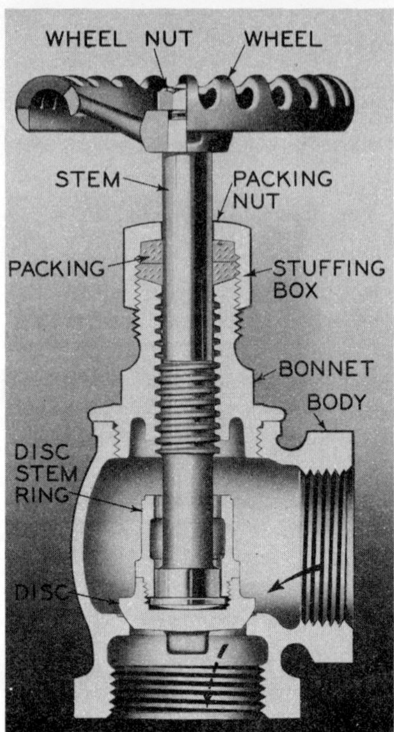

Fig. 24. Cutaway view of a brass angle valve.

ANGLE VALVES

Angle valves (Fig. 24) have the same operating characteristics as global valves and are available in a similar range of disc and seat designs. Used when making a 90-degree turn in a line, an angle valve reduces the number of joints and saves make-up time. Angle valves are available in brass, iron, and steel.

DISC DESIGNS IN GLOBE AND ANGLE VALVES

Plug-type Disc

The long, tapered *plug-type disc* and matching seat represent one of the advances in valve engineering (Fig. 25). Their

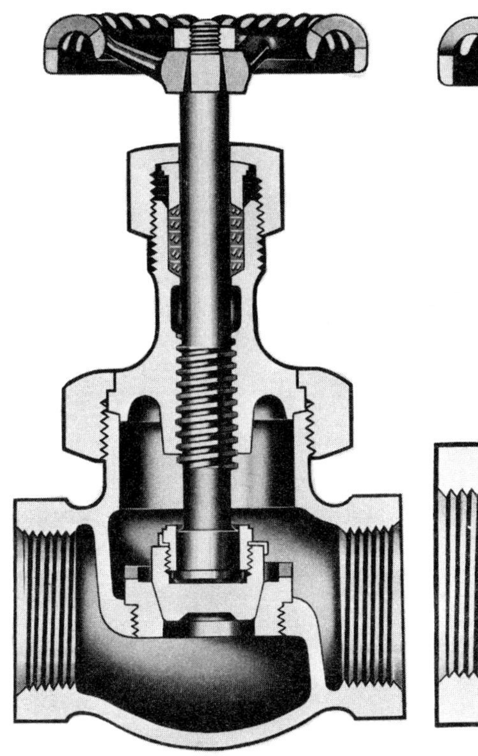

Fig. 25. Plug-type disc.

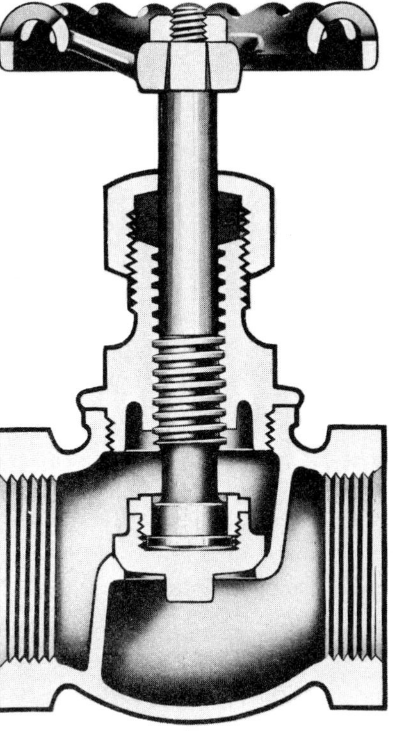

Fig. 26. Conventional disc.

wide bearing surfaces offer highest resistance to the cutting effect of dirt, scale, and other foreign matter present in pipe lines. This feature makes plug-type disc seating the best choice for the toughest flow control service.

In addition, the plug-type disc valve shown in Fig. 25 has a disc of special nickel alloy, and the seat is of stainless steel. This combination of metals provides the most effective protection against corrosive effects in most services.

Plug-type disc valves are made in globe and angle patterns, and highly recommended for services such as drip and drain lines, soot blowers, blow-off, boiler feed, and other severe services. Available in a wide variety of disc and seat materials to meet specific needs.

Conventional Disc

The *conventional disc* (Fig. 26) is so named because it is

so commonly known. Before the development of the plug type disc, it was considered best for the toughest services.

The main advantage of the conventional disc design is the ease of obtaining a pressure-tight bearing between the disc and seat. This is due to the thin line contact achieved by the taper of seat with face of the disc. The thin line bearing also helps break away hard deposits which form on the seats in some services.

The conventional disc is made in several seating styles, such as flat seating, ball seating, and with seating surfaces having varying degrees of taper.

Conventional disc valves are recommended for many cold and temperature services under moderate working conditions. They are available in brass, iron, and steel, with a wide variety of disc and seat materials for specific service needs.

Composition Disc

The *composition disc* (Fig. 27) works on the principle of a cap. The flat of its face is seated against the opening in the bridge wall. The disc unit consists of three parts: metal disc holder, composition disc, and retainer nut.

The main feature of the composition disc is the variety of disc material available for individual services, such as air, steam, hot and cold water, oil, gas, gasoline, and the like, and its quick changeover to any service, or quick renewal in case of leakage.

The composition disc is suited for all types of services, except throttling, at moderate pressures up to 400°F. with proper disc. While subject to the cutting effects of foreign matter, it will stand much embedding of dirt before leaking. The disc should be renewed as soon as a leak develops. These discs are available in brass and iron valves only.

CHECK VALVES

Check valves head up the family of non-return valves which are used to prevent back-flow in lines. They are made in two basic types—swing and lift. (*See* Figs. 28 and 29.) Each type has several variants, each with some particular feature suited for specific installations. However, in operating principle they all conform to one of the two basic patterns.

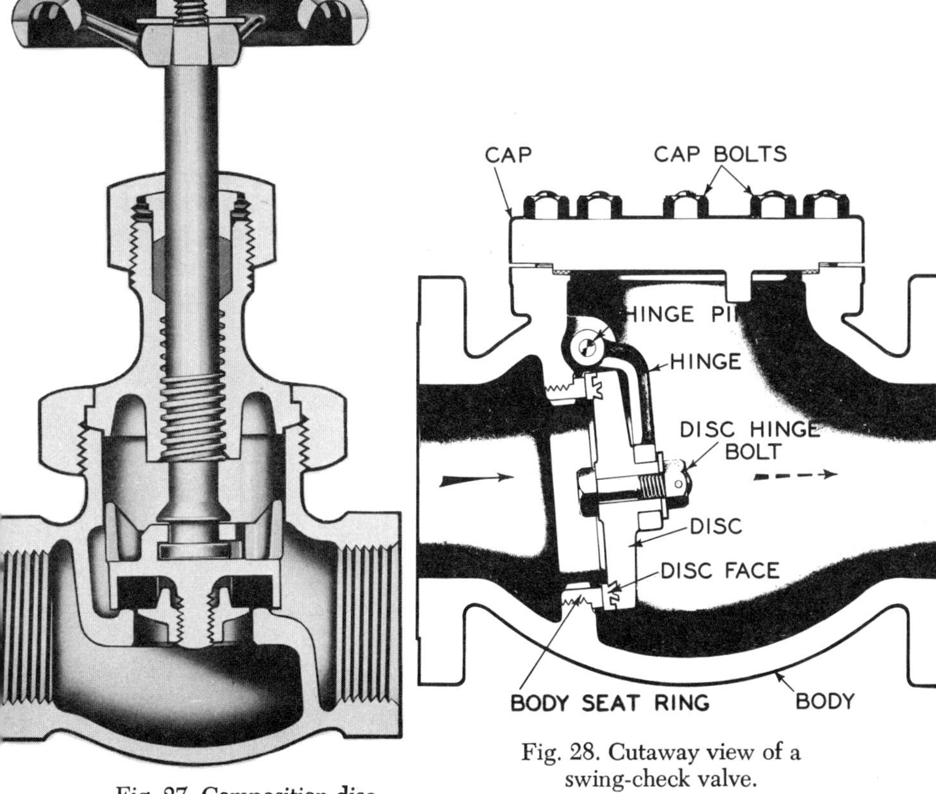

Fig. 27. Composition disc.

Fig. 28. Cutaway view of a swing-check valve.

Fig. 29. Cutaway view of a lift-check valve.

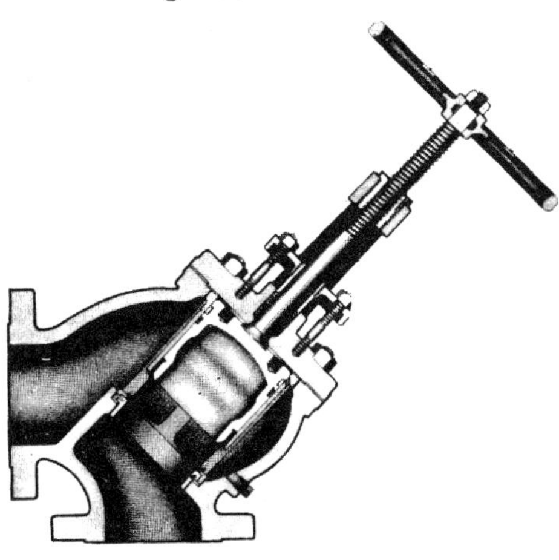

Fig. 30. Automatic stop-check valve.

Worthy of special note is the *automatic stop-check valve* shown in Fig. 30. A combination of stop valve and check valve, it is used almost exclusively in lines between boiler and header in multiple boiler installations. As a check valve, it prevents back-flow into a boiler and as a stop valve it is used to cut out a boiler when necessary. It belongs in the lift-check family.

Basic Swing-check Design

In the *swing-check valve*, flow moves through the body in a straight line as shown by the heavy arrows (Fig. 28). This is comparable to a gate valve. Similarly, it gives less resistance to flow than the lift-check.

A slightly tilted bridge wall crosses the line of flow, and the disc which is hinged at the top seats against a machined seat in the bridge wall opening. The disc is free to swing on its hinge pin in an arc from fully closed position to one parallel with the flow, thus permitting full flow. Flow keeps the valve open, the size of opening varying in direct relation to volume of flow. Gravity and reversal of flow seat the disc automatically.

Swing-back valves are widely used, mainly for low to moderate pressures. Because of low resistance to flow, they are best for liquids with low head pressure and velocity.

For years, it has been a general rule to use swing-checks in combination with gate valves.

Swing-checks are available in straightway patterns in brass, iron, and steel, with screwed or bolted caps. Some have a regrindable metallic disc, others a leather-faced disc, mainly for use in water lines. Outside lever and weight type is used when quick closing of disc with reversal of flow is necessary. All swing-checks must be installed so that disc is held to seat by gravity.

Basic Lift-check Design

Flow through a *lift-check* follows a turning course through a horizontal bridge wall on which the disc is seated, similar to the way it moves through a globe valve. Note the heavy arrows.

To insure proper seating, the disc is equipped with a short guide stem, usually above and below, which moves vertically in integral guides in the cap and bridge wall. It is seated by back-flow, or by gravity when there is no flow, and is free to rise and fall depending on the pressure under it.

In both lift- and swing-check valves shown in Figs. 28 and 29, the disc is the only moving unit, and both valves must be installed so that disc is fully seated by gravity alone.

The general rule on using lift-checks is to put them in lines in which globe valves are used. They are recommended for water, steam, air, gas, and general vapor service. They hold tighter than swing-checks, and their construction makes them better able to stand the punishment of high velocity flow.

Primary caution to be observed when installing any check valves is to see that flow enters at the proper end.

Lift-checks are made in globe, angle, and horizontal patterns; in brass, iron, and steel; with screwed, union ring, or bolted caps. Other types in the lift-check family are automatic stop-check, ball disc, vertical, and cushioned disc.

RANGE OF PIPE FITTINGS

Pipe fittings serve as the links in making up piping to a given layout. They join pipe into a continuous line or turn it at various angles. They are used to reduce or increase the size of pipe ends, to divert or divide flow, combine, or return it. The variety of fittings needed for these and the many other purposes they serve is almost amazing. (*See* Chap. 2.)

Fittings are made in the same variety of metals as valves—brass, iron, steel and various alloys—and they come in corresponding pressure classes. They are made with the same end connections as valves—screwed, flanged, and welding—and in all standard pipe sizes. But, as is true of valves, not all types of fittings are made in all sizes and materials.

Identification Marks

At first you may have a little trouble in telling cast and malleable iron materials apart. They look alike, but a common identification of *cast iron fittings* is their wider end bands than are put on malleable iron fittings. They are needed for greater strength and to resist damage to fittings on installation.

Brass fittings are readily recognized by their color. *Forged steel fittings* are marked as such, some with the letters "F.S."

Cast steel materials are usually marked "STEEL," and when new, they are generally painted in light gray or aluminum.

So great is even the regular range of fittings that to show them all here is impossible. But, most helpful to you will be an acquaintance with the typical patterns illustrated, and a general knowledge of the most common types and their uses which are described here.

Common Types of Fittings

An *elbow,* usually called an "ell," makes a 90-degree or right angle turn in a pipe line. For making wider turns, there are 60-degree, 45-degree and 22½-degree elbows.

To take a right angle line from a horizontal or vertical run of pipe, a *tee fitting*, shaped like a letter T, is used.

A *cross fitting* has four openings on the same plane at 90-degree angles to each other. It looks like a plus sign and is used to connect intersecting pipe lines.

A *reducing fitting* has smaller openings at one or more ends to permit joining of a given size line to one of smaller size. Elbows, tees, and crosses are also made in reducing patterns.

When you cannot get a reducing fitting of proper size, the fitting known as a *bushing* comes in handy. It screws into the end of another fitting or valve to reduce the size of the opening.

A *plug* is used to close off the female end of a line. *Caps* serve the same purpose on male ends.

The most frequently used fitting is a *coupling.* It connects two lengths of pipe.

Another type of elbow is the *street ell*, shaped like a regular ell but with one female end and one male end.

Nipples are standard short lengths of threaded pipe used to connect two other parts. They come in three types. The *close nipple* is used to make a close connection between parts, its entire surface being threaded with tapered threads cut from each end to the middle. The length of close nipples varies with pipe size; it is determined by the length of thread necessary to make a satisfactory joint.

A *short nipple* is longer than the close type and has a small unthreaded surface separating the threaded ends. A *long nipple* is any pipe up to 12 inches long, threaded on both ends. All longer pipe is called *cut pipe*.

The primary purpose of *unions* is also to connect pipe. But, they are designed for use in screwed lines to permit easy assembling, and easy opening or dismantling of the line.

A *flange union* serves the same purpose, but is bolted together instead of being held with a ring nut. It usually requires a gasket.

A *union fitting* is a union and fitting combined. It serves the purpose of both union and fitting with the added advantage of reducing the number of joints necessary in a line.

HANDLING VALVES, FITTINGS, AND PIPE

Good handling of piping materials begins with knowing the tools that are used on them. Let us start with wrenches—they are most important, but when used improperly, they can do a lot of damage. The types of wrenches you will be working with are pipe, monkey, strap, end, and pipe tongs, with the first two being your constant companions. (*See* Chapter 1, Plumbing Tools and How They Are Used.)

Before we go any further, let us learn this simple rule regarding the use of wrenches on every job: (1) Choose the right type, (2) choose the right size. Why this rule is so important, you will see in the following text.

Pipe Wrench

When working with a pipe wrench, the harder you pull, the tighter it squeezes. It was designed for use on pipe and screwed end fittings only. On parallel-sided objects, its efficiency is not up to that of a monkey wrench and, in addition, its squeezing action can do great damage. Many experienced piping workers have learned to their sorrow that using a pipe wrench on a valve may cripple or even crush its body.

Monkey Wrench

For hexagonal end valves and fittings, a *monkey wrench* with its smooth, square jaws is always your best bet. Not only does it fit better on the part to be turned, but it does not have the crushing effect of a pipe wrench. You hardly need to be told that a monkey wrench cannot be used on pipe or other round parts because its jaws have no means of gripping such shapes.

Open End Wrenches

For pulling up flange bolts and the like, you will find your *open end wrench* most suitable and fastest to work with. However, use the right size if you wish to avoid bruised knuckles due to slippage of the wrench, and to prevent wearing the bolt heads round.

Strap Wrench

The *strap wrench* is used mainly when working with plated or polished finish materials in order not to mar the surfaces. Also, it often comes in handy in tight places where you cannot get in with a pipe wrench.

Pipe Tongs

Pipe tongs are wrenches of the chain strap and lever type. They are made for handling pipe from ⅛-inch (.3175-cm.) size up, but are generally used for larger sizes, 6 inches (15.24 cm.) and above.

Right Size Wrench

Always use the right size wrench. A wrench that is too small makes it hard to pull up a joint. But there is more danger in picking one that is too large. It gives you more leverage than needed, and, unknowingly, you may pull up a joint so tight as to crack a fitting, twist a valve out of shape, or run the pipe clear into the seats. But, you will soon get the "feel" of wrenches and know what size is best for the job to be done.

Pipe Vise

Your next most frequently used tool will be the vise, and there are several cautions regarding its use that are worth noting. The *pipe vise* is used for pipe only. The *machinist's vise* has square jaws only, or a combination of square and gripper jaws, making it suitable for pipe as well as other work.

The screw action of a vise exerts a powerful force at the jaws. That is why experienced piping workers never put a valve or fitting in the vise when making up a joint at the bench. There is too much danger of squeezing the part so much as to distort it, or of putting the working parts of a valve out of line.

Fig. 31.

Fig. 32. Use of "soft" jaws.

The correct procedure shown in Fig. 31 is to put the pipe in the vise and to turn the valve or fitting. Note also, that the right place to put your wrench is on the pipe end of the valve or fitting. This gives you more direct leverage on the joint, and in the case of valves, prevents the possibility of twisting the unsupported valve body or hurting its working parts.

Copper or Lead Covers on Vise Jaws

When using a machinist's vise to hold a valve for any repair work, put it in with the jaws to the valve ends—not to the sides of the valve body. This is a precaution against damaging the valve body. When working on valves, or any valve parts, always use "soft jaws" (copper or lead covers over the jaws) to prevent damage to finished surfaces (Fig. 32).

Dirt in Pipe Lines

Dirt, sand, rust, scale, metal chips and burrs—any foreign matter—is a menace to valves or other equipment in a pipe line. Sooner or later it causes cutting and damaging of valve seats, with leakage resulting.

Valves and fittings should be stored where not exposed to foreign matter, weather, or harm in other ways—such as falling or being hit by other objects. The covers usually put on valve ends at the factory should not be removed until the valve is ready for use. Storing piping materials with the best protection possible will pay dividends after installation.

"Blow-out"

It is easy for sand, dirt, and scale to accumulate in pipe stored in a yard or shed. Never use pipe before blowing it out with compressed air or swabbing it out thoroughly.

When a new pipe line is installed, it is a good idea to flush it out completely with water to remove any loose scale or foreign matter.

Dirt in the threads can also get into lines when the joint is made up. Often it causes tearing of the metal when screwing up a connection; it increases friction and interferes with making a tight joint. The safest practice is to brush or wipe the threads of all pipe, valves, and fittings before you start to make a joint.

When pipe is threaded "on the job," particularly if a rotary type cutter is used, the ends should be filed or reamed to remove burrs before threading. Loose burrs in pipe lines may damage valve seats; those that hang on can cause obstruction to flow.

Flange Faces

The *faces of flanged end equipment* should be cleaned thoroughly before installing. Manufacturers usually coat them with a heavy oil or grease to prevent rusting. A solvent will easily remove this coating.

Special precaution should be taken against dirt on *gaskets*. They should be carefully wiped before using. Hard dirt particles in flanged joints can prevent proper seating of gaskets or cut into the machined flange faces deeply enough to cause leaks.

Dirt in Valves and Fittings

No matter how much care you have taken in storing valves and fittings, it is wise to give each one a careful looking over for dirt in the threads and body just before putting them in a line. If you have convenient facilities for blowing them out, do it. Never fail to take this precaution when using old materials.

Bruised Threaded Ends

Despite the care you give pipe on the job, occasionally a threaded end will get bruised in handling. When this happens, do not take a chance with a joint. If the end is badly damaged, cut it off and re-thread. More than likely, it will be repairable by running a die over the threads to clean up and straighten them. But do not fail to do this. *Bruised female threads* in pipe, valves, and fittings can usually be repaired by running a tap into the threads.

INSTALLATION OF PIPING

You cannot expect a valve, for example, to work at its best and stay on the job if it is put in wrong or installed in the wrong place. It is the responsibility of the worker to handle piping equipment with an eye to good performance, convenient operation, safe and durable service. A few cautions regarding the installation of piping for longer life and better service follow.

Installing Valves

Unless it is impossible, always install valves with the stem pointing straight up. Any position from straight-up to horizontal is satisfactory, but a compromise. When the stem points downward, the bonnet acts as a trap for sediment which may cut and damage the stem threads.

Using the Right Disc

You cannot beat *composition disc valves* for all-around work and quick disc repair. But you can save much disc-changing time by using the right quality disc for each service. A few extra disc holders with proper discs save more time when changing services. You just slip them on the stem.

Hammering in Steam Lines

Hammering in steam lines may be a signal that condensate is interfering with the flow of steam. This powerful force can pull piping from its anchors and cause erosion of valve seats by its high velocity. You cannot stop formation of condensate, but you can provide for its drainage by installing traps of suitable capacity.

Galvanized Materials

Galvanized materials need extra care in handling. Their coating is put on for extra protection against corrosion. But to be effective, it must be continuous. Rough handling when pulling up joints may injure the surfacing. Use the pipe wrench carefully.

Exposure of Valves

Valves cannot be abused and operate efficiently. Do not expose them to damaging blows. A bent stem not only cripples a valve but may cause a shutdown and result in costly repair.

Identification of Valves

All valves should be *identified* to show what lines they control. In addition, workers should be thoroughly familiar with valve use so that in emergencies valves can be quickly operated.

Working in Close Quarters

When you are putting a line into cramped space, it sometimes helps a lot to remove the bonnet assembly from the valves (Fig. 33). This gives you more clearance and protects the stems from possible damage.

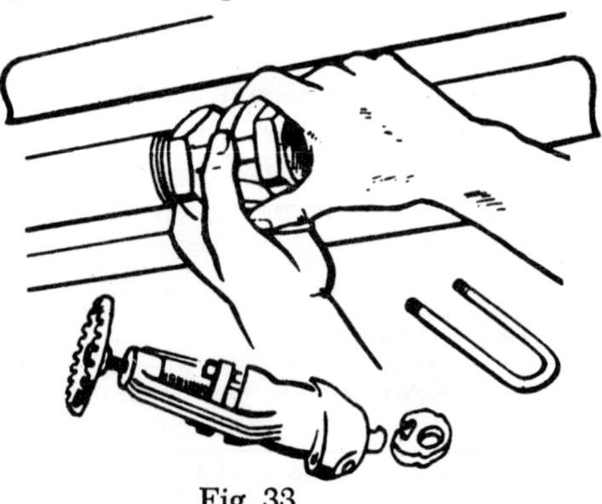

Fig. 33.

Where to Put Shut-off Valves

A standard practice of putting in *branch line shut-off valves* up close to the main line is safest. In an emergency, there is no time lost looking for valves. These valves should be fully identified and workers should be familiar on their use.

Use of Correct Gasket

It makes a big difference what kind of gaskets you use in flanged joints (Fig. 34). What is good for one service may not last in another. For durable, leak-proof joints, be sure the gaskets are right for every job.

Shutting Off Valves

Valves should be placed for convenient operation. There is danger to life and equipment any time a ladder must be used.

When hard-to-get-at valves must be frequently operated, it is best to relocate them for safety's sake, or equip them with a chain wheel, extension stem, or remote control device.

Use of Crosses

Crosses make cleaning easy. In lines where sediment is present, and in viscous fluid lines, there may be frequent clogging at the turns. By using *crosses* instead of elbows at turns, rodding out of lines is easily done by simply removing the plugs (Fig. 35).

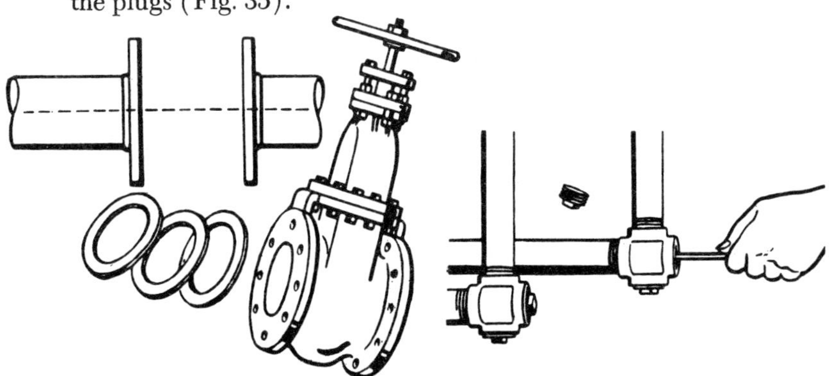

Fig. 34. Use of gaskets. Fig. 35. Use of crosses.

Placing Operating Valves

Place *operating valves* for convenient use. You cannot expect to open a valve fully, close it tight, or regulate it properly if you cannot do it conveniently and safely. Better care of valves is encouraged when they are installed so they can be operated without difficulty or danger.

Valve Seats

Quick wearing of *valve seats* is often due to valves that cannot be fully opened. Rising-stem valves must have plenty of clearance or non-rising-stem patterns should be used instead.

A *relief valve* makes this a safe hookup. Should the pressure regulator fail, the relief valve set at proper pressure limits will prevent damage to processing equipment. The bypass, installed with a plug disc valve for throttling, permits servicing of the pressure regulator without complete shutdown.

Water hammer can do serious damage to piping. When *quick-opening valves* are used in liquid lines, they may cause water hammer in varying degree unless proper protective measures are provided.

An adequate cushioning chamber should be installed ahead of quick-opening valves. An assembly of a tee, a length of pipe, and a pipe cap will serve this purpose efficiently and is easily put into any line.

HOW TO MAKE UP A SCREWED JOINT

The most common method of joining pipe is the *screwed connection*. All of us know that it consists merely of threaded male and female ends screwed together. Yet, the perfection that has been attained in standardizing the threading of piping materials is actually amazing. Pick up a valve made in Chicago, a fitting from Los Angeles, and join them to a pipe from Pittsburgh threaded with tools from Cincinnati. There is no question they will fit—perfectly.

This outstanding example of standardization practice began with the establishment of American Pipe Threads Standards. They rigidly guide the entire piping industry and are responsible today for the simplicity of making up a screwed joint.

The relation of male and female parts in a properly made-up

joint is controlled by standard dimensions for the pitch and depth of threads, number to the inch, length, and so on. Strict adherence to these dimensions when threading pipe is the secret of good screwed connections. For this, well-kept threading tools are necessary.

To get a good *metal-to-metal joint*, the first thing to do is to wipe both male and female threads clean. Dirt can make a good fit impossible. Wire brushing is recommended, especially if threaded pipe has been exposed to weather. Running a tap or die over the threads will usually straighten any that may be damaged.

A bit of good thread lubricant or "pipe dope," as it is frequently called, on the male threads only, comes next. Lubricant is used not as a cement to seal the joint, but to reduce friction when pulling up. It must not be expected to compensate for poor workmanship. Putting lubricant on the male threads only prevents any excess from squeezing into the pipe and causing harm to valve seats or other mechanisms.

Start the joint by hand, and "feeling" that thread engagement is right, turn it up as far as it will go. If the threading job was up to standard, and the threads free of dirt, you have an almost tight joint now. A few turns with the wrench is all that is needed.

Do not make the mistake of "leaning on the joint" with an oversize wrench. You must always remember the danger of crippling a fitting or running the pipe clear into the valve seats with too much pull.

No attempt should be made to run all the male threads into the joint. The lead of the die always leaves a few imperfect and unusable threads. And never should it be necessary to use a "hickey" (extension handle) on a wrench to make a joint tight. Like an oversize wrench (Fig. 36) it gives enough

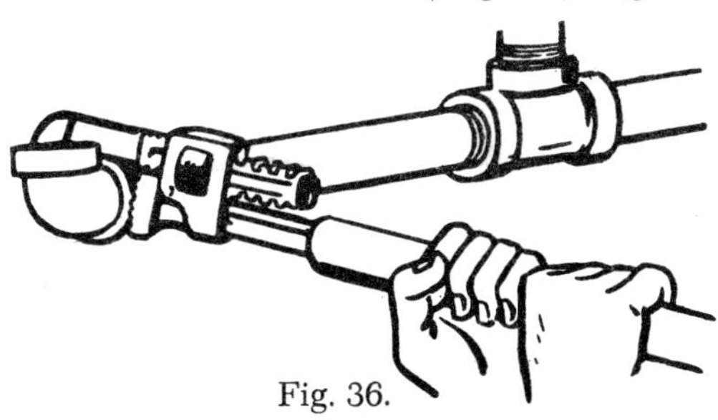

Fig. 36.

leverage to cripple a valve or stretch or crack a fitting. The only safe place for a hickey is on stubborn joints when taking down lines.

HOW TO MAKE UP A FLANGED JOINT

In the section on types of end connections used in piping previously discussed in this chapter, you saw the common methods of attaching flanges to pipe. Now, let us go through the process of joining this pipe to a flanged end valve. A *flanged joint* is the result of bolting the two flanges together, with a gasket between their machined faces.

Remember that the first step in making any joint is to clean all parts. So, with a solvent-soaked rag thoroughly remove the rust-preventing grease put on flanges at the factory. Clean off all dirt and grit particles that might cause trouble in the joint. Next, wipe off the gasket just to be sure there is nothing to keep it from seating properly.

You will have no trouble putting in a valve when pipe flanges are lined up properly and the piping is well supported. There is trouble ahead in making up a joint that is out of alignment. A severe stress on the valve flanges may distort the valve seats and prevent tight closure.

The pipe is in place, all properly supported. You know that a valve cannot be expected to hold up an unsupported length of pipe without causing a great strain. The pipe flanges have been accurately aligned to receive the valve—by checking with a spirit level, both horizontally along the pipe and vertically across flange faces—and are now ready to be bolted tight.

With the valve securely held in position, slip in half of the bolts through the flanges at the bottom. They will hold the gasket in place. Be sure to choose the best gasket for each service. But before inserting the gasket, coat its faces with a little graphite and oil or other recommended lubricant. This will make for easier removal when the joint must be opened. Then, slip in the bolts, all the way round, apply a little thread lubricant to each and turn up the nuts by hand as far as they will go.

Now, start pulling up with a wrench—not in rotation, but by the cross-over method to load the bolts evenly and eliminate any concentrated stress on the flanges. Tighten the bolts, over and across and keep going until the joint is uniformly tight, then to the other end to repeat the whole operation.

MAKING UP A SOLDER-JOINT

In piping, the term *solder-joint* is descriptive of a connection made by joining the pipe fitting or valve to the pipe with solder. Usually this type of connection is used with copper tubing, although some steel alloys can be soldered. The correct procedure for making such joints is illustrated in Figs. 37, 38, 39, 40, 41, and 42.

Fig. 37.

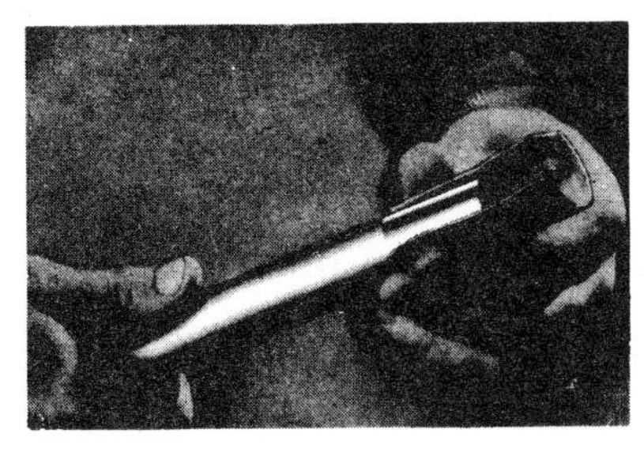

Fig. 38.

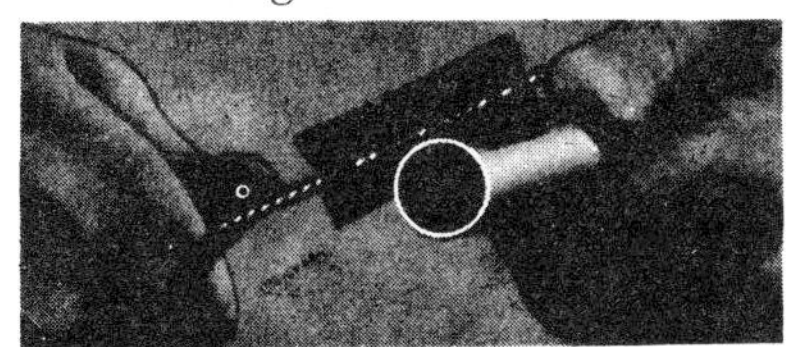

Fig. 40.

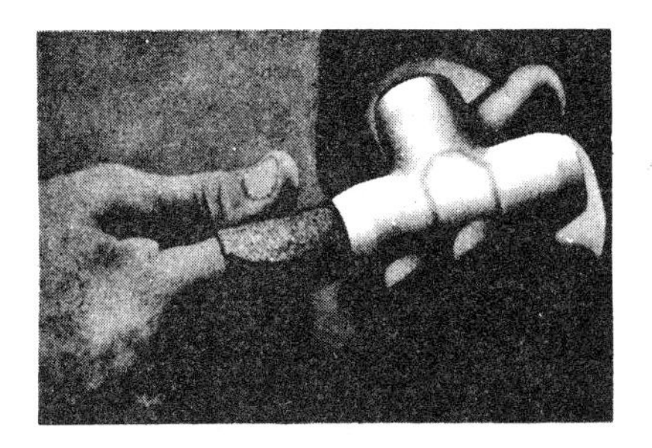

Fig. 39.

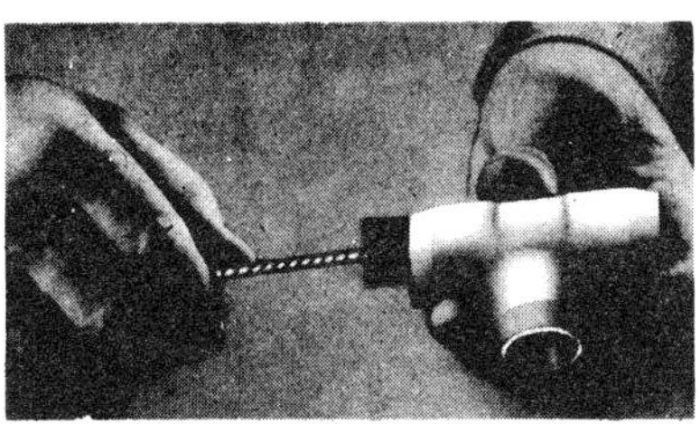

Fig. 41.

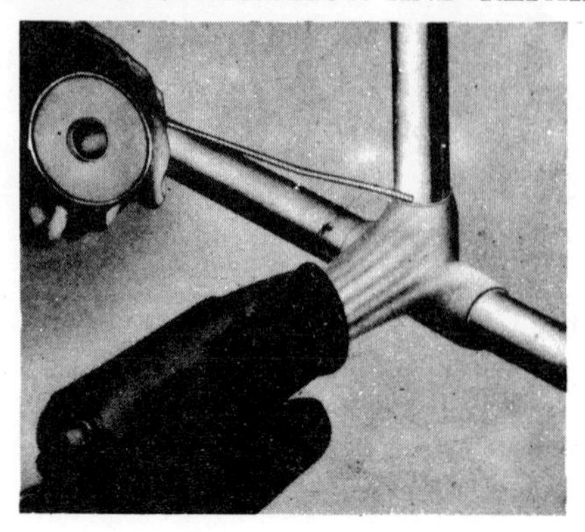

Fig. 42.

Cut Tubing to Length and Remove Burrs

Start by cutting the tubing to length, making sure the end is square (Fig. 37). Use a fine hacksaw blade. Be sure to remove all burrs with a file or a scraper.

Clean Tubing and Valve or Fitting

Clean outside end of tubing and inside end of valve or fitting thoroughly with medium grade sandpaper or emery cloth (Figs. 38 and 39). Next, wipe or brush away all emery dust and metal particles. A dirty solder joint will not hold.

Apply Flux and Heat

With a brush, apply a thorough and even coating of flux to the outside end of the tubing and the inside end of the fitting (Fig. 40). In cold weather, warm up the parts with a torch to 70° to 80°F. Slip the tubing in the fitting and turn back and forth once or twice to distribute the flux evenly (Fig. 41). The joint is now ready for soldering.

Solder the Connection

Heat the joint to good soldering temperature. Then feed solder around the edge of the fitting (Fig. 42). Fill the joint completely, wipe off any excess and the connection is made. Do not solder too hot or too cold. Solder flows freely at proper temperature, but burns when too hot.

Fig. 43. Butt-weld joint.

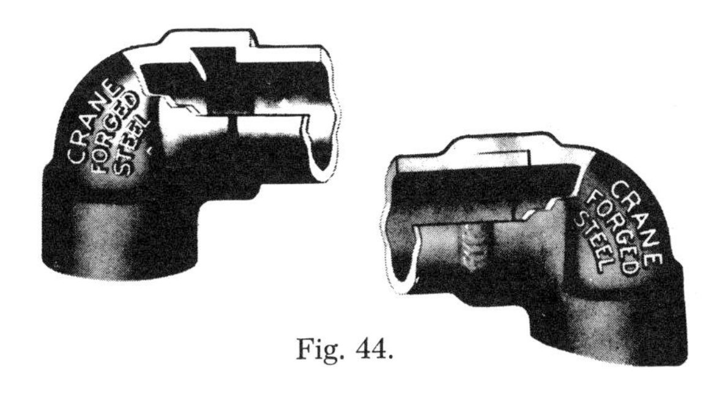

Fig. 44.

WELDED PIPE JOINTS

The practice of welding pipe joints was once limited to large piping. Now it is used for all sizes of high pressure-temperature lines, and is being used, more and more, for general service installations. In view of this trend, you will probably want to acquire a working knowledge of welding in order to know its uses and limitations in piping work.

Because the technique of welding is such a specialized operation, no effort is made here to describe actual procedures. Such information would take much more space than is available in this book. Instead, this section merely attempts to show the wide variety of welding materials—valves, fittings, flanges, and the like—which are regularly available.

Welded pipe joints fall into two general classifications: butt-weld and socket-weld. *Butt-welding* consists of beveling the two ends, lining up the two openings, and welding (Fig. 43).

Socket-welding fittings have deep sockets with ample "come and go." Pipe does not have to be cut accurately unless it must be butted against fitting shoulder (Fig. 44).

Fig. 45. Welded flange joints.

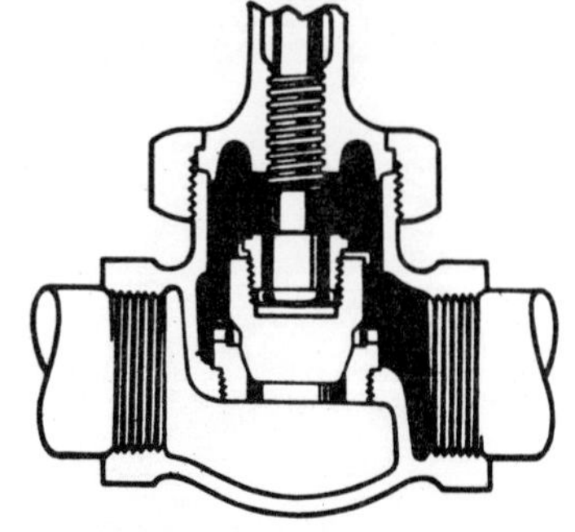

Fig. 46. Dark area shows pressure on top of disc.

Welded Flange Joints. Welding is also used in joining flanges to piping. Figure 45 shows typical methods for making such joints.

HOW TO INSTALL A GLOBE VALVE

The question of which is the better way to install a globe valve—with pressure above or below the disc—has had much discussion over the years. Sound reasons have been advanced for each preference. But, with the improvement made in trim materials, the method of globe valve installation appears to be of less significance. Nevertheless, there are still logical reasons for using one or the other depending on service conditions. The following examples should help you decide which method to use.

Suppose the disc of a globe valve in a boiler feed line which you know should have continuous flow, somehow became detached from the stem. Pressure from above the disc might cause it to seat and check the flow, if not shut it off completely. There is a case where pressure from below the disc is advisable.

Now, suppose the disc of a globe valve being used to regulate flow of steam to a steam-driven pump became detached from the stem. Pressure from below the disc might push the valve wide open and cause the pump to run wild. Here, pressure from above the disc would be safer. (*See* Figs. 46 and 47.) With pressure under the disc, cooling of the stem may

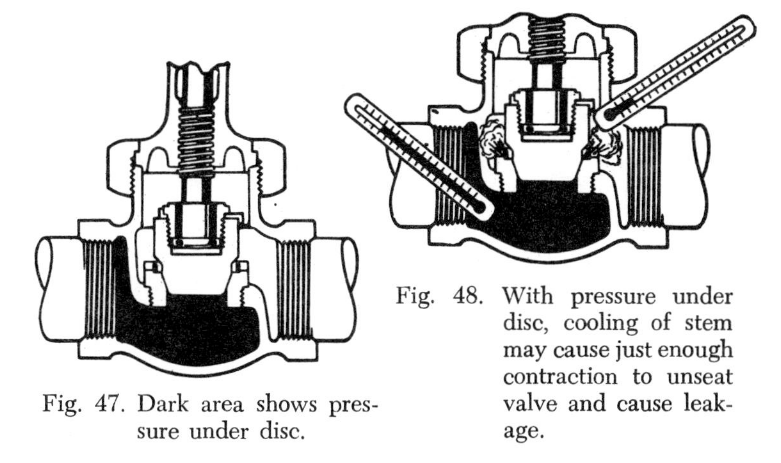

Fig. 47. Dark area shows pressure under disc.

Fig. 48. With pressure under disc, cooling of stem may cause just enough contraction to unseat valve and cause leakage.

cause just enough contraction to unseat the valve and cause leakage (Fig. 48).

Consider another case of temperature service with pressure under the disc. When the flow is shut off, the upper part of the valve is likely to cool. Cooling of the stem may cause sufficient contraction to unseat the valve just enough to cause leakage. The resulting extremely high velocity of flow may cause severe erosion of the disc and seat. And leakage of steam to empty kettles, for example, may cause severe damage. Here, too, pressure from above the disc would obviously be better. (*See* Fig. 49.)

Summary. Summarizing these applications, the general rule seems to be that unless pressure under the disc is definitely required, a globe valve will give more satisfactory service when installed with pressure above the disc. Follow the same rule for angle valves.

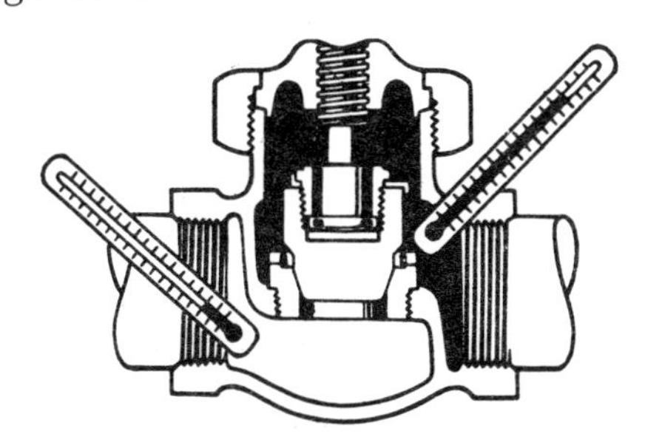

Fig. 49. Pressure and temperature above disc help insure tight seating.

General Cautions on Installing Gate, Globe, Angle, and Check Valves

Flow through angle valves is similar to flow through globe valves. Principles outlined for globe valves also apply to angle valve installations.

While most valves can be installed with the stem at any angle, the best position is with the stem pointing straight up. When the stem points downward, the bonnet acts as pocket for scale and other foreign matter in the line. Such matter may interfere with valve operation by cutting and eventually destroying inside stem threads.

In liquid lines subject to freezing temperatures, upside down position for valves is bad because liquid trapped in the bonnet may freeze and rupture it. Even when installed upright, valves in such lines should have drain plugs in the body as precaution against freeze-ups.

Recommended position for double disc gate valves is with the stem upright. The spreader mechanism in the disc has been known to jam in any other position. When the valve must be installed with the stem in a horizontal position, a wedge disc valve is to be preferred.

For severe throttling services, plug-type disc globe and angle valves are best. Gate valves are never recommended for throttling.

Special caution is advised when installing all check valves, so that the disc opens with flow.

To insure closing of the disc when back-flow occurs, the position of check valves in line must permit closure of the disc by gravity.

WHAT TO DO WHEN VALVES LEAK

Stopping Stuffing Box Leaks

In every gate, globe, and angle valve, the *stuffing box* is an important part. It holds the packing which seals the bonnet against leaks around the stem (Fig. 50). Leaks at this point are called *stuffing box leaks.* Pressure on packing is applied by a packing nut or gland flange, depending on valve design, which bears on a gland in the stuffing box.

Packing wears in direct relation to service condition. It

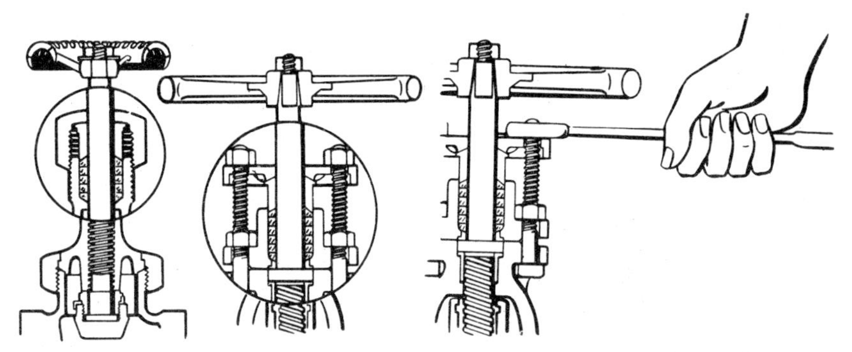

Fig. 50. Stuffing box leaks. Fig. 51. Tighten bolts evenly.

loses life with age, but wear is mainly due to rising and turning motion of the valve stem, combined with deteriorating effects of service conditions. In average services, packing lasts a long time and needs little attention. A few drops of oil on the stem, now and then, helps to reduce wear.

When New Valves Leak

Occasionally a stuffing box leak develops soon after a valve is installed. Unless the packing is unsuited for the service, the leak can usually be stopped by merely pulling down on the packing nut. On *gland flange type stuffing boxes*, you must be careful to tighten bolts evenly (Fig. 51) as cocking the gland will bind the stem.

When *leaks* show up on a *valve in service* for some time, repacking of the stuffing box may be necessary. The procedure is simple, and can usually be done without removing the valve from the line. There are valves, for example, that have a machined collar on the stem and a matched seating surface in the bonnet for back seating the valve when it is wide open. This feature permits repacking under pressure, provided that pressure is not too high.

To repack, first loosen all stuffing box parts to permit easy access. With a packing hook—usually a rattail file with bent end—take out all old packing and then clean up inside of stuffing box. To remove all particles adhering to the stem, or abrasions on it, polish the stem with any fine emery cloth.

After wiping all parts, inside and out, insert some of the new packing into the box. Most manufacturers use a split-ring

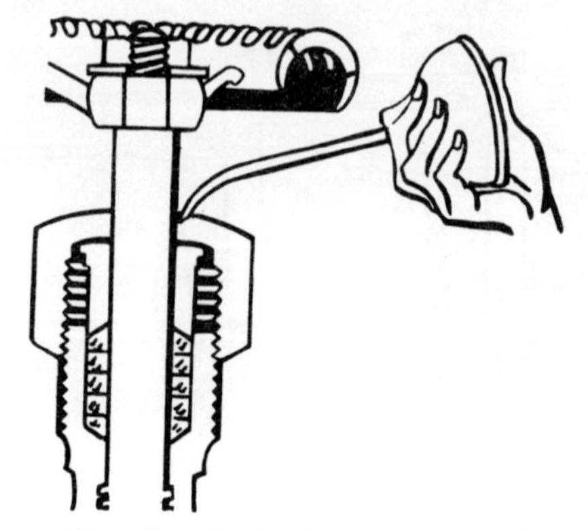

Fig. 52. A bit of oil helps work in packing.

type packing in new valves. If the same type is used for replacement, after putting in a few rings, tamp it well into place using the packing gland as a tamping tool. Then add to the packing to fill the box. Take care to stagger the ring splits so they are not all in line. Upon re-assembling the parts, a few turns of the handwheel and a few drops of oil will help work in the packing to the stem (Fig. 52).

USE OF GASKETS

To make flanged joints pressure tight, *gaskets* are usually inserted between the flanges. They come in three general types: (1) *Ring* gaskets which cover the face of the flange to the inside of bolt holes; (2) *full-face* gaskets, bolts are inserted through them; (3) a *metal ring* of elliptical section which fits into machined grooves in matching flange faces. This type of gasket is used only in very high pressure-temperature services.

Gaskets are made in various materials: *rubber, asbestos composition*, and *soft metals*. For low pressures and cold services, rubber gaskets are generally used. For cold service at higher pressures, and hot services up to 750°F., asbestos compositions are frequently used. Metallic gaskets, although suited for most services, are mainly used in steel piping.

Gaskets are regularly available in standard sizes. However, when there is a need for numerous gaskets of rubber and asbestos, you will find it economical to buy these materials in sheet form and cut gaskets as required. Good practice in cutting a gasket is to trace it from a template made first on heavy paper. This assures close and accurate fit. Consult manufacturers' recommendations for using various materials, also for treatment suggested before inserting gaskets in joints.

HOW TO REPAIR THE SEATS IN LEAKY GATE VALVES

When you definitely know that a valve leaks at the seats, do not delay in examining it to find out why. Neglected leaks will damage seating surfaces beyond repair. In new installations, leakage may be due to dirt or pipe scale. Immediate inspection and cleaning, or repairing if needed, may prevent serious damage.

To illustrate the procedure of making repairs, a solid wedge disc valve is used. The general method, however, applies to other types of gate valves which are repairable. The job is not an easy one, but it can be done in the following manner.

After removing the bonnet, clean and examine the disc and body thoroughly (Fig. 53). Carefully determine the extent of damage to body rings and disc. If corrosion has caused excessive pitting or eating away of metal, as in case of guide ribs in body, it may be impractical to attempt repairs.

A complete checkup and servicing of all parts of the valve is recommended. Remove the stem from the bonnet and ex-

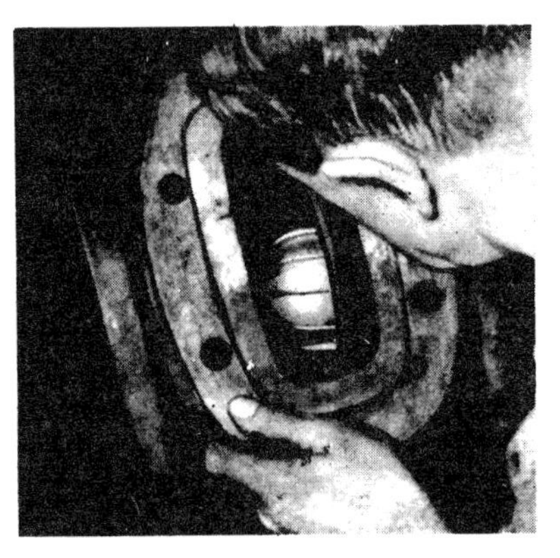

Fig. 53.

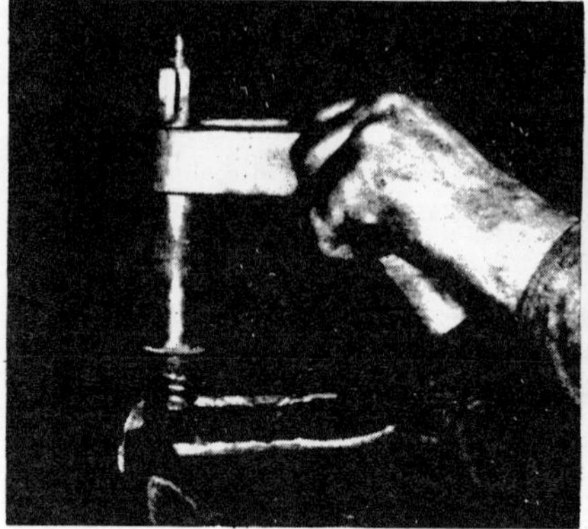

Fig. 54.

Fig. 55.

amine it for scoring and pitting where the packing bears. As a rule, light polishing with fine emery cloth is all that is needed to put the stem in good condition (Fig. 54). Use "soft" jaws if you put the stem in a vise.

Next, remove all the old packing and clean out the stuffing box (Fig. 55). At the same time, clean inside the valve bonnet and other parts to remove all dirt, scale, corrosion, and the like, to keep them from getting into valve seats.

It does not pay to salvage an old gasket. Remove it completely and replace it with one of proper quality and size (Fig. 56). After cleaning and examining all parts, if you see that the valve can be repaired by removing cuts from the disc and body seat faces, or by replacement of body seats, proceed as follows.

Set the disc in a vise with the face leveled. Wrap a piece of fine emery cloth around a flat tool and rub or "lap" the entire bearing surface (both sides of disc) to a smooth, even finish. Remove as little metal as possible.

The usual cuts and scratches found on body rings can also be repaired by "lapping." Use an emery block small enough to permit convenient rubbing of the rings all around (Fig. 57). Work carefully and watch closely; avoid removal of too much metal to prevent disc seating too low. When seating surfaces (disc and body rings) seem to be properly "lapped-in," coat faces of disc with Prussian blue and drop it in the body to check the "bearing." When a good continuous contact is obtained the valve will be tight—and is ready for reassembly.

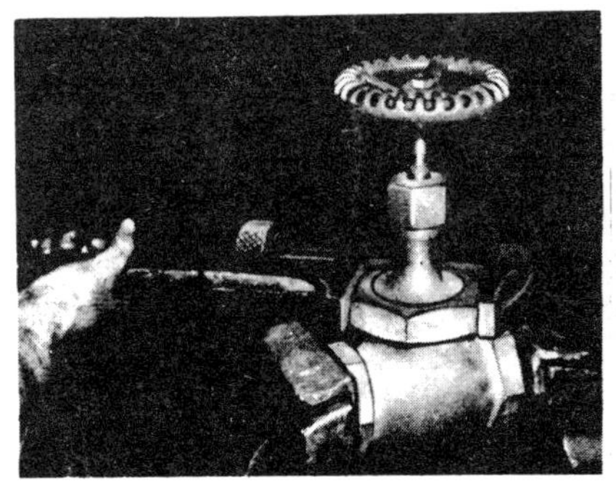

Fig. 56.

Fig. 57.

When assembling, first insert stem in bonnet, install new packing, assemble other parts, attach disc to stem and place assembly in body. Raise the disc to prevent contact with the seats so that the bonnet can be properly seated on the body before tightening the joint.

Good practice suggests testing the repaired valve before putting it back in the line. This is for assurance that repairs have been properly made.

When seat rings in gate valves must be replaced, they can be removed and replaced best with a power lathe. Chuck up body with rings vertical to the arbor and use a strong steel bar across the ring lugs to unscrew them. Removal by hand can be done with a diamond point chisel, being careful not to hurt the threads. The new rings must be "socked home" tight. A big wrench on a steel bar across the lugs will help, when rings must be put in by hand. Always be sure to coat the threads with a good lubricant before putting them in. "Lap-in" the rings (Fig. 57) for a perfect fit to disc.

HOW TO REGRIND OR RENEW DISC AND SEAT IN GLOBE VALVES

Repairs on globe and angle valves can frequently be made without removing them from the line. This feature should encourage immediate repair when leakage occurs, to prevent serious damage.

To illustrate the general procedure, a plug type disc valve is used. But the method applies to other globe and angle valves, except those with composition disc. The simple procedure for renewing composition discs is shown in Figs. 58, 59, 60, 61, 62, 63, and 64.

Working at a bench is most convenient. Set the valve firmly in the vise, straight up (Fig. 58). Use "soft" jaws to prevent damage to body. It is best not to use a pipe wrench on the bonnet or union rung. A monkey wrench will not damage these parts. While parts are disassembled, examine and clean them thoroughly. And it is also a good idea to renew the packing in the stuffing box at this time. The same method suggested for gate valves applies here.

Remove stem from bonnet—place disc in vise and unscrew disc stem ring (Fig. 59).

Next, lift out stem, insert a slug or coin inside of the disc (Fig. 60). Replace the stem and tighten disc stem ring. The slug takes up "play" between disc and stem. In some valves, a pin is inserted in hole provided to lock the disc on the stem. Now you are almost ready to grind.

Apply an emery base grinding compound to both disc and seat. A little compound will go a long way—too much is wasteful and likely to remove too much metal (Fig. 61).

Place bonnet and body together, giving bonnet or bonnet ring two or three turns by hand to provide a guide for the stem (Fig. 62). With firm hold oscillate the handwheel steadily until all pitting is removed and a continuous, smooth bearing on the seating surfaces is obtained. Do not grind more than is needed.

When grinding is completed, wipe the disc, seat, and body clean of all compound and dirt (Fig. 63). Before reassembling valve, be sure to remove slug from disc to give it free swivel action on stem.

When seat rings must be removed, it is desirable to use a tool specially designed for this purpose (Fig. 64). Doing so will prevent damage to the valve body.

HOW TO RENEW DISC IN COMPOSITION DISC VALVES

This procedure is simple and takes little time. And while

Fig. 58.

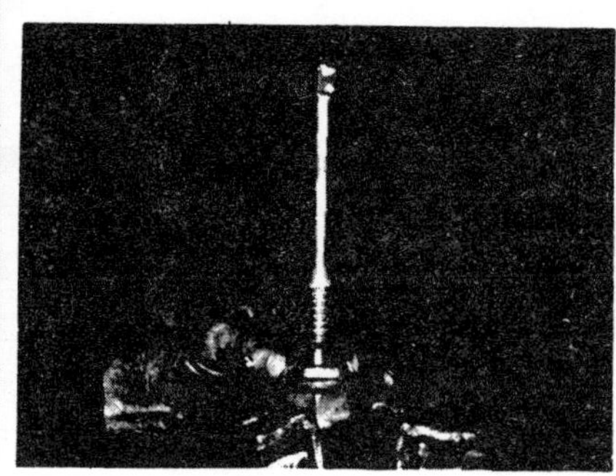

Fig. 59.

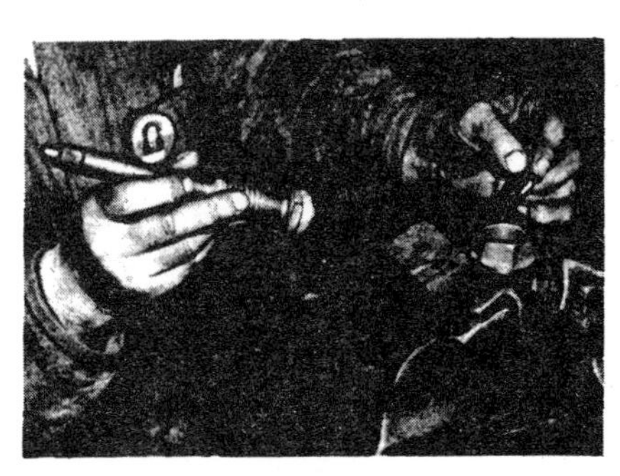

Fig. 60.

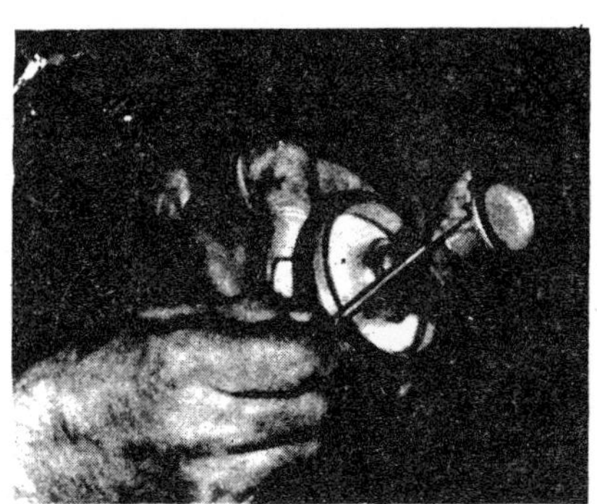

Fig. 61.

Fig. 62.

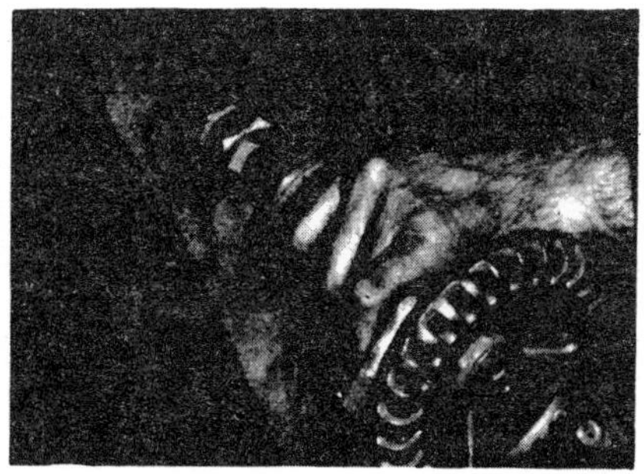

Fig. 63.

Fig. 64.

it may vary slightly for some valves, it is typical in principle.

Replacing a composition disc as soon as a seat leak is discovered will add much to the valve's life. Here is how it is done.

Turn the stem to partly open position, unscrew the bonnet joint, and lift out the bonnet assembly. Slip off disc holder. Remove disc retaining nut on under side of holder, and replace disc.

Where frequent disc changing is necessary, it is recommended that a few extra disc holders and discs be kept on hand. This saves time—it is also convenient when valves are changed to other services. Composition disc valves are ideal "all-purpose" valves, and discs are available for steam, hot and cold water, air, oil, gas, and gasoline. This feature simplifies maintenance and parts stocks problems.

Appendix I

Glossary of Terms

Absolute pressure. The gage pressure plus the atmospheric pressure is equal to the absolute pressure.

Absorption trench. A trench not over 36 inches in width with a minimum of 12 inches of clean, coarse aggregate and a distribution pipe, and covered with a minimum of 12 inches of earth cover.

Airway. A space between roof insulation and roof boards for movement of air.

All iron. Describes a valve all of whose parts are made of iron.

Anchor. A type of pipe support needed to hold piping rigid at a given point.

Angle valve. A variant of the globe valve design, having pipe openings at right angles to each other.

Atmospheric pressure. At sea level is 14.7 pounds per square inch.

Attic ventilators. In houses, screened openings provided to ventilate an attic space. They are located in the soffit area as inlet ventilators and in the gable end or along the ridge as outlet ventilators. They can also consist of power-driven fans used as as exhaust system. (*See also* Louver.)

Automatic stop-check valve. Combination check and shut-off valve designed primarily for use on multiple boiler installation.

Band. The raised collar put on the ends of certain screwed fittings and valves for reinforcement.

Blind flange. Solid plate-like fitting used to seal the end of a flanged-end pipe line.

Blow-off system. Piping hookup used for blowing scale and sediment from boilers, tanks, or receivers.

Blow-off valve. A valve designed specially for blow-off service and used in blow-off lines.

Boiler efficiency. The ratio expressed in percent of heat absorbed by the boiler to the heat released in the fire box.

Boiler horsepower. The equivalent evaporation of 34.5 pounds of water per hour from and at 212°F. This is equal to a heat output of $970.3 \times 34.5 = 33{,}475$ B.T.U. per hour.

Brass-to-iron. Designates a brass disc and iron seat, or vice versa, in a valve.

Brass trim or brass-mounted. Indicates that certain inside parts of a valve, such as stem, disc, and seat rings, are made of brass.

British thermal unit (B.T.U.) A measure of heat. A B.T.U. is the amount of heat necessary to raise the temperature of one pound of water one degree Fahrenheit or, roughly, the amount of heat produced by the burning of a wooden kitchen match.

Building (house) drain. That part of the lowest horizontal piping of a drainage system which receives the discharge from soil, waste, and other drainage piping inside a building and conveys it to the building (house) sewer beginning five feet outside the inner face of the building wall.

Building (house) sewer. That part of a drainage system which extends from the end of the building drain and conveys its discharge to a public sewer, private sewer, individual sewage disposal system or other point of disposal.

Building (house) storm drain. A building drain used for conveying rain water, ground water, sub-surface water, condensate, cooling water, or other discharge to a building storm sewer, or to a combined building drain or sewer.

Building (house) storm sewer. The pipe extending from the building storm drain to the public storm sewer, combined sewer, or other place of disposal.

Building (house) trap. A running trap installed in the building drain to prevent circulation of air between the drainage system of the building and the building sewer.

Bushing. Threaded and tapped fitting which is used to reduce size of end opening of a valve or fitting.

Bypass. An auxiliary loop in a pipe line, usually for diverting flow around a valve.

Calorie. One calorie is equal to 3.97 B.T.U.

Calorific value. The number of heat units available from the complete combustion of one pound of moisture-free coal (B.T.U. per pound).

Cesspool. A lined and covered excavation in the ground which receives the discharge of domestic sewage or other organic wastes from a drainage system, so designed as to retain the organic matter and solids, but permitting the liquids to seep through the bottom and sides.

Check valve. Valve designed to close automatically with reversal of flow in pipe line.

Clamp gate valve. Gate valve whose body and bonnet are held together by a U-bolt clamp.

Cock. Original form of valve. Has tapered plug with hole which is rotated to provide passageway for fluids.

CO. Symbol for carbon monoxide which denotes incomplete combination of the carbon in the fuel.

Co-efficient of heat emission. This term is usually applied to the heat emitted, or given off, by one square foot of actual radiator surface per hour for one degree temperature difference between the steam or water in the radiator and the air in the room.

Co-efficient of heat transmittance. The amount of heat (B.T.U.) transmitted per hour per square foot of surface per degree difference in temperature between the gases or liquid on the two sides of the surface. (This includes radiation, conduction, and convection.)

Combustion. The chemical union of the combustible of a fuel with the oxygen of the air resulting in heat or light or both.

Composition disc. Nonmetallic disc used in certain types of valves.

Condensate. Water in a steam line.

Condensation (house). Beads or drops of water (and frequently frost in extremely cold weather) that accumulate on the inside of the exterior covering of a building when warm, moisture-laden air from the interior reaches a point where the temperature no longer permits the air to sustain the moisture it holds. Use of louvers or attic ventilators will reduce moisture condensation in attics. A vapor barrier under the gypsum lath or dry wall on exposed walls will reduce condensation in them.

Conduction. By conduction, heat is transmitted from particle to particle of a body (internal conduction) or from one body to another, with which it is in contact (external conduction).

Continuous waste. A waste pipe from two or more fixtures connected to a single trap.

Convection heating. A form of heating in which the heat produced by the equipment is distributed through the natural movement of room air resulting from warm air rising.

Convector. A unit which gives off most of its heat directly to the air instead of through heat rays.

Conventional disc. Most commonly used design of disc for globe valves.

Corrosion. Effect of deterioration of materials due to chemical action.

CO_2. Symbol for carbon dioxide, which is the result of complete combustion of the carbon in the fuel.

Counterflashing. A flashing usually used on chimneys at the roofline to cover shingle flashing and to prevent moisture entry.

Coupling. Female-end pipe fitting used for joining two lengths of pipe.

Cross. A pipe fitting having four openings.

Cross-connection. A physical connection between piping systems, through which a supply of potable water could be contaminated.

Crawl space. A shallow space below the living quarters of a basementless house, normally enclosed by the foundation wall.

Dead end. A branch leading from a soil, waste, or vent pipe, building drain, or building sewer, that is terminated at a developed distance of two feet or more by means of a cap, plug, or other closed fitting.

Density. The weight per unit volume. (Generally used pounds per cubic foot.)

Disc. The part of a valve that actually closes off flow.

Dormer. An opening in a sloping roof, the framing of which projects out to form a vertical wall suitable for windows or other openings.

Downspout. A pipe, usually of metal, for carrying rainwater from roof gutters.

Drain. A drain or drain pipe is any pipe that carries waste water or water-borne waste in a building drainage system.

Drainage fitting. Type of fitting used mainly for plumbing drainage lines.

Drainage fixture unit value. A common measure of the probable discharge into a drainage system by various types of plumbing fixtures. This value for a particular fixture depends on its volume rate of drainage discharge, on the time duration of a single drainage operation, and on the average time between successive operations.

Drainage system. A drainage system, or drainage piping, includes all the piping within public or private premises that conveys sewage, rainwater, or other liquid wastes to a legal point of disposal, but does not include the mains of public sewage treatment or disposal plant.

Drip. (a) A member of a cornice or other horizontal exterior-finish course that has a projection beyond the other parts for throwing off water. (b) A groove in the underside of a sill or drip cap to cause water to drop off on the outer edge instead of drawing back and running down the face of the building.

Drip cap. A molding placed on the exterior top side of a door or window frame to cause water to drip beyond the outside of the frame.

Dry saturated steam. Steam containing no entrained moisture.

Ducts. In a house, usually round or rectangular metal pipes for distributing warm air from the heating plant to rooms, or air from a conditioning device or as cold air returns. Ducts are also made of asbestos and composition materials.

Eaves. The margin or lower part of a roof projecting over the wall.

Effect size. That size of sand of which 10 percent by weight is smaller.

Elbow. Fitting used for making a turn in direction of pipe line; also known as "ell."

End connection. Refers to the type of connection by which piping elements are joined together.

Estimated design load. The sum of the radiation, piping, domestic water and any other material or conditions, and expressed in terms of equivalent direct radiation: 240 B.T.U. steam; 150 B.T.U. water.

Expansion joint. Pressure-tight device that permits expansion or contraction of pipe lines.

Evaporation. The process of transforming a substance from the liquid state into a gaseous state.

Extra heavy. Denotes piping material suitable for working pressures up to 250 pounds.

Facing. Finish of the contact surface of flanged-end piping materials.

Female thread. Internal thread in pipe fittings and valves, for making screwed connection.

Fire-resistive. In the absence of a specific ruling by the authority having jurisdiction, applies to materials for construction not combustible in the temperatures of ordinary fires and that will withstand such fires without serious impairment of their usefulness for at least one hour.

Fire-retardant chemical. A chemical or preparation of chemicals used to reduce flammability or to retard spread of flame.

Fire stop. A solid, tight closure of a concealed space, placed to prevent the spread of fire and smoke through such a space. In a frame wall, this will usually consist of 2 by 4 cross blocking between studs.

Fitting. A piping element, other than valve or pipe, used in joining pipe lines. Also the work of installing piping as done by a pipefitter.

Fixture branch. The water-supply pipe between the fixture-supply pipe and the water-distributing pipe.

Fixture drain. The drain from the trap of a fixture to the junction of that drain with any other drain pipe.

Fixture-supply pipe. A water-supply pipe connecting the fixture with the fixture branch at the wall or floor line.

Flange. Run on the end of a pipe, valve, or fitting for bolting to another piping element.

Flanged end. End of a valve or fitting having flange for joining to other piping elements.

Flashing. Sheet metal or other material used in roof and wall construction to protect a building from water seepage.

Flue. The space or passage in a chimney through which smoke, gas, or fumes ascend. Each passage is called a flue, which together with any others and the surrounding masonry make up the chimney.

Flue lining. Fire clay or terra-cotta pipe, round or square, usually made in all ordinary flue sizes and in 2-foot lengths, used for the inner lining of chimneys with the brick or masonry work around the outside. Flue lining in chimneys runs from about a foot below the flue connection to the top of the chimney.

Fly rafters. End rafters of the gable overhang supported by roof sheathing and lookouts.

Foreign matter. General terms for scale, rust, and dirt in pipe lines.

Frost-proof closet. A closet that has no water in the bowl and has the trap and the control valve for its water supply installed below the frost line.

Fungi, wood. Microscopic plants that live in damp wood and cause mold, stain, and decay.

Fungicide. A chemical that is poisonous to fungi.

Gable. In house construction, the portion of the roof above the eave line of a double-sloped roof.

Gable end. An end wall having a gable.

Gage pressure. Pressure measured from atmospheric pressure as a base.

Gate valve. One of the basic valve types. Its name comes from the gate-like disc which regulates flow.

Globe valve. A basic valve type that gets its name from the globular shape of its body.

Grease interceptor (trap). A receptacle designed to collect and retain grease and fatty substances normally found in kitchen or similar wastes. It is installed in the drainage system between the kitchen or other point of production of the waste and the building sewer.

Ground joint. Denoting a connection in which two machined metallic surfaces are joined face to face.

Gutter or eave trough. A shallow channel or conduit of metal or wood set below and along the eaves of a house to catch and carry off rainwater from the roof.

Hanger. A device for supporting a pipe line.

Header. (a) A length of pipe or cast vessel to which two or more pipe lines are joined to carry fluid from a common source to various points of use. (b) A beam placed perpendicular to joists and to which joists are nailed in framing for chimney, stairway, or other opening. (c) A wood lintel.

Hearth. The inner or outer floor of a fireplace, usually made of brick, tile, or stone.

Heat. A form of energy believed to be a vibratory motion of the smallest particles (molecules) of which each body is composed.

The modes of propagation of heat are: conduction, convection, and radiation. Heat is transmitted from the steam or water side of a radiator to the air side by conduction. After heat is conducted to the outer surface of the radiator, approximately 80 percent of the total heat given off is transmitted by convection, and the balance by radiation.

Heat loss. The amount of heat that a home loses to the outdoors through walls, ceilings, windows, and the like, when the outdoor temperature is lower than the indoor temperature. This represents the amount of heat that must be added to a home by its heating system. (Heat gain is the opposite of heat loss and applies to cooling.)

Heat of liquid. The sensible heat of a quantity of liquid above an arbitrary zero.

Heat transmission. The passage of heat through any substance to a medium of lower temperature on the other side.

Hickey. A length of pipe or extension handle used on a wrench to get greater leverage.

Hip. The external angle formed by the meeting of two sloping sides of a roof.

Hip roof. A roof that rises by inclined planes from all four sides of a building.

Horizontal branch. A branch drain extending laterally from a soil or waste stack or building drain, with or without vertical sections or branches, that receives the discharge from one or more fixture drains and conducts it to the soil or waste stack or to the building (house) drain.

Hub end. Calked or leaded type of end connection used on valves, fittings, and pipe mainly for water supply and sewerage lines.

Humidifier. A device designed to increase the humidity within a room or a house by means of the discharge of water vapor. They may consist of individual room-size units or larger units attached to the heating plant to condition the entire house.

Hydronic heating. A type of heating in which heat is circulated through the house in the form of hot water or steam.

Hydronics. The science of heating and cooling with liquids.

ICC. A new system utilized in the Federal Housing Administration recommended criteria for impact sound insulation.

Increaser. Fitting with larger opening at one end, used to increase size of pipe opening. Common to drainage fittings.

Indirect waste pipe. A waste pipe that does not connect directly with the drainage system, but discharges into it through a properly trapped fixture or receptacle.

Individual sewage disposal system. A single system of sewage treatment tanks and disposal facilities serving only a single lot.

INR (Impact Noise Rating). A single figure rating which provides an estimate of the impact sound-insulating performance of a floor-ceiling assembly.

Installed radiation square feet. The sum of the heat output of all units installed for heating rooms or other purposes and expressed in square feet or 240 B.T.U. (steam) or 150 B.T.U. (water). This includes all types of heaters, such as radiators, convectors, unit heaters, pipe coils, heat exchangers, and the like.

Insulation board, rigid. A structural building board made of coarse wood or cane fiber in $\frac{1}{2}$- and $\frac{25}{32}$-inch thicknesses. It can be obtained in various densities, and with several treatments.

Insulation, thermal. Any material high in resistance to heat transmission that, when placed in the walls, ceiling, or floors of a structure, will reduce the rate of heat flow.

Interceptor. A receptacle designed and constructed to intercept or separate and prevent the passage of oil, grease, sand, or similar materials into the drainage system to which it is directly or indirectly connected.

Joint. The point of connection between two piping elements, whether screwed, bolted, or welded.

Joist. One of a series of parallel beams, usually two inches in thickness, used to support floor and ceiling loads, and supported in turn by larger beams, girders, or bearing walls.

Kilowatt hour (KWHR). A measure of the use of electric energy equal to 1000 watts of electricity used steadily for one hour. One KWHR will power a 100-watt light bulb for ten hours.

Lapping-in. Rubbing and polishing a surface such as a disc face to obtain a smooth bearing with body seat rings.

Latent heat. It is evident that a large quantity of heat or energy has been expended in evaporating water into steam, and this quantity of heat necessary to evaporate one pound of water from 212°F. to steam at 212°F. is approximately 970 B.T.U. This is called the "latent heat" of vaporization. When water is evaporated under pressure, the latent heat decreases as the pressure increases.

Latent heat available. When the steam is condensed into water.

Latent heat of evaporation. The heat required to change a liquid into a vapor without changing its temperature.

Latent heat of fusion. The quantity of heat to be added or extracted to change a solid into a liquid and vice versa without change of its temperature.

Leader. A leader or downspout is the water conductor from the roof to the storm drain or other means of disposal.

Long-sweep fitting. Fitting with a long-radius turn.

Louver. An opening with a series of horizontal slats so arranged as to permit ventilation but to exclude rain, sunlight, or vision. (*See also* Attic ventilators.)

Lubricant. Specially prepared compound used in making up screwed joints to reduce friction and tearing of threads.

Main sewer. The main sewer or public sewer is the sewer in a street, alley, or other premises under the jurisdiction of the city, town, or village.

Male thread. External thread on pipe fittings and valves, for making screwed connection.

Malleable fitting. Pipe fitting made of malleable iron.

Mantel. The shelf above a fireplace. Also used in referring to the decorative trim around a fireplace opening.

Master plumber. One who is licensed to engage in the business of installing plumbing as a contractor.

Mastic. A pasty material used as a cement (as for setting tile) or a protective coating (as for thermal insulation or waterproofing).

Needle valve. Globe-type valve with needle-point disc for extremely fine regulation of flow.

Nipple. Short length of pipe (up to 12 inches long) threaded on both ends, for joining piping elements.

Non-rising stem. Type of valve stem that merely turns and does not rise when valve is operated.

Offset. In a line of piping, a combination of elbows or bends that brings one section of the pipe out of line with but into a line parallel with another section.

Overall efficiency (also known as **Boiler and grate efficiency** for handfired boilers). The ratio of the heat absorbed by a boiler to the heat value of the fuel fired. A general definition of efficiency is output over input.

Packing. Material used in stuffing box of a valve to maintain a leakproof seal around stem.

Paper, sheathing. A building material, generally paper or felt, used in wall and roof construction as a protection against the passage of air and sometimes moisture.

Perm. A measure of water vapor movement through a material (grains per square foot per hour per inch of mercury difference in vapor pressure).

Pipe dope. (*See* Lubricant.)

Pipe scale. A hard scale-like material frequently found in new pipe. It is caused by heating operations in manufacture.

Pipe strap. Device for holding lightweight pipe to wall or ceiling.

Pipe support. A device for supporting pipe lines.

Piping. General term for materials used in plumbing, heating, and other pipe lines. A complete piping system. Also the act of making up a pipe line.

Pitch. The incline slope of a roof or the ratio of the total rise to the total width of a house, such as, an 8-foot rise and 24-foot width is a one-third pitch roof. Roof slope is expressed in the inches of rise per foot of run.

Plug. Screwed fitting for shutting off a tapped opening.

Plug-type disc valve. Globe or angle valve with tapered plug disc and cone-shaped seat having wide-bearing seating surface.

Ply. A term to denote the number of thicknesses or layers of roofing felt, veneer in plywood, or layers in built-up materials, in any finished piece of such material.

Pop valve. Spring-loaded safety valve which opens automatically when pressure exceeds limits for which valve is set. Used as a safety device on boilers and other equipment to prevent damage from excessive pressure. Pop safety valves are generally used with steam, air, or other gases.

Port opening. The pipe opening of a valve.

Potable water. Water from a public or private water-supply system or source, that is accepted by the proper governing authority as suitable for human consumption.

Pressure regulator. A valve used to automatically reduce and maintain pressure below that of the source for certain processing or heating devices.

R-Value. A measure of insulating quality. The higher the R-Value, the better the insulation and the less heating and cooling paid for and lost.

Radiant heating. A method of heating, usually consisting of a forced hot water system with pipes placed in the floor, wall, or ceiling; or with electrically heated panels.

Radiation. By radiation, heat passes through space by wave motion from one body to another without the aid of any material agency.

Radiator. A free-standing unit which gives off a substantial portion of its heat by radiation.

Rafter. One of a series of structural members of a roof designed to support roof loads. The rafters of a flat roof are sometimes called roof joists.

Rafter, hip. A rafter that forms the intersection of an external roof angle.

Rafter, valley. A rafter that forms the intersection of an internal roof angle. The valley rafter is normally made of double 2-inch-thick members.

Railing fitting. Type of fitting used for making up hand or guard railings.

Railroad union. Type of union. (*See also* Union.)

Rake. Trim members that run parallel to the roof slope and form the finish between the wall and a gable roof extension.

Rate of combustion. The number of pounds of coal burned per square foot of grate per hour.

Reducer. Fitting with smaller opening at one end for reducing size of opening.

Reflective insulation. Sheet material with one or both surfaces of comparatively low heat emissivity, such as aluminum foil. When used in building construction the surfaces face air spaces, reducing the radiation across the air space.

Regular. Denotes a piping item regularly catalogued by manufacturer.

Relative humidity. The amount of water vapor in the atmosphere, expressed as a percentage of the maximum quantity that could be present at a given temperature. (The actual amount of water vapor that can be held in space increases with the temperature.)

Relief valve. Safety valve similar to a pop safety valve in operation.

Return bend. A U-pattern fitting for reversing direction of pipe run. Used mostly in making up pipe coils.

Ridge. The horizontal line at the junction of the top edges of two sloping roof surfaces.

Ridge board. The board placed on edge at the ridge of the roof into which the upper ends of the rafters are fastened.

Rising stem. Type of valve stem that turns and rises when valve is opened.

Roll roofing. Roofing material, composed of fiber and saturated with asphalt, that is supplied in 36-inch wide rolls with 108 square feet of material. Weights are generally 45 to 90 pounds per roll.

Roof drain. A drain installed to receive water collecting on the surface of a roof and to discharge it into a leader (downspout).

Roof sheathing. The boards or sheet material fastened to the roof rafters on which the shingle or other roof covering is laid.

Saddle. Two sloping surfaces meeting in a horizontal ridge, used between the back side of a chimney, or other vertical surface, and a sloping roof.

Sand filter trenches. A system of trenches, consisting of perforated pipe or drain tile surrounded by clean, coarse aggregate containing an intermediate layer of sand as filtering material and provided with an underdrain for carrying off the filtered sewage.

Sanitary sewer. A sewer that carries sewage and excludes storm, surface, and ground water.

Saturated steam. Water vapor in the condition in which it is generated from the water with which it is in contact.

Screwed end. Type of end on valve, fitting, or pipe that is joined to other elements by screwed connection.

Scum. A mass of sewage matter which floats on the surface of sewage.

Scum clear space. Distance between the bottom of the scum mat and the bottom of the outlet device.

Seepage bed. A trench or bed exceeding 36 inches in width containing 12 inches at minimum of clean, coarse aggregate and a system of distribution piping, through which treated sewage may seep into the surrounding soil.

Seepage pit. A covered pit with lining designed to permit treated sewage to seep into the surrounding soil.

Sensible heat. When heat is applied to water at any temperature below the boiling point, the temperature of water will rise until the boiling point is reached. Heat which raises the temperature of water, and which affects the thermometer, is called "sensible heat."

Septic tank. A water-tight, covered receptacle designed and constructed to receive the discharge of sewage from a building (house) sewer, separate solids from the liquids, digest organic matter, and store digested solids through a period of detention, and allow the clarified liquids to discharge for final disposal.

Serial distribution. An arrangement of absorption trenches, seepage pits, or seepage beds so that each is forced to pond to utilize the total effective absorption area before liquid flows into the succeeding component.

Sewage. Any liquid waste containing animal or vegetable matter in suspension or solution, and may include liquids containing chemicals in solution.

Sheet metal work. All components of a house employing sheet metal, such as flashing, gutters, and downspouts.

Shingles. Roof covering of asphalt, asbestos, wood, tile, slate, or other material cut to stock lengths, widths, and thicknesses.

Shingles, siding. Various kinds of shingles, such as wood shingles or shakes and nonwood shingles, that are used over sheathing for exterior sidewall covering of a structure.

Side outlet. An ell or tee fitting with a side outlet.

Size of pipe or tubing. Nominal size by which pipe is commercially designated, unless otherwise stated.

Slip-on flange. Flange that slips onto pipe and is welded in place.

Sludge. The accumulated settled solids deposited from sewage and containing more or less water to form a semi-liquid mass.

Sludge clear space. The distance between the top of the sludge and the bottom of the outlet device.

Socket-welding fitting or valve. Socket-end type valve or fitting that slips over end of pipe and is made pressure-tight by welding.

Soft jaws. Copper or lead covers placed over vise jaws to prevent damage to materials held in vise.

Soil absorption field. A system of absorption trenches.

Soil absorption system. Any system that utilizes the soil for subsequent absorption of the treated sewage, such as an absorption trench, seepage bed, or a seepage pit.

Soil cover (ground cover). A light covering of plastic film, roll roofing, or similar material used over the soil in crawl spaces of buildings to minimize moisture permeation of the area.

Soil pipe. Any pipe that conveys the discharge of water closets or fixtures having similar functions, with or without the discharges from other fixtures.

Soil stack. A general term for the vertical main of a system of soil, waste, or vent piping.

Solder-joint. Type of end connection made by soldering. Generally used with copper tubing.

Special. Denotes an item that differs from manufacturer's standard design.

Specific gravity. The ratio of the weight of a given body of any substance to the weight of an equal volume of a substance used as a standard. (The standard for gases is generally air.)

Specific heat. The quantity of heat (B.T.U.) required to raise one pound of a substance one degree Fahrenheit.

Splash block. A small masonry block laid with the top close to the ground surface to receive roof drainage from downspout and to carry it away from the building.

Stack. A storm stack or sewer carries off surface or storm water from streets, roofs, or other areas (including street wash), but not sewage or liquid industrial waste.

Standard. Used in referring to piping materials for working pressures up to 125 pounds steam.

Standard absorption trench. A trench 12 inches to 36 inches in width containing 12 inches of clean, coarse aggregate and a distribution pipe, covered with a minimum of 12 inches of earth cover.

Steam. Water vapor.

Storm sash or storm window. An extra window usually placed on the outside of an existing one as additional protection against cold weather.

Street ell. Elbow fitting with male end and female end.

Subsoil drain. A drain installed for collecting subsurface or seepage water and conveying it to a place of disposal.

Subsurface sand filters. A wide bed, consisting of a number of lines of perforated pipe or drain tile surrounded by clear coarse aggregate, containing an intermediate layer of sand as filtering material, and provided with a system of underdrains for carrying off the filtered sewage.

Subsurface sewage disposal system. A system for the treatment and disposal of domestic sewage by means of a septic tank and a soil absorption system.

Sump. A tank or pit that receives the discharge for subdrains and from which the discharge is pumped or ejected into a drainage system.

Superheated steam. Steam heated higher than the temperature corresponding to its pressure.

Tap. Metal tool for forming internal or female thread.

Tee. A 3-way fitting shaped like the letter *T*.

Temperature. A measure of the intensity of heat.

Threader. Metal tool for cutting male thread.

Throttling. Regulation of flow through a valve.

Trap. A fitting or device so designed and constructed as to provide a liquid seal that will prevent the back passage of air without materially affecting the flow of sewage or waste water through it.

Trap seal. The vertical distance between the crown and the dip of the trap.

Trim. Term used in referring to bonnet, stem, disc, and seat parts of a valve.

Tubing. Light-weight pipe, usually copper, brass, or plastic.

Uniformity coefficient. A coefficient obtained by dividing that size of sand of which 60 percent by weight is smaller, by that size of sand of which 10 percent by weight is smaller.

Union. Fitting used to join lengths of pipe to permit easy opening of a line.

Union fitting. Fitting with union at one or more ends.

Valley. The internal angle formed by the junction of two sloping sides of a roof.

Valve body. Main part of valve. Contains passageway for fluid and seating surfaces for disc.

Vapor barrier. Material used to retard the movement of water vapor into walls and prevent condensation in them. Usually considered as having a perm value of less than 1.0. Applied separately over the warm side of exposed walls or as a part of batt or blanket insulation. (*See* Perm.)

Waste pipe. Any pipe that receives the discharge of any fixture, except water closets or similar fixtures, and conveys the same to the house drain, soil, or waste stack. When such pipe does not connect directly with a building drain or soil pipe it is termed an indirect waste pipe.

Weatherstrip. Narrow or jamb-width sections of thin metal or other material to prevent infiltration of air and moisture around windows and doors. Compression weather stripping prevents air infiltration, provides tension, and acts as a counterbalance.

Welding end. Type of end on valve, fitting, or pipe that is joined to other piping elements by welding.

Welding neck flange. Flange with integral extended neck for welding to pipe.

Wet saturated steam. Steam containing entrained moisture.
Wheel. Wheel handle for operating a valve.

Appendix II

Introduction of Terms

A = Branch Interval
B = Branch Vent
C = Building Drain
D = Building Sewer
E = Building Sub-drain
F = Circuit Vent
G = Continuous Waste & Vent
H = Dry Vent
H' = Double Offset
I = Dual Vent (Unit Vent)
J = Fixture Drain
J' = Group Vent
K = Horizontal Branch
L = Leader
M = Indirect Waste
N = Loop Vent
O = Offset
P = Primary Branch
Q = Relief Vent
R = Return Offset or Jumpover
S = Secondary Branch
T = Side Vent
U = Soil Stack
V = Vent Stack
W = Wet Vent
X = Stack Vent
Y = Yoke Vent
Z = Back Vent

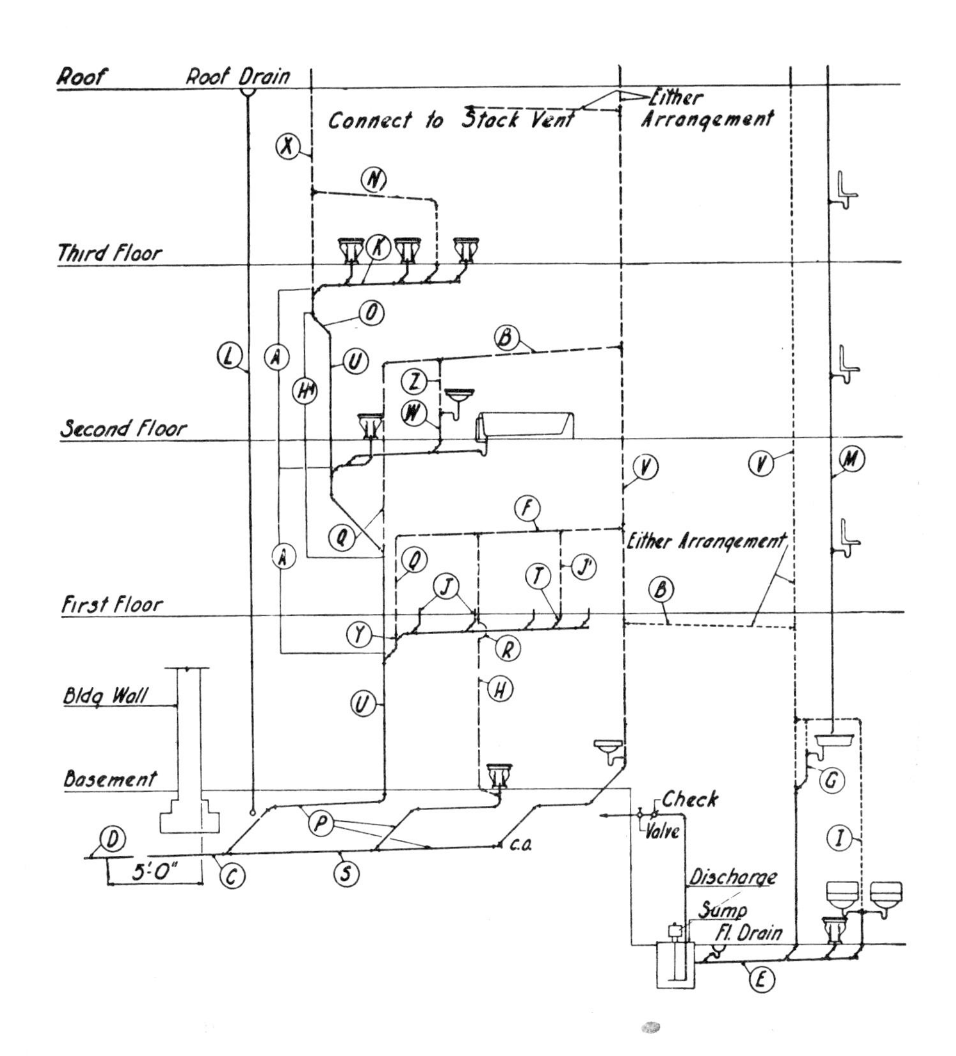
Roof
Roof Drain
Connect to Stack Vent
Either Arrangement
Third Floor
Second Floor
Either Arrangement
First Floor
Bldg Wall
Basement
Check
Valve
Discharge
Sump
Fl. Drain
5'-0"
C.O.
X
N
K
O
B
L
A
U
H'
Z
W
V
M
Q
F
J'
T
J
Y
R
H
G
P
D
C
S
I
E

Appendix III

Illustration of Terms

Symbol	Plan	Initials	Item
		D.	Drainage Line
		V.S.	Vent Line
			Tile Pipe
		C.W.	Cold Water Line
		H.W.	Hot Water Line
		H.W.R.	Hot Water Return
		G	Gas Pipe
		D.W.	Ice Water Supply
		D.R.	Ice Water Return
		F.L.	Fire Line
		I.W.	Indirect Waste
		I.S.	Industrial Sewer
		AW	Acid Waste
	A	A	Air Line
	V	V	Vacuum Line
	R	R	Refrigerator Waste
			Gate Valves
			Check Valves
CO CO		CO.	Cleanout
F.D		F.D.	Floor Drain
R.D		R.D.	Roof Drain
REF.		REF.	Refrigerator Drain
		S.D	Shower Drain
G.T.		G.T.	Grease Trap
S.C.		S.C.	Sill Cock
G		G.	Gas Outlet

Symbol		Abbr.	Term
VAC		VAC.	Vacuum Outlet
M		M	Meter
I×			Hydrant
H.R.		H R	Hose Rack
H.R.		H.R.	Hose Rack-Built in
L		L.	Leader
H WT		H.W.T.	Hot Water Tank
WH		W.H.	Water Heater
WM		W.M	Washing Machine
RB		R.B.	Range Boiler

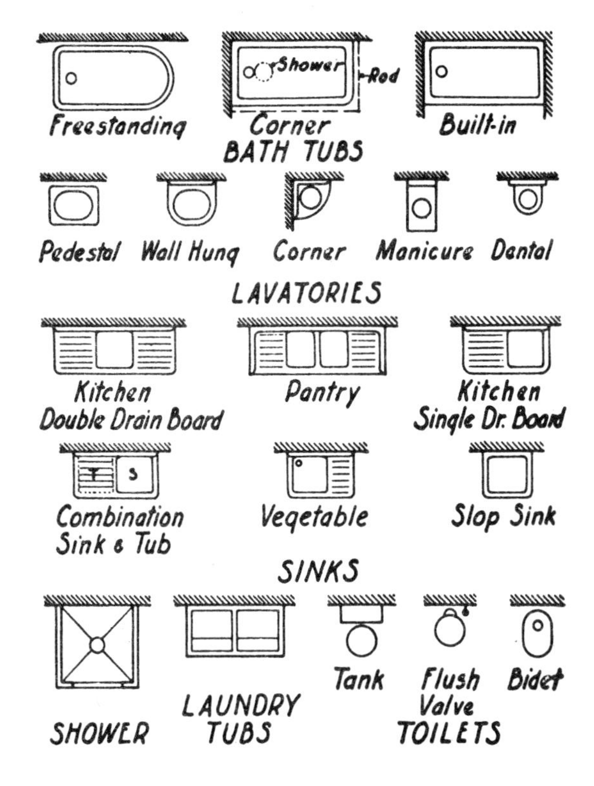

Appendix IV

Suggested Specifications for Watertight Concrete

1. *Materials*

Portland cement should be free of hard lumps caused by moisture during storage. Lumps from dry packing that are easily broken in the hand are not objectionable.[1]

Aggregates, such as sand and gravel, should be obtained from sources known to make good concrete. They should be clean and hard. Particle size of sand should range very fine to 1/4 inch. Gravel or crushed stone should have particles from 1/4 inch to a maximum of 1 1/2 inches in size. Water for mixing should be clean.

2. *Proportioning*

Not more than 6 gallons of total water should be used for each bag of cement. Since sand usually holds a considerable amount of water, not more than 5 gallons of water per bag of cement should be added at the mixer when sand is of average dampness. More mixing water weakens the concrete and makes it less watertight. For average aggregates, the mix proportions shown in the table below will give watertight concrete.

Average Proportions for Watertight Concrete

Max. Size Gravel (in.)	Cement (volume)	Water [1] (volume)	Sand (volume)	Gravel (volume)
1 1/2	1	3/4	2 1/4	3
3/4	1	3/4	2 1/2	2 1/2

[1] Assuming sand is of average dampness.

---

[1] Type V portland cement may be used when high sulfate resistance is required.

3. *Mixing and Placing*

All materials should be mixed long enough so that the concrete has a uniform color. As concrete is deposited in the forms, it should be tamped and spaded to obtain a dense wall. The entire tank should be cast in one continuous operation if possible, to prevent construction joints.

4. *Curing*

After it has set, new concrete should be kept moist for at least seven days to gain strength.

Appendix V

Drainage Fixture Unit Values

Type of Fixture or Group of Fixtures	Drainage Fixture Unit Value
Automatic clothes washer (2" standpipe)	3
Bathroom group consisting of a water closet, lavatory and bathtub or shower stall:	
Flushometer valve closet	8
Tank type closet	6
Bathtub (with or without overhead shower)	2
Combination sink and tray with food disposal unit	4
Combination sink and tray with one 1½" trap	2
Combination sink and tray with separate 1½" traps	3
Dishwasher, domestic (gravity drain)	2
Floor drains with 2" waste	3
Kitchen sink, domestic, with one 1½" waste	2
Kitchen sink, domestic, with food waste grinder	2
Lavatory with 1¼" waste	1
Laundry tray (1 or 2 compartments)	2
Show stall, domestic	2
Showers (group) per head	2
Sinks:	
Flushing rim (with valve)	6
Service (trap standard)	3
Service (P trap)	2
Pot, scullery, etc.	4
Wash sink (circular or multiple) each set of faucets	2
Water closet, tank operated	4
Water closet, valve operated	6
Unlisted fixture drain or trap size:	
1¼" or less	1
1½"	2
2"	3
2½"	4
3"	5
4"	6

Index